Arthur Sheridan Lea

The chemical basis of the animal body

Arthur Sheridan Lea

The chemical basis of the animal body

ISBN/EAN: 9783337228828

Printed in Europe, USA, Canada, Australia, Japan

Cover: Foto ©berggeist007 / pixelio.de

More available books at **www.hansebooks.com**

A TEXT BOOK

OF

PHYSIOLOGY.

BY

M. FOSTER, M.A., M.D., LL.D., F.R.S.,

PROFESSOR OF PHYSIOLOGY IN THE UNIVERSITY OF CAMBRIDGE,
AND FELLOW OF TRINITY COLLEGE, CAMBRIDGE.

SIXTH EDITION, REVISED.

PART IV.

COMPRISING THE REMAINDER OF

BOOK III.—THE SENSES AND SOME SPECIAL MUSCULAR MECHANISMS,
AND
BOOK IV.—THE TISSUES AND MECHANISMS OF REPRODUCTION.

London:
MACMILLAN AND CO.
AND NEW YORK.
1891.

PREFACE.

THE present Part IV. completes the work, with the exception of the Appendix, which differs so widely in character from the rest of the book that it seemed desirable to issue it separately; it will be published very shortly.

Besides receiving assistance from the friends who have helped me in former Parts, I have in this Part had the advantage of the criticisms of my friends, Dr. W. S. Griffith, Mr. Walter Jessop, and Dr. F. Semon, in the subjects with which they are respectively familiar, for which I tender them my best thanks. I also desire to thank my friends, Mr. James Ward and Prof. Postgate, of Trinity College, for valuable advice.

August, 1891.

CONTENTS OF PART IV.

BOOK III. (*continued*).

CHAPTER III.

SIGHT.

SECTION I.

ON THE GENERAL STRUCTURE OF THE EYE, AND ON THE FORMATION OF THE RETINAL IMAGE.

	PAGE
§ 702. Dioptic Mechanisms and Visual Impulses	1
§ 703. The General Structure of the Eye. The Formation of the Retinal Image	2
§ 704. A simple Optic System: its Cardinal Points. The Refractive Surfaces and Media of the Eye	6
§ 705. The Optic Constants of the Eye. The Diagrammatic Eye . . .	9
§ 706. The Paths of the Rays of Light through the Eye	10
§ 707. The Retinal Image in relation to the Sensations excited by it . .	12

SECTION II.

THE FACTS OF ACCOMMODATION.

§ 708. The Eye can accommodate for far and near Objects; far and near Limits of Accommodation	13
§ 709. Scheiner's Experiment	14
§ 710. Emmetropic, Myopic, Hypermetropic, and Presbyopic Eyes . .	16
§ 711. The Changes observed in the Eye during Accommodation . . .	17

SECTION III.

ON THE STRUCTURE OF THE INVESTMENTS OF THE RETINA.

		PAGE
§ 712.	The Sclerotic Coat	21
§ 713.	The Choroid Coat	21
§ 714.	The Ciliary Processes	23
§ 715.	The Iris	24
§ 716.	The Cornea	25
§ 717.	The Ciliary Zone	26
§ 718.	The Ciliary Muscle	28
§ 719.	The Lens	28
§ 720.	The Vitreous Humour and the Suspensory Ligament . . .	29

SECTION IV.

THE MECHANISM OF ACCOMMODATION AND THE MOVEMENTS OF THE PUPIL.

§ 721.	The Mechanism for Changing the Anterior Curvature of the Lens	31
§ 722.	The Evidence that such a Mechanism does effect the Result .	32

The Movements of the Pupil.

§ 723.	Circumstances leading to Constriction and to Dilation of the Pupil	34
§ 724.	How Constriction and Dilation are brought about	34
§ 725.	The Nerves supplying the Pupil	35
§ 726.	Constriction a Reflex Act by means of the Optic and Oculo-motor Nerves	37
§ 727.	Changes in the Pupil through the Action of the Cervical Sympathetic Nerve	40
§ 728.	The Nature of the Dilating Mechanism	41
§ 729.	Direct Action of Drugs and other Agencies on the Pupil . .	43
§ 730.	The Nervous Mechanism of Accommodation	45
§ 731.	The Association of the Movements of Accommodation and the Movements of the Pupil	46

SECTION V.

IMPERFECTIONS IN THE DIOPTRIC APPARATUS.

§ 732.	Imperfections of Accommodation	47
§ 733.	Spherical Aberration	48
§ 734.	Astigmatism	48
§ 735.	Chromatic Aberration	50
§ 736.	Entoptic Phenomena	51

CONTENTS. ix

SECTION VI.

The Structure of the Retina.

	PAGE
§ 737. The Optic Nerve	55
§ 738. The Layers of the Retina	56
§ 739. The Neuroglial Elements	57
§ 740. The Nervous Elements. The Rods and Cones with the Rod Fibres and Cone Fibres	59
§ 741. The Inner Nuclear Layer	62
§ 742. The Layer of Ganglionic Cells and Layer of Optic Fibres	64
§ 743. The Probable Connection of the Several Elements	64
§ 744. The Macula Lutea and Fovea Centralis	66
§ 745. The Blood-vessels of the Retina	67
§ 746. The Pigment Epithelium	68

SECTION VII.

On some General Features of Visual Sensations.

§ 747. The Relation of the Sensation to the Intensity of the Stimulus; Weber's Law	71
§ 748. The Relation of the Sensation to the Duration of the Stimulus	74
§ 749. Flickering and Continuous Sensations	75
§ 750. Sensations produced by various Changes in the Retina referred to some External Source of Light	76
§ 751. Localisation of Visual Sensations	77
§ 752. The Conditions of Discrete Visual Sensations	79
§ 753. The Region of Distinct Vision. The Limits of Distinct Vision	80
§ 754. Nature of the Discreteness of Visual Sensations; Retinal Visual Units	81

SECTION VIII.

On Colour Sensations.

§ 755. The Existence of many Kinds of Colour Sensations	84
§ 756. The Mixing of Colour Sensations	85
§ 757. The several usual Colour Sensations result from the Mixture of Simpler, Primary Sensations	86
§ 758. The Conditions which determine the Characters of Colour Sensations	88
§ 759. Complementary Colours	89
§ 760. Any Colour Sensation produced by the suitable Mixture of Three Colour Sensations	90
§ 761. The Young-Helmholtz Theory of Colour Sensations	91
§ 762. Hering's Theory of Colour Sensations. A Comparison of the Two Theories	93

CONTENTS.

		PAGE
§ 763.	Variations in Colour Vision. Colour Blindness. The different Kinds of Colour Blindness; Red Blind and Green Blind; the Young-Helmholtz Explanation of them	99
§ 764.	The Explanation of Colour Blindness on Hering's Theory . . .	102
§ 765.	The probable Subjective Condition of the Colour Blind	104
§ 766.	Blue or Violent Blindness; Absolute Colour Blindness	105
§ 767.	Colour Blindness in the Periphery of the Retina	106
§ 768.	The Influence of the Yellow Spot	106
§ 769.	Colour Sensations in Relation to the Intensity of the Stimulus .	107

SECTION IX.

ON THE DEVELOPMENT OF VISUAL IMPULSES.

§ 770.	The Blind Spot	110
§ 771.	Purkinjé's Figures; their Import	111
§ 772.	Possible Theories as to the Mode of Origin of Visual Sensations .	114
§ 773.	Photochemistry of the Retina; Visual Purple	115
§ 774.	Hypothetical Visual Substances; Insufficiency of our Present Knowledge	118
§ 775.	The Functions of the Layer of Rods and Cones. The Ophthalmoscope	119
§ 776.	Possible Differences of Function of Rods and Cones	121
§ 777.	Electric Currents in the Retina	122

SECTION X.

ON SOME FEATURES OF VISUAL SENSATIONS ESPECIALLY IN RELATION TO VISUAL PERCEPTIONS.

§ 778.	Simultaneous Visual Sensations; the Visual Field	123
§ 779.	The Psychological and Physiological Methods; Sensations and Perceptions; their Want of Agreement	123
§ 780.	Irradiation	126
§ 781.	Simultaneous Contrast	126
§ 782.	After-Images. Successive Contrast	127
§ 783.	The Phenomena of 'Contrast' in their Bearing on the Theories of Colour Vision	128
§ 784.	Recurrent Sensations. Ocular Phantoms or Hallucinations . .	132

SECTION XI.

BINOCULAR VISION.

§ 785.	The Movements of the Eye-Ball; their Limitations. Centre of Rotation, Visual Axis, Visual Plane	134

CONTENTS.

		PAGE
§ 786.	The Visual Field and Field of Sight of one Eye and of both Eyes	135
§ 787.	Corresponding or Identical Points	137
§ 788.	The Movements of the Eye-Ball; the Primary Position and Secondary Positions; the Kind of Movements which are possible .	139
§ 789.	Listing's Law; the Experimental Proof	141
§ 790.	The Muscles of the Eye-Ball	143
§ 791.	The Action of the Ocular Muscles	145
§ 792.	The Nervous Mechanism of the Movements of the Eye-Balls; the Co-ordination of the Movements	148
§ 793.	The Nervous Centres for the Movements of the Eye-Balls . . .	151
§ 794.	The Horopter	153

SECTION XII.

ON SOME FEATURES OF VISUAL PERCEPTIONS AND ON VISUAL JUDGMENTS.

§ 795.	On the Differences between the Objective Field of Sight and the Subjective Field of Vision	155
§ 796.	The Psychical Processes belonging to Visual Perceptions; Illusions and Visual Judgments	157
§ 797.	Appreciation of Apparent Size	158
§ 798.	Judgment of Distance and of Actual Size	160
§ 799.	The Judgment of Solidity	162
§ 800.	The Struggle of the Two Fields of Vision	163

SECTION XIII.

THE NUTRITION OF THE EYE.

§ 801.	The Arrangement of the Blood Vessels	165
§ 802.	The Vaso-Motor Changes in the Eye.	166

The Lymphatics of the Eye.

§ 803.	The Lymphatic Vessels and Lymph Spaces of the Eye	166
§ 804.	The Aqueous Humour; the Changes taking place in it; how effected	168
§ 805.	The Vitreous Humour; the Changes taking place in it	170

SECTION XIV.

THE PROTECTIVE MECHANISMS OF THE EYE.

§ 806.	The Eye-Lids and their Muscles	172
§ 807.	The Conjunctiva and its Glands. Tears. The Secretion of Tears	173

CHAPTER IV.

HEARING.

SECTION I.

ON THE GENERAL STRUCTURE OF THE EAR AND ON THE STRUCTURE AND FUNCTIONS OF THE SUBSIDIARY AUDITORY APPARATUS.

		PAGE
§ 808.	The Embryonic History of the Ear. The Otic Vesicle	176
§ 809.	The General Relations of the Parts of the Ear; Vestibule and Cochlea, Membranous and Bony Labyrinth, Tympanum, Auditory Ossicles, Membrana Tympani, and External Meatus	177
§ 810.	The General Use of the Several Parts	180

The Conducting Apparatus of the Tympanum.

§ 811.	The Auditory Ossicles	181
§ 812.	The Tympanum; its Structure and Relations	182
§ 813.	The Membrana Tympani	185
§ 814.	The Joints, Ligaments, and Positions of the Auditory Ossicles	187
§ 815.	The Chain of Ossicles as a Lever	189

The Conduction of Sound through the Tympanum.

§ 816.	Longitudinal and Transversal Sonorous Vibrations. The Vibrations of the Tympanic Membrane	190
§ 817.	The Conduction of Vibrations through the Chain of Ossicles	192
§ 818.	The Conduction of Vibrations through the Bones of the Skull	193
§ 819.	The Action of the Tensor Tympani and Stapedius Muscles	194
§ 820.	The Eustachian Tube	196

SECTION II.

THE STRUCTURE OF THE LABYRINTH.

§ 821.	The Parts of the Vestibule; Utricle, Saccule, and Semicircular Canals	198
§ 822.	The Perilymph Cavity of the Vestibule	200
§ 823.	The Auditory Nerve; the Maculæ and Cristæ Acusticæ	201
§ 824.	The General Structure of the Cochlea	202

The Vestibular Labyrinth.

§ 825.	The Minute Structure of the Parts free from Nerve Endings	206
§ 826.	The Structure of the Cristæ Acusticæ	206

CONTENTS. xiii

PAGE

§ 827. The Constituent Cells; Cylinder or Hair Cells, Rod or Spindle Cells   208
§ 828. The Structure of the Maculæ Acusticæ   210
§ 829. The Otoconia and Otoliths   210

The Cochlea.

§ 830. The Canalis Cochlearis, its several parts   211
§ 831. The Basilar Membrane   212
§ 832. The Organ of Corti   214
§ 833. The Rods of Corti   216
§ 834. The inner Hair Cells   217
§ 835. The outer Hair Cells   218
§ 836. The Endings of the Cochlear Nerve   219
§ 837. The Tectorial Membrane   221
§ 838. The Differences of the Organ of Corti in Different Parts of the Spiral   221
§ 839. Measurements of some Parts of the Cochlea   222

SECTION III.

ON AUDITORY SENSATIONS.

§ 840. Noises and Musical Sounds   223
§ 841. The Characters of Musical Sounds; Loudness, Pitch, and Quality;
        Fundamental and Partial Tones   223
§ 842. The Limits of Auditory Sensations   225
§ 843. Appreciation of Differences of Pitch   226
§ 844. The Number of Vibrations needed to excite a Sensation . . .   226
§ 845. The Characters of Noises   227
§ 846. The Effects of Exhaustion   228
§ 847. The Fusion of Auditory Sensations   229
§ 848. The Interference of Vibrations. Beats   229

SECTION IV.

ON THE DEVELOPMENT OF AUDITORY IMPULSES.

§ 849. The Transmission of Impulses through the Labyrinth; the Functions of the Hairs and Otoliths   232
§ 850. The Analysis of Complex Waves of Sound; Theories as to the Mode
        of Action of the Organ of Corti   234
§ 851. The Appreciation of the Nature of Sounds ultimately a Psychical
        Process   237
§ 852. Probable Functions of the Vestibular Labyrinth   238

CONTENTS.

SECTION V.

On Auditory Perceptions and Judgments.

		PAGE
§ 853.	Auditory Phantoms	241
§ 854.	The Appreciation of Outwardness in Sounds is connected with the Tympanum	242
§ 855.	Hearing binaural. The Judgment of the Direction of Sounds	243
§ 856.	Judgment of the Distance of Sounds	244

CHAPTER V.

Taste and Smell.

SECTION I.

The Structure of the Olfactory Mucous Membrane.

§ 857.	The Structure of Olfactory Mucous Membrane, Cylinder Cells and Rod Cells.	246
§ 858.	The Termination of the Olfactory Fibres	248

SECTION II.

Olfactory Sensations.

§ 859.	The Sensation due to Contact of Particles with the Membrane	250
§ 860.	The Chief Characters of Olfactory Sensations	251
§ 861.	Olfactory Judgments. The Olfactory Nerve the Nerve of Smell	252

SECTION III.

The Structure of the Organs of Taste.

§ 862.	The Structure of the Lining Membrane of the Mouth	254
§ 863.	Papillæ Foliatæ; Taste Buds, Rod Cells, and Subsidiary Cells. The Distribution of Taste Buds	255

SECTION IV.

Gustatory Sensations.

§ 864.	Sensations of Taste usually or frequently accompanied by other Sensations	259
§ 865.	The Different Kinds of Taste. Sensations of Taste provoked by Mechanical and Electrical Stimulation	259
§ 866.	The Conditions under which Taste Sensations are excited . . .	261
§ 867.	The Distribution of the Several Kinds of Tastes. Theories as to the Mode of Origin of Taste Sensations.	262
§ 868.	The Nerves of Taste; the Chorda Tympani ·	265

CHAPTER VI.

ON CUTANEOUS AND SOME OTHER SENSATIONS.

SECTION I.

THE NERVE ENDINGS OF THE SKIN.

		PAGE
§ 869.	General and Special Nerve Endings. End-bulbs	267
§ 870.	Pacinian Corpuscles	268
§ 871.	Grandry's Corpuscles	270
§ 872.	Touch Corpuscles	270
§ 873.	Nerve Endings in the Cornea	272
§ 874.	Nerve Endings in the Epidermis	272
§ 875.	'Touch Cells' in the Epidermis and elsewhere	273

SECTION II.

THE GENERAL FEATURES OF CUTANEOUS SENSATIONS.

§ 876.	Three Kinds of Cutaneous Sensations, of Pressure, of Heat and Cold, and of Pain	274

Tactile Sensations or Sensations of Pressure.

§ 877.	The General Characters of Tactile Sensations	274
§ 878.	The Localisation of Tactile Sensations	276

Sensations of Heat and Cold.

§ 879.	Sensations of Heat and Cold due to sudden Changes in the Temperature of the Skin	278
§ 880.	The General Characters of Temperature Sensations	279
§ 881.	Tactile and Temperature Sensations in Parts other than the External Skin	280

SECTION III.

ON PAINFUL AND SOME OTHER KINDS OF SENSATION.

§ 882.	Sensations of Pain distinct from other Sensations	281
§ 883.	Sensations of Pain are Extreme Degrees of Common Sensibility	281
§ 884.	Special Nerve Endings not Necessary for Sensations of Pain	284
§ 885.	Hunger and Thirst	285

SECTION IV.

ON THE MODE OF DEVELOPMENT OF CUTANEOUS SENSATIONS.

	PAGE
§ 886. The Specific Energy of Nerves. Special Terminal Organs necessary for the Sensations of Touch and Temperature as distinguished from Sensations of Pain	287
§ 887. The Terminal Organs for Sensations of Pressure different from those for Sensations of Temperature	290
§ 888. The Terminal Organs for Sensations of Heat different from those for Sensations of Cold	291
§ 889. The Importance of Contrast in Cutaneous Sensations	292
§ 890. The Nature of the Terminal Organs	293

SECTION V.

THE MUSCULAR SENSE.

§ 891. We possess a Sense of 'Movement,' of 'Position,' and of 'Effort'	295
§ 892. The Muscular Sense Distinguished from the Sense of Effort . .	296
§ 893. The Afferent Impulses forming the Basis of the Muscular Sense are Distinct from Cutaneous Impulses	297
§ 894. They are Derived from the Muscles, Ligaments, and Tendons . .	299

SECTION VI.

ON TACTILE PERCEPTIONS AND JUDGMENTS.

§ 895. The Ties between Touch and the Muscular Sense	302
§ 896. The Ties between Touch and Sight	303
§ 897. Cutaneous Sensations may arise otherwise than from Cutaneous Events	304
§ 898. Tactile Illusions	305

CHAPTER VII.

ON SOME SPECIAL MUSCULAR MECHANISMS.

SECTION I.

THE VOICE.

§ 899. Voice Produced by Vibrations of the Vocal Cords	306
§ 900. The Thyroid and Cricoid Cartilages	308
§ 901. The Arytenoid Cartilages	308

CONTENTS.

		PAGE
§ 902.	The Glottis. The Superior Aperture and the Several Features of the Larynx	309
§ 903.	The Minute Structure of the Parts of the Larynx	312
§ 904.	The Laryngoscopic View of the Larynx	313
§ 905.	The Fundamental Features of the Voice; Loudness, Pitch, and Quality. The Main Conditions of the Utterance of Voice; Adduction and Tightening of the Vocal Cords	316
§ 906.	The Muscles of the Larynx	317
§ 907.	The Action of the Muscles in Reference to narrowing and widening the Glottis and to tightening and slackening the Vocal Cords	322
§ 908.	The Nervous Mechanisms of the Larynx. The Respiratory Movements of the Larynx	323
§ 909.	The Nervous Mechanism of Phonation	325
§ 910.	The Cortical Area for Movements of the Larynx	325
§ 911.	The Different Kinds of Voice. Changes in the Glottis other than those of mere Adduction and General Tension	326
§ 912.	Chest-Voice and Head-Voice. The Registers of the Voice. The Complexity of the Laryngeal Movements	328
§ 913.	The Uses of the Ventricles and other Parts of the Larynx	332
§ 914.	The 'Break' in the Voice at Puberty	332

SECTION II.

SPEECH.

§ 915.	Speech, a Mixture of Musical Sounds and of Noises	333
§ 916.	Vowels and Consonants	333
§ 917.	The Manner of Formation of the Several Vowels	334
§ 918.	The Manner of Formation of the Several Groups of Consonants	337
§ 919.	The Manner of Formation of the more Important Individual Consonants	338

SECTION III.

ON SOME LOCOMOTOR MECHANISMS.

§ 920.	The General Characters of the Actions of Skeletal Muscles	342
§ 921.	The Erect Posture	343
§ 922.	Walking	343

BOOK IV.

THE TISSUES AND MECHANISMS OF REPRODUCTION.

	PAGE
§ 923. The General Features of Reproduction	349

CHAPTER I.

IMPREGNATION.

SECTION I.

ON SOME STRUCTURAL FEATURES OF THE FEMALE ORGANS.

§ 924. The General Structure of the Uterus	351
§ 925. The Minute Structure of the Uterus	352
§ 926. The Structure of the Vagina	353
§ 927. The Blood Vessels and Lymphatics of the Uterus	353

The Ovary.

§ 928. The Early History of the Ovary	353
§ 929. The General Structure of the Ovary	354
§ 930. The Structure of the Graaffian Follicles	355
§ 931. The Features of a Ripe Ovum	356
§ 932. The Graaffian Follicle nourishes the Ovum	357
§ 933. The Escape of the Ovum	358
§ 934. The Changes in the Ovary after the Escape of the Ovum; Corpus Luteum	358

SECTION II.

MENSTRUATION.

§ 935. The Transference of the Ovum from the Ovary to the Uterus	360
§ 936. The Changes in the Uterine Mucous Membrane	361
§ 937. The Causation of Menstruation	362

SECTION III.

THE MALE ORGANS.

§ 938. The Early History of the Testis	364
§ 939. The General Structure of the Testis and Epididymis	364
§ 940. The Seminal Tubules; the Spermatozoa	366

CONTENTS.   xix

		PAGE
§ 941.	The Formation of the Spermatozoa	366
§ 942.	The Movements of the Spermatozoa	368
§ 943.	The Chemical Constituents of Semen	369
§ 944.	The Lymphatics of the Testis	369

Accessory Organs.

§ 945.	The Vesiculæ Seminales and Prostate	369
§ 946.	Erectile Tissue	370
§ 947.	The Nature of Erection	371
§ 948.	The Emission of Semen	372

CHAPTER II.

PREGNANCY AND BIRTH.

SECTION I.

THE PLACENTA.

§ 949.	The Spermatozoon enters and unites with the Ovum	374
§ 950.	The Formation of the Decidua	374
§ 951.	The Decidua Serotina is transformed into the Placenta	376
§ 952.	The Structure of the Placenta	377
§ 953.	The Nature of the Vascular Events taking place in the Placenta	378
§ 954.	The Shedding of the Placenta	379

SECTION II.

THE NUTRITION OF THE EMBRYO.

§ 955.	The Embryo breathes by and feeds upon the Maternal Blood of the Placenta	382
§ 956.	The Blood and Blood-Flow in the Umbilical Arteries and Umbilical Vein	383
§ 957.	The Amniotic Fluid, its Nature and Origin, its Relations to the Nutrition of the Fœtus	385
§ 958.	The Transmission of Food Material from the Mother to the Fœtus	386
§ 959.	Glycogen in the Fœtus	388
§ 960.	The Movements of the Fœtus	388
§ 961.	The Digestive Functions of the Fœtus	389
§ 962.	The Fœtal Circulation towards the Close of Uterine Life	390
§ 963.	The Cause of the First Breath	392
§ 964.	The Changes in the Circulation taking place at Birth	393

SECTION III.

PARTURITION.

		PAGE
§ 965.	The Period of Gestation	395
§ 966.	The Events of "Labour"	396
§ 967.	The Reflex Nature of Parturition	399
§ 968.	The Nerves Concerned in the Act	400
§ 969.	The more Exact Nature of the Act	400
§ 970.	The Causes determining the Onset of Labour	401
§ 971.	The Inhibition of Parturition	402

CHAPTER III.

THE PHASES OF LIFE.

§ 972.	The Composition of the Babe as compared with the Adult	403
§ 973.	The Curve of Growth from Birth onwards	404
§ 974.	The Characters of the Nutrition of the Babe and Infant	405
§ 975.	The Nervous System of the Babe	407
§ 976.	Dentition	408
§ 977.	Puberty. Differences of Sex	409
§ 978.	Old Age	410
§ 979.	Periodical Events	412
§ 980.	Sleep	412
§ 981.	Other Diurnal Changes in the Functions	416

CHAPTER IV.

DEATH.

§ 982.	The General Causation of Death	417

LIST OF FIGURES IN PART IV.

FIG.		PAGE
134.	Diagrammatic Outline of a Horizontal Section of the Eye to illustrate the Relations of the Various Parts	3
135.	Diagram of simple Optical System	7
136.	Diagram of the Schematic or Diagrammatic Eye	10
137.	Diagram of the Formation of a Retinal Image	11
138.	Diagram of Scheiner's Experiment	15
139.	Diagram of Images reflected from the Eye	19
140.	Diagram of the Ciliary Muscle as seen in a Vertical Radial Section of the Ciliary Region	27
141.	Diagram to illustrate Accommodation	32
142.	Diagrammatic representation of the Nerves governing the Pupil	36
143.	Diagram illustrating Chromatic Aberration	50
144.	Diagram to illustrate Entoptical Images	52
145.	Diagram to illustrate the Structure of the Retina	60
146.	Diagram of three Primary Colour Sensations	92
147.	Diagram to illustrate Hering's Theory of Colour Vision	95
148.	Diagram illustrating the Formation of Purkinjé's Figures when the Illumination is directed through the Sclerotic	112
149.	Diagram illustrating the Formation of Purkinjé's Figures when the Illumination is directed through the Cornea	113
150.	Diagram to illustrate the Principles of a simple Form of Ophthalmoscope	121
151.	Figure to illustrate Irradiation	126
152.	The Visual Field of the Right Eye	136
153.	The Visual Fields (Fields of Sight) of the two Eyes when the Eyes Converge to the same Fixed Point	137
154.	Diagram illustrating Corresponding Points	138
155.	Figure to illustrate the Insertions of the Ocular Muscles	144
156.	Diagram to illustrate the Actions of the Ocular Muscles	146
157.	Diagram illustrating a simple Horopter	153
158.	Figure to illustrate the Appreciation of Apparent Size	158
159.	The Same	159
160.	Figure to illustrate an Optical Effect produced by Parallel Slanting Lines	159

LIST OF FIGURES IN PART IV.

FIG.		PAGE
161.	Figure to illustrate Binocular Vision	162
162.	Diagram to illustrate the General Structure of the Ear	178
163.	The Bony Labyrinth	179
164.	The Auditory Ossicles	182
165.	Diagram to illustrate the Relations of Auditory Passage, Tympanum and Eustachian Tube	183
166.	Diagram of the Median Wall of the Tympanum	183
167.	Diagram of the Outer Wall of the Tympanum as seen from the Mesial Side	184
168.	Frontal Section through the Tympanum	185
169.	The Membrana Tympani	186
170.	The Ossicles in Position	187
171.	The Ligaments of the Ossicles	188
172.	The Stapes in Position	189
173.	The Malleus and Incus in Position	190
174.	Diagram of the Outer Wall of the Tympanum as seen from the Mesial Side	195
175.	The Membranous Labyrinth as seen from above	198
176.	The Membranous Labyrinth and the Endings of the Auditory Nerve	199
177.	Diagram of a Transverse Section of a Whorl of the Cochlea	203
178.	Diagram to illustrate the Structure of a Crista or a Macula	207
179.	Diagram of the Organ of Corti	213
180.	Diagram of the Constituents of the Organ of Corti	215
181.	The Cartilages of the Larynx	307
182.	Diagram of the Superior Aperture of the Larynx	310
183.	Diagram of the Larynx in Vertical Section	310
184.	Diagram of the Larynx in Vertical Transverse Section	312
185.	Diagram of a Laryngoscopic View of the Larynx	314
186.	The Larynx as seen by means of the Laryngoscope in Different Conditions of the Glottis	315
187.	Diagram of the Transverse and Oblique Arytenoid and of the Posterior Crico-Arytenoid Muscles	318
188.	Diagram to illustrate the Thyro-Arytenoid Muscles	319
189.	The Internal Thyro-Arytenoid Muscle	319
190.	The Lateral Crico-Arytenoid Muscle	321
191.	The Crico-Thyroid Muscle	321
192.	Diagram to illustrate the Contact of the Feet with the Ground in walking	345
193.	Diagram to illustrate running	346
194.	Diagram to illustrate the Fœtal Circulation	391

CHAPTER III.

SIGHT.

SEC. 1. ON THE GENERAL STRUCTURE OF THE EYE, AND ON THE FORMATION OF THE RETINAL IMAGE.

§ **702.** In dealing with the brain we have been incidentally obliged to deal with some of the facts connected with the senses; but we must now study the details of the subject. And, for the very reason that it is the most highly developed and differentiated sense, it will be convenient to begin with the sense of sight; we shall find that the study of it throws more light on the simpler and more obscure senses than the study of them throws on it.

A ray of light entering the eye and falling on the retina gives rise to what we call a sensation of light; but in order that distinct vision of any object emitting or reflecting rays of light may be gained, an image of the object must be formed on the retina, and the better defined the image the more distinct will be the vision. Hence in studying the physiology of vision, our first duty is to examine into the arrangements by which the formation of a satisfactory image on the retina is effected; these we may call briefly the dioptric mechanisms. We shall then have to inquire into the laws according to which rays of light impinging on the retina give rise to nervous impulses, and into the laws according to which the sensory impulses thus generated, which we will call visual impulses, give rise in turn to visual sensations. Here we shall come upon the difficulty of distinguishing between the events which are of physical origin, due to changes in the retina and optic fibres, and those which are of psychical origin, due to features of our own consciousness; for many of our conclusions are based on an appeal to consciousness. We shall find our difficulties further increased by the fact, that in appealing to our own consciousness we are apt to fall into error by failing to distinguish between those affections of consciousness which are the primary and direct results of the stimulation of the retina and those secondary, more recondite, affections of consciousness to which the former, through the intricate working of the central nervous system, give rise, or, in familiar language, by confounding what we see with what we think we see. These two things we will briefly distinguish as visual sensations and visual judgments; and we shall find that even in vision with one eye, though more especially in binocular vision, visual judgments form a very large part of what we frequently speak of as our "sight."

§ 703. In the structure of the eye we may distinguish two parts: the one is the retina, in which visual impulses are generated; the other comprises all the rest of the eyeball, for all the other structures serve either as a dioptric mechanism or as a means of nourishing the retina. This distinction is readily seen when we trace out the early history of the eye.

The first of the three primary cerebral vesicles (§ 600), that which is the forerunner of the third ventricle, buds out on each side the stalked and hollow optic vesicle. The wall of this optic vesicle, like that of the rest of the medullary tube, consists of epithelium, and the cavity of the vesicle is at first continuous, through the canal of the hollow stalk, with that of the medullary tube. The whole is covered over by the layer of epiblast which, with scanty underlying mesoblast, is the rudiment of the future skin.

Very soon the vesicle is doubled back upon or folded in upon itself so that the originally more or less spherical hollow single-walled vesicle is converted into a more or less hemispherical cup with a double wall, one the hind or outer wall corresponding to the hind half, and the other the front or inner wall to the front half of the vesicle, the two walls of the cup coming eventually into contact so that the cavity of the vesicle is obliterated. The folding is somewhat peculiar, inasmuch as it takes place not only at the front but also and indeed chiefly at the side, forming at the side a cleft, the choroidal fissure, the edges of which ultimately unite. We cannot enter into the details of the matter here, and indeed only refer to the character of the folding in order to point out that it involves the stalk as well as the cup. The stalk is first flattened and then doubled up lengthwise, a quantity of mesoblastic tissue being thrust into the hollow of the fold; and eventually the originally hollow stalk becomes a solid stalk having within it a core of mesoblastic tissue, carrying blood vessels. This core of vascular mesoblast, the origin of the future central artery of the retina, is continuous with a quantity of mesoblast which enters into the hollow of the cup at the time of folding, and, as we shall see, the central artery of the stalk is up to a certain stage of development carried forward through the centre of the cup. The cup becomes what we may speak of broadly as the retina, and we may call it the *optic* or *retinal* cup; the solid stalk becomes the optic nerve; and the other parts of the eyeball are formed round this retinal cup, which remains as the essential part of the eye.

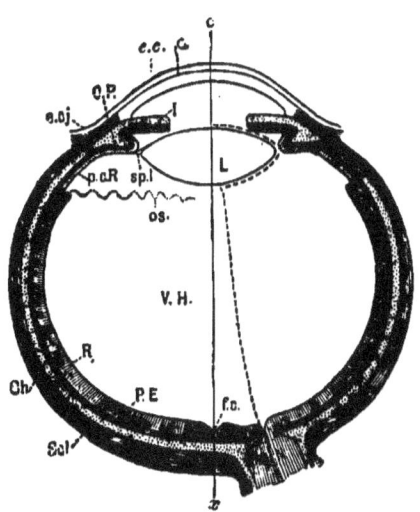

Fig. 134. Diagrammatic outline of a horizontal section of the eye, to illustrate the relations of the various parts.

The figure is to be regarded as very diagrammatic, more or less distortion of the relative sizes of the various parts and of the relative thickness of the coats being unavoidable in the effort to secure simplicity.

Scl. the sclerotic coat, shaded longitudinally, continuous with the (unshaded) body of the cornea. *e.c.* the epithelium of the cornea continuous with *e.cj.* the epithelium of the conjunctiva.

Ch. the choroid coat, with *C. P.* the ciliary process and *I.* the body of the iris, all stippled to indicate that they are all parts of the same vascular investment.

R. the retina or inner wall, and *P. E.* the pigment epithelium or outer wall of the retinal cup. In front of the wavy line *OS*, marking the position of the ora serrata, the retina proper changes into the pars ciliaris retinæ, *p.c.R.* Both the pigment epithelium and the pars ciliaris retinæ are represented as continued over the back of the iris as well as over the ciliary process.

L. the lens. *sp. l.* the suspensory ligament. The broken line round the lens, shewn on one side only, represents the membrana capsulo-pupillaris; and the straight continuation of it through *V. H.* the vitreous humor to *O. N.* the optic nerve indicates the embryonic continuation of the central artery of the retina.

o. x. the optic axis, in this case made to pass through the fovea centralis *fc.*

The front or inner wall of the retinal cup is from the first distinctly thicker than the hind or outer wall (Fig. 134); it soon consists of more than one layer of epithelium, and it alone, or, more strictly speaking, part of it alone, becomes the retina proper.

The hind or outer wall remains thin, and continues to consist of a single layer of epithelium, the cells of which are never developed into nervous elements but soon become loaded with pigment, and the greater part of it is known in the adult eye as the pigment epithelium of the retina, which, as we shall see, is in close functional connection with the nervous elements of the retina proper. At the time when the epithelial cells of the stalk of the retinal cup are developed into the fibres of the optic nerve, these become connected with the elements of the inner or retinal wall only of the cup; they pierce the outer wall of pigment epithelium, making no connections with the cells of that outer wall.

The retina then, in which by the action of light visual impulses are generated, is in reality a part of the brain, removed to some distance from the rest of the brain but remaining connected with it by means of the tract of white matter which we call the optic nerve; and, as we shall see, the retina is in structure similar to parts of the grey matter of the brain. The optic nerve is not like other nerves an outgrowth from the central nervous system, but like the olfactory tract (§ 674) a commissure of white matter between two parts of the brain, namely, between the outlying retina and the internally placed corpus geniculatum, pulvinar, and corpus quadrigeminum. We shall find accordingly that in structure it differs from ordinary cranial or spinal nerves.

Into the mouth of the retinal cup there is thrust a rounded mass of epithelium, an involution from the superficial epiblast; this becomes the lens. The hollow of the retinal cup is occupied, as we have said, by mesoblast; this ultimately becomes modified into the vitreous humour. The mesoblastic tissue surrounding the cup is developed into an investment of two coats; an inner, somewhat loose and tender, vascular and in part muscular coat, which on the one hand serves to nourish the retina, and on the other hand carries out certain movements of the dioptric apparatus, and an outer, firmer and denser coat, which affords protection to the whole of the structures within. The inner vascular coat, which may be compared to the pia mater, is called the *choroid* (Fig. 134 *Ch.*), and in the front part of the eye, at about the level of the lens, is thrown into a number of radiating folds or plaits, the ciliary processes *C.P.* The outer coat, which may be compared to the dura mater, is called the *sclerotic* (Fig. 134 *Scl.*). Over the greater part of the eyeball the two coats are in apposition, or separated only by narrow lymphatic spaces, which may be compared with the subarachnoid spaces, but towards the front they diverge; the choroid is bent inwards towards the central axis of the eye to form the diaphragm called the *iris* (Fig. 134 *I.*), while the sclerotic is continued forwards to form, beneath the epidermis into which the superficial epiblast is developed, the basis of the *cornea* (Fig. 134 *C.*). At the angle of divergence of the two coats is developed a small mass of muscular fibres, the ciliary muscle of which we shall speak in detail presently.

The inner or front wall of the retinal cup becomes as we have said thick, and is developed into the retina; but this takes place only over about the hind three-fourths of the cup. Along a meridian round the eye, at a wavy boundary line called the *ora serrata* (Fig. 134 *O.S.*), the retina proper ceases and the inner wall of the retinal cup in front of the ora serrata is continued on as a much thinner structure (Fig. 134 *p.c.R.*) consisting of a single layer only of cells; this is spoken of as the *pars ciliaris retinæ*. The outer or hind wall of the retinal cup consists throughout of a single layer of epithelium cells loaded with pigment. Behind the ora serrata, that is, in the region of the retina proper, these cells have, as we shall see, peculiar features, but in front of the ora serrata they lose these features and become ordinary cubical cells, though still loaded with pigment.

Hence the choroid may be described as having a double lining. Over the hind part of the eye, behind the ora serrata, it is lined by the single layer of pigment epithelium and the retina. In front of the ora serrata it, including the ciliary processes, is lined by the same layer of pigment epithelium representing the outer wall, and by the single layer of cells, free from pigment, representing the inner wall of the retinal cup, the latter being called, as we have said, the pars ciliaris retinæ. And as the ciliary part of the choroid passes on to form the iris, these two layers are also continued on to line the back of the iris, coming to an end at the margin of the pupil or central opening of the iris, which may accordingly be taken as marking the extreme lip of the retinal cup. Fig. 134. Here however, as we shall see, the two layers are not so easily and distinctly recognised as in the ciliary region; and the nature of the structures forming the back of the iris has been a matter of much controversy.

At an early stage the mesoblastic tissue, which fills up the hollow of the retinal cup and surrounds the lens, is continuous at the mouth of the retinal cup with the outer investment of the cup; it here forms around the lens the *membrana capsulo-pupillaris*, and at the margin of the iris the *membrana pupillaris* blocking up the future opening of the pupil. The arteria centralis retinæ, which during the folding of the cup and stalk is carried into the core of the optic nerve, does not at this early stage stop at the retina, but is continued forward through the middle of the vitreous humour to the membrana capsulo-pupillaris, and furnishes the developing lens with an abundant supply of blood. But neither layer of the retinal cup stretches over the pupillary membrane; they both stop, as we have said, at the margin of the iris. Before birth takes place, the membrana pupillaris is, in man, absorbed and the pupil is thus established, at the same time the central artery in the vitreous humour is obliterated beyond the retina, and the vascular membrana capsulo-pupillaris gives place to the non-vascular capsule of the lens and the suspensory ligament of which we shall speak hereafter.

Between the iris, which is the extreme front of the choroid investment, and the cornea, which is the extreme front of the sclerotic investment, the lymphatic spaces which over the rest of the eye are narrow and linear are developed into a large conspicuous chamber, the *anterior chamber* of the eye, which upon the establishment of the pupil by the absorption of the pupillary membrane becomes continuous with the smaller " posterior chamber " of the eye or space between the back surface of the iris and ciliary processes on the outside and the suspensory ligament with the lens on the inside. The cavity of the conjoined anterior and posterior chambers, being a continuation and enlargement of the flatter spaces between the choroid or pia mater of the eye, and sclerotic or dura mater of the eye, may be likened to the sub-arachnoid space, and like that space contains a peculiar fluid; this, which is called the aqueous humour, like the cerebro-spinal fluid, differs from ordinary lymph, and is probably, to a large extent, furnished by the ciliary processes in some such way as the cerebro-spinal fluid is furnished by the choroid plexuses (§ 695).

The Formation of the Retinal Image.

§ 704. The iris and choroid coat contain, as we have said, muscular elements, and by means of these muscular elements changes in the form and relations of some of the parts of the eye are brought about; hence we have to distinguish between the eye at rest, and the eye which is undergoing one or other of these changes. It will be convenient to reserve what we have further to say concerning the histological and other details of structure of these parts until we come to deal with these changes; and, simply premising that the cornea and lens are transparent bodies with surfaces of certain curvatures, we may at once pass to the consideration of the dioptrics of the eye at rest.

The eye is a camera, consisting of a series of surfaces and media arranged in a dark chamber, the iris serving as a diaphragm; and the object of the apparatus is to form on the retina a distinct image of external objects. That a distinct image is formed on the retina, may be ascertained by removing the sclerotic from the back of an eye, and looking at the hinder surface of the transparent retina while rays of light proceeding from an external object are allowed to fall on the cornea. To understand how such an image is formed, we must call to mind a few optical principles.

A dioptric apparatus in its simplest form consists of two media of different refractive power separated by a (spherical) surface; and the optical properties of such an apparatus depend upon (1) the curvature of the surface, (2) the relative refractive powers of the media.

Such a simple optical system is represented in Fig. 135, where *apb* represents, in section, a curved (spherical) surface separating a less refractive medium, on the left hand towards *O*, from a more refractive medium on the right hand towards *A*. The surface in question is symmetrically placed as regards the line *OA*, which falling normal (perpendicular) to the surface at *p* passes through the centre *n* of the sphere with whose surface we are dealing. This line is called the *optic axis*.

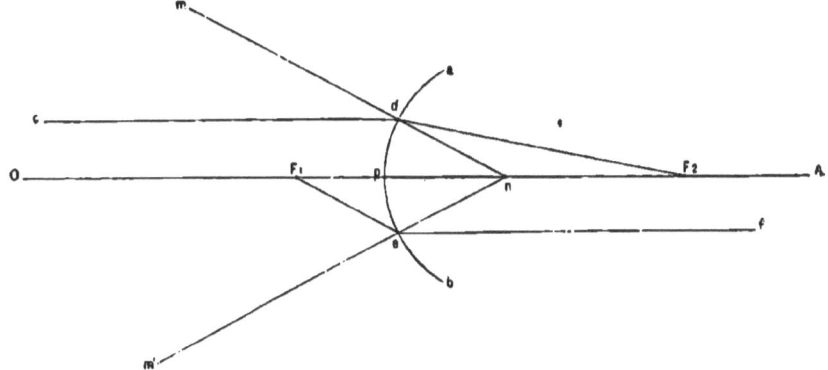

FIG. 135. DIAGRAM OF SIMPLE OPTICAL SYSTEM.

All rays of light which, in passing from the first less refractive to the second more refractive medium, cut the surface normally, such as the one, *Op*, in the line of the optic axis, and others, such as *md*, *m'e*, undergo no refraction; all such rays are continued on as straight lines, and all pass through *n* the centre of the sphere or *nodal point*. All other rays passing from the first to the second medium are refracted. Of these all those which lie in the first medium parallel to the optic axis, such as *cd*, are so refracted as to meet in the second medium at a point, F_2, on the optic axis; this is called the *principal posterior* (or second) *focus*. On the optic axis in the first medium there is another important point, F_1, the rays of light passing from which, such as F_1e, are so refracted in passing into the second medium as to become parallel, *ef*, to the optic axis; this point is called the *principal anterior* (or first) *focus*. The point at which the optic axis cuts the surface is, for reasons which we shall see presently, called the *principal point*. The above points, viz. the posterior and the anterior principal foci, the nodal point, and the principal point are the *cardinal points* of such an optical system.

Such a simple system, however, does not represent the optical conditions of the eye, for this consists of several media bounded by several surfaces, the latter differing from each other in curvature, though being approximately spherical. Rays of light in passing from an external object to the retina traverse in succession the following surfaces and media :—the anterior surface of the cornea, the substance of the cornea, the posterior surface of the cornea, the aqueous humour, the anterior surface of the lens, the substance of the lens, the posterior surface of the lens, and the vitreous humour; so that we have to deal with four surfaces, and, including the external air, four media. Indeed the matter is in reality still more complicated, for the structure of the lens, as we shall see, is such that the substance of the lens differs somewhat in refractive power in different parts, the central parts being more refractive than the peripheral parts; moreover the lens is covered in front by a capsule different in structure from the lens itself. We may, however, neglect, without fear of serious error, these smaller differences, and consider the lens as one medium of uniform refractive power bounded by an anterior and a posterior surface. The cornea again, as we shall see, is not absolutely uniform in structure, but this we may also neglect and consider the cornea as a medium, also of uniform refractive power, bounded by an anterior and a posterior surface. Moreover, the posterior surface of the cornea is parallel to (concentric with) the anterior surface or nearly so. Now when the two surfaces which bound a medium are parallel to each other we may, in dealing with refraction, neglect the thickness of the medium entirely, we may suppose it to be absent and treat the two surfaces as if they were one. We may therefore, without serious error, neglect the substance of the cornea, and consider the cornea as affording one surface, its anterior surface, bounding the air in front from the aqueous humour behind. Lastly, the aqueous humour differs in refractive power so little from the vitreous humour that we may consider the two as forming one medium.

We have therefore to deal with three surfaces separating three media, viz.:—first, the anterior surface of the cornea, at which considerable refraction takes place as the rays of light pass from the less refractive air into the more refractive aqueous humour; secondly, the anterior surface of the lens, at which again considerable refraction takes place as the rays pass from the less refractive aqueous humour into the more refractive substance of the lens; and lastly the posterior surface of the lens, at which refraction takes place as the rays pass from the more refractive substance of the lens into the less refractive vitreous humour. The three surfaces, differing in curvature, are all approximately centred, symmetrically disposed around, the optic axis of the system. This optic axis meets the retina, according to some authorities, not quite at the part of the retina which, under the name of *fovea centralis*, we shall hereafter speak of as the centre of the retina, but a little above and to the nasal side of that part; other authorities, however, maintain that it does cut the retina at the fovea centralis.

§ 705. The eye, therefore, even with the simplifications which we have introduced, presents a much more complex optical system than the one described above. It has, however, been shewn mathematically that a complex optical system consisting of several surfaces and media centred on one optical axis may be treated as if it were a more simple system consisting of two surfaces only. In such a simplified system each of the two (ideal) surfaces has its own nodal point and its own principal foci, anterior and posterior; moreover, the two points where the two surfaces cut the optic axis are called *principal points* (and vertical planes drawn through those points *principal planes*), first or anterior, and second or posterior. Hence the cardinal points of such a simplified complex system are six in number, namely, the anterior and posterior principal foci, the anterior and posterior principal points, and the anterior and posterior nodal points. (When such a system is, by removal of surfaces and media, converted into the still more simple system of one surface separating two media, the two nodal points become coincident in one point, namely, the centre of the sphere, and the two principal points become coincident in one point, namely, the point at which the optic axis cuts the surface.)

In order to effect such a simplification of a complex optical system, it is requisite to know :—(1) The refractive index of each medium. (2) The radius of curvature of each surface. (3) The distance along the optic axis between the first surface on which the rays fall and the succeeding surfaces. These can be and have been determined for the human eye, and the following table gives the several values usually adopted with some recent corrections, the latter being placed in brackets.

Refractive index of aqueous or vitreous humour... 1·3376 (1·3365)
Mean refractive index of lens .... 1·4545 (1·4371)
Radius of curvature of cornea .... 8 (7·829) mm.
    ,,   ,,   of anterior surface of lens . . 10   ,,
    ,,   ,,   of posterior ,,   ,, ... 6   ,,
Distance from anterior surface of cornea to anterior surface of lens .... 4 (3·6) ,,
Thickness of lens .... 4 (3·6) ,,

By means of these measurements the optical system of the eye may be simplified into an optical system of two surfaces. In this 'schematic, or diagrammatic, eye of Listing,' as it is generally called, the two (ideal) surfaces, and the principal points where these cut the optic axis (Fig. 136, p^1, p^2, the two surfaces being indicated by dotted lines), lie close together in the front part of the aqueous humour, and the nodal points, n^1, n^2, lie, also close together, in the back part of the lens.

Further, the two principal surfaces lie so close together that, for practical purposes, no serious error is introduced, if instead of two such surfaces we assume the existence of one surface lying midway between the two. In this way we arrive at the 'reduced diagrammatic eye,' or 'the reduced eye' as it is called, in which the several surfaces and media of the actual eye are replaced by

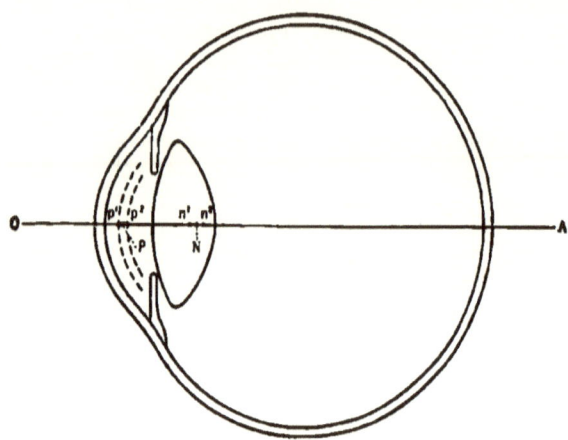

FIG. 136 DIAGRAM OF THE SCHEMATIC OR DIAGRAMMATIC EYE.

one (ideal) spherical surface (Fig. 136, P), having one nodal point, N; the two media which the surface separates are supposed to be air on the one side and water on the other.

The several positions of the cardinal points of this 'reduced eye' are as follows:

The *principal point*, where the one surface of the system cuts the optic axis, lies in the aqueous humour, 2·3448 mm. behind the anterior surface of the cornea.

The *nodal point* lies in the back part of the lens, ·4764 mm. in front of the posterior surface of the lens.

The *posterior principal focus* lies 22·647 (22·819) mm. behind the anterior surface of the cornea, that is to say, practically lies on the retina.

The *anterior principal focus* lies 12·8326 mm. in front of the anterior surface of the cornea.

The *radius of curvature* of the (ideal) surface is 5·1248 mm.; (that of the cornea is 8 mm. and of the interior surface of the lens 10 mm.).

§ **706.** By help of this 'reduced eye' we are enabled to trace out the paths of rays of light through the actual eye, and to study the formation of images on the retina. When an image of an external object, such as an arrow (Fig. 137), is formed in such an eye, each point of the object is considered as sending out a pencil of diverging rays, which by the system are made to converge again into the point in the image which corresponds to the point in the object. One such pencil of rays proceeds from the point at the extreme tip of the arrow, another from the extreme point at the other end, and other pencils from all the points between

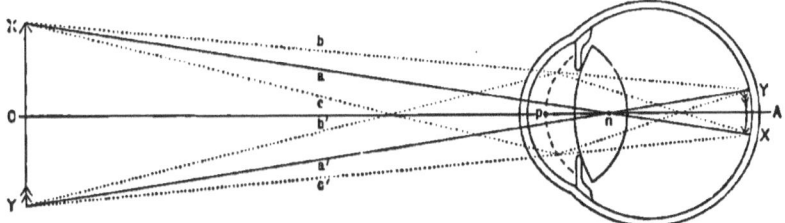

FIG. 137. DIAGRAM OF THE FORMATION OF A RETINAL IMAGE.

these two. Each such pencil has for its core a ray called the *principal ray*, a, a', around which are arranged, with increasing divergency, the other rays of the pencil, such as b, b', c, c'. When such a pencil of rays falls on the refracting surface, such as the 'principal surface' of the reduced eye, the principal ray of the pencil, a, being normal to that surface, is not refracted at all, but passes straight on through the nodal point n, while the other rays of the pencil, b, c, undergoing refraction according to their respective divergencies, are made to converge together at some point on the path of the principal ray, and thus form at that spot the image of the point from which the pencil proceeded. The exact position on the line of the principal ray, at which convergence takes place and at which the image is formed, will depend on the refractive power of the optical system in relation to the amount of divergence of the pencil; the refractive power of the system remaining the same, it will be nearer to, or farther from, the nodal point according as the rays are less or more divergent; and the divergence of the rays remaining the same, it will be nearer to, or farther from, the nodal point according as the refractive power of the system is greater or less.

Hence supposing the eye to be in that condition in which a distinct image of the arrow is formed on the retina, we can find the position on the retina of the image of the extreme point of the tip of the arrow, by simply drawing a straight line from that extreme point of the arrow X through the nodal point n of the 'reduced' eye. Such a straight line represents the path of the 'principal ray' of the pencil proceeding from the extreme tip of the arrow, and when an image is formed on the retina the other diverging rays of that pencil will be so refracted as to converge at the point x, where that line meets the retina; all the rays will form together there the image of the extreme point of the arrow. In a similar way a straight line drawn through the nodal point from the extreme point of the other end of the arrow, and continued until it meets the retina at y, will give us the position of the image of the other end of the arrow; and in like manner lines drawn from other points of the arrow through the same nodal point will give us the position on the retina of the images of those other points. In this way the construction of the reduced eye enables us to ascertain the position, magnitude and features of the retinal image of an object.

§ **707.** A ray of light, that is to say a series of waves of ether, falling upon a point of the retina stimulates certain structures in the retina and gives rise, as we have said, to visual impulses and so to a sensation of light; this we may consider as a visual sensation in its simplest form. When a number of different points of the retina are thus stimulated at the same time, as when an image of an external object falls in proper focus on the retina, the total result is a complex group of visual impulses and thus a complex sensation, by which we perceive, as we say, the object; and we frequently speak of this complex sensation as a visual image corresponding to the retinal image. The term is perhaps not a very desirable one, since it seems to imply an identity between the former which is a psychical matter, and the latter which is a physical matter; whereas, the one thing we may be sure about is that the psychical thing, though it is a sign and token of, is wholly different from the physical thing.

It will be as well perhaps thus early to call attention to the fact that, as indeed is shewn in Fig. 137, the image on the retina is an inverted one. What is the upper part of the object in the external world is represented in the lower part of the retinal image, what is on the right-hand side of the object is represented on the left-hand side of the image. In the visual judgment which is based upon the visual sensation, the retinal image is, as it were, reinverted; we take the left-hand side, or the bottom of the retinal image, as a token or sign of the right-hand side or the top of the object seen. We shall return to this matter later on; but in studying the dioptrics of the eye this inversion of the retinal image must always be borne in mind.

SEC. 2. THE FACTS OF ACCOMMODATION.

§ 708. When an object emitting or reflecting light, a lens, and a screen to receive the image of the object, are so arranged in reference to each other, that the image upon the screen is sharp and distinct, the rays of light proceeding from each luminous point of the object are brought into focus on the screen in a point of the image corresponding to the point of the object. If the object be then removed farther away from the lens, the rays proceeding in a pencil from each luminous point will be brought to a focus at a point in front of the screen, and, subsequently diverging, will fall upon the screen as a circular patch composed of a series of circles, the so-called *diffusion circles*, arranged concentrically round the principal ray of the pencil. If the object be removed, not farther, but nearer the lens, the pencil of rays will meet the screen before they have been brought to focus in a point, and consequently will in this case also give rise to diffusion circles. When an object is placed before the eye, so that the image falls into exact focus on the retina, and the pencils of rays proceeding from each luminous point of the object are brought into focus in points on the retina, the sensation called forth is that of a distinct image. When on the contrary the object is too far away, so that the focus lies in front of the retina, or too near, so that the focus lies behind the retina, and the pencils fall on the retina not as points, but as systems of diffusion circles, the sensation produced is that of an indistinct and blurred image. In order that objects both near and distant may be seen with equal distinctness by the same dioptric apparatus, the focal arrangements of the apparatus must be *accommodated* or adjusted to the distance of the object, either by changing the refractive power of the lens, or by altering the distance between the lens and the screen.

That the eye does possess such a power of accommodation or adjustment is shewn by every-day experience. If two needles be fixed upright, some two feet or so apart, into a long piece of wood, and the wood be held before the eye, so that the needles are nearly in a line, it will be found that if attention be directed to the far needle, the near one appears blurred and indistinct, and that, conversely, when the near one is distinct, the far one appears blurred. By an effort of the will we can at pleasure make either the far one or the near one distinct; but not both at the same time. When the eye is arranged so that the far needle appears distinct, the image of that needle falls exactly on the retina, and each pencil from each luminous point of the needle unites in a point upon the retina; but when the far needle is seen distinctly, the focus of the near needle lies *behind* the retina, and each pencil from each luminous point of this needle falls upon the retina in a series of diffusion circles; hence the image of the near needle is blurred. Similarly, when the eye is arranged so that the near needle is distinct, the image of that needle falls upon the retina in such a way, that each pencil of rays from each luminous point of the needle unites in a point on the retina, while each pencil from each luminous point of the far needle unites at a point *in front of* the retina, and then diverging again falls on the retina in a series of diffusion circles, and the far needle is now seen indistinctly. If the near needle be gradually brought nearer and nearer to the eye, it will be found that greater and greater effort is required to see it distinctly, and at last a point is reached at which no effort can make the image of the needle appear anything but blurred. The distance of this point from the eye marks *the near limit of accommodation* for near objects. Similarly, if the person be short-sighted, the far needle may be moved away from the eye, until a point is reached at which it ceases to be seen distinctly, and appears blurred; the *far limit of accommodation* is reached. In the one case, the eye, with all its power, is unable to bring the image of the needle sufficiently forward to fall on the retina: the focus lies permanently behind the retina. In the other, the eye cannot bring the image sufficiently backward to fall on the retina: the focus lies permanently in front of the retina. In both cases the pencils of rays from the needles strike the retina in diffusion circles.

§ **709.** The same phenomena may be shewn with greater nicety by what is called Scheiner's Experiment. If two smooth holes be pricked in a card, at a distance from each other less than the diameter of the pupil, and the card be held up, with the holes horizontal before one eye, the other being closed, and a needle placed vertically be looked at through the holes, the following facts may be observed. When attention is directed to the needle itself, the image of the needle appears single. Whenever the gaze is directed to a more distant object, so that the eye is no longer accommodated for the needle, the image of the needle appears double and at the same time blurred. It also appears double and blurred when the eye is accommodated for a distance nearer than that of the needle. When only one needle is seen, and the eye therefore is properly accommodated for the distance of the needle, the only effect produced by blocking up one hole of the card is that the needle and indeed the whole field of vision seems dimmer. When, however, the image is double on account of the eye being accommodated for a distance greater than that of the needle, blocking the left-hand hole causes a disappearance of the right-hand or opposite image, and blocking the right-hand hole causes the left-hand image to disappear. When the eye is accommodated for a distance nearer than that of the needle, blocking either hole causes the image on the same side to vanish. The following diagram will explain how these results are brought about

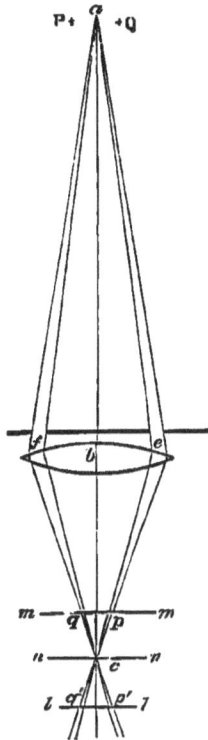

Fig. 138. Diagram of Scheiner's Experiment.

Let a (Fig. 138) be a luminous point in the needle, and ae, af the extreme right-hand and left-hand rays of the pencil of rays proceeding from the luminous point, and passing respectively through the right-hand e, and left-hand f, holes in the card. (The figure is supposed to be a horizontal section of the eye, and a forms part of a transverse section of the vertically placed needle.) When the eye is accommodated for a, the rays e and f meet together in the point c, the retina occupying the position of the plane nn; the luminous point appears as one point, and the needle will appear as one needle. When the eye is accommodated for a distance beyond a, the retina may be considered to lie[1] no longer at nn, but nearer the lens, at mm for example; the rays ae will cut this plane at p, and the rays af at q; hence the luminous point will no longer appear single, but will be seen as two points, or rather as two systems of diffusion circles, and the single needle will appear as two blurred needles. The rays passing through the right-hand hole e, will cut the retina at p, i.e. on the right-hand side of the optic axis; but, as we have already (§ 707) said, the image on the right-hand side of the retina is referred by the mind to an object on the left-hand side of the person; hence the affection of the retina at p, produced by the rays ae falling on it there, gives rise to an image of the spot a at P, and similarly the left-hand spot q corresponds to the right-hand Q. Blocking the left-hand hole, therefore, causes a disappearance of the right-hand image, and *vice versa*. Similarly when the eye is accommodated for a distance nearer than the needle, the retina may be supposed to be removed to ll, and the right-hand ae and left-hand af rays, after uniting at c, will diverge again and strike the retina in diffusion circles at p' and q'. The blocking of the hole e will now cause the disappearance of the image q' on the left-hand side of the retina, and this will be referred by the mind to the right-hand side, so that Q will seem to vanish.

If the needle be brought gradually nearer and nearer to the eye, a point will be reached within which the image is always double. This point marks with considerable exactitude the near limit of accommodation. With short-sighted persons, if the needle be removed farther and farther away, a point is reached beyond which the image is always double; this marks the far limit of accommodation.

The experiment may also be performed with the needle placed horizontally, in which case the holes in the card should be vertical.

The determination of the accommodation of the eye for near or far distances may be assisted by using two needles, one near and one far. In this case one needle should be vertical, and the other horizontal, and the card turned round so that the holes lie horizontally or vertically according to whether the vertical or horizontal needle is being made to appear double.

§ 710. In what may be regarded as the normal eye, the so-called *emmetropic* eye, the near limit of accommodation is about 10 or 12 cm., and the far limit may be put for practical purposes at an infinite distance. The 'range of distinct vision' therefore for the emmetropic eye is very great. In the *myopic*, or short-sighted eye, the near limit is brought much closer (5 or 6 cm.) to the

---

[1] Of course, in the actual eye, as we shall see, accommodation is effected by a change in the lens, and not by an alteration in the position of the retina; but for convenience sake, we may here suppose the retina to be moved

cornea; and the far limit is at a variable but not very great distance, so that the rays of light proceeding from an object not many feet away are brought to a focus in the vitreous humour instead of on the retina. The range of distinct vision is therefore in the myopic eye very limited. In the *hypermetropic*, or long-sighted eye, the rays of light coming from even an infinite distance are, in the passive state of the eye, brought to a focus beyond the retina. The near limit of accommodation is at some distance off, and a far limit of accommodation does not exist. The *presbyopic* eye, or eye of advanced years, resembles the hypermetropic eye in the near point of accommodation being at some distance, but differs from it inasmuch as the former is an essentially defective power of accommodation, whereas in the latter the power of accommodation may be good and yet, from the internal arrangements of the eye, be unable to bring the image of a near object on to the retina. When an eye becomes presbyopic, the far limit may remain the same, but since the power of accommodating for near objects is weakened or lost, the change is distinctly a reduction of the range of distinct vision. When no effort of accommodation is made, the principal posterior focus of the eye lies in the normal, emmetropic eye on the retina, in the myopic eye in front of it, and in the hypermetropic eye behind it.

§ 711. By what changes in the eye are we thus able, within the above mentioned limitations, to see distinctly objects at different distances? In directing our attention from a far to a very near object we are conscious of a distinct effort, and feel that some change has taken place in the eye; when we turn from a very near to a far object, if we are conscious of any change in the eye, it is one of a different kind. The former is the sense of an active exertion; the latter, when it is felt, is the sense of relaxation after exertion.

Since the far limit of an emmetropic eye is at an infinite distance, no such thing as active accommodation for far distances need exist. The only change which need take place in the eye in turning from near to far objects will be a mere passive undoing of the accommodation previously made for the near object. And that no such active accommodation for far distances takes place is shewn by the following facts; the eye, when opened after being closed for some time, is found adjusted not for moderately distant but for far distant objects; when the power of the eye to accommodate is impaired or abolished, as we shall see it may be, by atropin or nervous disease, the vision of distant objects may be unaffected; and we are conscious of no effort in turning from moderately distant to far distant objects. The sense of effort often spoken of by myopic persons as being felt when they attempt to see things at or beyond the far limit of their range seems to arise from a movement of the eyelids, and not from any internal changes taking place in the eye.

What then are the changes which take place in the eye, when we accommodate for near objects? It might be thought, and indeed once was thought, that the curvature of the cornea was changed, becoming more convex, with a shorter radius of curvature, for near objects. This is disproved by the fact that accommodation takes place as usual when the eye (and head) is immersed in water. Since the refractive powers of aqueous humour and water are very nearly alike, the cornea, with its parallel surfaces, placed between these two fluids, can have little or no effect on the direction of the rays passing through it when the eye is immersed in water. Moreover we have it in our power to detect any change in the curvature of the cornea which may take place. If a luminous body such as a candle be held in front of a convex surface like the cornea an image of the body is seen reflected from the surface; and, with the body at a certain distance, the image will be of a certain size. If now the curvature of the surface be increased, if the surface be made more convex, the image will diminish in size; if the curvature of the surface be diminished the image will increase in size. Indeed by measuring carefully the changes in the size of the image we may determine the amount of change in the curvature of the surface. And accurate measurements of the dimensions of an image on the cornea have shewn that these undergo no change during accommodation, and that therefore the curvature of the cornea is not altered. Nor is there any change in the form of the bulb; for any variation in this would necessarily produce an alteration in the curvature of the cornea, and pressure on the bulb would act injuriously by rendering the retina anæmic and so less sensitive. In fact, there are only two changes of importance which can be ascertained to take place in the eye during accommodation for near objects.

One is that the pupil contracts. When we look at near objects, the pupil becomes small; when we turn to distant objects, it dilates. This however cannot have more than an indirect influence on the formation of the image; the chief use of the contraction of the pupil in accommodation for near objects is to cut off the more divergent circumferential rays of light.

The other and really efficient change is that the anterior surface of the lens becomes more convex. If a light be held before the eye, three reflected images may, with care and under proper precautions, be seen by a bystander: one (Fig. 139 A, a) a very bright one caused by the anterior surface of the cornea, a second less bright, b, by the anterior surface of the lens, and a third very dim, c, by the posterior surface of the lens; when the images are those of an object, such as the flame of a candle, in which a top and bottom can be recognised, the two former images are seen to be erect, but the third inverted. When the eye is accommodated for near objects, no change is observed in the first, and none, or a very insignificant one, in the third of these images;

but the second, that from the anterior surface of the lens, is seen to become distinctly smaller, shewing that the surface has become more convex. When, on the contrary, vision is directed from near to far objects, the image from the anterior surface of

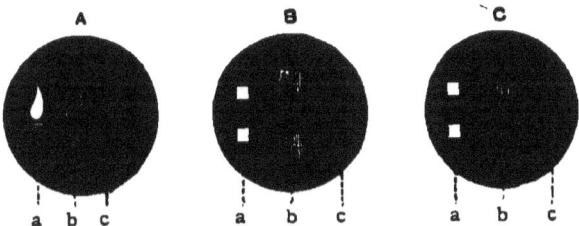

FIG. 139. DIAGRAM OF IMAGES REFLECTED FROM THE EYE.

In A are seen the three images of a candle reflected from *a*, the anterior surface of the cornea, *b*, the anterior surface of the lens, and *c* the posterior surface of the lens. *a* is bright and erect, *b* also erect, is larger but less bright, *c* inverted is small and dim.

B shews the images, two squares, as seen in the phakoscope when the eye is directed to a far object. C the same when the eye is accommodated for a near object. The pair *b* are in C, smaller and closer together than in B, shewing an increase of curvature.

the lens grows larger, indicating that the convexity of the surface has diminished, while no change takes place in the image from the cornea, and none, or hardly any, in that from the posterior surface of the lens. And accurate measurements of the size of the image from the anterior surface of the lens have shewn that the changes in curvature which do take place are considerable; the radius of curvature of the lens accommodated for near objects is 6 mm., for far objects 10 mm.; and this difference is sufficient to account for the power of accommodation which the eye possesses.

The observation of these reflected images is facilitated by the simple instrument introduced by Helmholtz and called a Phakoscope. It consists of a small dark chamber, with apertures for the observed and observing eyes; a needle is fixed at a short distance in front of the former, to serve as a near object, for which accommodation has to be made; and a lamp or candle is so disposed as to throw an image on each of the three surfaces of the observed eye. Since a change in the distance between two images is more readily appreciated than is a simple change of size of a single image, two prisms are employed so as to throw a double image in the form of bright squares on each of the three surfaces, Fig. 139 B, C. When the anterior surface of the lens becomes more convex the two images reflected from that surface approach each other, C, when it becomes less convex they retire from each other, B.

These observations leave no doubt that the essential change by which accommodation is effected, is an alteration of the convexity of the anterior surface of the lens. And that the lens is the agent of accommodation, is further shewn by the fact that after removal of the lens, as in the operation for cataract, the power of accommodation is lost. In the cases which have been recorded, where eyes from which the lens had been removed seemed still to possess some accommodation, we must suppose that no real accommodation took place, but that the pupil contracted when a near object was looked at, and so assisted in making vision more distinct.

To understand, however, the mechanism by which this change in the anterior surface of the lens is brought about, we must turn to some further details of the structure of the eye.

SEC. 3. ON THE STRUCTURE OF THE INVESTMENTS OF THE RETINA.

§ 712. *The Sclerotic Coat.* This coat is thickest at the hind pole of the eye, where it is pierced by the optic nerve at the *lamina cribrosa*, and becomes gradually thinner forwards, though at the extreme front, close to its junction with the cornea, it is again somewhat thickened by the insertion of the ocular muscles. It consists of bundles of ordinary white fibrillated connective-tissue, with which are mingled, especially on the inner surface next to the choroid, elastic fibres and connective-tissue corpuscles, the latter frequently carrying pigment. The bundles run for the most part in two directions, longitudinally, that is meridionally, and transversely, that is equatorially, the two sets of bundles crossing each other more or less regularly at right angles. The coat as a whole therefore presents itself as a thin but tough investment, with some but not very much extensibility and elasticity.

It is somewhat scantily supplied with blood vessels, and though it contains a large number of small lymph-spaces, lined with epithelioid plates, possesses no proper lymphatic vessels. On its inside next to the choroid is a large lymphatic space, the perichoroidal space, with which the above smaller spaces communicate; and a similar large space exists on the outer surface between the sclerotic itself and an investment of looser connective-tissue called Tenon's capsule; but we shall return to these when we come to speak of the lymphatics of the whole eye.

§ 713. *The Choroid Coat.* This, which around the optic nerve behind and at the junction of the cornea and sclerotic in front is closely attached to the sclerotic but loosely connected with it elsewhere, consists essentially of blood vessels and of certain muscular and nervous elements imbedded in connective-tissue possessing special features. It is, in the first instance, a vascular coat destined to nourish the all important retina; but muscular fibres, placed in certain parts of it, enable it to serve as a muscular mechanism as well.

The connective tissue, which forms the groundwork of the coat, has the special feature of being almost entirely destitute of white fibrillated bundles, at least in the choroid strictly so called.

It consists on the one hand of branched or irregularly polygonal corpuscles, generally loaded with pigment in the form of minute ellipsoidal crystals, and on the other hand of elastic fibres of various degrees of fineness, both elements being imbedded in a homogeneous ground substance, and for the most part arranged in lamellæ. The choroid coat therefore in contrast to the sclerotic coat is a conspicuously elastic coat.

The coat as a whole may be divided into three or more layers lying one above the other.

Immediately beneath the sclerotic the coat is loose in texture, and is made up of a series of lamellæ four or five in number, separated by lymph-spaces lined with epithelioid plates. Each lamella consists, as just described, of pigment cells and elastic fibres imbedded in a ground substance. This part of the coat, which is often called the *suprachoroidal* membrane, is separated by large irregular lymph-spaces from the inner surface of the sclerotic which being rich in pigment cells and of somewhat looser texture than the rest of the sclerotic is sometimes distinguished as the *lamina fusca*.

Below the suprachoroidal membrane lies the layer containing the larger blood vessels, often spoken of as the choroid proper. These blood vessels are on the one hand the trunks of the ciliary arteries, short and long, with the branches into which these break up, and on the other hand the ciliary veins, which gathered up from the ciliary processes and iris as well as from the choroid itself, are arranged along the equator of the eyeball in a number of venous whorls, *the venæ vorticosæ*, the issuing veins of which pierce the sclerotic. In a vertical section through a prepared and hardened choroid these larger veins and arteries are seen cut through in various ways, the relatively small spaces between them being occupied by pigment cells and elastic fibres; towards the inner surface the fibres are especially abundant and the pigment cells scanty. The blood vessels are surrounded by perivascular lymph-spaces and the tissue between the vessels contains numerous small lymph-spaces lined with epithelioid plates.

Beneath this layer of arteries and veins lies another layer consisting almost entirely of capillaries into which the arteries break up and from which the veins are gathered up. This is called the *chorio-capillary membrane*. The capillary network is exceedingly close set, almost as close as that of the pulmonary alveoli; and the tissue between the vessels, reduced to a minimum, consists of a homogeneous or finely dotted ground substance in which a few branched cells, devoid of pigment, may occasionally be seen, as well as, especially in myopic eyes, wandering leucocytes. Perivascular lymph-spaces are said to surround the capillaries.

This chorio-capillary membrane rests on a transparent homogeneous membrane, about 2μ thick, called the membrane of Bruch, or *basal membrane*, which separates the choroid from the underlying pigment epithelium of the retina, and so from the retina. Although the retina possesses, as we have said, vessels of its own, these as we shall see are largely confined to the anterior or inner layers of the retina adjoining the vitreous humour; the posterior or outer layers of the retina, and the adjoining pigment epithelium, are nourished by the blood vessels of the choroid. Hence the great development of the chorio-capillary layer; and the rest of the choroid, if we exclude the muscular and nervous elements of which we will speak presently, serves chiefly as a means to carry blood to and from the chorio-capillary plexuses, and as a bed for the lymphatic channels rendered necessary by the amount of lymph which must be continually furnished by such a close-set capillary network.

The pigment in the choroid may be regarded as serving in addition a dioptric purpose, absorbing the rays of light which have passed into it through the retina; but this cannot be of any great importance, since as we shall see such an absorption is chiefly carried out by the pigment epithelium of the retina.

In some animals part of the surface of the choroid when the eye is looked into shews various colours. The colouring is one not of pigments but of iridescence like the colouring of Newton's rings and thin films. It is due to the interference of light in a special layer, called the *tapetum*, intervening between the chorio-capillary membrane and the body of the choroid. In herbivora the interference of light causing iridescence is brought about by the peculiar arrangement of fine bundles of fibrillated connective-tissue; in carnivora the tapetum is composed of cells loaded with minute crystals and the interference is caused by the crystals.

§ 174. *The Ciliary Processes.* In front of the ora serrata, at which line the retina proper ceases, the choroid changes in character, being here thrown into the radiating plaits called the ciliary processes. These like the choroid, of which they are in fact a continuation, consist of blood vessels imbedded in a connective-tissue groundwork; but bundles of ordinary fibrillated tissue replace to a large extent the peculiar elastic lamellæ, and the capillary networks, though abundant, do not form a special close-set layer like the chorio-capillary membrane, but are more equally diffused through the bodies of the processes. The cells scattered throughout the connective-tissue bear pigment, especially in dark eyes.

The membrane of Bruch, or basal membrane, is continued over the processes, and here, often sculptured, rests on a layer of epithelium cells which do not maintain the features of the pigment epithelium of the retina, for these cease at the ora serrata, but are plain cubical cells simple in character and loaded, except in albino eyes, with black pigment. This pigment layer rests in turn on a layer of columnar cells transparent and free from pigment, into which the complex retina is suddenly transformed at the ora serrata, and which as we have already said is called the pars ciliaris retinæ.

§ 715. *The Iris.* Just as the ciliary processes continue forward in a modified form the choroid coat, so the iris continues forward the same coat still further, fresh modifications being introduced; the iris is like the choroid essentially composed of blood vessels lying in a bed of connective-tissue; but it has special features of its own.

The hind surface is covered with a layer of cells loaded in all except albino eyes with black pigment. The amount of pigment is so great that the outlines of the individual cells are greatly obscured and with difficulty distinguished; but we may probably regard the cells as representing two layers, both loaded with pigment, one the continuation of the pigment epithelium of the retina and the other of the pars ciliaris retinæ. That is to say, the cells together correspond to the two layers of the retinal cup, one of which, the outer or posterior, is pigmented over the whole of the extent of the cup, while the other, the inner or anterior, is pigmented only at the back of the iris, and elsewhere forms either the pars ciliaris retinæ or the retina itself. Fig. 134. The cells cease abruptly at the margin of the pupil, which thus, as we have said, forms the extreme lip of the retinal cup.

The front surface of the iris is also covered with a layer of epithelium, resting on an inconspicuous basement membrane; but this epithelium, which is easily detached and so overlooked, consists of flat polygonal epithelioid plates, and is really a lymphatic epithelium lining the large lymphatic space called the anterior chamber.

The body of the iris between the front and hind epithelium differs somewhat in the front and back parts. The front part or anterior layer contains relatively few blood vessels, and consists chiefly of a reticular form of connective-tissue furnished by spindle-shaped branched cells mingled with elastic fibres. The hind part of the body, or, counting the pigment behind as one layer, middle or vascular layer of the whole thickness of the iris, consists largely of blood vessels which are accompanied by imperfect sheaths of ordinary connective tissue, the intervals between the vessels being occupied by a reticular tissue like that of the anterior layer, save that the cells are more abundantly supplied with pigment. Bundles of nerve fibres, also accompanied by connective-tissue, are found in this layer but to a greater extent in the anterior layer.

Around the margin of the pupil, nearer the hind than the front surface, plain muscular fibres are gathered together in the form of a ring, the *sphincter iridis*, which compact on its inner edge towards the pupil becomes loose and frayed out on its outer edge, many of the fibres and small bundles curving away from the ring and taking a radial direction. The exact form and relative size of this sphincter muscle differs in different animals.

Resting upon the hind pigment epithelium, lying between it and the vascular layer, is a thin layer about which there has been much dispute. Seen *en face* the layer appears to consist of a number of long oval or even rod-shaped nuclei, arranged radially in a bed of material which is homogeneous save for some deposit of pigment and an obscure radiate striation. By some authors the nuclei are regarded as the nuclei of plain muscular fibres whose bodies form together the more or less homogeneous bed, and these authors accordingly speak of the layer as a radiate muscle, the *dilatator iridis*, the contractions of which would tend to widen the pupil. Other authors deny the muscular nature of this layer, and regard it as either a specially developed basement membrane, a continuation of the membrane of Bruch, or as a modified epithelial layer, the continuation of the pigment epithelium of the retina, the hind epithelium of the iris spoken of above corresponding, according to this view, to the pars ciliaris retinæ alone. We shall return to this question later on in speaking of the movements of the pupil.

§ 716. *The Cornea.* The sclerotic coat towards the front of the eye, at about the level of the attachment of the iris, is somewhat suddenly converted from an opaque membrane into the very transparent body of the cornea. The connective-tissue instead of being arranged in a feltwork by the interlacement of meridional and equatorial bundles as in the sclerotic, is in the cornea arranged in parallel, or rather concentric, layers of bundles all placed evenly in the same direction, the bundles of each layer and indeed the fibrillæ of the bundles being so united with cement substance that the whole is transparent. Moreover the connective-tissue corpuscles are distributed not irregularly but regularly in single layers between the layers of fibrillated material; they lie in spaces in the transparent cement substance uniting the layers together. Each cell is a broad flat thin plate with much-branched processes and a large, for the most part, oval nucleus. Since each cell lies with its broad surface parallel to the surface of the cornea and is very thin in the line of the rays of light, and since moreover the cell-substance and the nucleus is, in life, transparent, the presence of these cells does not interfere with the transparency of the organ.

The front surface of the cornea is covered with an epithelium of the same nature as the epidermis of the skin (§ 433) but somewhat modified. The cells form a few layers only. The lowermost layer of vertical cells is succeeded by two or three layers of cells corresponding to those of the Malpighian layer, but irregular in form and not bearing prickles. These are in turn covered by cells, two or three deep, which become flattened towards the surface, but retain their nuclei and are not so completely transformed into horny plates as are the corresponding cells of the epidermis. The substance of each cell is sufficiently transparent to render the whole epithelium transparent. Between the epithelium and the underlying connective tissue body which corresponds to the dermis, lies a thin cuticular sheet or basement membrane, the *anterior elastic lamina* or membrane of Bowman; and from the under surface of this, prolongations in the form of fibres arch downward into the substance of the cornea.

The concave hind surface of the body of the cornea is bounded by a conspicuous cuticular membrane possessing elastic properties. This, which is sufficiently thick to give in section an easily recognised double outline, is called the *posterior elastic lamina* or *membrane of Descemet* or *of Desmours*. Upon this lies a single layer of flat polygonal epithelioid plates, which, like the similar but less conspicuous cells covering the front surface of the iris, may be regarded as the lymphatic epithelium of the anterior chamber.

In the body of the cornea neither blood vessels nor lymphatic vessels are present. At the circumference are a few capillary loops and the beginnings of a few lymphatics; but within this circle the nutrition of the cornea is effected through the branched spaces in which the corneal corpuscles lie. These, communicating freely with each other by means of their branched prolongations, and being only partially filled by the substance of the cells, form a labyrinth of lymphatic spaces, along which a flow of lymph is continually taking place. We shall deal with the nerves of the cornea in connection with those of the skin in treating of touch.

§ 717. *The Ciliary Zone.* The region in the front part of the eye at which the sclerotic changes into the cornea is of great importance since here the choroid, represented by the outer parts of the ciliary processes and the outer margin of the iris, becomes joined to the sclerotic, and to the junction are attached the plain muscular fibres forming the ciliary muscle. This region which we may call "the ciliary zone" deserves especial attention.

When the transparent body of the cornea changes into the opaque sclerotic, the epithelium of the former becomes separated from the latter by the intercalation of ordinary loose connective tissue, which serves as a dermis to the epithelium and so forms the delicate skin covering the eyeball known as the conjunctiva; the epithelium of the cornea leaves the cornea to become the epithelium or epidermis of the conjunctiva (Fig. 140 *c.cj.*) This takes place on the outside of the ciliary zone.

On the inside the curved circumferential portion of the cornea makes a blunt angle, "iridic angle," with the outer edge of the more or less horizontal iris; and here several peculiar structures make their appearance. The thick membrane of Descemet with its epithelioid covering is in part continued on, greatly reduced in thickness, over the front surface of the iris to form the anterior basement membrane and overlying epithelioid layer of that structure. But in part only. At the angle the compact substance of the cornea is on its inner surface frayed out into a loose network of bundles of fibres, giving rise to a number of irregular spaces, of varying sizes, the *spaces of Fontana;* and the membrane of Descemet with its epithelioid covering is also split up and frayed out to supply a cuticular and epithelioid wrapping to the bundles of the network.

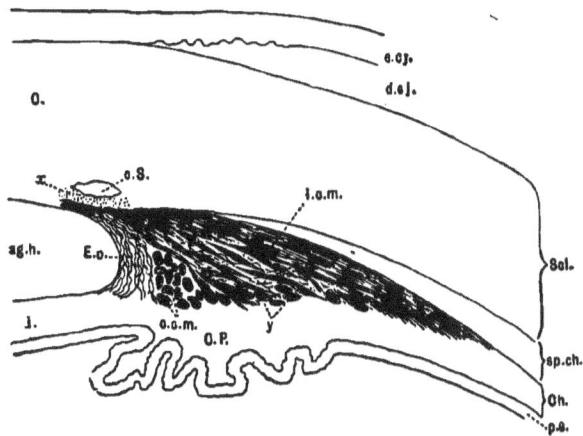

Fig. 140. Diagram of the ciliary muscle as seen in a vertical radial section of the ciliary region.

E.cj. epithelium of the conjunctiva. *d.cj.* dermis of the conjunctiva. *Scl.* Sclerotic. *sp.ch.* suprachoroidal layer. *Ch.* Choroid. *p.e.* pars ciliaris retinae and pigment epithelium represented as one layer. *C.P.* Ciliary processes. *I.* Iris. *ag.h.* anterior chamber. *E.p.* ligamentum pectinatum. *c.S.* canal of Schlemm, and *x* tissue to inside of it.

l.c.m. longitudinal, and *c.c.m.* circular ciliary muscle. *y* bundles of the longitudinal muscle cut across as they are taking a circular direction.

Thus, at the angle, a labyrinth of small irregular spaces is developed continuous at many points with the large anterior chamber of the eye as well as with each other; and the bars forming the walls of these spaces consist of bundles of the corneal substance passing on to the iris, coated with a continuation of the cuticular elastic membrane of Descemet and of the epithelioid covering of that membrane. The spaces, thus continuous on the inner side with the large lymphatic space of the anterior chamber, are small lymphatic spaces communicating on the outer side with the lymphatic vessels of the choroid and sclerotic, as we shall see when we come to deal specially with the lymphatics of the eye. The system of bars defining these spaces, conspicuous through their coating provided by the elastic membrane of Descemet, are spoken of as forming the *ligamentum pectinatum* (Fig. 140 *E.p.*). They are gradually lost in the peculiar choroidal stroma of the circumference of the iris.

In a vertical section of this region a somewhat large oval space, lined by epithelioid cells, is conspicuous to the outer side of the spaces of Fontana. This is the section of a circular canal, the *canal of Schlemm* (Fig. 140 *c.S.*) which, though properly a lymphatic channel, has direct connections with neighbouring veins, and under certain circumstances becomes filled with blood.

§ 718. *The Ciliary Muscle.* This occupies the space behind the ligamentum pectinatum between the sclerotic on the outside, and the ciliary processes and root or attachment of the iris on the inside. Just outside the ligamentum pectinatum the circular, equatorial bundles of the sclerotic are well developed, and just to the inside of the canal of Schlemm lies a small mass of denser tissue (Fig. 140 *x*); these structures serve as a point of attachment and sort of tendon for a number of small bundles of plain muscular fibres which run thence in a radiate meridional direction backwards. These interlace with other similar bundles having a similar direction and thus constitute all round the ring of the ciliary region a flat radiate meridional muscle, the *longitudinal* or *meridional ciliary muscle* (Fig. 140 *l.c.m.*), ending eventually in the front part of the choroid.

Between this and the ciliary processes and iris, is seen, at least in some eyes, a number of bundles of plain muscular fibres arranged circularly, equatorially, both ends of each bundle being attached to the loose connective-tissue forming the outer part of the ciliary processes and iris. These bundles constitute the *circular ciliary muscle*, or muscle of Müller (Fig. 140, *c.c.m.*). This is as a rule less conspicuous than the longitudinal muscle and in some animals is absent. Moreover the two muscles, longitudinal and circular, are not sharply defined from each other, many of the inner bundles of the longitudinal muscle curving round so as to take an equatorial direction (Fig. 140, *y*); and it is perhaps best to speak of the whole mass as forming one muscle, "the ciliary muscle."

§ 719. *The Lens.* This, as we have said, is of epithelial origin, and at an early stage presents itself as a sac with a wall consisting of a single layer of epithelium, supported by a scanty amount of mesoblast. The cavity, often from the first potential rather than actual, is soon obliterated by the elongation and growth of the cells of the hind half, and the sac is thus transformed into a solid more or less ellipsoidal mass in which one can distinguish an anterior part formed of short cubical and a posterior part formed of elongated epithelial cells; in a section the short anterior cells may at the edge of the mass be traced gradually lengthening into the long posterior cells.

The anterior cells remain throughout life as a single layer of cubical cells, often spoken of as the anterior epithelium of the lens. The whole of the lens except this thin layer is developed out of the posterior cells. These grow into long transparent flattened prismatic bands or fibres with a hexagonal transverse section, and with serrated edges by which each fibre locks into its neighbours. In the course of development the fibres assume a special disposition, being arranged in concentric lamellæ like the coats of an onion, the 'fibres' being so placed in the several lamellæ as to give rise to a star-shaped figure on the back of the lens. Most of the fibres still retain an elongated nucleus indicative of their epithelial nature but in many this is lost; in the adult as in the embryonic lens the transition from the characteristic fibres of the body of the lens to the cubical cells of the anterior layer may be observed in sections.

The lens developes around itself a cuticular membrane, structureless and elastic, called the *capsule of the lens*. This is much better developed in front than behind, where it is said to be partly wanting. It furnishes a distinct envelope for the lens, and when it is ruptured the lens may be turned out leaving the cavity of the capsule vacant.

The lens thus constituted is a transparent body, of a certain refractive power (§ 705), possessing considerable elasticity; its shape may be altered by pressure, but when the pressure is removed it regains its natural form.

The chemical nature of the lens is peculiar. The proteids which seem to form about 30 p.c. of the dry solids are of a globulin nature, being apparently nearly allied to the vitellin found in yolk of eggs and elsewhere; albumin seems to be absent. Even in healthy lenses a certain variable amount of cholesterin is present, and in lenses which have become opaque, forming cataracts, especially in what are called soft cataracts, this substance occurs in considerable quantity; it may amount to 5 p.c. of the dry solids.

§ 720. *The Vitreous Humour and Suspensory Ligament.* As we have said (§ 703), the mesoblast which is carried into the retinal cup at the involution of the lens is at first developed into the vascular pupillary and capsulo-pupillary membranes, which are supplied with blood by a continuation forwards of the central artery of the retina (Fig. 134). In the adult eye these membranes have been wholly absorbed, and the continuation of the central artery obliterated, so that all that remains of the mesoblast filling up the retinal cup is the jelly-like material known as the *vitreous humour*. This consists of little more than water, containing in solution, like the aqueous humour, about ·1 p.c. of proteids, namely, serum-albumin and globulin, as well as organic and inorganic salts; it seems chemically to resemble aqueous humour in spite of its different origin. A few scattered branched cells as well as wandering leucocytes are found in it.

Where it is in contact with the retina, the vitreous humour is defined by a structureless membrane, the *hyaloid membrane*, which is adherent normally to the overlying retina. This hyaloid membrane is continued forward beyond the ora serrata, and for some little distance is adherent to the pars ciliaris retinæ. A little farther forward, however, it leaves the ciliary processes and stretching inward as an independent, structureless, or faintly fibrillated, inelastic membrane, *the suspensory ligament* (Fig. 134) is attached to and becomes fused with the capsule of the lens on the anterior surface of that body.

During life the vitreous humour is in contact not only with the posterior surface of the lens, but also with the back surface of the suspensory ligament. After death, however, through changes in the vitreous humour, a space is developed, of a triangular form in section, between the suspensory ligament in front, the lens on the inside and the vitreous humour behind; this is often spoken of as the canal of Petit. According to some authors this canal exists during life and possesses a hind wall which is furnished by a membrane, distinct from the hyaloid membrane or suspensory ligament, which defines the front of the vitreous humour; and if, as asserted, the capsule of the lens be imperfect behind (§ 719), something of the nature of a membrane must exist in the front of the vitreous humour, since, when the lens is, in extraction of cataract, removed from the capsule, the vitreous humour does not escape into the vacant cavity. Since the suspensory ligament is attached on the outside, alternately to a projecting ciliary fold and to the depression between that and the next fold, the canal of Petit, when distended with air by blowing into it, has a beaded appearance. When the canal is thus blown out the suspensory ligament and its attachments are rendered very obvious; the ring thus formed by the suspensory ligament around the lens is sometimes called the zonule of Zinn.

We shall deal with the aqueous humour in speaking of the lymphatics of the eye.

that the impulses along the long ciliary nerves are ordinary motor impulses, leading to a contraction of this dilator muscle, whereby the constricting action of the sphincter is overcome. And it is further urged in support of this view that stimulation of one or some only of the long ciliary nerves, in contrast to an uniform stimulation of all of them together, such as happens when the cervical sympathetic is stimulated in the neck, leads to partial, uneven widening of the pupil; the long ciliary nerve stimulated acts only on a portion of the circle of the pupil and draws this part outward, forming more or less of a notch. Such a result it is argued could only be obtained by the direct local pull of some radial dilator fibres.

On the other hand, those who deny the existence of any such radial dilator muscle are led to explain the pupil-dilating influence of the sympathetic as due to the impulses along that nerve inhibiting the previously existing contraction of the sphincter. These argue that the sphincter may be compared to the heart, inasmuch as it possesses an automatic power of contraction, manifested however not in a rhythmic but in a tonic manner, and that like the heart its action may be either augmented or inhibited by nervous impulses; and we have seen (§ 429) that a similar view may be taken of the actions of the plain muscular fibres of the alimentary canal and of the bladder. According to this view the sphincter of the iris, when removed from all influences, is in a state of tonic contraction, pulling against the radiate strain of the elastic tissue of the iris and so maintaining a pupil of a certain size. Under the influence of light falling on the retina, impulses reaching the sphincter by the short ciliary nerves augment its contraction, and narrow the pupil in proportion to their intensity. On the other hand, impulses reaching the sphincter from the sympathetic by the long ciliary nerves inhibit the activity of the sphincter, diminish the force with which it is pulling against the elastic tissue of the iris, and so lead to a widening of the pupil, thus either diminishing the constriction which is being caused by the action of light on the retina or otherwise, or, in the absence of all external constricting influences, causing the pupil to become wider than it naturally would when removed from all extrinsic influences whatever. In support of such a view it is pointed out that the muscular tissue forming the sphincter is peculiar, since a slip of it when directly stimulated by a weak interrupted current elongates, in this respect also shewing its analogy with the heart whose activity may similarly be inhibited by the interrupted current. Again, in the extirpated eye, or even in the isolated iris, warmth dilates and cold constricts the pupil, the one relaxing, and the other increasing the contraction of the sphincter. Other arguments on the one side or on the other may also be brought forward, and the conflict between the two views cannot be regarded perhaps as definitely ended, though the influence of the cervical sympathetic may, as in the case of the vaso-constrictor action of the same nerve, be traced backwards down the neck to the upper thoracic ganglion, and thence to the spinal cord along the ramus communicans and anterior root of the second thoracic nerve, in the case of the dog, the monkey and man; in the frog it is the fourth spinal nerve. From thence the dilating influence may be further traced up through the bulb to a centre, which appears to be placed in the floor of the front part of the aqueduct not far from and apparently lateral to the centre for constriction of the pupil. Some authors have supposed that a part of the spinal cord in the lower cervical or upper thoracic region above the origin of the second thoracic nerve has a special share in carrying out the dilating action and hence have called this region the *centrum ciliospinale inferius*; but this seems very doubtful. Since, as a rule, a very decided amount of narrowing of the pupils follows upon mere section of the cervical sympathetic, we may infer that, unlike the case of the pupil-constrictor mechanism, tonic impulses habitually proceed from the pupil-dilator centre.

We may trace the path of dilating impulses in the other direction upwards along the cervical sympathetic, not to the sympathetic root of the ciliary ganglion and so to the short ciliary nerves, but to fibres which, passing over the Gasserian ganglion apply themselves to the ophthalmic division of the fifth nerve, and from thence along the nasal branch to the long ciliary nerves, and so to the iris; while the short ciliary nerves are the channels for pupil-constrictor impulses, the long ciliary nerves are the channels of pupil-dilator impulses.

§ 728. But while the mode of action of the pupil-constrictor impulses seems clear, since these have simply to throw into contraction, or increase the contraction of, the fibres of the sphincter, the mode of action of the pupil-dilator impulses is a matter which has been and still is disputed. We have already (§ 724) urged that the widening of the pupil cannot be simply the result of vaso-constrictor action on the blood vessels of the iris; and it is stated that the long ciliary nerves which act as pupil-dilators carry no vaso-constrictor impulses; it is said that stimulation of the long ciliary nerves while it widens the pupil produces no vaso-motor effects, and after the division of the long ciliary nerves stimulation of the cervical sympathetic, while it produces vaso-constriction in the eye as in other parts of the head and face, gives rise to no widening of the pupil. The impulses then which pass along the long ciliary nerves must affect in some manner or other the muscles of the iris other than those of its blood vessels.

Two views are held as to what that manner of action exactly is. Those who regard the radially disposed tissue lying immediately in front of the hind pigment layer of the iris as a muscular layer of radiating plain muscular fibres, maintain

stimulation of the retina; in other words a separate complete reflex mechanism exists on each side. If this be so there can be no question of any decussation of either afferent or efferent impulses, and the consensual pupil reaction, the constriction of one pupil caused by retinal stimulation of the other eye, must be carried out by some ties between the two centres of the two eyes. And indeed it has been suggested that the fibres passing from the nucleus of one side to that of the other (§ 623) furnish such a tie. If this be the state of things in the dog, it is difficult to believe that the arrangement in other animals is wholly different. We must not however carry on this discussion any further; but we may perhaps remark that, whichever view we take of the course of the afferent pupil constricting fibres, it is probable that they are not the same as those which subserve visual sensations, that light when it falls on the retina not only excites in certain fibres impulses which give rise to visual sensations, but also excites in other fibres impulses which go to govern the pupil.

Whatever be the exact nature of the central connections, the existence of such a reflex mechanism is an important fact, since the changes of the pupil which take place in actual life are to a large extent carried out by means of it; a constricted pupil indicates in the majority of instances an activity of the reflex mechanism, and a dilated pupil the absence of or diminution of that activity. In the normal, healthy organism the activity of the mechanism is in the first instance dependent on the amount of light falling on the retina; but even in the normal condition, and still more in an abnormal condition of the organism, other influences may become dominant. The activity of the centre may be exalted or depressed by nervous or other actions; the retina or optic nerves may be affected by the same amount of light to a degree less than or greater than the normal, and the efferent limb of the chain may be less or more effective.

§ 727. Besides, however, all the various changes which may thus be induced by playing upon the optic-oculo-motor reflex mechanism, there are other agencies capable of acting on the pupil quite apart from this reflex mechanism.

If the cervical sympathetic in the neck be divided, all other influences which could possibly affect the pupil being avoided, a constriction of the pupil will be seen to take place; this however is at times (in animals) not very well marked; but, whether it be so or no, if the peripheral portion of the nerve (*i.e.* the upper portion still connected with the head) be stimulated, a well-developed dilation is the result. The cervical sympathetic has, it will be observed, an effect on the pupil, the opposite of that which it exercises on the blood vessels of the head and neck; when it is divided, the pupil becomes constricted but the blood vessels dilate, and when it is stimulated the pupil is dilated while the blood vessels are constricted. This pupil-dilating

It is desirable to remember one important difference as to the behaviour of the pupil which obtains between man and some of the higher mammals on the one hand, and the lower mammals as well as other vertebrates on the other. In the former, the pupil-constricting nervous mechanisms of the two eyes are not completely independent; there is a functional communion between the two sides, so that when one retina is stimulated both pupils contract, and indeed, in man, as a rule, contract equally. Hence, when a change in the pupil of one eye is brought about by some means other than the one we are now considering, the pupil of the other eye is affected; when for instance one pupil is dilated with atropin, the larger amount of light thus admitted into that eye causes a narrowing of the pupil of the other eye, and thus increases the difference between the pupils of the two eyes. In the lower mammals and other vertebrates, the mechanisms in question are independent, stimulation of one retina produces no effect on the pupil of the other eye. This difference seems dependent on the character of the optic decussation, which, in turn, as we have seen (§ 667), is connected with the presence or absence of binocular vision. In man, with considerable binocular vision, the decussation is partial only, a considerable part of the optic nerve passes into the optic tract of the same side (Fig. 133); in the dog, with more limited binocular vision, a much smaller proportion of the optic nerve takes this course; and in the lower mammals, and other vertebrates which possess no binocular vision, the crossing is complete, the whole optic nerve of the one side passes into the optic tract of the opposite side, though the bundles of fibres may be interwoven at the decussation in various ways in different animals. If we reject the view mentioned above, that the optic fibres subserving the pupil mechanism leave the optic nerve at the decussation and suppose that these fibres pass with the other optic fibres into the optic tract, we must conclude that in such an animal as the frog, or bird, in which stimulation of the retina produces constriction of the pupil of the same eye and of that alone, a double crossing of impulses must take place; the afferent impulses of the pupil-constrictor mechanism must cross over to the opposite optic tract, and so reach the opposite centre, and the efferent impulses which these generate in the centre must cross back again to reach the pupil of the eye whose retina was stimulated. But such a conclusion is opposed to the results which have been obtained on the dog, in which, as we have just said, the decussation is partial. In this animal, after division of the floor of the third ventricle and aqueduct lengthways in the median line, not only does unilateral direct stimulation of the pupil-constrictor centre, or at least of the region of the nucleus of the third nerve, produce constriction in the pupil of the same side alone, but it is stated that, under these circumstances, reflex constriction of the pupil may be obtained, on either side, by

shifted from the front backwards; first movements of accommodation, next constriction of the pupil, and then contractions of the ocular muscles. Now in this region lies the elongated nucleus of the third nerve (§ 623); and it would appear that while the fibres of the third nerve concerned in accommodation arise from the extreme front of the nucleus, those which act upon the pupil start from a succeeding part, the remaining hinder part giving rise to the fibres which govern the ocular muscles. It seems therefore natural to regard the part of the nucleus from which the pupil-constricting fibres spring, as the centre of the reflex pupil-constricting mechanism, as the pupil-constrictor centre.

There is no difficulty as to the connection of the centre with the efferent limb of the reflex chain. The pupil-constrictor fibres pass from the nucleus to the trunk of the third nerve of the same side, and so by the short root to the ciliary ganglion (Fig. 142 $r.b.$), whence they reach the pupil by the short ciliary nerves; section of the short ciliary nerves breaks the reflex chain of which we are speaking, and stimulation of them or of their peripheral ends causes narrowing of the pupil.

But considerable difficulties are met with in determining the connection of the optic fibres, the afferent limb of the chain, with the centre. We should perhaps naturally suppose that the afferent nervous impulses which affected the pupil were the same as, or at least took the same course as those which gave rise to visual sensations, and that there was some connection between the part of the third nucleus serving as the pupil-constrictor centre and one or other or all of the three bodies in which, as we have seen (§ 669), the optic fibres end, namely, the lateral corpus geniculatum, the pulvinar, and the anterior corpus quadrigeminum, the connection being perhaps more especially with the latter. But we have no exact or satisfactory knowledge of such a connection; and indeed that the connection between the optic fibres and the pupil-constrictor centre is not furnished by any of these bodies is shewn by the fact that the pupil may still react to light after their removal, though they, or at all events the anterior corpora quadrigemina, are so far connected with the pupil mechanism that interference with them or stimulation of them gives rises to changes in the pupil. The tie, whatever be its nature between the optic fibres and the pupil-constrictor centre, must leave the optic path, if we may use the expression, before these bodies are reached. It has indeed been maintained that the afferent fibres of the optic nerve which affect the pupil do not pass into the optic tract at all, but leave the optic nerve at the decussation, being some of those fibres which pass directly from the optic decussation to the floor of the third ventricle (§ 669), and so make their way directly to the nucleus of the third nerve, through the substance of the floor of the third ventricle. But this has been disputed, and other connections have been suggested; these however we cannot discuss here.

long ciliary nerves, piercing the sclerotic somewhat nearer the front of the eye, are distributed to the muscles of the iris, and probably to the ciliary muscle.

The third or oculo-motor nerve we may trace back, as we have seen (§ 623), to its nucleus in the floor of the aqueduct; the sympathetic root we may trace back along the cervical sympathetic to the spinal connections of that nerve, on which we have so often dwelt; the remarkable ophthalmic division of the fifth nerve we may trace back to the nucleus of the fifth nerve, which we have seen (§ 621) to be exceedingly complex, and indeed we have reason to consider this ophthalmic division as an independent nerve, which in the course of evolution has become annexed to other nerves to form what we call 'the fifth' nerve.

§ 726. We may now make the broad statement, qualifications of which we will consider later on, that constriction of the pupil, brought about by light falling on the retina, is a reflex act, of which the optic is the afferent nerve, the third or oculo-motor the efferent nerve, and the centre some portion of the brain lying in the front part of the floor of the aqueduct at the level of the anterior corpora quadrigemina. This is shewn by the following facts. When the optic nerve is divided, light falling on the retina of that eye no longer causes a constriction of the pupil: we are supposing that the observations are confined to one eye. When the third nerve is divided, stimulation of the retina or of the optic nerve no longer causes constriction; but direct stimulation of the peripheral portion of the divided third nerve causes constriction of the pupil which may be extreme. If the region of the brain spoken of above as the centre be carefully stimulated, constriction of the pupil will take place even when no light falls on the retina or after the optic nerve has been divided. After destruction of the same region stimulation of the retina is ineffectual in narrowing the pupil. But if the centre and its connections with the optic nerve and third nerve be left intact and in thoroughly sound condition, constriction of the pupil will occur as a result of light falling on the retina, though all other parts of the brain be removed.

It might be imagined that this cerebral centre acted as a tonic centre, whose action was simply increased, not originated, by the stimulation of the retina; but this is disproved by the fact that if (still dealing with one eye) the optic nerve be divided subsequent section of the third nerve produces no further dilation.

When the rootlets of the third nerve are separately divided as they leave the brain, it is found that section of those placed more anteriorly interferes with accommodation and constriction of the pupil, while section of the hinder ones affects the ocular muscles. Moreover if the hind part of the floor of the third ventricle and front part of the floor of the aqueduct be carefully explored (in the dog) by means of the interrupted current, the following movements may be observed in succession as the electrodes are

ciliary ganglion (*l.c.*) which is connected by means of its three roots, (1) through the so-called 'short root' with the third nerve (*r.b.*), (2) with the cavernous sympathetic plexus and so, along the internal carotid artery, with the cervical sympathetic nerve (*sym.*), and (3) through the so-called 'long root' with the nasal branch of the ophthalmic division of the fifth nerve (*r.l.*). Besides the short ciliary nerves, the eyeball is supplied by the long ciliary nerves (*l.c.*) coming direct from the nasal branch of the ophthalmic division of the fifth nerve. The short ciliary nerves, which are the

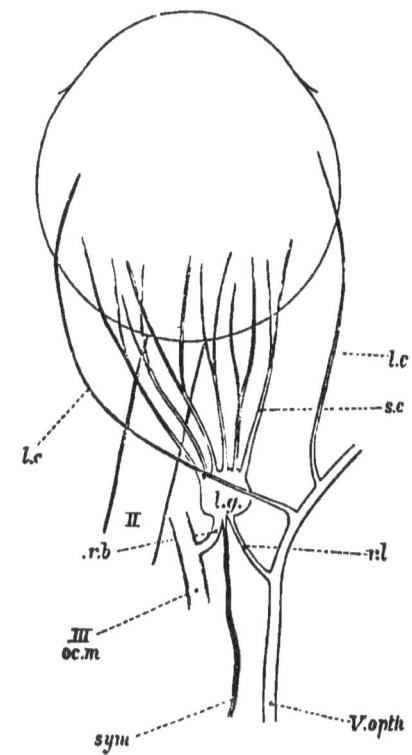

FIG. 142. DIAGRAMMATIC REPRESENTATION OF THE NERVES GOVERNING THE PUPIL.

II. Optic nerve. *l.g.* ciliary ganglion. *r.b.* its short root from *III.* *oc.m.*, third or oculo-motor nerve. *sym.* its sympathetic root. *r.l.* its long root from *V. ophthm.* the nasal branch of the ophthalmic division of the fifth nerve. *s.c.* the short ciliary nerves from the lenticular ganglion. *l.c.* the long ciliary nerves from the nasal branch of the ophthalmic division of the fifth nerve.

most numerous, pierce the sclerotic at the hind part of the eyeball and are distributed on the one hand to the blood vessels of the choroid, ciliary processes and iris, and on the other hand to the ciliary muscle and to the sphincter of the pupil. The less numerous

versely, the body of the iris being elastic as well as extensible, a relaxation of the muscular fibres of the sphincter will lead to a widening of the pupil. We may therefore in the first instance at all events consider the constricted pupil as the result of a contraction of the sphincter muscle, and the dilated pupil as the result of a diminution of that contraction; whether the dilated pupil is merely a negative result, whether it is simply due to a lessening of the activity of the sphincter, or whether there is in addition an active dilator muscle, we will discuss later on.

We may here, however, remark that, considering how vascular the iris is, it does not seem unreasonable to interpret some of the variations in the condition of the pupil as the results of simple vascular turgescence or depletion brought about by vaso-motor action or otherwise. When the blood vessels are dilated and filled they will cause the iris to encroach on the pupil, making the latter small and narrow, and conversely a constricted and emptied condition of the blood vessels would lead to the pupil being large and wide. And indeed slight oscillations of the pupil, due to greater or less fulness of the blood vessels, may be observed synchronous with the heart-beat, and others synchronous with the respiratory movements. But the variations in the pupil are, as a rule, too marked to be merely the effects of vascular changes; and indeed that turgescence of the vessels of the iris is not the only cause of constriction of the pupil, nor depletion the only cause of dilation, is shewn by the facts that both these events may be witnessed in a perfectly bloodless eye, and that the movements of the pupil when brought about by agents which also affect the blood vessels begin some time before the changes in the calibre of the blood vessels begin, and indeed may be over and past before these have arrived at their maximum. Moreover those fibres of the sympathetic which are concerned in causing dilation of the pupil, are said to run a somewhat different course from those which govern the blood vessels; it is stated that we can experimentally, by stimulating one or the other set, either dilate the pupil without any marked change in the blood-supply, or affect the blood-supply without materially changing the pupil. We may therefore adhere to the view that the main changes of the pupil in the direction of narrowing and widening are brought about by means of plain muscular fibres in the iris apart from those of the blood vessels.

§ **725.** Of all conditions affecting the size of the pupil, the one most important and most frequently at work is the falling of light on the retina; and to this we may now return. But before doing so it will be desirable to recall to mind the nervous supply of the eyeball, omitting for the present the nerves governing the six ocular muscles which move the eyeball as a whole.

The eyeball is supplied, in the first place, by the short ciliary nerves (Fig. 142 *s.c.*) coming from the ophthalmic or lenticular, or

assist in slackening the suspensory ligament. But no very decisive explanation has been given why the circular fibres are often largely developed in some eyes, it is said hypermetropic or long-sighted eyes, and scantily present in others, myopic or short-sighted eyes. And indeed there are several points in the whole action of accommodation which still require to be cleared up.

Accommodation is in a certain sense a voluntary act; we can by looking at near or far objects bring about the change whenever we please. Since, however, the change in the lens is always accompanied by movements in the iris, it will be convenient to consider the latter before we speak of the nervous mechanism of the former.

The Movements of the Pupil.

§ 723. Although by looking at near or far objects, and so voluntarily bringing about changes in the accommodation mechanism, we can call forth the accompanying changes in the iris, and can thus at pleasure produce a constriction, narrowing, or a dilation, widening, of the pupil; it is not in our power to bring the will to act directly on the iris by itself. This fact alone indicates that the nervous mechanism of the pupil is of a special character, and such indeed we find it to be.

The pupil is *constricted*, contracted, narrowed, (1) when the retina (or optic nerve) is stimulated, as when light falls on the retina, the brighter the light the greater being the contraction; (2) when we accommodate for near objects. The pupil is also constricted when the eyeball is turned inwards, when the aqueous humour is deficient, in the early stages of poisoning by chloroform, alcohol, and similar substances, in nearly all stages of poisoning by morphia, physostigmin, and some other drugs, in the early part of the day, in deep slumber, in the epileptic seizure, and in certain nervous diseases. The pupil is *dilated*, widened, (1) when stimulation of the retina (or optic nerve) is diminished or arrested, as in passing from a bright into a dim light or into darkness, (2) when the eye is adjusted for far objects. Dilation also occurs when there is an excess of aqueous humour distending the anterior chamber, during dyspnœa, during violent muscular efforts, as the result of stimulation of sensory nerves, as an effect of emotions, as an effect of fatigue, in the later stages of poisoning by chloroform, alcohol and similar substances, in all stages of poisoning by atropin and some other drugs, and in certain nervous diseases.

§ 724. Constriction of the pupil is caused by contraction of the circularly disposed muscular fibres which form within the iris the sphincter muscle (§ 715). The more or less spongy body of the iris being extensible, the shortening of the fibres and bundles of fibres of the sphincter must necessarily narrow the ring of the pupil of which the sphincter forms the almost immediate margin. Con-

compressed elastic lens to bulge forward. And experimental evidence shews that this is what does take place. The ciliary muscle is governed, as we shall presently see, by the ciliary nerves. If in a living animal (dog) or in an eye immediately after removal from the body, an opening be made in the sclerotic in order to watch the choroid, it may be seen that when the ciliary nerves are stimulated the choroid does move forward at the same time that the front surface of the lens becomes more convex; a needle, the point of which is carefully lodged in the choroid, moves in such a way as to shew that the choroid moves forward, though no appreciable movement can be seen in a needle thrust into the front part of the ciliary muscle itself. If the cornea be cut away so as to leave only at one place a small fragment still connected to the junction of the sclerotic and cornea, this piece moves backward when the ciliary nerves are stimulated, shewing that the ciliary muscle does pull on the point of junction of the sclerotic with the cornea. When, however, the cornea is intact, or even when a sufficiently large part of it is left, the junction becomes a fixed point, at least relatively to the moveable choroid. Moreover not only the contraction of the ciliary muscle and movement of the choroid, but the actual slackening of the suspensory ligament and change in the curvature of the lens may be observed to follow upon stimulation of the ciliary nerves. We may conclude, therefore, that the possible explanation given above is the actual one.

One or two additional points are worth mentioning. During accommodation for near objects the pupil is narrowed; we shall speak of this presently. A narrowing of the pupil means that the edge and inner part of the iris moves over the front surface of the lens towards the centre of the pupil. In becoming more convex, the front surface of the lens, especially the central portion, projects further forward into the anterior chamber, and in so doing carries with it the pupillary edge and inner part of the iris; for the iris lies close upon and indeed in contact with the anterior surface of the lens. And when the eye is carefully watched sideways this projection forwards of the pupillary margin of the iris may be observed. While the edge of the pupil thus moves forward, and the body of the iris increases in a radial direction, becoming correspondingly thinner (cf. Fig. 141), the circumferential edge of the iris is carried slightly backwards, owing to the giving way to a certain extent of the elastic ligamentum pectinatum on which the ciliary muscle pulls; and thus additional space is afforded in the anterior chamber for the aqueous humour driven aside by the projection of the anterior surface of the lens.

The action of the circular, equatorial fibres of the ciliary muscle, and of the fibres intermediate between these and the longitudinal meridional fibres, is not quite so clear. We may, however, suppose that the circular fibres acting in concert with the longitudinal fibres would bring the ciliary processes nearer to the lens, and so

The cavity of the eyeball behind the suspensory ligament is filled with the vitreous humour. If this is sufficiently abundant it will distend the cavity and render the suspensory ligament tense. But since the suspensory ligament passes obliquely forwards, all round, from the ciliary processes to the front of the lens, tension of the ligament will tend to flatten the lens, altering its shape but not its bulk.

The choroid, of which, as we have seen, the ciliary processes form the forward continuation, is loosely attached to the sclerotic along the line of the lamina fusca and suprachoroidal membrane; the one can to a certain extent be slipped backwards and forwards beneath the other.

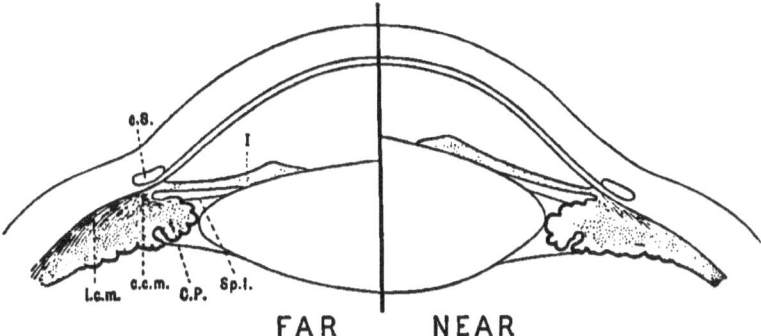

FIG. 141. DIAGRAM TO ILLUSTRATE ACCOMMODATION. (After Helmholtz.)

C. P. Ciliary process. *I.* Iris. *Sp. l.* suspensory ligament. *l. c. m.* longitudinal ciliary muscle. *c. c. m.* circular ciliary muscle. *C. S.* canal of Schlemm.

The left half represents the arrangement for viewing far objects and the right half that of viewing near objects.

The (longitudinal) ciliary muscle is, as we have seen, attached on the one hand to the junction of the sclerotic and cornea, and on the other hand to the front part of the choroid. If we suppose the former to be a fixed point, the contraction of the muscle would pull the moveable choroid and ciliary processes somewhat forward. But the pulling forward of these structures would slacken the suspensory ligament by bringing its ciliary attachment more forward. And a slackening of the suspensory ligament by relieving the pressure on the elastic lens would allow the front surface to become more convex. This is shewn diagrammatically in Fig. 141, one-half of which, the left half, is intended to represent the eye directed towards distant objects, while the other half represents the change taking place during accommodation for a nearer object.

§ 722. It seems possible then that the accommodation for near objects may be brought about by a contraction of the (longitudinal) ciliary muscle dragging forwards the choroid and ciliary processes, thus slackening the suspensory ligament, and so permitting the

SEC. 4. THE MECHANISM OF ACCOMMODATION AND THE MOVEMENTS OF THE PUPIL.

§ 721. We have seen (§ 711) that the essential change in the eye during accommodation for near objects is an increase in the curvature of the anterior surface of the lens. How is this brought about?

It has been supposed to be due to a compression of the circumference of the lens by a contraction of the sphincter muscle of the iris, which, as we shall see, is the cause of the narrowing of the pupil attendant upon accommodation for near objects; but this is disproved by the fact that normal accommodation may take place in eyes from which the iris is congenitally absent or has been wholly removed by operation. It has also been attributed to vasomotor changes, to increased fulness of the vessels of the iris or ciliary processes, surrounding and pressing upon the lens; but this also is disproved, not only by the fact just mentioned but as well by the fact that accommodation may be effected, after death, in an eye which is practically bloodless, by stimulating the ciliary ganglion or short ciliary nerves with an interrupted current or by other means; as we shall see, these nerves govern the accommodation mechanism. The real nature of the mechanism seems to be as follows:

The lens, as we have said, is a body of considerable elasticity. When the curvature of the anterior surface of the lens is determined, as may be done by appropriate means (by measurements of images seen by reflection from it), in its natural position in the eye at rest, and then again determined, after the lens has been removed from the eye, the anterior surface is found to be more convex in the latter than in the former case. There seems to be in the eye in its natural condition at rest some agency at work, keeping the anterior surface of the lens somewhat flattened. All that is needed is some means of counteracting this agency, and thereby allowing the lens through its elasticity to assume its natural form. And the arrangements of the suspensory ligament described in a previous section afford an explanation of what is the agency in question, and how it is counteracted.

balance of evidence seems to be against the existence of a distinct radial dilator muscle.

Whatever be the view adopted as to the exact mode of action of the sympathetic there remains the broad fact that the pupil is under the dominion of two antagonistic mechanisms: one a constricting mechanism, reflex in nature, the third nerve serving as the efferent, and the optic as the afferent tract; the other a dilating mechanism, apparently tonic in nature, but subject to augmentation from various causes, and of this the cervical sympathetic is the efferent channel. Hence, when the third or optic nerve is divided, not only do constricting impulses cease to be manifest, but the effect of their absence is increased, on account of the tonic dilating influence of the sympathetic being left free to work. When, on the other hand, the sympathetic is divided, this tonic dilating influence falls away, and constriction results. When the optic or third nerve is stimulated, the dilating effect of the sympathetic is overcome, and constriction results, and when the sympathetic is stimulated, any constricting influence of the third nerve which may be present is overcome, and dilation ensues.

The former, optic oculo-motor mechanism is the instrument by means of which the pupil is adapted to the amount of light, the latter, sympathetic mechanism appears to be employed when other influences are brought to bear on the pupil. Thus the characteristic pupil-dilating effects of emotions such as fear, of the painful stimulation of sensory nerves, of dyspnœa, and in part of some drugs, appear to be carried out through the sympathetic mechanism.

§ 729. In the case of many drugs, however, the effect produced is either in part or wholly independent of both these nervous mechanisms. A small quantity of atropin introduced into the system, or even directly into the eye, causes a dilation of the pupil which may be so great that the iris is reduced to a mere rim, while physostigmin (eserin) similarly introduced into the system or eye produces a constriction of the pupil which may be so great that the pupil is narrowed to a mere pin's point. Since both these drugs may produce their full effects after division of the optic-oculo-motor and the sympathetic nerves, and indeed may produce their effects in an extirpated eyeball, it is obvious that those effects are not due to the drugs acting on the central parts of the above mechanisms. Their action is a local one. They do not act by means of the ciliary ganglion, for both drugs continue to produce their effects to a most marked degree after the ganglion has been excised. Nor have we any evidence that their action is dependent on any other local nervous mechanism, such as might be afforded by the nerve cells lying in the choroid or even in the iris. They appear to act directly on the sphincter, atropin paralyzing it or producing relaxation, and physostigmin increasing or producing contraction, both often of an extreme character. Whether the drugs act on the actual muscular tissue itself or on the endings of the nerve fibres in the muscular tissue, or on both together, is a question which we cannot discuss here. Nor need we stay to consider the difficulties which are added, if we accept the view of the existence of a special dilator muscle. The important point is that the action of both these drugs is a local one; hence, when they have produced their full effects, the normal nervous mechanisms on which we have been dwelling are of little or no use; even an abundance of light leads to no constriction in the full atropinized eye, and removal of light produces little or no dilation in an eye fully under the influence of physostigmin.

We may here mention the fact that in certain animals at all events, for instance the eel, light falling into the eye, even into an extirpated eye, will cause constriction of the pupil; and this seems to be brought about by means of some nervous connection between the retina and the iris, for the effect ceases when the retina is destroyed. But this peculiar action has not yet been satisfactorily explained.

The share of the fifth nerve in the work of the iris seems to be chiefly at least a sensory one; the iris is sensitive, and the sensory impulses which are generated in it pass from it along the fibres of the fifth nerve. Some observers maintain that in addition to the dilating fibres of the sympathetic which reach the iris after having joined the fifth nerve, the latter nerve also contains fibres of its own which passing by the ophthalmic branch and long ciliary nerves are able to dilate the pupil; but this is doubtful.

We may sum up the nervous mechanism of the pupil then somewhat as follows. The salient and most frequently repeated event, the constriction of the pupil upon exposure to light, is a reflex act, the centre of which is placed in the brain; and the correlative widening of the pupil upon diminution of light is due to the tonic action of the sympathetic making itself felt upon the waning of its antagonist. The dilating effects of emotions, of sensory impressions, especially painful ones, and of dyspnœa appear to be brought about by an increased activity of the dilating centre, assisted possibly in the latter instance by a depression of the constricting centre. The constriction of the pupil in the earlier stages of the action of alcohol and chloroform and in slumber is probably due to an increased action of the constricting centre, but the narrow pupil caused by such a drug as physostigmin is due, chiefly if not exclusively, to a local action. The constricted pupil of morphia appears to be due partly to central and partly to local action. The dilating effects of such a drug as atropin are chiefly if not exclusively due to a local action, but in the widened pupil of the later stages of alcohol poisoning and of some other drugs we can probably trace the effects of an exhaustion of the constricting centre, assisted possibly by an increased activity of the dilating centre.

§ 730. *The nervous mechanism of accommodation.* The ciliary muscle which brings about accommodation is governed in this action by fibres which may be traced, through the short ciliary nerves and ciliary ganglion, along the third nerve, to a centre which lies (in dogs) in the extreme front of the floor of the aqueduct, or rather perhaps in the extreme hind part of the floor of the third ventricle, and which is especially connected with the extreme front of the nucleus of, and so with the most anterior bundles of the root of, the third nerve. As we have already said stimulation of this centre, or of the third nerve, or of the short ciliary nerves, leads to a contraction of the ciliary muscle and to accommodation for near objects.

This nervous mechanism, unlike that for the pupil, is under the command of the will, though the will needs to be assisted by visual sensations; it is moreover only brought into play by the direct action of the will; we are not led to accommodate by any other influence than the desire to see distinctly near or far objects. The mechanism may, however, be affected by the local action of drugs. Such drugs as atropin and physostigmin which have a special action on the pupil, also affect the mechanism of accommodation, and that in a corresponding way. Atropin paralyses it, so that the eye remains adjusted for far objects; and physostigmin throws the eye into a condition of forced accommodation for near objects. This double action has been explained by the supposition that, by acting on the muscular fibres, or on the nerve endings, or on both, atropin inhibits the contraction of or paralyses, while physostigmin throws into contraction or augments the contraction of the ciliary muscle. But the phenomena, on further inquiry, are found to be more complicated than they appear to be at first sight. For instance, we have no clear evidence that the mechanism is, like that of the pupil, a double one; when we pass from accommodation for a near object to that for a far object, we simply 'let go' the previous effort; we cease to contract the ciliary muscle, and, so far as we know at present, the return of the suspensory ligament and other parts is simply the passive result of the cessation of the contraction of the ciliary muscle. We have no evidence of antagonistic impulses passing along the long ciliary nerves for instance, and undoing the work of the previous impulses. It has, however, been stated that when the eye is brought in forced accommodation for near objects by the action of physostigmin, stimulation of the long ciliary nerves abolishes or diminishes the accommodation, bringing about a flattening of the anterior surface of the lens by inhibiting the previous contraction of the ciliary muscle. We may add that, were the change from near to far a mere passive relaxation of a previous contraction we should, judging from our experience of ordinary muscular contractions, expect the time taken up by it to be greater, or at least not less than the time taken up by the change from far to near;

but as a matter of fact it is very much shorter, indeed the act is an exceedingly rapid one. This and other facts indicate that our knowledge of the mechanism of accommodation is far from being complete.

§ 731. There remains a word to be said concerning the constriction of the pupil which takes place when the eye is accommodated for near objects, and when the pupil is turned inwards (the two being closely allied, since the two eyes converge to see near objects), and the return to the more dilated condition when the eye returns to rest and regains the condition adapted for viewing far objects. These are instances of what are called "associated movements." A similar instance is afforded by certain cases of blindness of one eye due to atrophy of the optic nerve; in such cases the pupil of the blind eye may be wholly insensible to light, and yet becomes narrowed when the subject looks at a near object with the sound eye. In so doing he throws into action the accommodation mechanism, and with that the pupil-constricting mechanism of both eyes. Two movements are thus spoken of as "associated" when the special central nervous mechanism employed in carrying out the one act is so connected by nervous ties of some kind or other with that employed in carrying out the other, that when we set the one mechanism in action we unintentionally set the other in action also. In this constriction of the pupil associated with accommodation the nervous ties between the parts of the central nervous system concerned in the generation of the will, the centre for accommodation, and the centre for the constriction of the pupil, are such that whenever the will stirs up the impulses necessary to carry out accommodation, it at the same time stirs up corresponding impulses in the pupil-constrictor mechanism. More than this we cannot at present say.

We can, as we have said, accommodate at will; few persons only can effect the necessary change in the eye unless they direct their attention to some near or far object, as the case may be, and thus assist their will by visual sensations. By practice, however, the aid of external objects may be dispensed with; and it is when this is achieved that the pupil may seem to be made to dilate or contract at pleasure, accommodation being effected without the eye being directed to any particular object.

SEC. 5. IMPERFECTIONS IN THE DIOPTRIC APPARATUS.

§ 732. *Imperfections of accommodation.* The emmetropic eye, in which the principal posterior focus lies on the retina, may, as we have said, be taken as the normal eye. The myopic, in which the principal posterior focus lies in front, and the hypermetropic eye, in which it lies beyond the retina, may be considered as imperfect eyes, though the former possesses an advantage over the normal eye in so far that it can see minute objects more distinctly than can the normal eye, since these can be brought so near the eye as to give a relatively large retinal image and yet remain within the limits of accommodation. An eye may be myopic from too great a convexity of the cornea, or of the anterior surface of the lens, or from permanent spasm of the accommodation-mechanism, or from too great a length of the long axis of the eyeball. The last appears to be the usual cause. Similarly, the cause of hypermetropism is in most cases the possession of too short a bulb. In presbyopia the failure or loss of accommodation may be due either to a loss of elasticity of the lens, or to increasing weakness of the ciliary muscle, or to the parts becoming rigid; the first appears to be the more common cause; the change, which may affect not only normal but also other eyes, generally begins in the fifth decade of life.

These several defects may be remedied by the use of appropriate lenses, by wearing proper spectacles. The myopic eye needs for distant objects the rays of which fall parallel on the cornea (or at least so little divergent that they still are brought to a focus in front of the retina) a concave glass, of such a refractive power, of such a focal length, as to give to parallel rays sufficient divergence before they fall on the cornea to enable the dioptric mechanism of the eye to bring them to a focus on, and no longer in front of, the retina.

The hypermetropic eye needs a convex glass of such a focal length as will give to parallel rays sufficient convergence before they fall on the cornea to enable the eye to bring them to a focus on the retina.

The presbyopic eye similarly needs a convex glass the focal

length of which must depend on the amount of accommodation still possessed by the eye; it must give the rays just so much convergence that the weakened mechanism is able to bring them to a focus on the retina, the convexity or refractive power of the glass being increased, that is to say its focal length diminished, as the loss of accommodation increases.

§ 733. *Spherical aberration.* In a spherical lens the rays which are refracted by the circumferential parts are brought to a focus sooner than those which pass through the more central parts; in consequence the rays proceeding from a luminous point are no longer brought to a single focus at one point, but form a number of foci at different distances. Hence, when rays are allowed to fall on the whole of the lens, the image formed on a screen placed in the focus of the more central rays is blurred by the diffusion-circles caused by the circumferential rays which have been brought to a premature focus. In an ordinary optical instrument spherical aberration is obviated by a diaphragm which shuts off the more circumferential rays. In the eye the iris is an adjustable diaphragm; and when the pupil contracts in near vision the more divergent rays proceeding from a near object, which tend to fall on the circumferential parts of the lens, are cut off. The lens however, as we have seen, is not uniform in structure, and the refraction which it exercises does not, as in the case of the ordinary lens, increase regularly and progressively from the circumference to the centre, but varies most irregularly; hence the purpose of the narrowing of the pupil cannot be simply to obviate spherical aberration; and indeed the other optical imperfections of the eye are so great, that such spherical aberrations as are actually caused by the lens produce no obvious effect on vision.

§ 734. *Astigmatism.* We have hitherto treated the eye as if its dioptric surfaces were all parts of perfect spherical surfaces. In reality this is rarely the case, either with the lens or with the cornea. Slight deviations from the spherical shape do not produce any marked effect, but there is one deviation, known as regular astigmatism, which, present to a certain extent in most eyes and very largely developed in some, frequently leads to very imperfect vision. This defect is due to one or other of the dioptric surfaces being not spherical but more convex along one meridian than another, more convex, for instance, along the vertical than along the horizontal meridian. When this is the case with the dioptric surface of an optical system the rays proceeding from a luminous point are not brought to a single focus at a point, but possess two linear foci, one nearer than the normal focus and corresponding to the more convex surface, the other farther than the normal focus and corresponding to the less convex surface. If the vertical meridians of the surface be more convex than the horizontal, then the nearer linear focus will be horizontal and the farther linear focus will be vertical, and *vice versa*. (This can be shewn much more effectually on a model than in a diagram in which we are limited to two dimensions.) Now, in order to see a vertical line distinctly, a needle held vertically for instance, it is much more important that the rays which diverge from the line in a series of horizontal planes should be brought to a focus properly than those which diverge in the vertical plane of the line itself; for the former contribute to a far greater extent than do the latter to the sum of rays which go to form the retinal image of and so to excite the sensation of the line. Similarly, in order to see a horizontal line distinctly it is much more important that the rays which diverge from the line in a series of vertical planes should be brought to a focus properly than those which diverge in the horizontal plane of the line itself. When a horizontal line is held before an astigmatic dioptric surface, more convex in the vertical meridian, it will give rise to a strong image of a horizontal line at the nearer focus where the many vertical rays diverging from the line are brought to a linear horizontal focus, and to a weak image of a vertical line at the farther focus where the fewer horizontal rays are brought to a linear vertical focus. Similarly, a vertical line held before the same surface will give rise to a strong image of a vertical line at the farther focus where the horizontal rays diverging from the vertical line are brought to a linear vertical focus, and to a weak image of a horizontal line at the nearer focus. But in the case of an astigmatic eye trying to see a horizontal or vertical line, such as a horizontal or vertical needle, the eye will neglect the weaker image, and take the stronger image as the only image of the object. Hence an astigmatic eye, more convex in the vertical meridian, will see a horizontal needle distinctly when the nearer, and a vertical needle distinctly when the farther of the two foci falls on the retina; it will require a different accommodation to see the one and the other distinctly. If the astigmatism is such that the horizontal meridian be the more convex, the vertical needle will be seen most distinctly at the nearer, and .the horizontal at the farther focus. In both forms of astigmatism the horizontal and the vertical lines which go to make up the features of the surface of an object will fail of being seen distinctly at the same time; and the vision of the object will be imperfect.

Rays of light proceeding from a line, which is neither vertical nor horizontal but oblique, give rise in an astigmatic system to a number of foci arranged in so complex a manner that no distinct image can be formed on the retina; the presence of these lines accordingly adds to the imperfection of the vision of any object.

Most eyes are thus more or less 'regularly' astigmatic, and generally with a greater convexity along the vertical meridian. If a set of horizontal or vertical lines be looked at, or if the near point of accommodation be determined by Scheiner's experiment, for the needle placed first horizontally and then vertically, the distance from the eye at which the horizontal lines or needle are seen distinctly will be found, in most cases, to be appreciably and in many cases considerably shorter than that at which the vertical lines or needle are seen with equal distinctness. In other words, in the case of most eyes, a vertical line must be farther from the eye than a horizontal one, if both are to be seen distinctly at the same time. The cause of astigmatism is, in the great majority of cases, the unequal curvature of the cornea; but sometimes the fault lies in the lens, as was the case with the philosopher Young.

Regular astigmatism may be remedied by the use of cylindrical glasses, that is to say, glasses which are convex along one meridian but plane along the other. Thus the ordinary astigmatic eye with the greater curvature along the vertical meridian will be benefited by a cylindrical glass, plane in the vertical plane but possessing such convexity in the horizontal plane as will make up for the relatively deficient horizontal curvature of the cornea.

When the curvature of the cornea or of the lens differs not in two meridians only but in several, *irregular* astigmatism is the result. A certain amount of irregular astigmatism, due to the cornea or lens, exists in most eyes, thus causing the image of a bright point, such as a star, to be not a round dot but a radiate figure; in some cases the irregularity is so great that several imperfect images are formed of every object.

§ 735. *Chromatic aberration.* The different rays of the spectrum are of different refrangibility, those towards the violet end of the spectrum being brought to a focus sooner than those near the red end. This in optical instruments is obviated by using compound lenses made up of various kinds of glass. In the eye we have no evidence that the lens is so constituted as to correct this fault; still the total dispersive power of the instrument is so small, that such amount of chromatic aberration as does exist attracts little notice. Nevertheless some slight aberration may be detected by careful observation. When the spectrum is observed

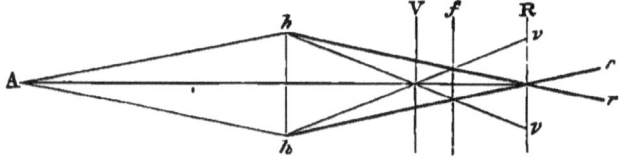

Fig. 143. Diagram illustrating Chromatic Aberration.

hh is the dioptric surface, *hv* represents the blue, and *hr* the red rays; *V* is the focal plane of the blue, *R* of the red rays.

at some distance off the violet end will not be seen in focus at the same time as the red end. Again, if a luminous point be looked at through a narrow orifice covered by a piece of violet glass, which while shutting out the yellow and green allows the red and blue rays to pass through, there will be seen alternately an image having a blue centre with a red fringe, or a red centre with a blue fringe, according as the image of the point looked at is thrown on one side or other of the true focus. Thus supposing f (Fig. 143) to be the plane of the mean focus of A, the violet rays will be brought to a focus in the plane V, and the red rays in the plane R. If the rays be supposed to fall on the retina between V and f, the diverging or blue rays will form a centre surrounded by the still converging red rays; whereas if the rays fall on the retina between f and R, the converging red rays will form a centre with the still diverging blue rays forming a fringe round them. If the rays fall on the retina at f, the two kinds of rays will be mixed together; as will be seen from the figure, the circumferential still converging red ray hr as it cuts the plane of the retina is, in ordinary vision, accompanied by the diverging violet ray hv, and thus by a sort of compensation, we see together, though not in absolutely proper focus, even the rays which differ most in refraction. The experiment may be varied by blocking up one half of the pupil with a piece of card and using the uncovered half of the pupil to look through a piece of red glass at a white surface or a candle flame. The red strip will be seen to have a blue edge.

§ **736.** *Entoptic phenomena.* The various media of the eye are not uniformly transparent; the rays of light in passing through them undergo local absorption and refraction, and thus various shadows are thrown on the retina, of which we become conscious as imperfections in the field of vision, especially when the eye is directed to a uniformly illuminated surface. These are spoken of as entoptic phenomena, and are very varied, many forms having been described.

Tears on the cornea, or temporary unevenness of the anterior surface of the cornea after the eyelid has been pressed on it, may give rise to retinal images and so to visual sensations; but in these cases the cause lies outside the eye and the result can hardly be spoken of as entoptic.

Changes in the margin of the pupil appear in the shadow of the iris which bounds the field of vision. If we look at a bright object or luminous surface through a pin-hole in a card placed close in front of the eye (in order to get the best image on the retina, the pin-hole should occupy the position of the principal anterior focus), the dark circle which bounds the field of vision is the image caused by the shadow of the margin not as might at first be supposed of the pin-hole but of the iris. This is at once shewn by the changes which it can be made to undergo, while the pin-hole remains motionless, by alternately closing and opening the other eye; the field of vision of the eye which is looking through the pin-hole may be observed to contract when light enters, and to expand when the light is shut off from the other eye; for as we have seen (§ 726) light falling on one retina leads to consensual narrowing of the pupil of the other eye. Other changes or irregularities in the iris may be observed by this method.

Imperfections in the lens or in its capsule may also give rise to entoptic images. Not unfrequently a radiate figure corresponding to the arrangement of the fibres of the lens makes its appearance.

The most common entoptic phenomena are those caused by the presence of floating bodies in the vitreous humour, the so-called *muscæ volitantes*. These are readily seen when the eye is turned towards a uniform surface, and are frequently very troublesome in looking through a microscope. They assume the form of rows and groups of beads, of single beads, of streaks, patches and granules, and may be recognised by their almost continual movement especially when the head or eye is moved up and down. When an attempt is made to fix the vision upon them they immediately float away.

Since the images on the retina are in these cases shadows and since the strongest shadows are cast by parallel rays, the images are best seen when the rays of light giving rise to the shadows on the retina traverse the vitreous humour in parallel lines; hence the best illumination for examining the phenomena is one placed in the principal anterior focus, the rays diverging from which fall parallel on the retina (§ 704, Fig. 135). The sharpness of the images is also increased by using a small but bright source of light, as by looking at a bright light through a small hole in a screen.

The sensations which these objects in the vitreous humour excite by means of the retinal images to which they give rise do not

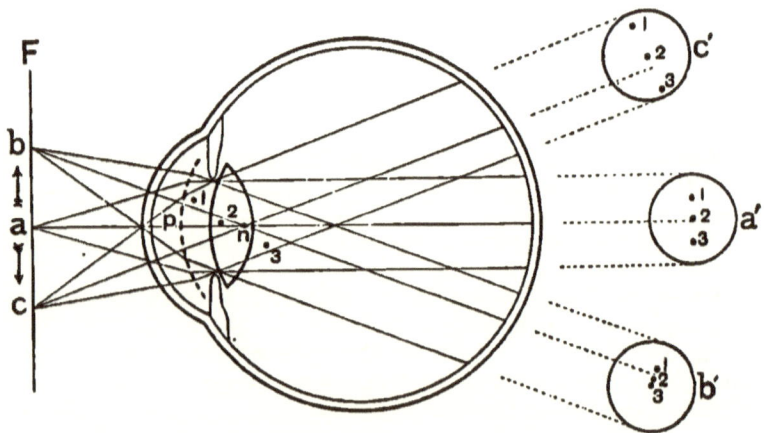

Fig. 144. Diagram to illustrate Entoptical Images.

tell us that the objects are in the vitreous humour. As we shall see we refer all affections of the retina, all visual sensations to some changes in the external world; and if we trusted to our sensations alone in the cases of these entoptic phenomena we should suppose that the causes existed outside ourselves. It is only by means of inferences drawn from the features and behaviour of the sensations that we arrive at the conclusion that the causes lie in the vitreous humour.

The accompanying diagram (Fig. 144) illustrates how the position of objects in the eye may be determined by the movements of their shadows on the retina. It represents the reduced diagrammatic eye seen in vertical section, with n the nodal point, p the principal plane, and F the plane of the principal anterior focus. 1 represents an object in the anterior chamber, 2 another in the substance of the lens, and 3 a third in the vitreous humour. If a bright light be looked at through a pin-hole in a card placed in the plane of the principal anterior focus F so that the hole is at the principal anterior focus a, the rays of light may be considered as diverging from a, and we may draw them as refracted at the principal plane p, and then passing parallel through the vitreous humour. The image on the retina in this case may be represented by a'. The field of vision, limited by the shadow of the iris, will be circular; the shadow of 2 will lie close to the optic axis, that of 1 a little above it, and that of 3 some little way below it. It will of course be remembered that in the *apparent* image all the features will be inverted (§ 707). If now the card be moved upward so that the light emanates from the pin-hole at b, and the paths of the rays of light be drawn as before, the image resulting will be that shewn at b'. The shadow of 2 has changed very little in position; but that of 1 has moved downwards, while that of 2 has moved upwards so that all three lie closer together. If, on the contrary, the card be moved downward to c the result will be that shewn in c'; the shadow of 1, as before, has moved but little, while that of 1 has moved upward, and that of 3 downwards, so that the three shadows are farther apart.

Thus while the shadows of objects in the anterior chamber move in a direction the opposite to that of the movement of the source of an illumination placed in the plane of the principal anterior focus, the shadows of objects in the vitreous humour move in the same direction as the source of illumination. Hence, by observing the direction of the movement of an entoptic image resulting from the movement of the illumination, the position in the eye of the object giving rise to the image may be determined, regard of course always being had to the so-called mental inversion of the retinal image. Stated more strictly the rule would run thus. The shadows of objects in front of the nodal point (§ 705) in the lens move in a direction contrary to and those of objects behind the nodal point in the same direction as the movement of the illumination; moreover the more distant the object from the nodal point the greater the movement of the shadow caused by the same movement of the illumination.

In this connection we may refer to one or two matters which however cannot be called dioptric imperfections.

If a white sheet or white cloud be looked at in daylight through a Nicol's prism, a somewhat bright double cone or double tuft, with the apices touching, of a faint blue colour, is seen in the centre of the field of vision, crossed by a similar double cone of a somewhat yellow darker colour. These are spoken of as Haidinger's brushes; they rotate as the prism is rotated, and are supposed to be due to the unequal absorption of the polarised light in that part of the retina which we shall study presently as "the yellow spot." The prism must be frequently rotated, since when the prism remains at rest the phenomena fade. We may here remark that the media of the eye are fluorescent: a condition which favours the perception of the ultraviolet rays. There are other entoptic phenomena due to features of the retina, of which we shall speak in treating of the development of visual impulses.

Lastly, returning to dioptric imperfections, we may add that the optical arrangements are also to a certain extent imperfect inasmuch as the dioptric surfaces are, according to most observers, not truly centred on the optic axis.

SEC. 6. THE STRUCTURE OF THE RETINA.

§ 737. We have now to inquire how the rays of light thrown on to the retina, by means of the dioptric mechanism, in the form of an optical image give rise to visual sensations and so to a perception of the object sending forth the rays. For this purpose we must turn to the structure of the retina, including with it the pigment epithelium derived from the outer, as the retina proper is from the inner, layer of the retinal cup.

The optic nerve, as we have already said, is not so much an ordinary nerve as a strand of white matter extending from the brain; and several of its features shew this. Its outer wrapping is not an ordinary perineural sheath but a prolongation of dura mater, and within this lies a 'pial' sheath, a continuation of the pia mater. Between the two may be recognised a membrane corresponding to the arachnoid membrane with scanty sub-dural and sub-arachnoid spaces. The pial sheath sends supporting septa into the interior of the nerve, and at about 15 or 20 mm. from the eyeball a large process of the pial sheath passing obliquely forwards carries the central retinal artery and vein into the middle of the nerve, and thence onwards along the axis of the nerve to the retina.

The nerve fibres, for the most part of very small diameter 2 μ, though a few (possibly "pupil" fibres) are larger, 5 μ or 10 μ, are up to the eyeball medullated fibres, but as in the brain and spinal cord possess no neurilemma. They are supported by neuroglia very similar to that of the spinal cord, and continuous with the pial septa, which like those of the spinal cord are irregular, the arrangement, so common in an ordinary nerve, of definite longitudinal bundles, each with its own sheath, being absent.

The number of fibres in the whole nerve has been calculated to be about 500,000 ; but higher estimates have been made.

Where the nerve joins the eyeball the dural sheath becomes continuous with the sclerotic coat and the pial sheath with the choroid coat, fine bundles from the sclerotic and also to some extent from the choroid passing transversely into the nerve and forming a network, the "lamina cribrosa." At this level the fibres of the nerve suddenly lose their medulla and pass on to the retina as naked axis cylinders supported by neuroglia; owing to the loss of medulla the thickness of the nerve is greatly diminished.

The medullaless fibres pass on to the level of the anterior, inner surface of the retina, forming the *optic disc* or optic papilla, in the centre of which lie the central artery and vein. From the optic disc the fibres radiate in curves, giving a pattern not unlike that known as 'engine-turned,' over the retina as far as the ora serrata, and form on the anterior, inner side of the retina, next to the vitreous humour, a layer which we shall describe presently as 'the layer of optic fibres.' Thus the optic nerve spreads itself out as a thin film lining the interior of the retinal cup next to the vitreous humour; and the other structures of the retina, in which the fibres end, lie outside or behind this film, between it and the choroid.

§ 738. *The Layers of the Retina.* Vertical sections of the retina, which has an average thickness of about ·15 mm., shew that it is made up of a series of layers superimposed, the one on the other; and the broad features of the layers are very much the same over the whole extent of the retina except at one part, the *macula lutea* containing the depression called the *fovea centralis*. The structure of this part differs materially from the rest of the retina, and we must consider it by itself, but we may treat of all the rest of the retina surrounding the optic disc as one.

The layer of optic fibres (Fig. 145, I.) lies, as we have said, next to the vitreous humour and forms what we may henceforward call the innermost layer. Next to this comes a layer in which relatively large branched nerve cells are present; this is the *layer of ganglionic corpuscles* (Fig. 145, II.). It is succeeded by a peculiar layer, very closely resembling the molecular layer of the cerebellum (§ 648) and the ground substance of the cortex (§ 649), and hence called the *molecular layer* or *reticular layer* (Fig. 145, III.), or to distinguish it from another somewhat similar layer, the *inner molecular layer* or *inner reticular layer*. Next comes a layer characterized by the presence of conspicuous nuclei closely packed together (Fig. 145, IV.), and still farther outwards lies a similar but somewhat different second layer (Fig. 145, VI.) of closely packed nuclei. The first is called the *inner nuclear layer*, or sometimes the "inner granular layer," the second the *outer nuclear layer*, or sometimes "the outer granular layer." The two layers in question are separated from each other by a layer (Fig. 145, V.) often very thin, which since in some of its features it resembles the inner molecular layer, is called the *outer molecular layer*, or "outer reticular layer;" it is sometimes called the "fenestrated layer or membrane." Outside the outer nuclear layer comes the remarkable *layer of rods and cones* (Fig. 145, VII.) which is the last of the layers of the retina proper; this is in turn succeeded by the *layer of pigment epithelium*, beyond which we come at once upon the limiting membrane of the choroid, the so-called membrane of Bruch (§ 713).

As we shall see, there is a functional connection, even if not actual continuity of structure, between the optic fibres on the inside and the rods and cones on the outside. As we shall see, the processes through which the rays of light are able to give rise to visual impulses begin in the region of the rods and cones; the rays of light have to pass through the whole or nearly the whole of the thickness of the retina before they begin their work; their work begun in the rods and cones is carried back through the thickness of the retina to the optic fibres and gathered up from the layer of optic fibres to the optic disc and so to the optic nerve. It is a necessity therefore that all the several elements of the retina should be very transparent.

We have already called the retina a piece of the brain; and even the brief statement which we have just made helps, from the likeness to cerebral structures which it suggests, to justify such a view. It is not surprising to find that in the retina as in the brain purely nervous elements are mixed with neuroglial elements, and that further, as in the brain, great difficulty is often met with in determining exactly which structure is really nervous and functional, and which merely neuroglial and sustentative. The broad distinctions however are easy.

§ 739. *Supporting* or *Neuroglial Elements.* We apply the term neuroglial to all those elements of the retina which, though not nervous in nature, are derived from epithelial cells and not of mesoblastic origin. The inner surface of the retina, that which lies in contact with the hyaloid membrane investing the vitreous humour, is defined by a thin transparent membrane of a cuticular nature, the *inner limiting membrane* (Fig. 145, *m.l.i.*). Between the outer nuclear layer and the layer of rods and cones lies a somewhat similar thin transparent membrane, the *outer limiting membrane* (*m.l.e.*), pierced with holes for the passage of the inner parts of the bodies of the rods and cones, the greater part of the bodies of which lie outside the membrane plunged, as we shall see, in the pigment epithelium. These two membranes correspond to the inner and outer surface of the inner (anterior) wall of the retinal cup, which we have said (§ 703) alone furnishes the retina proper. The rods and cones therefore may be said to project beyond the retina, and indeed, when the development of the retina is traced out, we find that the rods and cones do really grow out from the inner wall of the retinal cup and thrust themselves into the outer wall, which is wholly transformed into the pigment epithelium.

Stretching radially from the inner to the outer limiting membrane in all regions of the retina, and therefore seen in a vertical section as vertically disposed structures, are certain peculiar shaped bodies known as the *radial fibres of Müller* (Fig. 145). Each fibre is the outcome of the changes undergone by what was at first a simple columnar epithelial cell. The changes are in the main that the columnar form is elongated into that of a more or less prismatic fibre, the edges of which become variously branched, and that while the nucleus is retained the cell-substance becomes converted into neurokeratin (see § 563). And indeed at the ora serrata the fibres of Müller may be seen suddenly to lose their peculiar features and to pass into the ordinary columnar cells which form (§ 714) the pars ciliaris retinæ.

At the inner limiting membrane, the fibre of Müller begins with a broad expanded foot fused with the membrane; indeed the inner limiting membrane may be regarded as in reality formed out of the coalesced broad more or less polygonal bases of the fibres of Müller. From this broad base the fibre narrows to a thin columnar body stretching radially outwards through the layer of optic fibres and that of ganglionic cells. From the edges of the body numerous fine processes extend in various directions, affording support to the optic fibres which wind in and out between the feet of, and to the ganglionic cells which are lodged in the spaces between the bodies of these fibres of Müller. Stretching still radially outwards the fibre as it passes through the inner molecular layer gives off processes which either divide into or at least are lost in the numerous neuroglial fibres forming part of the fibrillar structure of that layer. In the inner nuclear layer broader lateral processes become especially abundant, forming a neurokeratinal basketwork in the spaces of which the nervous elements of this layer are lodged; and here the body of the fibre bears the nucleus (Fig. 145 *n*) of the fibre itself, a somewhat elongated oval nucleus placed vertically. At the level of the outer molecular layer the body of the fibre seems to end, its processes contributing to form that layer in some such manner as in the inner molecular layer, but in reality it breaks up into a basketwork or spongework of delicate neuro-keratinal shreds, supporting the nervous structures of the outer nuclear layer, and terminating at the outer limiting membrane or, possibly, even passing beyond it.

Each fibre of Müller may be regarded as a brace or tie between the inner and the outer limiting membrane, its nucleus appearing among the nuclei of the inner nuclear layer, and its numerous processes on the one hand contributing to form the inner and outer molecular layer and on the other hand affording throughout the thickness of the retina a support for the nervous elements.

Besides these conspicuous fibres of Müller, which we may regard as large neuroglial elements, small, more ordinary neuroglial cells are also present in the various layers, especially perhaps in the outer molecular layer and in the layers of optic fibres and ganglionic cells. In addition to these neuroglial elements the retinal blood vessels as we shall presently see carry with them a small amount of actual connective-tissue.

§ 740. *The Nervous Elements*. As we have said, there is at least a functional continuity between the rods and cones on the outside and the optic fibres on the inside. The structures of the outer nuclear layer are on the one hand closely and obviously continuous with, indeed form part of, the rods and cones, and on the other hand are if not actually continuous at least functionally connected across or by means of the outer molecular layer with the structures of the inner nuclear layer; while these in turn make connections, across or by means of the inner molecular layer, with the optic fibres either indirectly through the nerve cells of the ganglionic layer or in a more direct manner.

The rods and cones. Each rod consists of at least two quite distinct parts, of wholly different nature, called respectively the outer and the inner limb. The *outer limb* (Fig. 145, *r.o.*) is a cylinder, in man about 30 μ in length by 2 μ in diameter, which when seen in a natural condition is transparent, though highly refractive, and also doubly refractive. In prepared specimens it often appears fluted or grooved lengthwise and is very apt to cleave transversely into discs or fragments of varying thickness. It is probably made up of a number of excessively thin discs ·6 μ or less in thickness, superimposed on each other and united by some kind of cement substance. The material of which the limb, as a whole, is composed stains deeply with osmic acid, but not at all with carmine and similar staining reagents, and is of a peculiar nature, in many respects resembling but still differing from the medulla of a medullated nerve. It is sensitive to light, swelling up when in the living eye it is exposed to light and shrinking again when the light is removed. During life it is coloured with a peculiar pink colouring matter of which we shall treat later on, called visual purple, and which as we shall see is bleached by the action of light.

The *inner limb* (Fig. 145, *r.i.*) is an elongated ellipsoidal or fusiform body, about as long as, and at its broadest part slightly broader than, the outer limb, being truncated at its outer end where by a flat surface it lies in contact with the inner end of the outer limb. It is of a wholly different nature from the outer limb, staining with carmine and other staining reagents, and having the ordinary optical features of protoplasmic cell-substance. The outer part lying next to the outer limb exhibits a longitudinal striation or fibrillation, but the inner part is faintly and finely granular, the whole however being very transparent. This slight difference of marking indicates a division of the inner limb into an outer and an inner moiety differing in nature from each other; and in some animals the outer part is occupied by a distinctly differential structure, called the "ellipsoidal body."

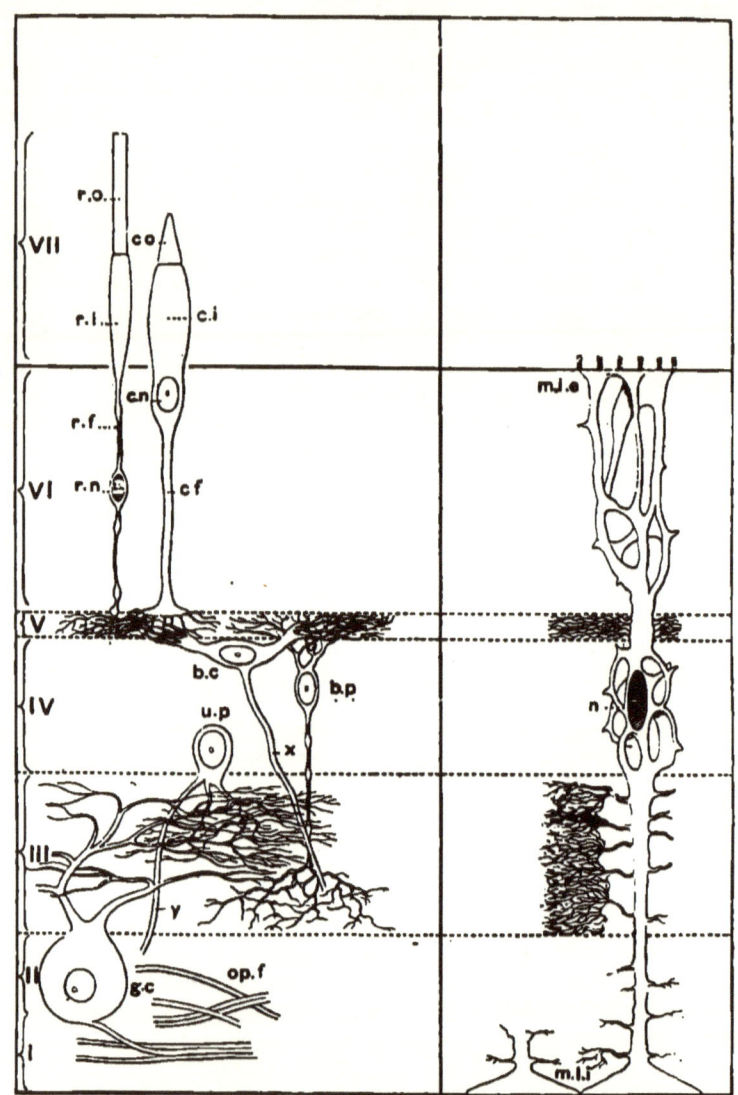

Fig. 145. Diagram to illustrate the Structure of the Retina.

The figure is quite diagrammatic and is introduced simply to assist in the description of the probable structure of the retina.

The several "layers" are indicated by the numeral I, &c. in order from within (vitreous humour) outwards.

On the right-hand side is shewn a fibre of Müller reaching from the inner limiting membrane $m.l.i.$, formed by the coalescence of its feet, to the outer limiting membrane $m.l.e.$; the lateral processes of the fibre are indicated, as well as the neuroglial fibrils (possibly independent of the fibres of Müller), in III. the inner molecular, and V. the outer molecular layer. *n.* the nucleus of the fibre of Müller.

On the left hand side the nervous elements are represented. A rod with its *r.o.* outer, and *r.i.* inner limb, continued on as the rod fibre *r.f.* with its banded nucleus *r.n.*; also a cone with its outer *c.o.* and inner *c.i.* limb, and cone fibre *c.f.* with the nucleus *c.n.*

In IV., the inner nuclear layer, are represented, (1) one of the bi-polar cells, *b.p*, with its outer process lost in the outer molecular layer, and its inner process ending in a tangle in the inner zone of the inner molecular layer; (2) one of the so-called unipolar cells *u.p.*, with its single process branching into the outer zone of the inner molecular layer, the possible axis cylinder process of such a cell being also shown at *y*; (3) one of the cells *b.c.* in the outermost tier of the nuclear layer, branching on the outer side into the outer molecular layer, and sending inwards an 'axis-cylinder process' *x* to join the layer of optic fibres. It will be remembered that by far the greater number of the nuclei of the inner nuclear layer belong to the first of these three kinds of cells.

In II. is shown a 'ganglionic' cell *g.c.* with its axis cylinder process becoming an optic fibre, and the branches of the cell body lost to view as fine processes in the middle zone of the inner molecular layer. Optic fibres are also shown at *op.f.*

The inner end of the inner limb, piercing the external limiting membrane, narrows rapidly to a delicate thread *r.f.*, which is directed straight inwards towards the outer molecular layer, is often varicose, and has otherwise much the appearance of a fine nervous filament. This process of the inner limb, known as the *rod-fibre*, bears at some part of its course, sometimes nearer the external limiting membrane, sometimes nearer the outer molecular layer, an oval nucleus *r.n.* over which the fibre expands to form a very thin layer of cell-substance. The nucleus is peculiar inasmuch as its staining, chromatin, constituents are, at least frequently, arranged in transverse bands, giving the stained nucleus a banded appearance, like that of the planet Jupiter.

At the outer molecular layer the rod-fibre is lost to view; whether, as some think, it ends abruptly at the layer in a knob-like termination, or changing its direction and running transversely ends in fine fibrils, thus contributing to the network of the layer, or again whether, as others think, it crosses the layer and passes into the inner nuclear layer cannot at present be regarded as settled.

Obviously we may consider the 'rod,' taken as a whole, to be an elaborated epithelium cell, whose nucleus lies on its inner thread-like continuation or rod-fibre on the inner side of the external limiting membrane and contributes to the nuclei of the outer nuclear layer, and whose main body, lying outside the external limiting membrane, is sharply divided into an inner part, inner limb, which remains more or less protoplasmic in nature, and into an outer part, outer limb, which has undergone a differentiation having a certain resemblance to cuticular differentiation but still of a distinctly peculiar kind. Between the outer and inner limb there appears to be a distinct break; the two are in apposition but are not continuous; and indeed in some creatures, such as birds, the two are not even in apposition, being separated from each other by the intercalation of a spherical globule of a fatty nature often coloured. We may infer that the outer limb serves in some way or other as a purely physical dioptric apparatus, and that the strictly physiological changes, those which initiate the visual nervous impulses, begin in the inner limb.

A cone, like a rod, consists of an outer and an inner limb. The outer limb (Fig. 145, *c.o.*) is conical, not cylindrical in form, and, though much shorter than the outer limb of the rod, being in man about 10 μ in length, is in nature in every way similar to it save that it contains no visual purple. The inner limb *c.i.* is almost in all ways like the inner limb of a rod, save that it is usually broader, the diameter of the cones (outside the fovea) being about 6 μ. Piercing the external limiting membrane it narrows to a fibre, *cone fibre, c.f.*, which however is broader than a rod fibre, and the oval nucleus *c.n.* which it bears, usually close under the external limiting membrane, is not banded and contains a conspicuous nucleolus. On reaching the outer molecular layer the cone fibre, expanding into a sort of foot, breaks up into a number of fibrils, which like the corresponding end of the rod fibre cannot be satisfactorily traced beyond their entrance into the layer.

Over the retina (we are now, it will be remembered, excluding the macula lutea) the rods are much more numerous than the cones, there being about two or three rods in the line joining two cones, and since the outer limbs of the cones are much shorter than those of the rods, a surface view of the retina seen from the outside shews nothing but rods when the tops of the rods are in focus; if the focus be carried lower down, inwards, the cones will come into view and a mosaic of rods with cones interspersed between them will be seen. Towards the extreme periphery of the retina the cones become relatively more numerous, and close to the ora serrata are alone present. The total number of cones has been calculated to be more than three million.

To complete the account of the layer of rods and cones, we may add that the outer limbs and also, to a certain extent and under certain conditions of which we shall speak later on, the inner limbs are imbedded in the layer of pigment epithelium, and that between the inner limbs a number of fine acicular cilia-like processes, apparently of cuticular nature, start up from the external limiting membrane, forming a sort of basketwork support for those structures; on the other side of the limiting membrane, as we have already said, all the nervous structures of the outer nuclear layer, that is to say the rod fibres and cone fibres with their respective nuclei, are supported by neuro-keratinal spongework proceeding from the fibres of Müller.

§ **741.** The nuclei of the *inner nuclear layer*, with the exception of the nuclei of the fibres of Müller, which, as we have said, are placed in this layer, belong to cells in which the amount of cell-substance is in most cases small compared to the nucleus. These cells are arranged in several tiers and, though possessing a general resemblance to each other, are not all of the same kind.

In the cells (Fig. 145, *b. p.*) which form the greater number of the tiers, all in fact except the innermost and outermost tiers, the nucleus, round and highly refractive, is surrounded by a thin layer of cell substance, which is prolonged in a radial direction from the opposite poles of the nucleus into an outer or peripheral and an inner or central process. The peripheral process, directed straight towards the outer molecular layer, soon branches and dividing into fine fibrils is lost to view in that layer; in some animals, however, the process is said to give off a branch which, continued undivided through the outer molecular layer and the outer nuclear layer ends abruptly just inside the external limiting membrane in a club-shaped swelling. The central process, thinner and more delicate than the peripheral one, and frequently varicose, is directed straight inwards to the inner molecular layer, and, after traversing that layer for some distance without dividing, is lost to view, or, according to some observers, ends in a peculiar tangle of fine nervous fibrils in the inner zone of the layer (cf. Fig. 145). From their possessing two obvious processes running in two opposite directions these cells have been called the *bipolar* cells of this layer.

The cells of the innermost tier (Fig. 145, *u. p.*), whose nuclei are also round and refractive with conspicuous nucleoli, vary more in size than do those of the outer rows, and are on the whole larger. The cell-substance around each nucleus is continued not into two processes, but into a single one only, which is directed inwards into the inner molecular layer and is there lost to view, or, according to some observers, divides into fine fibrils in the outer zone of the layer (Fig. 145); the cells have therefore been called *unipolar* to distinguish them from the bipolar cells just described. They have also been called "spongioblasts" and have been supposed to be concerned in the maintenance of the neuroglial framework; but they are probably nervous in nature. Some observers have described them as giving off besides the branched process just described an undivided axis cylinder process (Fig. 145 *y*), which, running through the whole thickness of the inner molecular layer, joins the layer of optic fibres and indeed becomes an optic fibre.

A third kind of cell has been described as forming or contributing to the outermost tier of nuclei of this inner nuclear layer just inside the outer molecular layer. The body of a cell of this kind (Fig. 145, *b. c.*), somewhat flattened in the plane of the retina, is said to give off on the outer side processes which, running transversely and soon branching, are lost to view as fine fibrils in the outer molecular layer, and on the inner side a single process which has very much the appearance of an axis cylinder process (Fig. 145, *x*.) but which cannot be traced beyond the inner molecular layer, though it has been supposed to pass on and become an optic fibre.

Thus while the nuclei of the outer nuclear layer are nuclei of the rod fibres and cone fibres, the nuclei of the inner nuclear layer are the nuclei ·of cells of a peculiar character; most of them belong to cells called bipolar, some of them, the innermost, belong to so-called unipolar cells, while some of the outermost are described as belonging to a third kind of cell. In several respects this inner nuclear layer resembles the nuclear layer of the superficial grey matter of the cerebellum.

§ 742. The *layer of ganglionic cells* is closely connected with that of the *optic fibres*. The latter (Fig. 145, *op. f.*) consists, as we have said, of nerve fibres, that is of naked axis cylinders without medulla and without neurilemma, radiating in all directions from the optic disc. Over the retina generally these fibres exist as a single layer of bundles, arranged in a plexiform manner, winding between the feet of the fibres of Müller, and supported by a neuroglia to which the processes of the fibres of Müller contribute, as well as by a scanty amount of connective-tissue belonging to the retinal blood vessels. The fibres and bundles of fibres become less numerous, and the plexuses more open, from the optic disc towards the ora serrata, and at the latter line cease altogether. The diminution is due to the optic fibres continually passing away from the layer into the other, outer layers.

The *layer of ganglionic cells* is over the retina generally a single layer of nerve cells. Each cell (Fig. 145, *gc.*), 30 μ or so in diameter, consists of a cell body which is transparent or nearly transparent, though sometimes containing pigment, and a relatively large spherical nucleus. The cell-body sends inwards a single undivided axis cylinder process which becomes an optic fibre, and outwards a number of branched processes, which, passing into the inner molecular layer and dividing, chiefly in the middle zone of the layer, into delicate filaments, are lost to view; owing to these branched processes the cells are spoken of as multipolar. These ganglionic cells are far less numerous than the optic fibres, so that only some of the latter are continued into the former; the rest of the fibres are either connected as suggested with some of the cells of the inner nuclear layer or possibly end without the intervention of cells, by branching in the inner molecular layer in a manner similar to that of the processes of the nerve cells.

§ 743. We may then consider the retina as made up of three main layers of cells; (1) the rods and cones with the outer nuclear layer, (2) the cells forming the inner nuclear layer, and (3) the ganglionic cells with the optic fibres. On theoretical grounds we may conclude that these three layers are functionally continuous, that changes set going in the (inner limbs of the) rods and cones sweep through the inner nuclear layer, and issue along the fibres of the optic nerve as nervous impulses; but, with at least our present knowledge, we cannot demonstrate a structural continuity between them. Two conspicuous breaks occur at the two molecular layers. We can trace the rod fibres and the cone fibres to the outer molecular layer, and there we lose them. We can trace the optic fibres through or apart from the ganglionic cells to the inner molecular layer, and there we lose them too. We can trace the processes of the cells of the inner nuclear layer on the one hand to the inner and on the other to the outer molecular layer, and there these too are lost. These two molecular layers, which repeat in the outlying part of the brain which we call the retina some of the characteristic features of the brain itself, are obviously of no little importance in the development of visual impulses; but for a proper understanding of their nature we must await the results of further inquiry. At present all we can perhaps say is that each layer consists of a network of fine nervous fibrils imbedded in neuroglia, but that, as in corresponding cerebral structures, we cannot accurately distinguish neuroglial from nervous elements, much less trace out the exact disposition of the latter. In the outer molecular layer among the tangle of fibrils, nervous and neuroglial, we may distinguish small branched cells, lying flatwise in the plane of the layer; these are probably neuroglial cells whose branched processes become neuroglial fibrils. In the inner molecular layer such cells are absent or at least inconspicuous; the layer seems to consist on the one hand of nervous fibrils derived from the branching processes of nerve-cells and on the other hand of neuroglial fibrils, all imbedded in a peculiar ground-substance which stains deeply with osmic acid, and indeed is of a nature in some respects allied to the medulla of a nerve fibre.

We have reason to think that the molecular changes which light induces in the inner limbs of the rods and cones differ very considerably in character from the molecular changes in the fibres of the optic nerve which constitute a nervous impulse, and that the transformation from the one set of changes to the other is effected through some or other of the retinal structures which we have described. But we cannot attribute definite functions to the several elements; and here, as in the case of the brain and spinal cord, we may hesitate to assign too much to cellular elements. We may, perhaps, in conformity with what we have urged elsewhere (§ 579), regard the cells of the ganglionic layer as being largely concerned in nutritive labours, and may even apply the same view to the nuclear layers; if this be so, no small part of the work of the retina in transforming the first crude effects of the impact of light into true nervous impulses, may be looked upon as being carried out by the tangle of nerve fibrils, in the two molecular layers and elsewhere.

§ **744.** *The Macula Lutea and Fovea Centralis.* On the temporal side of the optic disc, at a distance of about 4 mm. from, and a little below the horizontal level of, its centre an oval area, with its long axis of about 2 mm. placed horizontally, is distinguished from the rest of the retina by its yellowish or brownish tint; this is the *macula lutea* or 'yellow spot.' At the edges of the macula the retina is thickened but in the centre becomes very thin. So that the macula presents a central depression, about ·3 mm. in diameter, the *fovea centralis*, surrounded by a raised rim.

A vertical section through the macula lutea shews that, in contrast to the retina generally, in which the rods are more numerous than the cones, in the very centre of the fovea cones alone are present; their outer limbs, however, are very elongated, 60 μ long by 2 μ or 3 μ broad, and indeed are rod-like. Moreover, all the several layers of the retina described above are here absent except the layer of cones, and the outer nuclear layer or layer of cone fibres. Between the pigment epithelium and the vitreous humour are found only these elongated cones, with the nucleated cone fibres belonging to them; the latter supported by a delicate reticular neuroglia curve away towards the periphery of the fovea; next to the vitreous humour lies an exceedingly thin layer of neuroglia. About 7000 cones are supposed to be crowded together in the very centre of the fovea.

A vertical section through the thickened rim of the macula shews all the layers present, the layer of ganglionic cells being exceedingly prominent and consisting not as elsewhere of a single layer or at most of two layers but of several, eight or nine, layers of cells; indeed the greater thickness of the retina at the rim of the macula is chiefly due to an increase in the layer of ganglionic cells, the inner nuclear layer being somewhat increased, but not to any great extent, and the others hardly at all. In the layer of rods and cones, the cones are more numerous than outside the macula, and their outer limbs are somewhat elongated; but otherwise the features of the several layers are the same as in the retina generally; we may add, however, that the ganglionic cells appear bipolar rather than multipolar, one process being continued as elsewhere into an optic fibre and the other directed obliquely towards the central parts of the macula; the latter process, however, sooner or later divides into fine branches.

If in a vertical section taken across the whole macula we examine the features of the section from the periphery towards the centre, we find the layer of optic fibres rapidly thinning out and soon disappearing, and the layer of ganglionic cells, after its temporary thickening, also thinning out and disappearing. The inner molecular layer and the inner nuclear layer extend a little farther towards the centre, but these also eventually disappear, all that is left of them and of the outer molecular layer being the thin layer of neuroglia mentioned above. At the same time the rods, relatively numerous at the peripheral parts of the macula, gradually grow scanty and finally disappear. In this way nothing is left in the fovea itself but the cones and the cone fibres.

Over the retina generally the several elements seem, as seen in a vertical section, to be disposed vertically the one over the other. During the thinning out and disappearance just mentioned, this vertical disposition of the several elements gives way, except so far as the cones themselves are concerned, to an oblique disposition, which becomes the more marked as the centre is approached. It is as if the parts of the inner nuclear layer, of the inner molecular layer and of the ganglionic layer belonging to the cones and cone fibres of the central region of the macula were dragged on one side towards the periphery. We may imagine that each (rod or) cone with its (rod or) cone fibre is connected in some way or other with one or more of the cells of the inner nuclear layer, and so with one of the ganglionic cells; over the retina generally these are placed in the same vertical line, and before a ray of light can act on the rod or cone it must pass through these several other structures. In the fovea these other structures are pulled on one side, so that in the very centre of the fovea the light can gain access to the outer limb of the cone, without having to pass through any other structures than the cone itself and its cone fibre. This oblique disposition is also obvious in the cone fibres, which elsewhere short and vertical are in the fovea much elongated, and radiate obliquely from the centre of the fovea to their more peripherally placed inner nuclei or other structures with which they have to make connections of some kind or another.

The colour of the macula which is said to be in the living eye brown rather than yellow is most intense in the thickened rim, fading gradually away peripherally and centrally, and being wholly absent from the central fovea. The colour is due to a pigment diffused through the layers lying to the inside of the outer nuclear layer, being absent from this layer as well as from the cones themselves; hence the absence of the colour from the very centre. The pigment seems to be attached at least as much to the neuroglial as to the nervous elements. It thus presents a contrast to the visual purple which is limited to the outer limbs of the rods, not being present elsewhere, not even in the outer limbs of the cones.

§ 745. *The Blood Vessels of the Retina.* The central artery and vein of the retina running, as we have seen, with their sheath of connective-tissue in the middle of the optic nerve reach the surface at the centre of the optic disc, which is excavated so that the blood vessels seem, on a surface view of the retina, to arise out of a minute crater. Both artery and vein divide into two main trunks, one directed upwards and the other downwards, the division of the vein taking place while it is still imbedded in the nerve; each trunk divides again into two, and these into several branches which are distributed over the retina as far as the ora serrata, the branches on the nasal side radiating in a more or less straight direction, while those on the temporal side arch round above and below the macula lutea, veins and arteries taking much the same course.

Both arteries and veins run close to the internal limiting membrane in the layer of optic fibres, being accompanied by sheaths of delicate connective-tissue enclosing perivascular lymphatic spaces. The capillaries into which the arteries break up, and from which the veins are gathered up, form in the first instance a somewhat close capillary network between and among the optic fibres and ganglionic cells; but the vessels also extend a certain distance outward and form a second outer also fairly close capillary network in the inner nuclear layer. Some of the loops may reach the outer molecular layer, and in some few animals capillaries may be seen in the outer nuclear layer. In no case do they extend beyond the external limiting membrane, and as a rule the outer molecular layer may be taken as marking the limit beyond which they do not extend.

The arteries and veins sweep, as we have said, round the macula lutea, which however is largely supplied by two small arteries and veins coming straight from the optic disc. Capillaries extend into the margin of the macula, reaching as far as the layers of ganglionic cells and the inner nuclear layer; from the fovea itself, blood vessels are wholly absent.

§ **746.** *The Pigment Epithelium.* This is a single layer of epithelium cells lying between the rods and cones of the retina proper on the inside and the limiting membrane of the choroid, membrane of Bruch, on the outside. It is wanting where the optic nerve passes forward to join the retina, but is present, and exhibits the same features over the whole of the retina up to the ora serrata, at which line it passes into the more ordinary pigment epithelium of the ciliary processes.

It very readily separates from the choroid, and frequently comes away with the retina when the latter is removed from the eyeball. In such cases the layer is, in a surface view, seen to be composed of cells which have a polygonal outline and are loaded with black pigment granules, the nucleus of the cell being more or less obscured by the pigment, and the outlines of the cells being well defined by clear lines of cement material, apparently neurokeratinal in nature, free from pigment; a similar appearance is presented by the pigment epithelium of the ciliary processes though the outlines are not so well marked.

A vertical section however taken through the layer in position shews that the cells possess peculiar features absent from the simpler epithelium beyond the ora serrata. Each cell is seen to consist of a more or less cubical body, the outer part of which next to the choroid is free from black pigment, though often containing small irregular masses of yellowish material of a fatty nature allied to the medulla of nerves. The inner part, that next to the retina, is on the other hand loaded with black pigment granules, which on closer examination are found to be minute crystals; the substance of which these are composed is called *fuscin*. The nucleus lies either wholly in the clear part or partially imbedded in the pigment.

The most important feature of the cell however is that the inner surface next to the retina, is not smooth and even, but frayed out into a number of rod-like or filamentous processes, each carrying a load of pigment crystals. Further when eyes are subjected immediately before death to different conditions as regards light, these processes are found in different states.

If an eye be fully exposed to light before removal and examination, the processes carrying pigment are found to stretch a long way inwards between the outer limbs of the rods and cones, investing these outer limbs with a sheath of pigment, and even reaching between the inner limbs. If on the contrary the eye be kept in the dark before removal and examination, the processes are found to be short and to stretch a little way only inwards, not reaching much farther than the tops of the outer limbs of the rods and cones. The substance of the cell has in fact the power of amœboid movement, at one time throwing out long filamentous processes inwards between the rods and cones, and at another time retracting the processes into the body of the cell. As they move to and fro these processes carry with them the crystals of pigment with which they are studded; hence in the extended condition much of the pigment is carried away from the body of the cell inwards between the rods and cones, leaving the nucleus less covered with pigment, while in the retracted condition the pigment is carried back to the body of the cell and the nucleus becomes obscured. Further, while various circumstances may determine whether the processes are extended or retracted, the falling of light on the retina has the most marked and potent effect. When light falls on the retina the processes hurry inwards and envelope the outer limbs of the rods and cones with pigment; when the light is shut off from the retina the processes carry back the pigment to the body of the cell.

Hence in an eye exposed to light the processes and pigment being largely jammed in between the outer limbs of the rods, and these outer limbs at the same time swelling, the pigment epithelium adheres closely to the retina, and when the retina is removed is carried away with it. In an eye kept in the dark, the processes being withdrawn, and the outer limbs of the rods shrinking again, the attachment of the retina to the epithelium is much less, and the retina can be more readily removed so as to leave the pigment epithelium adherent to the choroid.

Urari has an effect on these cells of the pigment epithelium of such a kind that they cease to throw out their processes; they seem to be paralysed. Hence in the eye of a urarized animal the pigment epithelium readily separates from the retina.

We may add that in frogs at least, this shifting of the pigment may be seen to be accompanied by a change of form in the inner limbs of the cones. Under the influence of light the inner limb becomes shorter and broader, in fact contracts, and when the influence of the light is removed elongates to its original length. Moreover these changes in the cones may be induced, not only by light falling on the retina but also, through a mechanism not at present fully understood, as the result of stimulation of the skin, by light or otherwise; in these latter cases the change of form of the cone is not necessarily accompanied by migration of the pigment.

SECT. 7. ON SOME GENERAL FEATURES OF VISUAL SENSATIONS.

§ 747. When light falls upon the retina it produces, under favourable circumstances, a change in our consciousness which we call a sensation of light, a visual sensation. The immediate effect of the light is to stir up certain changes in the retina; these retinal changes give rise in turn to nervous changes in the optic fibres; these latter, which we have called 'visual impulses,' start in the brain a further series of events, one effect of which is a change in our consciousness; and it is this change in our consciousness which we call a sensation. We may, and often do, speak of light as a 'stimulus' to the retina, the result of the stimulation being visual impulses; but we may also speak of light as a stimulus to the whole visual apparatus, central as well as retinal, regarding the sensation as if it were the direct and immediate, instead of being the indirect and ultimate effect of the stimulus. We may, by observing certain general features of visual sensations, such as can be ascertained by means of a direct and simple appeal to our own consciousness, study the relations which obtain between the characters of the stimulus on the one hand and those of the sensation on the other. There are certain advantages indeed in doing this before we proceed to discuss the nature of the changes in the retina through which rays of light give rise to visual impulses in the optic fibres. But in taking this course we must bear in mind how complex is the whole process through which the stimulus gives rise to the sensation. We must remember that, as we have already said, though some of the characters of a visual sensation are impressed upon it while it is as yet immature, as yet in the stage of visual impulses, others are introduced later on in the course of the cerebral changes. Since we are now dealing for the first time with sensory impulses studied in this way, we may venture to enter into some details, for the deductions which may be drawn concerning visual sensations will apply to a large extent to other sensations.

To simplify matters we will in the first instance suppose that the luminous object, the object emitting or reflecting light, is so small that the image of it on the retina may be considered as a mere point; we may speak of it as a luminous point. If for the sake of illustration or otherwise we have to consider a larger luminous object, we shall do so without regard to the size of the image on the retina unless this is specially mentioned.

We may begin with the preliminary remark that in dealing with light as a stimulus of visual sensations, we have to consider not only the intensity of the stimulus, but also its duration. A luminous point may appear dim and feeble, that is to say, the waves of light from it have a small amplitude and so bring little energy to bear on the retina, or it may appear bright and strong, that is to say, the waves of light have a large amplitude and so bring much energy to bear on the retina. Whether dim or bright, the luminous point may act on the retina for a longer or a shorter time; and, moreover, during its action may remain steady, not varying in intensity, or may vary in intensity and become unsteady or flickering. In estimating the total visual effect of a luminous point, we have to consider both these features, its intensity or brightness and its duration.

Neglecting for the present the feature of duration, we find that a luminous point must possess a certain amount of brightness in order to produce any conscious sensation at all, in order to be visible. If the waves of light fall on the retina with less than a certain amplitude, if their energy sinks below a certain minimum, they fail to give rise to visual impulses, or at least to such as can affect consciousness; for we may suppose that visual impulses might be generated and yet be so feeble as not to produce in the cerebral centre changes sufficiently great to affect consciousness. It will be understood, of course, that the exact degree of brightness at which the luminous point becomes visible depends on the greater or less irritability, on the sensitiveness, of the retina. The same amount of luminous energy which, falling on one retina or on one part of a retina, produces a distinct sensation, may, falling on a less sensitive retina or on a less sensitive part of the same retina, produce no sensation whatever.

From the minimum onwards the intensity of the sensation increases with the luminous intensity of the object; a wax candle appears brighter than a rushlight, and the sun brighter than any candle; we are dealing now with the intensity of the light quite apart from the size of the luminous object. The ratio, however, of the sensation to the stimulus is not a simple one. If the luminosity of an object be gradually increased from a very feeble stage to a very bright one, it will be found that, though the corresponding sensations likewise gradually increase, the increments of the sensations due to increments of the luminosity gradually diminish, and at last an increase of the luminosity produces no appreciable increase of sensation; a light, when it reaches a certain brightness, appears so bright that if it becomes brighter we do not recognize that it is brighter. Hence it is much easier to distinguish a slight difference of brightness between two feeble lights than the same difference between two bright lights; we can easily tell the difference between a rushlight and a wax candle; but two suns, or even two bright lamps, one of which compared with the other gave out just that additional amount of light, just that additional quantity of luminous energy, which a wax candle gives out in addition to that given out by a rushlight, would appear to us to have exactly the same brightness. In a darkened room an object placed before a candle will throw what we consider a deep shadow on a sheet of paper or any white surface. If, however, sunlight be allowed to fall on the paper at the same time from the opposite side, the shadow is no longer visible. The difference between the total light reflected from that part of the paper where the shadow was, and which is illuminated by the sun alone, and that reflected from the rest of the paper which is illuminated by the candle as well as by the sun, remains the same; yet we can no longer appreciate that difference because the whole surface has become so bright.

On the other hand, when we carefully compare the visual sensations excited by measurable differences of luminosity, we come upon the following remarkable result. If we place two candles so as to throw two shadows of some object on a white surface, the shadow caused by each light will be illuminated by the other light, and the rest of the surface will be illuminated by both lights. If now we move one candle away we shall reach a point at which the shadow caused by it ceases to be visible, that is to say, we fail at this point to appreciate the difference between the surface illuminated by the near light alone and that illuminated by the near light and the far light together. If now, having noted the distance to which the candle had to be moved, we repeat the same experiment with two bright lamps, moving one lamp away until the shadow it casts ceases to be visible, we shall find that the lamp has to be moved just as far as the candle; that is to say the least difference between the illumination of the bright lamps which we can appreciate is the same as in the case of the dimmer candles. Many similar examples might be given shewing a similar result; in fact, it is found by careful observation that, within tolerably wide limits, the smallest difference of light which we can appreciate by visual sensations is a constant fraction (about $\frac{1}{100}$th) of the total luminosity employed. As we shall see, the same relation holds good with regard to the other senses as well. It may be put in a general form, as a law of sensation, often called Weber's law, somewhat as follows: The smallest change in the magnitude of a stimulus which we can appreciate through a change in our sensation always bears the same proportion to the whole magnitude of the stimulus. It should however be stated that this law holds good within certain limits only; it fails when the stimulus is either above or below a certain range of intensity.

Hence, if we take the smallest difference which we can appreciate in a stimulus as a measure of our sensibility to differences in the stimulus, we may say that on the one hand in respect to absolute differences, such as that between two lamps and that between two rushlights, our sensibility varies inversely as the magnitude of the stimulus; we are more sensible of the same absolute difference when that is a difference between two rushlights than when it is a difference between two lamps. On the other hand, in regard to relative differences, our sensibility is independent of the magnitude of the stimulus; the difference of which we are sensible in the case of the lamp bears the same proportion to the whole luminosity of the lamp as the difference of which we are sensible in the case of the rushlight bears to the whole luminosity of the rushlight.

§ 748. Returning now to consider the duration of the stimulus, as distinguished from its intensity, we find that a stimulus of extremely brief duration may give rise to a distinct sensation; the flash of an electric spark, for instance, is readily visible. There is probably a limit in respect to duration within which the stimulus fails to produce a sensation; it is probable, for instance that a certain number of undulations in succession must fall on the retina in order to give rise to a visual sensation, and that a single undulation of the ether falling on the retina, if such a thing were possible, would produce no visual effect; but the exact limit will depend on the intensity and nature of the light, and we need not enter upon these details here.

It is of more importance to note that the visual sensation caused by a very brief stimulus lasts a considerable time; the sensation has a duration much greater than that of the stimulus. The sensation of a flash of light, for instance, lasts for a much longer time than that during which luminous vibrations are falling on the retina. In this respect, we may roughly compare a visual sensation to a simple muscular contraction caused by such a stimulus as a single induction shock. We might indeed construct a "visual sensation curve" very much after the fashion of a "muscle curve." We should find that after a very obvious latent period the sensation began, rose to a maximum and then declined. This latent period forms an important part of the "reaction period," on which we dwelt in a former part of this work (§ 691). As we have said, in all the sensations with which we are now dealing, we have to distinguish at least two parts, the peripheral part, the events taking place in the retina, and the central part, the events taking place in the brain, the two being united by means of the visual impulses passing along the optic nerve. And within the latent period are comprised the changes in the retina which start the visual impulses, the passage of these impulses along the optic fibres, and the changes in the brain antecedent to consciousness beginning to be affected; of these the retinal changes probably take up the most time, but into this point we cannot enter now.

The length of the sensation, as compared with that of the stimulus, is illustrated by viewing objects in motion under a very brief illumination, such as that of a single electric spark. In such a case the light reflected from the object is sufficient to generate a distinct sensation, to give rise to a distinct image of the object, but it ceases before the object can make any appreciable change in its position, and the image accordingly is that of a motionless object. When a moving body is illuminated by several rapid flashes in succession, several distinct images corresponding to the positions of the body during the several flashes are generated; this, as we shall see presently, is because the images of the body corresponding to the several flashes fall on different parts of the retina.

The duration of the stimulus remaining the same, the characters of the sensation and the form of the sensation curve will, in accordance with what was stated above, vary with the intensity of the stimulus; a bright flash will produce a sensation greater and of longer duration than that produced by a feeble flash, the curve will be higher and more extended. We have reason to think, too, that the form of the curve is dependent on the intensity of the stimulus in such a way that the decline from the maximum begins earlier and at all events in the first part of its course, is more rapid with the stronger than with the feeble stimulus.

When the stimulus is not a mere flash, but is of some duration leading to a prolonged sensation, we can readily distinguish between that part of the sensation which is going on while the light is still falling into the eye, and that part which goes on after the light has ceased to fall on the retina; this latter part is often spoken of as the *after-image*. When the light is very bright this "after-image" frequently becomes very prominent even after a very brief exposure. Thus, if we look, even for a moment only, at the sun, and then immediately shut the eye, an intense visual sensation, a bright visual image of the sun, remains for some considerable time. After images, especially as they are vanishing, are marked by certain features, which we shall study later on, and which, as we shall see, are related to the fatigue or exhaustion of the retina; for the retina, or rather the whole visual apparatus, is, we need hardly say, subject to fatigue.

§ **749.** From the prolonged duration of visual sensations it results that when two or more stimuli, such as two or more flashes of light, follow each other at a sufficiently short interval, the two sensations or the several successive sensations are fused into one more or less uniform sensation. Thus a luminous point moving rapidly round in a circle gives rise to the sensation of a continuous circle of light. We might, in a very general manner, compare this with the way in which a series of simple muscular contractions resulting from rapidly repeated induction shocks are fused into a fairly uniform tetanus. When the stimuli succeed each other so rapidly that each sensation begins before its predecessor has had time to appreciably decline, the total sensation is as completely uniform as if the stimulus were constant. If the interval between each two stimuli be just so long that each sensation in turn has had time to distinctly diminish before the next sensation begins, the result is a "flickering;" and there are of course many degrees of flickering between a perfectly steady and an obviously intermittent light. The interval at which fusion takes place, that is, the interval between successive stimuli which must be exceeded in order that successive distinct sensations may be produced, varies according to the intensity of the light, being shorter with the stronger light; with a faint light it is about $\frac{1}{10}$ sec., with a strong light $\frac{1}{30}$ or $\frac{1}{50}$ sec. This may be shewn by rotating rapidly before the eye a disc arranged with alternate black and white sectors of equal width. With a faint illumination, the flickering indicative of the successive sensations from the white sectors not being completely fused, ceases when the rotation becomes so rapid that each pair of black and white sectors takes only $\frac{1}{10}$ sec. in passing before the eye. When a brighter illumination is used the rapidity must be increased before the flickering disappears; this is owing to the decline of the stronger sensation, as stated above, beginning earlier and being more rapid than that of the weaker sensation.

§ 750. When a luminous point excites the retina, we recognize in the sensation not only the features of intensity, duration and constancy or steadiness, but also a character which is dependent on the position in the retina of the image of the luminous point. We recognize the sensation caused by a luminous point whose image falls on the temporal side of the retina, as different and distinct from the sensation caused by a luminous point whose image falls on the nasal side of the retina, and so with other positions of the images; indeed, as we shall see presently, we are able to distinguish, to recognize as different and distinct, two sensations excited by two luminous points, the images of which lie very close indeed to each other on the retina. We distinguish the sensations, however, not by reference to the parts of the retina affected, but by reference to the relations in space of the luminous points giving rise to the sensations. Since this is a matter of some importance we may treat of it in some detail.

In the vast majority of cases the changes in the retina which give rise to visual impulses, and so to visual sensations, are brought about by light falling on the retina. But the retina may be stimulated by other agencies than that of light. When this is the case the changes in the retina, however produced, if they are able to affect consciousness at all, give rise to visual sensations, and to visual sensations only. A mechanical stimulation of the retina, as when a blow is struck on the eye, produces a visual sensation, a sensation of light; pressure exerted on the eyeball so as to produce pressure on the retina gives rise to visual sensations in the form of rings of light, of coloured light, known as 'phosphenes;' and when the retina is subjected in various ways to stress or strain, as by rapid accommodation, or by rapid movement of the eyeball from side to side, there often result visual sensations in the form of light of some kind or other, best appreciated when objective light is cut off from the retina and when the retina has by long repose been rendered unusually sensitive. Electrical stimulation also gives rise to visual sensations; not only is the induced current, or the break and make of a constant current, thus able to excite the retina, but during the whole time of the passage of a constant current of adequate strength, even though it remain of uniform intensity, visual impulses, and thus visual sensations, are being generated; in this respect the retina resembles sensory and differs from motor nerves. It is stated that when the current is directed from the layer of optic fibres to the layer of rods and cones, the sensation is a positive one, a sensation of light or of increased light, but that a current in the reverse direction gives rise to a negative sensation, a sensation of diminished light, a sensation of blackness.

That the stimulation of retinal structures by other agents than light may thus give rise to visual sensations, and apparently to visual sensations alone, may be verified by experiment at any time. The occasions on the other hand are rare in which evidence can be gained as to whether stimulation of the optic nerve apart from the retina, whether stimulation of the optic fibres themselves, and not of their special endings in the retina, also gives rise to visual sensations and to visual sensations alone. In certain cases of removal of the eye it has been stated that when the optic nerve was divided in the absence of anesthetics, the patient "saw a great light" accompanied by no more pain than could be accounted for by the filaments of the fifth nerve which are distributed to the optic nerve as nervi nervorum. Such experiences are urged in support of the view that all impulses passing along the optic nerve however generated, whether by retinal changes or by other means, are visual impulses and visual impulses only; they give rise to visual sensations and to visual sensations alone. On the other hand, in other cases of removal of the eye in the absence of anesthetics, neither section of the optic nerve nor subsequent stimulation of the stump has given rise to visual sensations. We shall return to this question later on when we have to speak of what is known as the "specific energy of nerves," and have only referred to it incidentally now.

§ 751. Visual sensations then may be produced in many

other ways than by the falling of light on the retina; and the point to which we wish to call attention now is that we are unable to distinguish a sensation thus produced from the visual sensation produced by light itself. We cannot by the help of the mere sensation alone recognize the nature of the agency which has produced the changes in the retina giving rise to the sensation. The identity of sensations due to mechanical stimulation with those due to luminous stimulation may be illustrated by the story of the witness in a case of assault, who swore that he recognized his assailant by help of the flash of light produced by a blow on his eye. Since light emitted or reflected from external objects is the normal stimulus for visual sensations, all our visual sensations seem to us to be produced by rays of light proceeding from external objects; we look for their cause not in the retina itself, but in the external world; and when we wish to know why we have felt the sensation of a flash of light, we ignore the retina and seek at once in the external world for some source of the rays of light corresponding to the sensation. Hence, also, when in a particular part of the retina, in a spot for instance on the nasal side of the right eye, changes take place such as would be produced by the image of a luminous point falling on that spot, though we recognize the sensation which results as having a certain feature, owing to its being started in that particular spot, we do not through the sensation learn anything about the retina itself, we do not through it recognize that the nasal side of the retina or any particular spot in the nasal side has been affected; what we do recognize, or infer, is the existence in the external world of such a luminous point as would give rise to the sensation in question. The dioptric arrangements of the eye are, as we have seen (§ 707), such that a luminous point in order to give rise to an image in the spot in question, and so to the sensation in question, must occupy a particular position on what we call the right-hand side of the external world. We accordingly recognize the sensation as having been caused by, or refer the sensation to, a luminous point having that position on our right hand. And so with the sensations similarly generated in all other spots in the retina; we recognize them as caused by luminous points occupying such positions in the external world that their images fall on those spots. In each case we ignore the retina itself, and the changes taking place in it are to us simple tokens of luminous events in the external world. When with the right eye we see a luminous point on our right-hand side, if we know that changes are taking place on the nasal side of the retina of that eye, it is not because we are directly aware that the nasal part of the retina is being affected, but because our knowledge of the dioptrics of the eye teaches us that the image of the luminous point is falling on the nasal side of the retina. If we are suffering from right-sided hemiopia (§ 670) all that our sensations can of themselves tell us is that we cannot see things on the right-hand side; they do not tell us anything about the retina itself; they cannot even tell us whether the deficiency of vision is due to changes failing to be set up in the retina or to the cerebral centres failing to be affected by the retinal changes; such questions we have to decide by some other means than a simple examination of our sensations, and by a similar roundabout way only are we able to conclude that in such a hemiopia it is the nasal side of the right retina, and the temporal side of the left retina, which fail to give rise to visual sensations. Our sensations, in fact, tell us of themselves nothing about the optical image on the retina; they do not tell us whether the retinal image is inverted or no; the fact that the retinal image is an inverted one does not in itself influence our visual sensations, and hence the inversion needs no correction on our part.

§ 752. As we have just said, if the images of two luminous objects, two luminous points, fall on the retina at a certain distance apart, the consequent sensations are distinct. If, however, the two objects are made to approach each other, a point will be reached at which the two sensations are fused into one. Two stars at a certain distance apart may be seen distinctly as two stars, while two stars nearer each other appear to be one star; we cannot analyse the latter sensation into its constituent parts.

Similarly, when the images of a number of luminous points, of equal luminosity, fall on the retina sufficiently near each other, the effect is not a number of sensations of luminous points, but one sensation, that of a luminous surface. This introduces a new feature of visual sensations, namely, that of size. If the luminous points be few, so as to involve a small area of the retina, the sensation is that of a small surface; if the luminous points, equally near to each other as before, be numerous, so as to involve a large area of the retina, the sensation is that of a large surface. Moreover, such a sensation of a surface will be referred to some position in space corresponding, as we have just seen, to the region of the retina affected, and will possess features determined by the relative positions and so the figure formed by the luminous points; it will be the sensation of a surface of a certain form, round, square or the like; thus the retinal area stimulated supplies data, which are used, in a manner which we shall study later on, for judging the size and form as well as the position of visible objects.

When the images of two luminous points are at a certain distance apart on the retina, the two sensations may have no effect whatever on each other; but when they are within a certain distance from each other, the sensations do affect each other, in a manner which we shall study later on. Meanwhile we will merely say that when two images approach so closely that the two sensations become fused into one, such a mutual influence is exerted that the intensity of the total sensation produced is greater than that of either of the sensations caused by the single image, though less than the sum of the two. A number of luminous points scattered over a wide surface would appear each to have a certain brightness; each would give rise to a sensation of a certain intensity. If they were all gathered into one spot, that spot would appear brighter than any of the previous points; the intensity of the sensation would be greater.

§ 753. *The region of distinct vision.* The distance at which two images must be apart from each other in order that the two sensations may be separate is not the same for the whole area of the retina. If two luminous points lie near the optic axis, so that their images fall on the fovea centralis or on the yellow spot, they will be seen as two distinct points, even when their images lie very close indeed to each other. If the luminous points be moved aside, so that the images fall on the retina outside the yellow spot, the two luminous points, though at the same distance apart from each other, will give rise to one sensation only, and be seen as one point; they may be moved even farther apart from each other and still give rise to one sensation; and if the two points be placed so much on one side that their respective images fall on the extreme peripheral parts of the retina near the ora serrata, the two images may be separated from each other a very considerable distance and yet give rise to one sensation only. We may vary the experiment by making use of a negative sensation, and take two black dots on a white surface only just so far apart that they can be seen distinctly as two when placed near the axis of vision so that their images fall on or near the fovea, and then, keeping the axis fixed, move the two points outwards, so that their images travel outwards from the fovea; it will be found that the two soon appear as one. The two sensations become fused, as they would do if brought nearer to each other in the centre of the field. The farther away from the centre of the field, the farther apart must two points be in order they may be seen as two.

It is obvious that the more sharply we can distinguish the several sensations produced by the images of the several points of which any external object may be supposed to be made up, the more distinct will be our vision of the object. In the fovea centralis our power of thus distinguishing sensations is at its maximum; in the outer parts of the yellow spot around the fovea it is less; just outside the yellow spot it is much less; and thence diminishes more gradually towards the periphery of the retina. Hence we speak of the fovea centralis, including more or less of the whole yellow spot, as the "region of distinct vision;" and when we wish to examine closely the features of an external object, we so direct the eye, we so 'look' at the object, that its image falls as far as possible on the fovea centralis. The diminution of distinctness does not take place equally from the centre to the circumference along all meridians. The outline described by a line uniting the points where two spots at a certain distance apart cease to be seen as two when moved along different radii from the centre, is a very irregular figure; it differs very much in different individuals; is often not the same in the two eyes of the same person, and does not necessarily correspond to the figure of "the field of vision" to which we shall later on refer. We may add that the power of distinguishing two points in the peripheral parts of the retina is much increased by practice.

As we have just said, when we look intently at an object such as a star in the heavens we so direct the eye that the image of the object falls on the fovea centralis. In the case of most people, two stars so looked at appear to become one when the angle subtended by the distance between them becomes less than 60 seconds or one minute; when they are nearer than this the two sensations become one. And similar measurements are obtained when other images are made to fall on the fovea, such as those of parallel white streaks on a black ground or black streaks on a white ground. In the case of an acute and trained observer this minimum distance may be diminished to 50 seconds; in many cases, on the other hand, it is not less than 73 seconds and may be more. Now the distance between two points subtended by an angle of 50 seconds, corresponds in the diagrammatic eye (§ 705) to a distance of 3·65 μ in the retinal image, and one of 73 seconds to 5·36 μ. Hence in the fovea centralis the elements of the retina excited by light, must lie 3·65 μ or 5·36 μ apart, or in round numbers about 4 μ apart, in order that the two sensations, excited at the same time, may remain distinct.

In the periphery of the retina the distance must be much greater; thus at the extreme periphery, two black dots distant apart about 15 mm. viewed at a distance of 20 cm. and therefore giving a distance of more than a millimeter in the retinal image, are still seen as one point.

§ 754. In accordance with the above, we may suppose the retina to be divided into areas, stimulation of the retina within which gives rise to a single sensation; we might speak of these as visual areas, and of the stimulation of a visual area as a sensational unit. The areas are very small, and the sensational units very numerous in the fovea centralis and yellow spot; the areas are larger, and the sensational units fewer, over the rest of the retina, increasingly so towards the periphery. The smaller or larger the areas, the more numerous or fewer the sensational units in any retina or in any part of a retina, the more or less distinct will be the vision.

Now in the human eye 50 cones may be counted along a line of 200 μ in length drawn through the centre of the yellow spot; this would give 4 μ for the distance between the centres of two adjoining cones in the yellow spot, the average diameter of a cone at its widest part being here about 3 μ and there being slight intervals between neighbouring cones. Hence if we take the centre of a cone as the centre of an anatomical retinal area, these anatomical areas correspond very fairly in the region of distinct vision to the physiological visual areas just spoken of. If two points of the retinal image are less than 4 μ apart, they may both lie within the area of a single cone; and it is just when they are less than about 4 μ apart that they cease to give rise to two distinct sensations. It must be remembered, however, that the fusion or distinction of the sensations is ultimately determined by the brain. The retinal area must be carefully distinguished from the sensational unit, for the sensation is a process whose arena stretches from the retina to certain parts of the brain, and the circumscription of the sensational unit, though it must begin as a retinal area, must also be continued as a cerebral area, the latter corresponding to, and being as it were the projection of, the former. Two points of the retinal image less than 4 μ apart might lie both within the area of a single cone; but the reason why, under such circumstances, they give rise to one sensation only is not because one cone-fibre only is stimulated. For, two points of a retinal image might lie, one on the area of one cone and another on the area of an adjoining cone, and still be less than 4 μ apart; in such a case two cone-fibres would be stimulated; and yet only one sensation would be produced.

In the case where the two points lie entirely within the area of a single cone, it is exceedingly probable that, even if the adjacent cones or cone-fibres in the retina are not at the same time stimulated, impulses radiate from the cerebral ending of the excited cone into the neighbouring cerebral endings of the neighbouring cones; in other words, the sensation-area in the brain does not exactly correspond to and is not sharply defined like the retinal area, but gradually fades away into neighbouring sensation-areas. We may imagine two points of the retinal image so far apart that even the extreme margins of their respective cerebral sensation-areas do not touch each other in the least; in such a case there can be no doubt about the two points giving rise to two sensations. We might, however, imagine a second case where two points were just so far apart that their respective sensation-areas should coalesce at their margins, and yet that, in passing from the centre of one sensation-area to the centre of the other, we should find on examination a considerable fall of sensation at the junction of the two areas; and in a third case we might imagine the two centres to be so close to each other that in passing from one to the other no appreciable diminution of sensation could be discovered. In the last case there would be but one sensation, in the second there might still be two sensations if the marginal fall were great enough, even though the areas partially coalesced.

That the ultimate differentiation of the sensations rests with the brain is still more clear in the case of sensations started in the periphery of the retina; two points of a retinal image might stimulate two cones a considerable distance apart, or several cones.

to say nothing of the intervening rods, might be stimulated, and yet one sensation only result.

Thus, the distinction or fusion of visual sensations is ultimately determined by the disposition and condition of the cerebral centres. Hence the possibility of increasing by exercise the faculty of distinguishing two sensations, since by use the cerebral sensation-areas become more and more differentiated, though the mosaic of rods and cones fixes for the power of discrimination of each individual a limit beyond which exercise cannot carry improvement. This effect of exercise is however shewn in touch even more strikingly than in sight.

SEC. 8. ON COLOUR SENSATIONS.

§ **755.** The sensation excited by a luminous point possesses still another character besides those of intensity, duration, constancy, and localisation, namely the one which we speak of as colour.

When we allow sunlight reflected from a white cloud or from a sheet of white paper to fall into the eye, we have a sensation which we call that of white light. When we look at the same light through a prism and allow different parts of the spectrum to fall in succession into the eye, we have a series of sensations, differing in character from the sensation of white light and from each other; these we call 'colour sensations,' sensations of red, yellow, and the like. In the latter case the luminous undulations are dispersed in a linear series according to their wave-lengths, from the short waves of the extreme violet to the long waves of the extreme red; and we learn from the spectrum, on the one hand, that undulations having different wave-lengths produce different sensations, and on the other hand that undulations having wave-lengths longer than that of the extreme red, about λ 760[1], or shorter than that of the extreme violet, about λ 390, are unable to excite the retina and are therefore invisible. When we look directly at a white object all this dispersion is absent, and the retina is excited at the same time by undulations of all the above wave-lengths. A sensation of 'colour' then is a sensation evoked by undulations of particular wave-lengths, a sensation of 'white' is the sensation which results when the retina, or a part of it, is simultaneously excited by undulations of all wave-lengths which are able to affect it, that is by the whole visible spectrum. When we direct our eyes to an object in such a way that the rays of light proceeding from it might fall on the retina when we bring the object within our field of vision, and yet experience from it neither any sensation of white nor any of the various colour sensations, we call the resulting affection of consciousness a sensation of 'black,' we say that we see 'black.' Sometimes the word 'colour' is confined to the sensations other than those

[1] λ signifies a millionth of a millimeter or ·001 μ.

of white and black, sometimes it is used to comprise these two sensations as well.

When we examine the spectrum we are able to perceive a very large number of different colours, we experience a multitude of sensations, no two of which are exactly alike. There are certain broad differences which we express by common names, such as red, orange, yellow and the like. But we can go much further than this. If we take any part of the spectrum, the green for instance, we find that a very slight change in the wave-length produces a change in the character of the sensation. For convenience sake we call a whole group of sensations green; but we are obliged to admit that there are several kinds of green, several distinct kinds of sensations, though we do not possess names for all of them; a trained eye will recognize that within the green of the spectrum, the sensation produced by one part is a different sensation from that produced by an adjoining part differing in wave-length from the former by an exceedingly small amount. The same is the case with other parts of the spectrum. And in general we may say that any change in the wave-length will produce a change in the sensation, so that we might speak of almost each wave-length as producing a separate sensation.

On the other hand we also easily recognize that the sensations produced by the spectrum are not all wholly unlike, that some are allied to others, and that in some cases one sensation is intermediate between two other sensations and partakes of the nature of both. We recognize the sensation produced by the part of the spectrum lying between the green and the yellow as partaking on the one hand of the nature of the sensation of green and on the other hand of yellow, and call it yellowish green or greenish yellow; we similarly recognize a greenish blue or a bluish green, and so on. This suggests that our colour sensations are in reality mixed sensations, that the multitude of different sensations to which the spectrum gives rise are brought about not by each wave-length giving rise to a separate and independent sensation, but by means of a certain smaller number of primary sensations excited in different degrees by different wave-lengths and mixed in various proportions.

§ 756. This view is confirmed when we study in a systematic manner the results of mixing or fusing together colour sensations.

The best method of fusing colour sensations is that of allowing two different parts of the spectrum to fall on the same part of the retina at the same time. We may make use of surfaces coloured with pigments, but in doing so we must bear in mind the nature of the colour of pigments. A pigment possesses colour because when white light falls upon it some of the rays are absorbed while others are reflected. Thus gamboge absorbs the blue rays very largely as well as to a slight extent the red rays, but reflects the yellow rays and with these many of the green rays; indigo on the other hand absorbs the red and yellow but reflects the blue and a good deal of the green. Hence when we look at a yellow gamboge patch our retina is excited not by those rays alone which form the yellow of the spectrum, but by many other rays as well; the colour is not a 'pure' colour, does not correspond to one of the colours of the spectrum, but is a mixture of more than one. And this is the case with most pigments; hence when they are employed in experiments on the mixture of sensations, difficulties and even errors arise which are avoided by the use of the colours of the prism. We may here incidentally remark that mixing the sensations excited by looking at pigments gives very different results from mixing the pigments themselves. Thus when gamboge and indigo are mixed the mixture is green because the gamboge absorbs the blue and the indigo absorbs the red and yellow, while both reflect the green. We shall see presently that when the sensation excited by gamboge is mixed with the sensation excited by indigo the result is a sensation not of green but of white; and we shall see why this is. What we have just said with regard to surfaces coloured with pigments applies also to glasses stained with pigment, it being understood that the colour of stained glass, seen as a transparent object, corresponds to the rays which it does not absorb. When pure pigments, *i.e.* pigments corresponding as closely as possible to the prismatic colours, are used, satisfactory results may be gained, either by using the reflected image of one pigment, and arranging so that it falls on the retina at the same spot as the direct image of the other pigment, or by allowing the image of one pigment to fall on the retina before the sensation produced by the other has passed away. The first result is easily reached by the simple method of placing two pieces of coloured paper a little distance apart on a table, one on each side of a glass plate inclined at an angle. By looking with one eye down on the glass plate the reflected image of the one paper may be made to coincide with the direct image of the other, the angle which the glass plate makes with the table being adjusted to the distance between the pieces of paper. In the second method, the 'colour top' is used; sectors of the colours to be investigated are placed on a disc made to rotate very rapidly, and the image of one colour is thus brought to bear on the retina so soon after the image of another that the two sensations are fused into one.

§ 757. When by any of the above methods sensations corresponding to the red and yellow of the spectrum are mixed together in certain proportions the result is a sensation of orange, quite indistinguishable from the orange of the spectrum itself. Now the latter is produced by rays of certain wave-lengths, whereas the rays of red and of yellow are respectively of quite different wave-lengths. The orange of the spectrum cannot be made up by any mixture of the red and the yellow of the spectrum in the sense that the red and yellow rays can unite together to form rays of the same wave-lengths as the orange rays; the three things are absolutely different. It is simply the mixed sensation of the red and yellow which is indistinguishable from the sensation of orange; the mixture is entirely and absolutely a subjective one. In the same way we may by appropriate mixtures produce the sensations corresponding to other parts of the spectrum. Now we must suppose that rays of different wave-lengths affect the retina in different ways and so give rise to different visual impulses, that, for instance, the visual impulses generated by orange rays are different from those generated by red rays or by yellow rays. Hence we are led by the fact of mixed sensations being identical with other apparently simple sensations to infer that the visual impulses and hence the visual sensations which any ray originates are of a complex character. We conclude, for instance, that the impulses which a ray in the middle of the orange gives rise to are not simple impulses answering exclusively to the colour of that ray, but complex impulses, parts of which may be excited by rays other than the particular orange ray in question. In saying this we must bear in mind that we possess no direct information of the nature of visual impulses, our knowledge of these being limited to what we learn through the sensations to which they give rise; the complexity of the sensation may be, and indeed probably is, of a different order from that of the visual impulse.; to this point we shall return.

The view that our ordinary colour sensations are mixtures of simpler sensations is further confirmed by an examination of the colours of external nature. For, though we see around us very many colours besides those present in the spectrum, yet we find that the sensations of all these colours may be reproduced by mixtures of sensations excited by various parts of the spectrum. Thus the colour purple, which is so abundant in the external world and yet so conspicuous by its absence from the spectrum, may be at once reproduced by fusing in proper proportions the sensations of red and of blue. And very many other colours present in the external world but not seen in the spectrum itself may be produced by mixing various spectral colours in various proportions.

Other colours in nature may be reproduced by mixing spectral colours with white or with black. When by means of a slit we allow a certain limited part of the spectrum, say in the green, to fall on a certain area of the retina, the rays exciting that area have certain wave-lengths, lying within certain limits, say from λ 525 to λ 535; no rays but these are affecting the retina at the time, and the result is the sensation which we call spectral green. But we might easily so arrange matters that a certain amount of white light, that is of light of all wave-lengths of the visible spectrum, should fall on the area in question at the same time that the green is falling upon it; the result would be a mixed sensation, a sensation of spectral green mixed with the sensation of white, and we should recognize this sensation as different from the sensation of spectral green. Further by varying the proportion of white to green falling on the area in question at the same time we should have a whole series of different sensations from a green in which there was hardly any white to a white in which there was hardly any green. In such a series of colour sensations we recognize a hue supplied by the spectral colour, and we use the phrase more or less "saturated" to express the proportion of white light; when very little white is present, we speak of the colour as being highly saturated. It need hardly be said that not only individual spectral colours, but all mixtures of these also, may be thus "mixed with white."

Again, taking a given area of the retina, we may, on the one hand, throw on to the area a small amount of a spectral colour in such a way that all the elements of the retina in the area are excited, to a slight degree, giving rise to a feeble sensation of that colour; but we may, on the other hand, so scatter a few rays over the area that while some elements are excited others remain at rest and yet in such way that the excitation of the whole area still gives rise to one sensation only. We may speak of each of these sensations as one in which the sensation of the spectral colour is mixed or fused with the sensation which we call black; or we may distinguish the former as merely a feeble sensation and the latter as more strictly mixed with black. Many of the colours of the external world are of this nature; thus the colours which we call "browns" are mixtures of yellow or of red or of both (and possibly of other spectral colours also) with more or less black. In a similar way we may mix, not a spectral colour, but white with black, various mixtures forming various "greys."

§ 758. Putting aside these more or less peculiar cases of mixture with black, we may say that the character of a colour depends (1) on the wave-lengths of the particular rays which, either alone or in excess of other rays, are falling on a given area of the retina; (2) on the amount of this coloured light falling on that area in a given time; and (3) on the amount of white light falling on that area at the same time. The first determines what we call the hue, the second the intensity, and the third the amount of saturation. Our common phrases do not distinguish with sufficient accuracy these three conditions, which obviously may exist under various combinations. On the one hand we frequently use wholly unlike names for colours which differ only in degree of saturation, such as carmine and pink; on the other hand we often use the same adjectives for quite different conditions. It is desirable to employ the word 'pale,' to mean little saturated, largely mixed with white, and the word 'deep' or 'rich' to mean highly saturated, slightly mixed with white. The word 'tint' might be used to express various degrees of saturation, the word 'hue' being reserved to denote the dominant wavelength. 'Tone' is frequently employed to express variations of wave-length within a named colour, as for instance different tones of red. The word 'bright' is often used somewhat loosely, but it is desirable to employ it exclusively as identical with 'luminous,' that is to say, as indicating the intensity of the sensation; a colour is more or less bright according to the amount of luminous energy which is being expended on the retina. We may remark, in passing, that while we can easily compare the brightness or luminosity of two white lights or of the same part of the spectrum under a feeble and under a strong illumination, we may feel some difficulty in comparing the amount of brightness of one colour with that of another, the brightness for instance of a given yellow with that of a given red. Conversely the word 'dark' is used to denote feeble intensity, or admixture with black. Lastly, our appreciation of the colours of external objects is modified by the nature of the surface which is coloured, and features so arising receive various names; but these are in reality outside actual colour sensations.

§759. Admitting that our colour sensations may be considered to be much fewer in number than those which we appear to have when we look on the colours of the spectrum or of nature, admitting that rays of light awake in us certain "primary" colour sensations, which mixed in various proportions reproduce all our colour sensations, we have now to ask the question, What is the nature or what are the characters of these primary colour sensations?

In view of the answer to this question we must call attention to certain results which may be obtained by a further study of the mixing of colours, meaning by that the mixing of colour sensations.

We have seen that all the colours of the spectrum mixed together make white. We have now to add that white may also be produced by mixing two colours only, provided that these are properly chosen. If we take a part of the red of the spectrum, and by any of the methods given in § 756, mix it with successive parts of the spectrum, we shall find that the mixture with a particular part of the green or blue green gives white. These two colours are said to be *complementary* to each other. In order to get a complete white, that is a white free from all colour, a certain proportion between the relative amounts of red and green light, that is to say between the intensities of the two sensations, must be observed. And it will be understood that the white thus produced by two small parts of the spectrum is not equal in intensity to the white which would be produced by the combined effect of the whole of the same spectrum. The following may be taken as characteristic complementary colours, the respective wave-lengths being given:

Red, λ 656,	Blue Green, λ 492,
Orange, λ 608,	Blue, λ 490,
Gold Yellow, λ 574,	Blue, λ 482,
Yellow, λ 564,	Indigo-blue, λ 462,
Greenish Yellow, λ 564,	Violet, λ 433.

It will be understood that the above are not the only complementary colours; as we pass from the red end of the spectrum towards the green, each successive part of the spectrum has its complementary part on the other, blue side of the spectrum, each wave-length on the red side has its complementary wave-length on the blue side. When we reach the greenish yellow at λ 564, the complementary colour is on the very margin of the violet end of the visible spectrum. But we may go, so to speak, outside the spectrum, for the green of the spectrum has for its complementary colour, purple. Or, to put it in another way, while each end of the spectrum has its complementary colour at the other end, the complementary colour of the middle of the spectrum is a combination of the two ends.

The bearing of these facts on the theory of primary colour sensations is obvious. Two complementary colours excite between them all the primary sensations which are excited by white light. though not to the same intensity.. Rays of the wave-length λ 656 falling on the retina give rise to the sensation which we denote as a particular kind of red; they do this however, not by the simple and exclusive stimulation of a particular red sensation, but by exciting all the primary sensations which are not excited by the wave-length λ 492. Conversely rays of the wave-length λ 492, produce the sensation of blue green by exciting all the primary sensations which are not excited by λ 656. Similarly complex is the effect of other wave-lengths. We may roughly describe each of two complementary wave-lengths as stirring up about half the whole of the primary sensations which can be excited by rays of all wave-lengths.

§ **760.** To produce white out of two colours, out of two parts of the spectrum, we are limited to certain pairs; if we take one colour, we are limited to one other colour, to its pair; we have no choice. If however we are allowed three colours instead of two, we have a much greater range. If we take any three colours, provided only that they lie a certain distance apart along the spectrum, we can produce white by mixing them in certain proportions. If we take any red, green and blue, we can by adjusting the amount of each, that is the intensity of each, produce white.

We may go further than this. By adjusting the amounts of each of the three colours we can reproduce all the colours of the spectrum. If we take, for instance, a red of a certain wave-length, a green of a certain wave-length, and a blue of a certain wave-length, we can, without calling to our aid any other wave-lengths, by varying the relative intensities of the three, produce not only white light, but also orange, yellow, and violet, with all the intermediate tints, that is to say, produce all the colours of the spectrum; and we may in the same way produce the non-spectral purple. Our choice however is to a certain extent limited; the three colours which we choose must be spread over the spectrum, for we cannot obtain these results with three colours taken from the red and yellow alone, or from the green and blue alone. Moreover, the result is not a complete one; the colour which we thus produce by combining three spectral colours differs from a true spectral colour in not being saturated; it is "mixed with white," more so in some cases than in others; in relation to this deficiency of saturation, the green region of the spectrum behaves differently from the red end and the blue end.

§ 761. These results shew that the primary colour sensations out of which our recognized colour sensations originate, may be reduced to three in number. If we suppose that we possess three primary sensations so disposed in reference to the spectrum, so arranged so to speak along the spectrum, that a ray of light affects each of the three differently according to its wave-length, we can understand how all our multitudinous colour sensations may arise from the varied excitation of these primary sensations. There may be more than three of these primary sensations, but if so they must behave as if they were three; they cannot be less, since as we have seen the results of mixing two sensations only are extremely limited. We may therefore speak of our vision as *trichromic*, as based on three, or the equivalent of three, primary sensations.

When we attempt to inquire further into the nature of these primary sensations, we find ourselves in the face of two rival theories.

The one, propounded by Young but more fully elaborated by Helmholtz and Maxwell, and known as the Young-Helmholtz theory, teaches that there are three and only three such primary sensations. As we have just seen, any three parts of the spectrum, with certain restrictions, might be chosen as corresponding to these three primary sensations so far as concerns the reproduction, by means of them, of all other colour sensations; hence in determining the nature of the primary sensations we must have recourse to other considerations. We may for instance very naturally suppose that two of the three correspond to the two ends of the spectrum, and may therefore be spoken of as more or less closely corresponding to our recognized sensations of red, and of violet. If red and violet be thus two of the sensations the third one must correspond to green, for only a sensation corresponding to green would give white when mixed with the other two sensations. Or again, choosing green in the first instance as one of the primary sensations for the reason that it stands apart from the others in its complement, purple, not being a spectral colour, we may decide that the two other primary sensations ought to differ as much as possible from each other, and therefore choose red and blue rather than red and violet since violet is obviously more allied to red than is blue, indeed we may perhaps regard violet, on account of its relations to red, as the beginning of a second spectrum the greater part of which is invisible. The decision between these two forms of the same theory rests on a number of considerations, into the discussion of which we cannot enter here. Unless we specially call attention to the difference between them, which acquires importance on certain occasions only, we shall treat them as identical, and use the words blue and violet in this connection indifferently.

Such a view of three primary colour sensations is represented in the diagram (Fig. 146). Thus the red primary sensation, excited to a certain extent by the rays at the extreme red end,

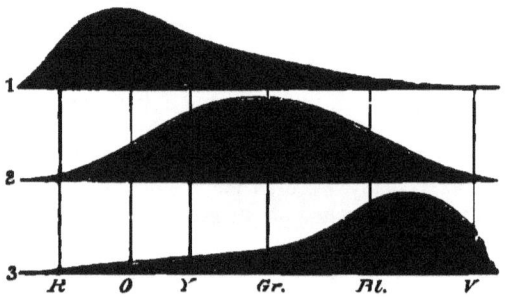

FIG. 146. DIAGRAM OF THREE PRIMARY COLOUR SENSATIONS.

1 is the so-called 'red,' 2 'green,' and 3 'violet' primary colour sensation. R, O, Y, &c. represent the red, orange, yellow, &c., colour of the spectrum. The diagram illustrates, by the height of the curve in each case, how the several primary colour sensations are respectively excited to different extents by vibrations of different wave-lengths. But, in this, and also in Fig. 147, the curves are to be understood not as careful curves of actual variations in the intensity of the several changes, but as simply serving to illustrate roughly the nature of the theory.

is most powerfully affected by the rays at a little distance from that end, the rays from this point onwards towards the blue end producing less and less effect. The curve of the green primary sensation begins later and reaches its maximum in the green of the spectrum, while the blue or violet primary sensation is still later and only reaches its maximum towards the blue end of the spectrum. Each ray calls forth each primary sensation though to a different degree, and the total result of each ray, or of each group of rays, is determined by the proportionate amount of the three sensations. Thus the sensation of orange (O in the figure) is brought about by a mixture of a great deal of the primary red with much less of the primary green, and hardly any of the primary blue; the orange sensation is converted into a yellow sensation by diminishing the primary red and largely increasing the primary green, the primary blue undergoing also some slight increase. And similarly with all the other sensations. When all the three primary sensations are together excited, each to its whole extent, as when ordinary light falls on the retina, the result is a sensation of white. According to this theory, black is simply the absence of sensation from the visual apparatus.

It will be understood that the pure primary red sensation need not necessarily appear in any of our actual sensations of red; we may suppose that it is always more or less mixed with the primary green and even with the primary blue. So also we may suppose that we never actually experience the primary sensations of green or of blue; to this point we shall return.

In the view, as originally put forward by Young, the three primary sensations were supposed to be represented by three sets of fibres, each set of fibres being differently affected by different rays of light, and the impulses passing to the brain along each set awakening a distinct sensation. No such distinction of fibres can be found in the retina; but an anatomical basis of this kind is not necessary for the theory; we can easily conceive of the same fibre transmitting three distinct kinds of impulses; and indeed, as we shall see later on, there are more ways than one by which we can imagine the sensations to be differentiated.

§ **762.** Another theory, that of Hering, starts from the observation that when we examine our own sensations of light we find that certain of these seem to be quite distinct in nature from each other, so that each is something *sui generis*, whereas we easily recognize all other colour sensations as various mixtures of these. Thus red and yellow are to us quite distinct: we do not recognize any thing common to the two, but orange is obviously a mixture of red and yellow. Green and blue are equally distinct from each other and from red and yellow, but in violet and purple we recognize a mixture of red and blue. White again is quite distinct from all the colours in the narrower sense of that word, and black, which we must accept as a sensation, as an affection of consciousness, even if we regard it as the absence of sensation from the field of vision, is again distinct from everything else. Hence the sensations, caused by different kinds of light or by the absence of light, which thus appear to us distinct, and which we may speak of as 'native' or 'fundamental' sensations, are white, black, red, yellow, green, blue. Each of these seems to us to have nothing in common with any of the others, whereas in all other colours we can recognize a mixture of two or more of these.

This result of common experience suggests the idea that these fundamental sensations are the primary sensations, concerning which we are inquiring. And Hering's theory attempts to reconcile, in some such way as follows, the various facts of colour vision with the supposition that we possess these six fundamental sensations. The six sensations readily fall into three pairs, the members of each pair having analogous relations to each other. In each pair the one colour is complementary to the other; white to black, red to green, and yellow to blue.

The little we know about the actual nature of sensations leads us to believe that the nervous processes which are at the bottom of sensations are, like other nervous processes, the outcome of metabolic changes in nervous substance. We shall presently call attention to the view that vision originates in the metabolic changes of a certain substance (or substances) in the retina, that the metabolism of this substance, which has been called visual substance, is especially affected by the incidence of light, and that the metabolic changes so induced determine the beginnings of visual impulses and thus of visual sensations. In the metabolism of living substance, we recognize (§ 30) two phases, the upward constructive anabolic phase, and the downward destructive katabolic phase; we may accordingly, in the absence of any distinct leading to the contrary, on the one hand suppose that different rays of light, rays differing in their wave-length, may affect the metabolism of the visual substance in different ways, some promoting anabolic, others promoting katabolic changes, and on the other hand that different changes in the metabolism of the visual substance may give rise to different sensations. We say 'in the absence of distinct leading to the contrary,' because though in our study of muscular contraction we were led to regard the effect of a stimulus as a katabolic one, as of the nature of an explosive decomposition, we cannot take a muscular contraction as the exclusive type of the effect of a stimulus; and indeed even in the case of muscular tissue we saw, in the instance of the augmentor and inhibitory cardiac nerves, that nervous impulses, in acting as stimuli, might have contrary effects, effects moreover suggesting that the one, augmentor, were associated with katabolic and the other, inhibitory, with anabolic changes. In all probability, we ought to regard the study of sensations as more likely to throw light on the molecular changes involved in muscular contraction, than to take the little we know about the latter as a guide to our views concerning the former.

We may therefore regard ourselves as at liberty to suppose that there may exist in the retina a visual substance of such a kind that when rays of light of certain wave-lengths, the longer ones for instance of the red side of the spectrum, fall upon it, katabolic changes are induced or encouraged, while anabolic changes are similarly promoted by the incidence of rays of other wave-lengths, the shorter ones of the blue side. But, as we have already said, it is difficult in these matters of sensation, to distinguish between peripheral, retinal, and central, cerebral events; we may accordingly extend the above view to the whole visual apparatus, central as well as peripheral, and suppose that when rays of a certain wave-length fall upon the retina, they in some way or other, in some part or other of the visual apparatus, induce or promote katabolic changes and so give rise to a sensation of a certain kind, while rays of another wave-length similarly induce or promote anabolic changes and so give rise to a sensation of a different kind.

The theory of Hering, of which we are now speaking, applies this view to the six fundamental sensations, and supposes that

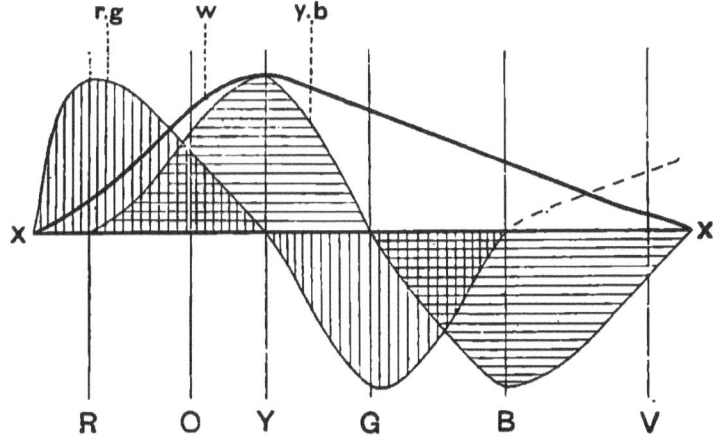

Fig. 147. Diagram to illustrate Hering's Theory of Colour Vision.

The lines *R.O.Y.G.B.V.* indicate, as in Fig. 146, the position on the spectrum of red, orange, yellow, green, blue and violet.

The line *r.g.*, which includes a space, shaded vertically, is intended to represent the effect of rays of different wave-lengths on the red-green visual substance. In the red, orange and yellow up to the line Y, the effect is katabolic, one of dissimilation (red sensation). *Y.* marks the position of equilibrium; beyond this the effect is anabolic, one of assimilation (green sensation). Beyond the blue, *B* the effect (indicated by a broken line) is represented as once more katabolic.

The line *y.b.* similarly represents the behaviour of the yellow-blue substance, shaded horizontally, katabolic (yellow) up to *G*, anabolic (blue) beyond.

The line *w.* similarly indicates the white-black substance, unshaded, katabolic (sensation of white) along the whole length of the spectrum.

each of the three pairs is the outcome of a particular set of katabolic and anabolic changes; these we may provisionally speak of as changes in a distinct visual substance, without attempting to decide whether the changes are retinal or cerebral or both. The theory supposes the existence of what we may call a red-green visual substance, of such a nature that so long as its metabolism is normal, katabolic and anabolic changes being in equilibrium, we experience no sensation, but that when katabolic changes (changes of dissimilation is Hering's own term), are increased, we experience a sensation of (fundamental) red, and when anabolic changes (changes of assimilation) are increased we experience a sensation of (fundamental) green. A similar yellow-blue visual substance is supposed to furnish through katabolic changes, a yellow, through anabolic changes a blue sensation; and a white-black visual substance similarly provides for a katabolic sensation of white and an anabolic sensation of black. The two members of each pair are therefore not only complementary but also antagonistic. Further these substances are of such a kind that while the white-black substance is influenced in the same way though to different degrees by rays along the whole range of the spectrum, the two other substances are differently influenced by rays of different wave-length (see Fig. 147). Thus in the part of the spectrum which we call red, the rays promote a large katabolism of the red-green substance with comparatively slight effect on the yellow-blue substance; hence our sensation of red. In that part of the spectrum which we call yellow the rays effect a large katabolism of the yellow-blue substance but their action on the red-green substance does not lead to an excess of either katabolism or anabolism, this substance being neutral to them; hence our sensation of yellow. The green rays, again, promote anabolism of the red-green substance, leaving the anabolism of the yellow-blue substance equal to its katabolism; and similarly blue rays cause anabolism of the yellow-blue substance, and leave the red-green substance neutral. Finally at the extreme blue end of the spectrum, the rays once more provoke katabolism of the red-green substance, and by adding red to blue give violet. When orange rays fall on the retina, there is an excess of katabolism of both the red-green and the yellow-blue substance; when greenish-blue rays are perceived there is an excess of anabolism of both these substances; and other intermediate hues correspond to varying degrees of katabolism or anabolism of the several visual substances.

When all the rays together fall on the retina, the red-green and yellow-blue substance remain in equilibrium, but the white-black substance undergoes great katabolic changes; and we say the light is white.

Such are the two main theories of colour vision; and much may be said in favour of both of them; at the same time both of them present difficulties. We may perhaps regard as the distinctive feature of Hering's theory the view that white is an independent sensation, and not, as according to the Young-Helmholtz theory, the secondary result of the mixture of primary sensations. In Hering's theory rays of all wave-lengths (within the range of the visible spectrum) give rise to the sensation of white, whatever may be the colour sensation produced at the same time; a fully saturated colour, one wholly unmixed with white, according to this view does not exist. This assumption enables us to explain much more readily than does the Young-Helmholtz theory the occurrence under certain circumstances of white sensations replacing or accompanying, that is to say diminishing the saturation of, colour sensations. On the other hand it introduces what appears to many minds a grave difficulty in reference to black. The theory supposes that the sensation of black is the result of the predominance of anabolic changes in the white-black substance. But what name are we to give to the sensation when the white-black substance is in a condition of equilibrium? We cannot investigate the corresponding conditions of equilibrium in the red-green, or in the yellow-blue substance, because we can never study these by themselves. When either of them occurs, as when rays limited to certain wave-lengths are falling on the retina, we are by hypothesis at the same time subject to changes in the white-black substance; we may therefore leave these two conditions of equilibrium on one side. But we are constantly experiencing the condition of equilibrium of the white-black substance, unaccompanied by any stimulation of either the red-green or yellow-blue substance; we do so when the influence of light has for some time been wholly removed from the eye, or again taking the view, which is the more probable one, that the changes of which we are speaking are cerebral changes, when the retina by disease or injury has become insensible to light. Under such circumstances we must suppose that the previous katabolic excitement of the white-black substance has died away, and that the substance is in equilibrium. Now when we examine our sensation under these circumstances, we find that though it is one of darkness it is one which differs from a sensation of intense blackness. So distinct is the difference that the sensation in question has been spoken of under the phrase "the intrinsic light of the retina." And that we may experience sensations of black different from this sensation due to the retina being at rest may be shewn in several ways. When we close and shade the eyes after they have been exposed to a very bright sunlight, we first experience a sensation of blackness, but this soon gives way to the sensation of mere darkness corresponding to the "intrinsic light of the retina." Again if we stare for some time at a white disc on a black field and then close the eyes, what we shall speak of presently as a negative after image is developed; the part of the field of vision corresponding to the white disc appears as a black disc, which by its blackness stands out in fairly strong contrast to the rest of the field of vision, which corresponding to the area of the retina previously free from the stimulus of light, now yields the sensation of the "intrinsic light of the retina."

And other examples of a similar kind might be given. Admitting then that the "intrinsic light of the retina" corresponds to a condition of equilibrium of the white-black substance, we may speak of this as the neutral condition on one side of which we have sensations of white and on the other side sensations of black. Such a neutral condition has been spoken of as a "neutral grey," but the word grey is so often associated with a mixture of white and black sensations coexisting at the same time rather than with a neutral condition, that the term seems unsuitable. Many minds find it difficult to realize that the condition of which we are speaking is a true neutral condition, the various degrees of blackness being insignificant compared with the various degrees of intensity of white, and accordingly find it difficult to accept Hering's theory.

Both theories conform to the conclusion (§ 761) that normal vision is trichromic in the sense of being made up of three factors; for the three pairs of fundamental sensations of the one theory (the two members of each pair being reciprocally antagonistic, the positive and negative phase of the same thing), play the same part in the equations of mixtures as the three primary sensations of the other theory. Indeed it will be found on examination that all the results of the mixtures of colours are equally explicable on both theories. In comparing the two theories, however, especially in reference to the results of mixtures, we must bear in mind that "brightness" "or luminosity" does not possess the same meaning in the two theories. In the Young-Helmholtz theory brightness is dependent on the extent to which the primary sensation is excited, on the amount of energy expended in the physical substratum, whatever that may be, of the primary sensation. The red of the extreme red end of the spectrum has a minimum of brightness since the extreme red rays excite the red sensation to a minimum and the other two sensations hardly or not at all. As we pass bluewards the brightness increases, partly because the red sensation is more powerfully excited, but also because to the brightness of the red sensation there is now added the brightness of the green sensation. And the brightness of a saturated yellow, such as that of the spectrum, is the sum of the brightnesses of the red and green sensations and nothing else; we neglect for the sake of simplicity the minute adjunct of the blue sensation. In Hering's theory the case is different. The lack of brightness at the red end of the spectrum is due not merely to the feeble development of the red sensation, to the feeble (katabolic) excitation of the red green substance, but also to the feeble development of the white sensation, to the feeble (katabolic) excitation of the white black substance; and the brightness of the yellow of the spectrum is due not merely to the large development of the yellow sensation but also to the large increase of the white sensation.

We may here remark when the extreme red end of the spectrum is examined it is found that along a certain length, between λ 760 and λ 655, there is no change in the sensation as regards hue but only as regards luminosity; the red remains exactly the same kind of red, it only becomes brighter and more readily seen. Similarly at the other end from λ 430 to λ 390 the sensation of violet remains of the same hue though differing in luminosity. And these facts have been brought forward on the Young-Helmholtz theory in support of violet being a primary sensation; it is urged that the red and violet which thus do not change in hue but only in luminosity correspond to the actual primary sensations. The behaviour at the red end is quite intelligible on Hering's theory, since, as the waves shorten in length both the red and the white sensations are supposed to increase, though probably in this part of the spectrum the white sensation is very feeble, rapidly increasing a little farther on. The behaviour at the violet end presents difficulties, since if the violet be due to admixture with a second octave so to speak of red, the violet should change in hue, become more red, as the rays shorten. But the same difficulty presents itself to the Young-Helmholtz theory if blue be accepted as a primary sensation. Moreover observations on this part of the spectrum are exceedingly difficult. We cannot however, attempt to discuss the contending theories properly; this would carry us beyond the limits of this book. We must content ourselves with incidental reference to some conclusions, which are suggested by the study of some other features of colour sensations as well as of abnormal colour vision, and to these we may now turn.

§ 763. *Variations in Colour Vision. Colour-Blindness.* Persons differ very much in their power of appreciating and discriminating colours, and that quite independently of their ability to give expression to their colour sensations, that is to say, of their skill in naming colours. One person will regard as identical two colours which another person recognizes as different. In many cases such differences in the power of discriminating colours are slight, but in some cases they are great. Certain persons are met with who regard as quite alike, or nearly alike, colours which to most people are glaringly distinct; such persons are said to be "colour-blind."

The most common token of "colour-blindness" is the inability to distinguish, or the difficulty in distinguishing, red and green. The great chemist Dalton, who was colour-blind, found great difficulty in recognizing at a distance his red (Glasgow) college gown when it was lying on the college grass plot; the colour-blind can tell a cherry among the leaves on a tree much more by its form than by its colour, and when such persons are asked to 'make matches' between coloured objects, such as skeins of coloured wools, they will put together a red skein and a green skein as being of the same colour. Most colour-blind people more or less confound red and green; but when a number of such colour-blind persons are tested in making matches either between skeins of wool or otherwise, it is found that they do not all make the same matches; they do not agree as to the particular red and green which they regard as identical, and they disagree in various other matches. But they all agree in this that when they are tested by the method of mixing colours it is found that all the colour sensations which they experience, including white, may be reproduced by mixtures of two colours only, whereas as we have seen (§ 761) normal vision requires three. For instance all the colours which they see may be reproduced by varying mixtures of yellow and blue. The vision of these colour-blind people is therefore dichromic not trichromic. All their colour sensations are compounded of two not three (or two, not three pairs of) primary sensations.

On further examination it is found that these ordinary colour-blind persons may be more or less successfully divided into two classes. The members of one class have the following characters. The spectrum seems to them shortened at the red end; that is to say they fail to receive any visual sensation from the rays of extremely long wave-length which still give to the normal eye a distinct sensation of red. The blue-green of the spectrum seems to them less deeply coloured than the rest of the spectrum either on the red or on the blue side; this part gives rise in them to a sensation like that caused by feeble white light; they have a difficulty in recognizing any hue in it and they often speak of it as grey, while in the remainder of the spectrum both to the blue and to the red side they have distinct sensations of colour. We may call this region of the spectrum the 'neutral band'; and it is one of the characters of this class that they see such a neutral band in the blue-green. They confound, as we have said, reds and greens, but when asked to make an exact match between a red and a green they choose a bright red and a dark green; they are more or less uncertain about all colours containing red or green, and when asked to match a purple they generally select a blue or a violet.

To the members of the other class, the spectrum is not shortened; they receive sensations as far to the red side as does the normal eye. They also see a 'neutral' band, but this is placed in the green, that is to say, nearer the red end than is the case with the first class. When asked to make an exact match between red and green they choose a dark red and a bright green; when asked to match a purple they generally select a green or a grey.

Persons whose vision belongs to one or other of these classes are sometimes spoken of as 'totally' colour-blind; for there are grades of difference between such a kind of vision and normal vision, and some eyes may be called 'partially' colour-blind.

Moreover, even among these 'totally' colour-blind persons individual differences occur in each class; indeed not a few cases are met with which do not seem to fit into either class, since they unite in themselves some of the characters of each class. But even if we make allowance for these exceptions, the existence of the two classes with their respective features seems to offer a strong support to the Young-Helmholtz theory. In both classes vision is dichromic not trichromic, that is to say according to that theory in both classes one of the three primary sensations is missing. Since the characteristic mistake which they both make is to confound red and green, we may infer that the missing primary sensation is not blue but either red or green. If we further suppose that in the first class red is missing, in the second green, all the features of the two classes seem intelligible.

On this view all the visual sensations which the first class experience are made up of green and blue; and their vision might be represented by Fig. 146 with the upper curve (1) omitted. Owing to the absence of the red sensation, the extreme red rays hardly affect them at all. Since all their visual sensations are made up of various mixtures of the primary green and primary blue sensations, and since the sensation which they call white light (whatever it may be when compared subjectively with that of the normal eye) is the sensation produced when rays of all the wave-lengths of the visible spectrum are falling on the retina at the same time, that is to say when both of the two primary sensations are being equally excited at the same time, it follows that any particular wave-lengths which equally excite both the two sensations should also produce a sensation which to them is identical with that of white-light. Now the blue-green rays do excite equally both the green and the blue sensation (cf. Fig. 146); and it is just at this part of the spectrum that these persons see the 'neutral band' spoken of above. Further, the matches which eyes of this class make are such as we might imagine would be made if the sensation of red were absent, and the two remaining sensations when mixed together made white. Hence members of this class are spoken of as being "red-blind."

In eyes of the second class, since red is present though green is wanting, the spectrum extends redwards as far as in the normal eye; the least coloured part of the spectrum, the 'neutral band', occupies about the same position as the green seen by the normal eye, for here the red sensation and blue sensation are excited to about the same extent; and the matches made by eyes of this class are such as might be expected in the absence of the green sensation. Members of this class are accordingly spoken of as "green-blind."

It might appear at first sight that the lack of a primary sensation, that is to say, the want of a third of all visual sensations, would lead to a general deficiency of vision; for the lack of one-third of visual sensations would be equivalent to a diminution of the total illumination of external objects to the extent of one-third, and this, unless we suppose that the normal eye lives in a superfluity of light must, especially in feeble light, lead to dim vision; moreover a vision which has to trust to two-fold differences must be less sure than one based on three-fold differences. But this does not necessarily follow; the two remaining sensations might become more highly developed, might so to speak expand in the absence of the third. And as a matter of fact the general vision of colour-blind people seems to be as good as that of normal eyes; moreover, within the range of the colours which they can see, colour-blind people are if anything more acute than most people; though they regard as more or less alike two colours which seem to the normal eye wholly unlike, they can more easily detect minute differences such as those of shade or tone, within each of the two colours.

§ 764. The phenomena, however, of these two classes of colour-blind eyes can also be interpreted on Hering's theory. In both of them the red-green substance may be supposed to be missing, and their dichromic vision to be made up exclusively of changes in the yellow-blue and white-black substances. Since they are thus supposed to have neither red nor green sensations, they must necessarily confound red and green, and the smaller differences, which, as we have seen, divide into two classes all those which confound red with green may be explained as follows.

Even in eyes which may be considered normal as regards colour vision, eyes which certainly cannot be called colour-blind, considerable differences will, on closer examination, be found in regard to sensations of yellow. If by means of a special arrangement we bring a certain amount of the red part of the spectrum and a certain amount of the green part of the spectrum on to the eye at the same time, the result is a sensation of yellow; according to the Young-Helmholtz theory yellow is a mixture of red and green. By the same arrangement we can bring on to the eye at the same time a certain amount of the actual yellow of the spectrum. In this way we can make a match between a mixture of spectral red and green on the one hand, and spectral yellow on the other, comparing the mixed sensation derived from two parts of the spectrum with the sensation derived from a single (yellow) part. We have to adjust the quantities of red light and green light until the mixture seems of the same hue and the same brightness as the yellow, not shewing either a reddish or a greenish tone. When this is done it is found that different people differ very materially as to the proportion of red and green, the proportion of the intensities of the two sensations, necessary to make the match with yellow; with the same quantity of red some need more green, others less green, to make the match. This, on the Young-Helmholtz theory, is interpreted as meaning that the development of the red and green primary sensations differs even in people whose colour vision is considered normal. But on Hering's theory, in which yellow is a fundamental sensation, it may be interpreted as meaning that in passing along the spectrum toward the red-end, the point at which the yellow-blue substance ceases to be affected by rays of light, is placed much nearer the red end in some people than in others. By Hering's hypothesis the green of the spectrum affects not only the red-green substance, cf. Fig. 147 (in way of anabolism), but also to some extent the yellow-blue substance (in way of katabolism); the red rays on the other hand affect the yellow-blue substance very slightly, while the (pure) yellow rays are neutral to the red-green substance producing neither katabolic red nor anabolic green, but simply yellow by katabolic action on the yellow-blue substance. This at least represents the condition of the majority of eyes. If, however, we suppose that in other eyes the yellow-blue substance is considerably affected by red rays, if in Fig. 147, we suppose the curve representing the yellow sensation to be considerably extended towards the red end, in these eyes the red rays would give rise to a sensation of yellow at the same time that they excited a sensation of red, the red would be mixed with yellow; hence in such eyes a certain amount of red being already mixed with yellow would need less green (with its necessarily accompanying yellow) to produce a certain amount of yellow as the result of the mutual neutralisation of the red and green. In such cases we may suppose not that the whole relation of the yellow-blue substance to wave-lengths is altered, but merely that the sensitiveness to long wave-length is increased; the curve of the yellow-blue is not shifted bodily along the spectrum, but the form of the curve is altered so that, the maximum of yellow remaining the same, the yellow end of the curve extends further into the red. Not only this match of red and green with yellow, but other matches of a similar nature, show that in different eyes the yellow sensation (in Hering's sense) is more prominent in some people than in others, that some people so to speak are more yellow sighted than others.

The application of this fact to the colour-blind cases is obvious. In the one class, the red-blind of the Young-Helmholtz theory, the relations of the primary sensations, the distribution along the spectrum of the visual substances are the same as in the normal eye save that the red-green substance and the corresponding sensations are missing; and since the visibility of the red end of the spectrum is chiefly effected by the red sensation, the white-black substance being as compared with the red-green substance but slightly sensitive to the extreme rays, the spectrum is shortened. The feeble white visual impulses excited are insufficient to affect consciousness unless supported by red visual impulses. In the second class, the yellow-blue substance has undergone an expansion similar to but probably greater than that which obtains in the yellow-sighted but otherwise normal eyes mentioned above, it is sensitive to even the rays at the red end of the spectrum; hence the spectrum to eyes of this class seems of the ordinary visible length.

§ 765. So far then both theories may be made to explain the ordinary phenomena of colour-blindness; but it is obvious that the subjective condition of the colour-blind must be different according to one theory from what it is according to the other. According to the Young-Helmholtz theory the red-blind person does not experience in any degree the sensation of either red or yellow; from the green of the spectrum to the red end he only sees some sort of green. Indeed along the whole spectrum, the sensations which he experiences are only various kinds of green and blue, with various amounts of the sensation whatever it be, whether white or simply green-blue, or some other sensation unknown to the normal eye, which results from the mixture of the green and blue sensations. The green-blind person, according to the same theory, has only the sensations of red and blue, with the sensation whatever it may be derived from the mixture of these two, he never has the sensation of either green or yellow. Obviously the sensations of the two classes ought to differ very widely.

According to Hering's theory, both classes agree in seeing neither red nor green, all their sensations are made up of yellow and blue, with white and black, and the only difference between the two classes is that the one, the green-blind of the other theory see more yellow than do the other.

We cannot of course tell what are the actual sensations of the colour-blind; no man can tell what are the sensations of his fellow man; but a person who having normal vision in one eye was colour-blind in the other eye could act as an interpreter. Such cases are recorded; and of them it is said (the cases were of the red-blind class) that the sensation which they experienced in their colour-blind eye from the red side of the spectrum resembled not the green but the yellow of their normal eye. Moreover intelligent colour-blind persons who have studied their own cases are confident that something corresponding to the yellow sensation of the normal eye enters largely into their vision, and is not, as according to the Young-Helmholtz theory ought to be the case, wholly absent. No great stress of course can be laid on this, but as far as it goes it supports the conclusion that what we can learn about actual subjective conditions is in favour of Hering's theory. We may add that according to Hering's theory we should expect the sensations of the so-called red-blind and green-blind not to be wholly unlike since they differ only in the amplitude of their sensations of yellow, whereas according to the Young-Helmholtz theory they ought to be wholly unlike, since red is wholly unlike green. It is true that when the curves of the red and green primary sensations are carefully worked out they lie much closer to each other than does either to that of the primary blue sensation; but this does not do away with the fact that subjectively the primary sensation of red must be something quite different from that of green. Hence further support is given to the former theory by the fact that while there is no difficulty in finding out whether a person is colour-blind or no, it needs much greater care to determine to which class he belongs; indeed, as seems easy on the one theory, but difficult on the other, cases occur which by different observers are placed now in the one class, now in the other.

§ 766. We have treated of the colour-blind as if they were confined to those who confounded red with green. According to the Young-Helmholtz theory, another class of colour-blind is possible, the blue- or violet-blind, those who while possessing red and green sensations lack the third, blue or violet sensation. And indeed cases of such blue-blindness have been described; but none of them are free from doubt. The drug santonin has also been said to produce violet-blindness; the effect of the drug is first to excite sensations of violet, and then annul them; but careful observations shew that vision under the influence of santonin is not truly dichromic and that the effects of the drug therefore are not due to its abolishing a primary sensation. We may add that the peculiar effect of the drug is not due to any coloration of retinal structures, but appears to be the result of cerebral or at least of central changes.

Lastly we may remark that absolute colour-blindness, a condition in which shades of black and white alone indicate the features of external objects, while possible on Hering's theory, is impossible on the Young-Helmholtz theory. According to the latter a person reduced to one primary sensation must see either red, green or blue; this one sensation is excited in him both by objects which we called coloured and by objects which we call white. He would probably call it white; but it would be either red, green, or blue. According to Hering's theory he might still see white and black in the total absence of both the red-green and the yellow-blue substance. A case has been recorded in which only black and white were seen; it would be hazardous to insist much on a single case and that one of obvious disease; but such a case if indubitably established would seem to afford an almost complete refutation of the Young-Helmholtz theory. It might indeed be reconciled to that theory by help of the supposition that in such a case the primary sensations were not wanting but shifted so to speak in their relative positions along the length of the spectrum, brought into the same position on the spectrum, so that each ray of light affected them all equally;

or, what amounts to the same thing put in another way, that in such a case there had been a return to a primitive condition, in which light produced one kind of visual sensation only, not yet differentiated into colour sensations. But such a supposition in laying one difficulty would raise many others.

§ 767. What we have said concerning colour vision refers to the central parts of the retina only. If a coloured object be moved so that its image travels from the central to the peripheral parts of the retina, the colour sensations change and the peripheral parts may be spoken of as colour-blind. In studying the changes of colour which are thus undergone, some results are obtained which seem to favour Hering's theory rather than the Young-Helmholtz theory. Thus the sensation of red is lost towards the periphery, which may be spoken of as red-blind, while in the same region other sensations, at all events that of blue, are still felt. If we suppose this peripheral red-blindness to be due to the loss of the primary sensation of red, the image of a white object ought to give in this region a sensation compounded of the two remaining primary sensations, that is blue-green; but the sensation actually felt is white. Moreover at the extreme periphery even blue is wanting, that is, all the primary sensations are wanting, and yet we receive by it uncoloured sensations, sensations of black and white. But the phenomena of peripheral colour vision need a fuller discussion than we can afford to give them here. We may however add that in certain diseases of the eyes the central parts of the retina may become more or less colour-blind, while the peripheral parts suffer comparatively little; but in these cases, though red and green disappear first, there seems to be rather an unequal diminution of all the primary sensations than a loss confined to any particular one, and the failure to recognize colour is accompanied by failure to recognize form.

§ 768. *Influence of the pigment of the yellow spot.* In the macula lutea, or yellow spot, the yellow pigment which (§ 744) is diffused through the retinal structures in this region absorbs some of the greenish-blue rays of the light which falls upon it. We may use this feature of the yellow spot for the purpose of making the spot, so to speak, visible to ourselves, by the following experiment. A solution of chrome alum, which only transmits red and greenish-blue rays, is held up between the eye and a white cloud. The greenish-blue rays are absorbed by the yellow spot, and here the light gives rise to a sensation of red; whereas in the rest of the field of vision, the sensation is that ordinarily produced by the purplish solution. The yellow spot is consequently marked out as a rosy patch. This very soon however dies away.

Though, when we wish our vision to be most acute, we use the fovea centralis in which the pigment is extremely scanty or absent owing to the thinness or absence of all retinal layers except the cones and cone fibres, still in ordinary vision we make large use of the whole yellow spot, and our sensations of the colour of external objects must be to a certain extent influenced by the pigment of the spot. The light which reaches the rods and cones of this region from objects which we call white, is in reality more or less tinged with yellow; in other words what we call white is more or less yellow. Indeed variations in the amount of pigment present in the yellow spot have been offered in explanation of some of the differences in colour vision discussed above; but the explanation does not seem satisfactory, since there is not such a difference between the sensations derived through the yellow spot and those derived through the colourless retina immediately around the yellow spot as is required for the explanation; in fact the presence of the yellow pigment seems to affect our vision much less than might be imagined.

§ 769. In speaking of the relation between a visual sensation and the intensity of the stimulus (§ 747) we were confining our remarks to white light; when we inquire into the behaviour of our colour sensations under variations in the intensity of the stimulus, we come upon results which are in many ways complicated. We must be content with pointing out one or two only of these.

Each of our colour sensations, when the light giving rise to it reaches a certain intensity, ceases to be a colour sensation and becomes a sensation of white. The theory of three primary colour sensations may be used to explain this. Thus, taking violet as a primary sensation, a violet light of moderate intensity appears violet because it excites the primary sensation of violet much more than those of green and red. If the stimulus be increased the maximum of violet stimulation will be reached, while the stimulation of green will continue to be increased and even that of red to a slight degree. The result will be that the light appears violet mixed with green, that is to say, appears blue. If the stimulus be still further increased while the green and violet are both still largely excited the red stimulation may be increased until the result is violet, green, and red in the proportions which make white light. And so with light of other colours. But the same facts may also be explained on Hering's theory, for this supposes that the stock, so to speak, of white-black substance is far greater than that of either of the other two visual substances; hence under violent stimulation the white sensation wholly overpowers any accompanying colour sensation.

Conversely when the intensity of the stimulus is diminished, colour sensations may disappear before all sensation of light is lost. When the light is very dim we cease to recognize the colour of coloured objects though we continue to see the objects. And this is not merely because the white light reflected from the object, (and it is through this that we chiefly become aware of the form of an object,) is more powerful than the particular rays which give the object colour; since even a saturated colour behaves in the same way. If with a feeble illumination we allow a very small part of the spectrum to fall on the retina, we are much more distinctly conscious of a sensation of light than of any particular colour sensation; indeed the minimum sensation thus felt has been called a 'grey' for all parts of the spectrum. Moreover the colour which is first recognized upon gradually increasing the illumination, appears less saturated, that is to say apparently more mixed with white than when a large amount of light of the same refrangibility falls on the retina; and such distinct colour sensation as may be felt at the first moment of looking at such a light soon diminishes, giving way to a mere sensation of light. These results are perhaps on the whole more intelligible on Hering's theory than on the other.

When we attempt to compare one colour sensation with another in reference to their behaviour towards variations in the intensity of the stimulus we find the results to a certain extent conflicting. When we diminish the intensity of the stimulus by diminishing general illumination, when we look for instance at objects in nature under light of varying intensity, we find that the colours change unequally as the light diminishes; as is well known the colours of flowers look very different when night is falling from what they do under bright daylight. In particular we find that as the light diminishes red sensations and also yellow sensations disappear earlier than blue sensations. Hence in dim lights, as those of evening and moonlight, blues preponderate, reds and yellows being less obvious, whereas in bright lights yellows and reds become prominent.

On the other hand, if we test our sensitiveness to different colours in a different way we get results which are opposed to the above. If for instance we determine the distance at which we cease to recognize the colour of a piece of coloured paper, say 1 cm. square, we find that the blue goes first, then green and next yellow, red being recognizable at the longest distance, though the difference between red and yellow is not very great. It will be understood of course that in this experiment we are dealing not only with diminished energy, with diminished amplitude of the luminous waves, but also with a diminished area of retinal stimulation.

Or again, if we take the heating effects of rays of different wave-lengths as a measure of their energy, we may determine the amount of energy needed, in the case of the several colours, to produce a given visual effect. When this is done it is found that the rays in the green, about wave-length λ 530 are the most effective; from this part of the spectrum the efficiency declines both towards the violet and the red.

The three several methods lead to three different results, the one teaches that blue, the other that red or yellow, and the third that green is the colour to which the eye is most sensitive. It would be hazardous to found important conclusions on any of them.

There are several other facts of considerable importance bearing on the theory of colour vision, but it will be best to consider these in connection with certain modifications of visual sensations with which we shall have presently to deal. Meanwhile having acquired some general notions of visual sensations, we may turn from the study of the little we know concerning the way in which these sensations originate through retinal changes, to the study of the way in which light falling on the retina gives rise to visual impulses.

SEC. 9. ON THE DEVELOPMENT OF VISUAL IMPULSES.

§ 770. We have already called attention to the important fact that the changes which give rise to visual impulses begin on the outer side of the retina, that the rays of light pass through the inner layers of the retina without, as far as we know, producing any effect, and do not begin their work until they reach the region of the rods and cones. It is in this region that the energy of light is transformed into energy of another kind; and the processes here started travel back to the layer of fibres in the inner surface of the retina and thence pass as visual impulses along the optic nerve. That on the one hand the optic fibres themselves are insensible to light and that on the other hand visual impulses do begin in the region of the rods and cones is shewn by the phenomena of the blind spot and of Purkinjé's figures respectively.

The Blind Spot. There is one part of the retina on which rays of light falling give rise to no sensations; this is the entrance of the optic nerve, and the corresponding area in the field of vision is called the blind spot. If the visual axis of one eye, the right for instance, the other being closed, be fixed on a black spot in a white sheet of paper, and a small black object, such as the point of a quill pen dipped in ink, be moved gradually from the black spot sideways over the paper away towards the outside of the field of vision, at a certain distance the black point of the quill will disappear from view. On continuing the movement still farther outward the point will again come into view and continue in sight until it is lost in the periphery of the field of vision. If the pen be used to make a mark on the paper at the moment when it is lost to view and at the moment when it comes into sight again, and if similar marks be made along the other meridians as well as the horizontal, an irregular outline will be drawn circumscribing an area of the field of vision within which rays of light produce no visual sensations. This is the blind spot. The dimensions of the figure drawn vary of course with the distance of the paper from the eye. If this distance be known, the size as well as the position of the area of the retina corresponding to the blind spot may be calculated from the diagrammatic eye (§ 705). The position thus determined coincides exactly with the entrance of the optic nerve, and the dimensions (about 1·5 mm. diameter) also correspond; the exact size and shape of the blind spot differs however in different individuals. While drawing the outline as above directed the indications of the large branches of the retinal vessels as they diverge from the entrance of the nerve can frequently be recognized. The existence of the blind spot is also shewn by the fact that an image of light, sufficiently small, thrown upon the optic nerve by means of the ophthalmoscope, gives rise to no sensations.

The existence of the blind spot proves that the optic fibres themselves are insensible to light, that light can stimulate them only through the agency of the retinal structures in which they end.

§ 771. *Purkinjé's Figures*. If one enters a dark room with a candle and while looking at a plain (not parti-coloured) wall, moves the candle up and down, holding it on a level with the eyes by the side of the head, there will appear in the field of vision of the eye of the same side, projected on the wall, an image of the retinal vessels, similar to that seen on looking into an eye with the ophthalmoscope. The field of vision is illuminated with a glare, and on this the branched retinal vessels appear as shadows. In this mode of experimenting the light enters the eye through the cornea, and an image of the candle is formed on the nasal side of the retina; it is the light emanating from this image which throws shadows of the retinal vessels on to the rest of the retina. In Fig. 149 the light a forms an image on the retina at b; the light reflected from this spot casts a shadow of the retinal vessel ν on to another part of the retina at c, and the image of this shadow appears in the field of vision at d. A far better method is for a second person to concentrate the rays of light, with a lens of low power, on to the outside of the sclerotic where this is thin (§ 712) just behind the cornea; the light in this case emanates from the illuminated spot on the sclerotic and passing straight through the vitreous humour throws a direct shadow of the vessels on to the retina. Thus the rays passing through the sclerotic at b, Fig 148, in the direction $b\nu$, will throw a shadow of the vessel ν on to the retina at β; this will appear as a dark line at B in the glare of the field of vision. This proves that the structures in which visual impulses originate must lie behind the retinal vessels, otherwise the shadows of these could not be perceived.

If the light be moved from b to a, the shadow on the retina will move from β to a, and the dark line in the field of vision will move from B to A. If the distance BA be measured when the whole image is projected at a known distance, kB from the eye, k being the nodal point (§ 705) of the reduced diagrammatic eye, then, knowing the distance $k\beta$ in the diagrammatic eye, the distance βa can be calculated. But if the distance βa be thus estimated,

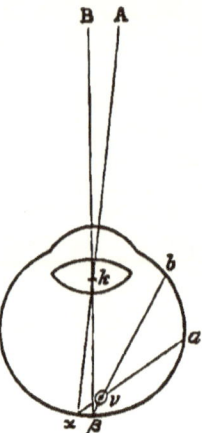

FIG. 148. DIAGRAM ILLUSTRATING THE FORMATION OF PURKINJÉ'S FIGURES WHEN THE ILLUMINATION IS DIRECTED THROUGH THE SCLEROTIC.

and the distance ba be directly measured, the distance $\beta\nu$, $a\nu$, $b\nu$, $a\nu$ can be calculated; and if the appearance in the field of vision is really caused by the shadow of ν falling on β, these distances ought to correspond to the distances of the retinal vessels ν from the sclerotic b on the one hand, and from that part of the retina β where visual impressions begin, on the other. When this is done it is found that the distance $\beta\nu$ thus calculated corresponds fairly well to the distance of the retinal vessels from the layer of rods and cones. Thus Purkinjé's figures prove in the first place that the sensory impulses which form the commencement of visual sensations originate in some part of the retina behind the retinal vessels *i.e.* somewhere between them and the choroid coat; and calculations based on the movements of the shadows following movements of the illumination, even if they do not give absolutely exact results, at least go far to shew that these impulses originate at the outermost part of the retina, viz. the layer of rods and cones.

In the second method of experimenting, where the light passes through the sclerotic, the image always moves in the same direction as the light, as it obviously must do; when the spot of light on the sclerotic is moved from a to b (Fig. 148) the shadow on the retina moves from a to β, and the (inverted) image moves from A to B. In the first method, where the light enters through the cornea, the image moves in the same direction as the light when the light is moved from side to side, provided the movement does not extend beyond the middle of the cornea, but in the opposite direction to the light when the latter is moved up and down. In Fig. 149, which represents a horizontal section of an eye, if a be moved to a, b (the illuminated spot on the retina, the light reflected from which casts a shadow of v on to c) will move to β, the shadow on the retina c to γ, and the image d to δ. If on the other hand a be supposed to move above the plane of the paper, b will move below, in consequence c will move above, and d will appear to move below, *i.e.* d will sink as a rises.

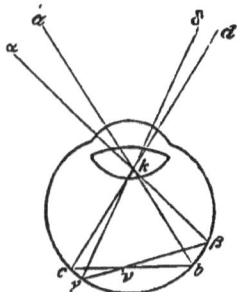

FIG. 149. DIAGRAM ILLUSTRATING THE FORMATION OF PURKINJÉ'S FIGURES WHEN THE ILLUMINATION IS DIRECTED THROUGH THE CORNEA.

It is desirable in these cases to keep moving the light to and fro, especially in the first method, since the retina soon becomes tired, and the image fades away. To give rise to a conscious sensation of the slight difference between shadow and absence of shadow the retina must be extremely sensitive; if the shadow remains motionless, the sensitiveness rapidly decreases in the parts which are not in shadow, until the visual sensations from these parts are no stronger than those from the parts in shadow; when the light is moved the parts which were in shadow, not having been so much stimulated, are sufficiently sensitive to the light which now falls on them, while those parts which had been previously fatigued recover their sensitiveness by resting in the shadow. The experiment, like the experiment by which the yellow spot (§ 768) is made visible, is incidentally useful as shewing how extremely sensitive and how soon fatigued are the retinal structures.

Some observers can recognize in the axis of vision a faint shadow corresponding to the edge of the depression of the fovea centralis.

The retinal vessels may also be rendered visible by looking through a small orifice such as a pin-hole in a card placed close to the eye, in the position of the principal anterior focus, at a bright surface such as a white cloud, and moving the orifice very rapidly from side to side or up and down. If the movement be from side to side, the vessels which run vertical will be seen; if up and down, the horizontal vessels. In this case, as in the similar instance of shadows cast by objects in the vitreous humour

(§ 736), the shadow is cast by the rays passing parallel through the vitreous humour; hence the change from shadow to absence of shadow is more marked with the vertical vessels when the movement is sideways and with the horizontal vessels when it is up and down. The fine capillary vessels are seen more easily in this way than by Purkinjé's method. The same appearances may also be produced by looking through a microscope from which the objective has been removed and the eye-piece only left (or in which at least there is no object distinctly in focus in the field), and moving the head rapidly from side to side or backwards and forwards. Or the microscope itself may be moved; a circular movement of the field will then bring both the vertically and horizontally directed vessels into view at the same time.

§ 772. It being admitted that the processes which give rise to visual impulses begin somewhere in the region of the rods and cones, we have to ask the question, How do they begin and what is their nature? We are accustomed to consider light as the undulations of an ether; a nervous impulse is, so far as we can understand, a molecular change propagated along the substance of the axis cylinder of a nerve fibre; and, though as we have seen our knowledge of the subject is very limited, still the analogy of a muscular contraction, and of other responses of living substance to a stimulus, lead us to conclude that chemical changes play a part in this molecular change. By what steps does the undulation of the ether give rise to the material molecular change? In attempting to answer this question we may adopt one or other of two views.

On the one hand we may suppose that the vibrations of the ether are able, through the means of the retinal apparatus of the rods and cones for example, to give rise in some more or less direct manner to the molecular vibrations which are the beginnings of the nervous impulses in the optic nerve. And the rapidity with which events must come and go in the retina in order that the eye may be, what it is, an instrument for appreciating rapidly repeated minute changes, lends support to this view. But the present state of our knowledge of physical phenomena does not afford us an adequate explanation of how such a direct transformation can be effected. The recent progress of science tends, it is true, more and more to lay bare the close relations which obtain between optical and electric phenomena, and the latter, as we have so often seen, play an important part in the generation of nervous impulses. Then again many of the phenomena of fluorescence seem to supply a bridge between the vibrations of ether, and the vibrations of molecules. But in neither of these directions is it possible, at present at all events, to frame a hypothesis which can be satisfactorily applied to retinal processes.

On the other hand we may perhaps more naturally turn to a chemical explanation. We are familiar with the fact that rays of light are able to bring about the decomposition of very many chemical substances; and we accordingly speak of these substances as being sensitive to light. All the facts dwelt on in this book illustrate the great complexity and corresponding instability of the composition of living matter. And we might reasonably suppose that living matter itself would be sensitive to light; that is to say that rays of light falling on even undifferentiated protoplasmic substance might set up a decomposition of that substance and so bring about a molecular disturbance; in other words, that light might act as a direct stimulus to living matter. As a matter of fact, however, we meet with very little evidence of this, especially when we make a distinction between thermic rays, rays which though they produce physical results are to us invisible, and luminous rays which alone when they fall on our retina give rise in us to the sensation of light. Nor can we be surprised at this apparent indifference of living matter towards light when we reflect that living matter in what we may call its purest form is remarkable for its transparency, that is to say the rays of light pass through it with exceedingly little absorption. But in order that light may produce chemical effects, it must be absorbed; its energy must be spent in doing the chemical work. Accordingly the first step towards the formation of an organ of vision, that is to say an organ through which the body of a living being reacts towards light, is the differentiation of a portion of the substance of the body into a pigment at once capable of absorbing light, and sensitive to light, *i.e.* undergoing decomposition upon exposure to light. An organism, a portion of whose body had thus become differentiated into such a pigment, would be able to react towards light. The light falling on the organism would be in part absorbed by the pigment, and the rays thus absorbed would produce a chemical action and set free chemical substances which before were not present. We have only to suppose that the chemical substances thus produced are of such a nature as to induce other chemical changes, or in some way or other to act as a stimulus to other parts of the organism, (and we have manifold evidence of the exquisite sensitiveness of living matter in general to chemical stimuli,) in order to see how rays of light falling on the organism might excite movements in it, or modify movements which were being carried on, or might otherwise affect the organism in whole or in part. A comparatively simple illustration of this is afforded by some of the lowly organisms called bacteria, especially by the one which has been called *bacterium photometricum*. This organism is remarkably sensitive to light, and especially reacts towards certain rays of light. It is coloured with a purple pigment, apparently allied to chlorophyll; and the rays of light, to which it is especially sensitive, are just those which are absorbed by the pigment.

§ **773.** *Photochemistry of the Retina.* Such considerations as the foregoing may be applied to even the complex organ of vision of the higher animals. If we suppose that the actual terminations of the optic nerve are surrounded by substances sensitive to light, then it becomes easy to imagine how light falling on these sensitive substances should set free chemical bodies possessed of the property of acting as stimuli to the actual nerve-endings and thus give rise to visual impulses in the optic fibres. We say "easy to imagine," but we are, at present, far from being able to give definite proofs that such an explanation of the origin of visual impulses is the true one, probable and enticing as it may appear.

One of the most striking features in the structure of the retina is the abundance of black pigment, fuscin (§ 746), in the retinal epithelium. It is difficult to suppose that the sole function of this pigment is to absorb the superfluous rays of light, and that the rays thus absorbed are put to no use and simply wasted. And indeed it has been shewn that the pigment is sensitive to light; but the changes in it induced by light are excessively slow. Moreover its presence cannot be of fundamental importance, since vision is not only possible but fairly distinct with albinos in which this pigment is absent.

Then again, in the vast majority of vertebrate animals, the outer limbs of the rods are suffused with a purplish-red pigment, the so-called visual purple, which is so eminently sensitive to light that images of external objects may by appropriate means be photographed in it on the retina. When the eye of a frog or of a rabbit is examined in an ordinary way, with full exposure to light, the retina appears colourless. But if the eye be kept in the dark for some time before it is examined, the retina, if removed rapidly, will be found to be of a beautiful purplish-red or pink colour. Upon exposure to light the colour changes to yellow and then fades away, leaving however the retina, not only white but more opaque than it was before. Upon examination with the microscope it is found that the purple colour is confined exclusively to the rods and to the outer limbs of the rods, the inner limbs being wholly devoid of it.

The colour of the rods is due to the presence of a distinct pigment, the "visual purple," diffused through the substance of the outer limbs; and this may be extracted from the rods by dissolving these in an aqueous solution of bile salts. A clear purple solution is thus obtained, which is capable of being bleached by the action of light, and in its general features and behaviour is similar to the pigment as it naturally exists in the retina.

Visual purple is found as we have said exclusively in the outer limbs of the rods; it has never yet been found in the cones, and it is accordingly absent from (or exceedingly scanty in) the retinas (such as those of snakes) which are composed of cones only (or contain very few rods), and from the greater part of the macula lutea and the whole of the fovea centralis of the retinas of man and the ape. The intensity of the coloration varies in different animals, and the retinas even of some animals possessing rods (bat, dove, hen) seem to be wholly devoid of the visual purple; it is generally well marked in retinas in which the outer limbs of the rods are well developed. Its absence or presence is not dependent on nocturnal habits, since the intense colour of the retina of the owl is in strong contrast to the absence of colour in the bat. It has been found in the retina of the embryo.

The visual purple is bleached not only by white but also by monochromatic light. Of the various prismatic rays the most active are the greenish-yellow rays, those to the blue side of these coming next, the least active being the red. Now it is precisely the greenish-yellow rays which are most readily absorbed by the colour itself. A natural coloured retina or a solution of visual purple gives a diffuse spectrum without any defined absorption bands, and according to the amount of colouring material through which the light passes, absorption is seen either to be limited to the greenish-yellow part of the spectrum or to spread thence towards the blue and, to a much less extent, towards the red. Thus the various prismatic rays produce a photochemical effect on the visual purple in proportion as they are absorbed by it. Under the action of light the visual purple, whether in solution, or in its natural condition in the rods, passes through a purplish orange to a yellow, and finally becomes colourless; and we appear to be justified in speaking of a "visual yellow" and "visual white" as products of the photochemical changes undergone by the visual purple.

For the restoration of the visual purple, after it has been destroyed by light, the maintenance of the circulation of the blood through the tissues of the eye is not essential. The retinal epithelium has by itself, provided that it still retains its tissue life, the power of regenerating the purple. If a portion of the retina of an excised eye be raised from its epithelial bed, bleached, and then carefully restored to its natural position, the purple will return if the eye be kept in the dark.

If the image of some bright object such as a lamp or a window be thrown on to the retina, either of an eye in its natural position or of one recently excised, care having been taken to keep the retina for some time previous away from all rays of light, the portion of the retina on which the rays have fallen will be found to be bleached, the rest of the retina remaining purple. In fact an "optogram" of external objects may be thus obtained; and if the retina be removed and treated with a 4 p.c. solution of potash alum before the retinal epithelium has had time to obliterate the bleaching effects, the retina may remain permanently in that condition: the photochemical effect may, as the photographers say, be "fixed."

It seemed very tempting, especially upon the first discovery of it, to suppose that this visual purple is directly concerned in vision. If we suppose that visual purple itself is inert towards, produces no effect on, the endings of the optic nerve, but that either visual yellow or visual white, *i.e.* some product of the action of light on visual purple, may act as a stimulus to those endings, the way seems opened to understanding how rays of light can give rise to sensory impulses in the optic nerve. And such a view receives incidental support from the fact that the visual efficiency of rays of different wave-lengths corresponds very closely to their photochemical efficiency towards visual purple; the greenish-yellow rays which are most active towards visual purple are precisely those which seem to us the brightest, most luminous, which produce the greatest effect on our consciousness. But visual purple is absent from the cones, it is in ourselves absent from the fovea centralis, the region of most distinct vision; it is further entirely wanting in some animals which undoubtedly see very well; and lastly animals such as frogs, naturally possessing the pigment, continue to see very well and even apparently to see colours when their visual purple has been absolutely bleached, as it may be by prolonged exposure of the eyes to strong light. We cannot therefore, at present at least, explain the origin of visual impulses by the help of visual purple. It is difficult to suppose that it plays no part in the origination of visual impulses; but even in a photochemical theory of vision we cannot allot to it more than a subsidiary function, possibly something analogous to the "sensitizer" of the photographer. At the same time its history suggests that some substances, sensitive like it to light, but unlike it, colourless and therefore escaping observation, may exist, and by photochemical changes be the means of exciting the optic nerves; but if so we must suppose that these substances, though colourless, are capable of absorbing light, since otherwise they would not be acted upon by it.

§ **774.** If the eyeball be steadily compressed so as to arrest or greatly interfere with the circulation, the anæmic retina becomes insensitive to light, so that the eye becomes temporarily blind. If while the pressure is being applied the eye is directed to a white and black surface, so arranged that half the field of vision is white and the other half black, that is to say half the retina is stimulated while the other half is at rest, and if as soon as the white half becomes invisible, the black half is made white (by the withdrawal for instance of the black sheet which furnished the black half) then for an instant that new white half becomes visible, though the sensation soon fades away. This result has been interpreted as shewing the existence in the retina of a "visual substance" nourished by the previous blood supply but ceasing to be replenished when the blood supply was interfered with. We may suppose that over the half of the retina stimulated by the light, over the white half of the field of vision, this visual substance was wholly exhausted, while over the half not so stimulated some remained, sufficient to give rise to a fugitive visual sensation when that half was stimulated by light. But the result may be explained without reference to any special visual substance; it is only natural that some part or the whole of the retinal nervous apparatus should be sooner exhausted when stimulation is added to deprivation of nourishment than when stimulation is absent. We may here remark that pressure on the eye-ball gives rise to temporary colour-blindness like that existing in the periphery of the retina; as the pressure is continued red and green pass through yellow into white, while yellow and blue pass directly into white.

We have then no satisfactory evidence of the existence of any visual substance or substances, of a photochemical or other nature, lying outside the nervous elements. There may be such, but we have at present no proof of their existence. And it must be remembered, with regard to the hypothetical 'visual substances' of Hering's theory, or of other theories involving the existence of visual substances, that it is not necessary that these should lie outside the nervous conducting elements, or indeed should lie exclusively in the retina. The visual substance is merely supposed to be the basis of visual sensations, it is introduced to explain these sensations; but those sensations are developed in the brain, and as we have already more than once insisted, it is always difficult, and in many cases impossible to distinguish in a sensation between central and peripheral events. All our knowledge goes to shew that sensations, like other nervous processes, are the outcome of the metabolism of nervous substance; and it is at present quite open to us to suppose that the visual substances, changes in which we recognize as the basis of colour and other visual sensations, are substances forming part of the nervous material of the central organs of vision, each substance being affected in its own way by nervous impulses generated in the retina by rays of light of certain wave-lengths. The various exhaustions, mixtures and the like, supposed by the theories, would on this view take place in the central organ. On the other hand the visual substances may reside in the retina and the central organ may do little more than, so to speak, record the changes which had already taken place in the retina. Or perhaps we ought to make no such sharp distinction between retina and central organ. Our present knowledge is unable to decide these matters; but the acceptance or rejection of the theories of colour vision is quite independent of any view as to the exact nature or position of the visual substances.

§ 775. Whatever view we adopt, whether photochemical or other, as to the changes which lead to stimulation of the real endings of the retinal nervous mechanism, we cannot at present state anything definite concerning those nerve-endings or the manner of their stimulation.

Each outer limb of a rod is a cylinder of highly refractive material, closely packed round with the black pigment of the retinal epithelium. When an image of an external object, such as a candle-flame, is formed on the retina, at or near the layer of rods and cones, the rays of light diverge again beyond the focal plane in the form of pencils of rays from each point of the image. Of these some passing between the rods are absorbed by the pigment, while others pass into the outer limbs of the rods; of these latter some traversing the whole length of the limb, are absorbed by the pigment beyond, while others undergo "total reflection" at the sides, or are absorbed by the pigment after reflection. Hence of all the rays which fall on the layer of rods and cones, a small number only are reflected back into the vitreous humour and so through the pupil; hence the eye when looked into usually looks black. In the case of the conical outer limbs of the cones the amount of light thus thrown back into the vitreous humour must be still less. We may fairly assume that the light which thus disappears, partly in the actual outer limbs of the rods and cones, partly in their immediate surrounding, sets up changes which, whatever be their exact nature, either are or in some way assist the very beginnings of visual impulses. It also seems probable that these changes, so long as they are confined to the region of the outer limbs, ought not to be considered as nervous in nature, it seems probable that they do not take on a nature analogous to that of a nervous impulse, until they have passed the conspicuous break which divides the outer from the inner limbs. But on these matters we have no certain knowledge.

We may here turn aside for a moment to remark that when an image of a candle-flame is formed on the retina the rays reflected back, as stated above, from the retina through the pupil form a second image in the position of the candle-flame; hence to see an image of an illuminated retina the observing eye must be placed in the position of the source of illuminaton. This is the principle of the ophthalmoscope.

There are many forms of this instrument, but the accompanying diagram (Fig. 150) will illustrate its essential features. The rays from the lamp L (or other source of illumination) are reflected by the concave mirror M, M, and brought to a focus at a. The rays diverging from a are, by means of the lens l, rendered parallel, and thus, through natural dioptric arrangements of the observed eye B, are brought to a focus on the retina at a'. The rays reflected back from the part a' of the retina thus illuminated, will, as stated above, follow the same path as on entering, and so return to the focus a. Hence the rays reflected from a number of points on the retina, such as those forming the arrow at a', will be brought to a focus in a corresponding number of points at a, i. e. will from an (inverted) image of the arrow at a. And the observing eye placed at A

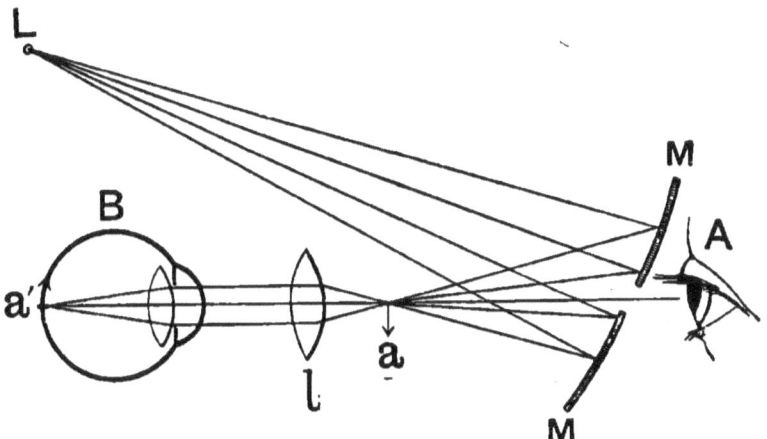

Fig. 150. Diagram to illustrate the Principles of a simple Form of Ophthalmoscope.

behind the hole in the mirror will see at a an inverted image of the illuminated retina.

§ 776. As to the meaning of the difference between rods and cones·no satisfactory statement can be made. It has, it is true, been suggested that the cones subserve the vision of colour and the rods that of form only. This, however, is in flagrant contradiction to both the theories of colour vision discussed above. For colourless vision of form is the appreciation of differences in black and white; and according to the Young-Helmholtz theory, white is simply a combination of colour sensations. Sensations of white, apart from colours ordinarily so called, are only possible on Hering's theory, and an extension of this theory in the direction that the rods are connected exclusively with the white and black substance, and the cones exclusively with the red-green and yellow-blue substances, lands us at once in absurdity. Moreover since it is in the fovea centralis that we have the most acute vision of both form and colour, the cones alone must be able to serve as the instruments of all visual sensations. The argument that in nocturnal animals the rods are developed almost to the exclusion of cones, because such animals do not need colour sensations, is one which can be turned against itself, since it may be urged that the dim light in which these creatures move calls for increased and not diminished appreciation of small differences of colour. The coloured globules intercalated between the outer and inner limbs of cones in some of the lower animals, such as birds and reptiles, have probably no closer relation to colour vision than has the yellow pigment of our own macula lutea.

The close resemblance in their general features, apart from form, between the rods and cones, suggests that their functions differ in degree rather than in kind, and this view is supported by the rod-like character assumed by the cones in the macula lutea and especially in the fovea centralis. But we can hardly expect to be able to differentiate the functions of the two, so long as we know so little about either.

With regard to what goes on in the other layers of the retina our ignorance is complete. We may fairly suppose that the events which take place in the inner limbs of the rods and cones are different from those which take place in the optic fibres. We may conclude that the latter are of the nature of nervous impulses, though we may here repeat what we have already urged, namely, that it is hazardous to infer that the little we know of motor nervous impulses may be applied with little or no modification to sensory nervous impulses; but as to the nature of the events in the inner limbs of rod and cones, or as what happens in the intervening layers of the retina, we know nothing. The double method of connection of the optic fibres with the other retinal layers, on the one hand by the mediation of the ganglionic cells, on the other hand by some more direct way (§ 742) with either the inner molecular, or the inner nuclear layer, a like double connection being observed in other parts of the brain, the cerebral cortex and the cerebellar superficial grey matter for example, suggests that the retinal processes are exceedingly complex, and that visual impulses are so to speak largely moulded before they find their way into the optic nerve.

§ 777. The little objective knowledge which we possess concerning retinal processes is almost limited to the detection of electric currents. The retina and optic nerve like other nervous structures develope electric currents which may be spoken of as currents of rest and currents of action. They may be shewn by placing one electrode on the retina of a bisected eye, or on the cornea of a whole one, and the other on the optic nerve, or hind part of the eye-ball or on the cortical visual centre or even on some distant part of the body. They are also manifested by the isolated retina itself. The phenomena appear somewhat complicated by the appearance now of positive, now of negative variations; but this fact comes out clearly that the incidence of light on the irritable retina developes an electric change, the magnitude of which is to a certain extent proportionate to the intensity of the light acting as a stimulus. The changes gradually diminish and disappear as the retina gradually loses its irritability. We may add that these electric phenomena appear to be quite independent of the condition of the visual purple.

SEC. 10. ON SOME FEATURES OF VISUAL SENSATIONS ESPECIALLY IN RELATION TO VISUAL PERCEPTIONS.

§ 778. In our previous study of visual sensations we dealt chiefly with the more simple and fundamental characters of sensations; we considered each sensation by itself and discussed its features irrespective of the influence of other sensations excited at the same time, except so far as it became necessary, in treating of the localisation of sensations, to speak of the circumstances which determined the fusing of two neighbouring sensations into one. It very rarely occurs however that any object or event in the external world gives rise to a simple sensation such as those on which we have dwelt; each part of the external world, each external object such as a tree, is the source of many distinct sensations differing from each other in intensity and other characters. In looking at a tree we are conscious of many sensations of different colours and intensities, each having a definite localisation; but these are coordinated in our consciousness into a whole and we say we "see a tree." The effect which the whole visible world has upon us is not that of a multitude of single sensations each separate from and independent of the other, but of a smaller though still large number of groups of sensations corresponding to what we call the objects of nature. And we have now to turn our attention to certain facts concerning vision which become especially prominent when we are the subject not of an isolate single visual sensation but of complex groups of simultaneous visual sensations. The sum of visual sensations and groups of sensations which are excited by images falling on the retina at any one time, we call, as we have already said (§ 666), the 'field of vision,' or 'visual field.'

§ 779. Before we proceed any further however it will be well to call to mind that in studying vision as we are now doing by means of an appeal to our own consciousness, we are deserting the ordinary methods of physiology for the methods which are more strictly speaking those of psychology. Or rather in our study of vision we are using both methods, suddenly turning from one to the other. We are using ordinary physiological methods when we are studying how the various rays of light proceeding from a tree form an image of the tree on the retina, and how these rays thus falling on the retina give rise to visual impulses. But when we study the change in our consciousness which is brought about by the visual impulses thus excited through the image of the tree falling on the retina, we are dealing with psychological problems. The object, the tree itself, and our vision of it, the one being commonly spoken of as the cause of the other, are connected by a chain of events; one end of the chain we study by physiological, the other end by psychological methods; and the difficulty of our task arises from the fact that we have to use these two different methods for a common purpose, namely that of explaining how the tree gives rise to the vision of it.

When we turn to the physiological side of the problem we cannot at present say much more than that the rays of light proceeding from the tree give rise to the changes in the optic fibres which we have called visual impulses. We have seen in dealing with the brain reason to think (§ 643) that visual impulses, like other sensory impulses, may influence the working of the central nervous system without producing any such change of consciousness as can be studied by psychological methods; and we further suggested (§ 673) that in the structures of the mid-brain which we called the primary visual centres a visual impulse underwent a development by which it became no longer a mere impulse but something more, and that the changes in these primary visual centres transmitted to the occipital cortex gave rise there to the changes with which the distinct affection of consciousness is associated. It is undesirable to speak of the events in the primary visual centres as "sensations," since it is convenient to reserve this term for the psychical events, the changes of consciousness of which we can become aware by examining our own minds; nor is there at present any need to give them any name at all; but it is important when we are using the psychological method to remember that between the physiological visual impulses and the psychological sensation there are events which must not be ignored.

Turning now to the psychological side of the problem we find that the psychical events are also complex, and that the psychical effects due to the same visual impulses are not all of the same kind. This is seen even in the case of simple and isolated visual sensations. Taking the effect of a luminous point, shining for a moment only, as a simple form of visual sensation, we must distinguish what we may call the mere change of consciousness, the mere sensation of light, from the further psychical effect of which we have already spoken and through which we associate the sensation with a luminous point occupying a particular position in external nature. Though the latter always accompanies the former, though whenever we experience a visual sensation we refer it to its cause in the external world, we can dissociate the two in our minds, and can speak of the mere sensation independently of the further psychical action. When we have vision not of such a simple object as a luminous point, which we may consider as giving rise to a single sensation, but of a tree which gives rise to a complex group of sensations, the psychical actions which accompany the mere sensations are manifold and become prominent in the total visual effect produced by the tree. That total visual effect is determined not only by the sensations to which the retinal image of the tree is at the time giving rise, but also by various psychical events dependent on the previous knowledge of the nature of trees which we have gained by touch as well as by sight, and on other circumstances. In common language we distinguish between the mere sensation and the further psychical visual effect by saying that we 'feel' a sensation and 'perceive' an object; and, though the term 'perception' has been employed in different meanings by different writers, we may here make use of it, in what is perhaps its most usually accepted meaning, to denote the further visual effect to which we have just called attention as distinguished from the immediate sensation. We feel a sensation of light, and we may feel at one and the same time a number of such sensations of different intensity and quality; we perceive an object, it may be a simple object such as a mere transient flash of light or a complex object such as a tree or a scene.

From what we have said above it follows that, although it is perfectly true as we have insisted (§ 702), that our perception of external objects is based on the optical sharpness of the retinal image, and on the distinctness of the several sensations which the retinal image excites, we should be wrong in supposing that when an image of an object is formed on the retina the visual impulses correspond exactly to the retinal image, the sensations correspond exactly to the impulses, and the perception corresponds exactly to the sensations, so that the perception is as it were a "mental image" corresponding exactly to the retinal image and hence to the object itself. The truth lies in the contrary direction; things are not what they look, or, since the same applies to other senses besides vision, what they seem; and one object of philosophy is to ascertain the exact relations between things as they are and things as we think them to be. We must of course confine ourselves here to pointing out, in regard to vision, some of the more salient differences which obtain between the actual features of an object and our perception of the object.

Of these differences some are clearly of psychical origin. Our perception of a tree is in part determined by events other than the actual sensations, by psychical processes arising out of our previous experiences of trees, and in other ways. Some of these psychical processes we shall consider a little later on.

Other differences are either clearly or possibly of physiological origin; the view may at least be argued that they arise either during the retinal changes through which visual impulses are developed or during the subsequent cerebral changes, spoken of above, through which the visual impulses give rise to visual sensations; and it is to some of these that we with first to call attention.

§ 780. *Irradiation.* A white patch on a dark ground appears larger, and a dark patch on a white ground smaller, than it really is. In Fig. 151, the white square on the right hand side looks

Fig. 151.

larger than the black square on the left hand side though both are exactly of the same size. So also neighbouring white surfaces tend to melt together. The effect is increased when the object is somewhat out of focus, and may be then partly explained by the diffusion circles which, in each case, encroach from the white upon the dark. But over and beyond this, any sensation coming from a given retinal area occupies a larger share of the field of vision, when the rest of the retina and central visual apparatus are at rest, than when they are simultaneously excited. It is as if the neighbouring, either retinal or cerebral, structures were sympathetically thrown into action at the same time. In this way a certain difference is established between the retinal image and the perception.

§ 781. *Simultaneous contrast.* If a white strip be placed between two black strips, the edges of the white strip, near to the black, will appear whiter than its median portion; and if a white cross be placed on a black background, the parts close to the black will appear sometimes so white, compared with the centre of the cross, that the latter will seem dim or even shaded. This effect which occurs even when the object is well in focus, is spoken of as one of 'simultaneous contrast'; the increased sensation of light which causes the apparent greater whiteness of the borders of the cross is regarded as the result of the 'contrast' with the black placed immediately close to it. Still more striking results are seen with coloured objects. If a book, or pencil, be placed vertically on a sheet of white paper, and illuminated on one side by the sun, and on the other by a candle, two shadows will be produced, one from the sun which will be illuminated by the yellowish light of the candle, and the other from the candle which will in turn be illuminated by the white light of the sun. The former naturally appears yellow; the latter, however, appears not white but blue; it assumes, by contrast, a colour complementary to that of the candle-light which surrounds it. If the candle be removed, or its light shut off by a screen, the blue tint disappears, but returns when the candle is again allowed to produce its shadow. If, before the candle is brought back, vision be directed through a narrow blackened tube at some part falling entirely within the area of what will be the candle's shadow, the area, which in the absence of the candle appears white, will continue to appear white when the candle is made to cast its shadow, and it is not until the direction of the tube is changed so as to cover part of the ground outside the shadow, as well as part of the shadow, that the latter assumes its blue tint. If a small piece of grey paper be placed on a sheet of pale green paper, and both covered with a sheet of thin tissue paper, the grey paper will appear of a pink colour, the complementary of the green. This effect of contrast is far less striking, or even wholly absent, when the small piece of paper is white instead of grey, and generally disappears when the thin covering of tissue paper is removed. It also vanishes if a bold broad black line be drawn round the small piece of paper, so as to isolate it from the ground colour. And many other instances of this kind of contrast might be given. It is obvious that whenever in vision this effect intervenes, a discrepancy is introduced between the features of an object and our perception of them. But before we attempt to point out the exact manner in which the effect is produced, it will be convenient to turn to some other effects which are also sometimes spoken of as those of "contrast."

§ **782**. *After-images. Successive Contrast.* As we have already (§ 748) seen the visual sensation lasts much longer than the stimulus, and under certain circumstances the sensation is so prolonged that it is spoken of as an after-image. Such after-images are best developed when an eye, which has for some time been removed from the influence of light, is momentarily exposed to a somewhat strong stimulus. Thus if immediately on waking from sleep in the morning the eye be directed to a window for an instant and then closed, an image of the window with its bright panes and darker sashes, the various parts being of the same colour as the object, will remain for an appreciable time.

When, however, the eye has been for some time subjected to a stimulus, the sensation which follows the withdrawal of the stimulus is of a different kind; the result is what is called a *negative after-image*, or *negative image*, to distinguish it from a *positive after-image*, like the one mentioned above, which is simply a continuation of the sensation primarily excited with all its characters unchanged except that of intensity. If, after looking stedfastly at a white patch on a black ground, the eye be turned to a white ground, a grey patch is seen for some little time. A black patch on a white ground similarly gives rise when the eye is subsequently turned towards a grey ground to a negative image in the form of a white patch. This may be explained as the result of exhaustion. When the white patch has been looked at steadily for some time, that part of the retina on which the image of the patch fell has become tired; hence the white light, coming from the white ground subsequently looked at, which falls on this part of the retina, does not produce so much sensation as in other parts of the retina; and the image, consequently, appears grey. And so in the other instance; in this case, the whole of the retina is tired, except at the patch; here the retina is for a while most sensitive, and hence the white negative image. In speaking of the retina being tired we are using these words for simplicity's sake. We have no right to suppose that the exhaustion takes place in the retinal structures only; it may occur in the central cerebral structures during the development of visual impulses into sensations; indeed the chief part of it is probably of such a cerebral origin.

When a red patch is looked at, and the eye subsequently turned to a white or to a grey ground, the negative image is a greenish blue; that is to say, the colour of the negative image is complementary to that of the object. Thus also orange produces a blue, green a pink, yellow an indigo-blue, negative image, and so on; the negative image is in each case complementary to the primary one.

Similarly, when the eye, after looking at a coloured patch, is turned not to a white or grey but to a coloured ground, the colour of the negative image is a mixture of the colour complementary to the primary image with the colour of the ground; if a yellow ground be chosen after looking at a green object, the negative image will appear as a mixture of red and yellow, a reddish yellow; and so on.

Though these negative images only becomes striking after a prolonged or intense excitation of the retina, such as rarely occurs in ordinary vision, still the effect must intervene, even if to a slight extent only, in our daily sight, and proportionately contribute to the discrepancy between the perception and the object.

§ 783. The phenomena of 'simultaneous' and 'successive contrast' are further of interest in relation to the theory of colour vision; and we may venture for a little while to consider them in this connection. The mere occurrence of the negative images can be explained as a result of exhaustion on either hypothesis of colour vision. According to the Young-Helmholtz theory when the coloured patch is looked at, one of the three primary colour sensations is much exhausted, and the other two less so, in varying proportions, according to the exact nature of the colour of the patch; and the less exhaustive sensations become prominent in the after-image. Thus, the red patch exhausts the red primary sensation, and the negative image is made up chiefly of green and blue sensations, that is, appears to be greenish-blue, or bluish-green, according to the particular hue or tone of the red. So also the yellow patch exhausts both the red and green sensations leaving the blue only to make itself felt. On Hering's hypothesis, we may suppose that, owing to the continued effect of looking at the red patch, the katabolic changes of the red-green substance become less and less, leading to a prominence and indeed to an actual increase of anabolic changes in the same substance; hence, the sensation of green dominating in the negative image; and we may suppose that like events occur in the yellow-blue substance.

So far the facts suit both theories, but Hering's theory offers a more ready explanation than does the rival theory of the fact that it is easier to produce a negative green image after positive exposure to red, or negative blue after positive yellow than, vice versa, red after green, or yellow after blue, in other words, that the red and yellow sensations are more readily exhausted than the green and blue sensations, as indeed is shewn by general experience. For all living substances are more prone to katabolic than to anabolic changes, destruction is easier than construction; and, as we have already seen (§ 769), the fact that blue sensations preponderate in dim lights must not be taken as shewing that the blue or green sensations are more readily excited than the yellow or red. Further, several phenomena of colour vision, to some of which we have already alluded, seem explicable on the view that the stock so to speak of red-green substance in the retina is sooner exhausted than the stock of yellow-blue substance, and this in turn than the white-black substance.

The Young-Helmholtz theory does not explain so readily as does the other why negative images often follow upon positive images without any stimulation of the retina subsequent to the primary one. As we have already said, if a white patch on a black ground be looked at for some time, and the eyes be then shut, a negative image of the spot will be seen on the ground of the 'intrinsic light' of the retina much blacker than the ground, and having in its immediate neighbourhood a sort of bright corona. Conversely a black patch on a white ground will give rise to a patch of exaggerated 'intrinsic light' in contrast to the blackness of the rest of the field. So also, if a window be looked at and the eyes then closed, the positive after-image with bright panes and dark sashes gives way to a negative after-image with bright sashes and dark panes. On Hering's theory all this is readily intelligible as a mere physiological process. On this theory the retina, or rather the visual apparatus, has to return to equilibrium after every exposure to light of any kind; the part of the retina giving rise to the sensation of black, whether it be the patch or the general ground, is not in a condition of equilibrium at the moment of shutting the eyes, the white-black substance is here undergoing anabolism in excess; and, when the light is wholly removed, it passes into equilibrium by katabolic changes, and in doing so developes a sensation, a feeble sensation it may be but still a sensation, of white. The Young-Helmholtz theory cannot offer any such physiological explanation; black being the effect of the absence of all stimulation from the visual apparatus cannot be followed by any physiological rebound; and the theory has to seek a psychological explanation. The parts of the visual apparatus which had been stimulated by light, give rise, when the eyes are shut to a sensation of black, and the parts which had not been stimulated, appear in contrast with these to yield a sensation of light; but only appear, the effect is a psychological not a physiological one.

Hering's theory also offers a physiological explanation of the fact that not only in the case of black and white, but also in the case of colours, the negative after-image with its black, green, &c., corresponding to the white, red, &c., of the positive image, may give way to a return of the positive image with all its original features, to be succeeded by a second negative image like the first, and thus often by a whole series of alternate positive and negative images, each gradually becoming fainter and more obscure. For such rhythmic oscillations from one sensation to its complementary or correlative and back again, pointing to katabolism and anabolism alternately gaining the upper hand, are not without analogies in other common instances of the metabolism of living substance. We may of course apply a like hypothesis to the three primary sensations, but the explanation of the phenomena, as thus given, is not so direct as that afforded by Hering's theory, especially when we consider that each occurrence of the negative image of black has to be accounted for on psychological grounds.

On somewhat the same line of argument, the phenomena of simultaneous contrast may be appealed to in favour of Hering's theory. The explanation of these effects, given by the supporters of the Young-Helmholtz theory is, like that offered for the negative image of black, a psychological one. In the case for instance of the grey patch seen as pink in the midst of a green field, it is argued that the patch does not actually excite a sensation of pink but that we think it is pink because we attribute the greenness of the whole field to the covering tissue paper, and seeing the patch shine through this judge the patch to be reflecting just those rays, namely pink, which mixing with the green would give rise to white, that is to a colourless grey. And a similar psychological explanation has been given of the other cases of simultaneous contrast. Such an explanation is in itself not very convincing; and against it may be urged the fact that in the cases quoted the predominance of the primary colour over the field is not necessary for the effect; and yet the interpretation is based on this. Moreover an experiment may be so arranged that a marked effect of simultaneous contrast should present itself in the vision of one eye but not of the other; now we can hardly imagine that, when both eyes are being used, we can interpret in different ways the sensations derived through the two eyes.

Hering's theory on the other hand offers a direct physiological explanation of the effect; it supposes that when one part of the retina is stimulated, the neighbouring portions of the field of vision are affected at the same time in a manner which may be roughly but only roughly compared to electric induction, so that they undergo changes antagonistic or complementary to those going on in the part of the field of vision corresponding to the portion of the retina actually stimulated. Thus in the case of the grey patch on the green field, the anabolism of the red-green substance in the green field surrounding the grey patch leads to a certain amount of katabolic action of the red-green substance within the grey patch, and so gives rise to a red sensation. In a similar way the bright corona seen round the black negative image developed by shutting the eyes after staring at a white patch on a black ground is due to the anabolism in the black negative image inducing katabolism of white-black substance, that is a sensation of (white) light, in its immediate surroundings. It will of course be understood that the theory does not maintain that the effect is necessarily produced in the retina itself; it may be developed in the visual centres, or in them and in the retina together. We must not go into further details, but we may add that many of the details of these effects of simultaneous contrast are more easily explained on Hering's theory when the possibility of such an inductive action is admitted than by any psychological hypothesis.

We have, contrary to our wont, dwelt so long on two contending theories, and have here renewed our discussion of them in connection with their effects of contrast, partly because of the intrinsic interest of the matter, but also, and not least, because an attempt to decide between the two views opens up important lines of thought and leads to considerations which must have great influence on our conceptions not only of visual sensations but also of all sensations and indeed of nervous processes in general. We have not attempted anything like a full discussion of the subject; we have only ventured to indicate some of the leading criticisms which may be made on each theory; and so far as we are aware no crucial test between the two has as yet been brought forward. We may now leave the matter with the remark that while the

Young-Helmholtz theory tends to lead us direct from the retinal image to the psychological questioning of the sensations, and seems to offer no bridge between the first step and the last, Hering's theory is distinctly a physiological theory, and at least holds out for us the promise of being able to push the physiological explanation nearer and nearer home before we are obliged to take refuge in the methods of psychology.

§ **784.** We have seen (§ 750) that visual sensations may be produced in other ways than by light falling on the retina. In such cases the effect which is produced upon our consciousness is wholly misleading. A mechanical or electrical stimulation of the retina may give rise to a visual sensation identical with that which would be produced by the rays from a flash of light falling upon a part of the retina. In both cases we should have a perception of a flash of light occurring in a certain part of the field of vision; and so far as the perception itself is concerned we could not distinguish between the latter which is a real and the former which is a false perception.

Not only single and simple sensations, but also complex groups of sensations may be excited by other means than that of light falling on the retina, and we may thus experience varied and intricate perceptions which have no objective reality at all. Many people when they close their eyes at night, or indeed at other times, see images of faces or other objects; and though under such circumstances it is easy to recognize the subjective origin of the perception, that conclusion is reached by reasoning upon the circumstances, and not because the perception itself differs in character from a like perception caused by looking at an external object. In such cases it is probable that some causes or other of a physiological nature give rise either in the lower visual centres or in the cerebral cortex to just such changes as would be induced by corresponding visual impulses, though those impulses are wholly wanting; in other words the causes in question give rise to visual sensations, in the physiological meaning of that word, which produce a psychological effect identical with that of visual sensations produced in the ordinary way through the action of light on the retina. In some cases perhaps the process may begin even in the retina itself; abnormal changes in one or other of the retinal structures may lead to the development of complex coordinate visual impulses.

Sometimes the sensations and perceptions thus occurring, especially those which are met with on closing the eyes at night, may be recognized as revivals, more or less altered, of sensations experienced during the day; something sets going again the series of cerebral events which were set going by actual rays of light. These are generally spoken of as "recurrent sensations."

At other times, there is no history of any like sensation having been felt in the immediate past; the psychical effect appears to have no objective cause at all. Moreover such false sensations and perceptions having a distinctness which gives them an apparent objective reality quite as striking as that of ordinary visual perceptions, may occasionally be experienced not only when the eyes are closed, but even when the eyes are open, and when therefore ordinary visual perceptions are being generated, with which they mingle and with which they are often confused. They are then spoken of as *ocular phantoms* or *hallucinations*. They sometimes become so frequent and obtrusive as to be distressing, and form an important element in some kinds of delirium, such as delirium tremens.

It is probable, as we have just suggested, that these false perceptions may be started by events, which in ordinary language may be called physiological; but the whole chain of events between the visual impulse or even the immediate effect of the impulse which we may consider as the physiological sensation, and the terminal psychological perception is long and complex; the discordance between the perception and its apparent cause, in other words, the falsity of the perception, may be introduced in the later, psychological, links of the chain. And an hallucination may have such an origin that it may fitly be spoken of as purely psychological.

This naturally leads to the remark that a perception may be revived in the mind, without the usual physiological antecedents, as the result of purely psychological processes; it is then generally spoken of as an 'idea.' And we find, upon examination, that each new perception which we experience is more or less modified by memories and ideas resulting from bygone perceptions of a like kind. But we have already determined to defer the consideration of these and other more or less distinctly psychical modifications of perceptions until we have studied certain results arising from the use of two eyes.

SEC. 11. BINOCULAR VISION.

§ 785. So far we have treated of vision as if it were carried out by means of one eye and have only incidentally referred to our possessing two eyes. Our ordinary vision is, however, carried out by means of two eyes, our vision is binocular not monocular; and to the characters of this binocular vision we must now turn. In dealing with monocular vision we rarely had occasion to refer specially to the movements of the eyeball; but in binocular vision these play an important part; and even before we go into details, it will be desirable to point out not only certain general facts, but also the meaning of certain terms which we shall have to use.

The eye is virtually a ball placed in a socket, the bulb or eyeball and the orbit forming a ball-and-socket joint. In its socket joint the eyeball is capable of various movements, but these are limited to those of rotation within the socket; the eyeball cannot by any voluntary effort be moved out of its socket. It is stated that by a very forcible opening of the eyelids the eyeball may be slightly protruded; but this trifling locomotion may be neglected. By disease, however, the position of the eyeball in the socket may be materially changed.

The movements of rotation to which the eyeball is thus limited are carried out round a centre in the eye which is termed the *centre of rotation*, and which has been determined to lie in the vitreous humour about 13·5 mm. behind the anterior surface of the cornea, not quite 2 mm. behind what, though the eyeball is not a sphere, may be considered as the geometric centre of the eyeball; it is of course quite different from the optical centre or nodal point of the diagrammatic eye (§ 705).

When we, in looking, direct our vision to a point, a line drawn from such a point, which we may call the *fixed point* of vision, to the centre of rotation, is called the *visual axis;* prolonged past the centre of rotation it meets the retina in the centre of the fovea centralis; hence in the view of those who hold that the optic axis, the line on which the dioptric surfaces of the eye are centred, meets the retina on one side of the fovea, the visual axis does not coincide with, and is different from the optic axis. When with both eyes we look straightforwards to the far distance, the visual axes of the two eyes are parallel; when we direct the two eyes to the same fixed point, the two visual axes converge to the fixed point, the amount of convergence being the greater the nearer the fixed point to the observer.

The horizontal plane in which the two visual axes lie is called the *visual plane;* and a vertical plane at right angles to this, midway between the two eyes, or more exactly bisecting a line, sometimes called the "base line" or "fundamental line" joining the nodal points of the two eyes, is called the *median plane.*

§ 786. As we have seen, the sum of the sensations which we can receive from the retina at the same time is spoken of as the "visual field," or "field of vision." The term therefore has properly a subjective meaning, but it is sometimes used in an objective sense to denote the space or area of the external world, rays of light from which are capable of exciting the retina at any one time; where we wish to distinguish between the two, we may call the latter the "field of sight." The dimensions of the field of sight for one eye will even in the same individual vary with the width of the pupil and other dioptric arrangements of the eye; individual variations are also considerable; but the ordinary dimensions may be stated as subtending an angle of about 145° in the horizontal and about 100° in the vertical meridian, the former being distinctly greater than the latter. When an external object lies outside the area subtending these angles we say that it is outside the field of sight for that position of the eye; it may of course be brought into the field of sight by moving it or by moving the eye. The outline of the field is an irregular one, and stretches farther towards the temporal side of the fixed point, that is, towards the nasal side of the retina, than on the other side; it is somewhat larger and of a different form when the eye is turned towards the temporal side than when the eye is directed straight forwards, cf. Fig. 152. It will be understood that the two visual fields of the two eyes are unlike, cf. Fig. 153.

When we use both eyes a large part of the visual field of each eye overlaps that of the other; that is to say, the rays of light proceeding from a large part of the field of sight of each eye fall upon and affect both retinas. But at the same time a certain part of each visual field does not so overlap any part of the other. If the right hand be held up above the right shoulder and brought a little forward it soon becomes distinctly visible to the right eye, it enters into the field of sight of the right eye. But if the right eye be closed, the right hand kept in the former position is not visible to the left eye; it is outside the field of sight of that eye; it has to be brought much further forward until it comes into the field of sight of the left eye; the profile of the face and especially of the nose prevent

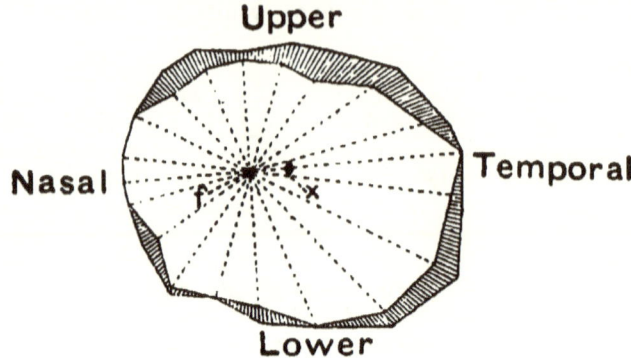

FIG. 152. THE VISUAL FIELD OF THE RIGHT EYE. (Aubert).

The figure represents the visual field projected into space and therefore corresponds to the objective field of sight; the temporal side of the figure corresponds to the nasal side of the retina. The shaded part indicates the increase gained by looking outwards towards the temporal side. f, fovea; x, blind spot.

the rays reflected from the hand gaining access to the left retina until the hand is brought a certain distance forward. The right-hand side of the objective field of sight of the right eye, corresponding to the nasal side of the retina of that eye, extends much farther to the right than does the right-hand side of the field of sight of the left eye, which corresponds to the temporal side of the retina of that eye. Cf. Fig. 153. Similarly, the left hand side of the field of sight of the left eye extends farther to the left than does that of the right eye. Hence on the one hand the total field of sight of the two eyes together is increased in the horizontal diameter, subtending on an average an angle of 180° instead of 145°; and on the other hand while a certain right-hand and left-hand part of the united fields of sight belong respectively to the right and left eye only, the remainder of the field is common to the two eyes. The area common to the two eyes when the visual axes converge to the same fixed point, is shewn as the shaded part in Fig. 153. Rays of light from objects in the common part affect the retinas of both eyes at the same time, vision is here binocular; rays of light from objects at the extreme right and left affect only the right and left retina respectively, vision in these parts of each eye is never binocular, always monocular. The amount of each retina which is thus cut off from binocular vision is determined by the prominence of the nose and profile between the eyes; in some of the lower animals the position of the eyes is so completely lateral that no rays of light proceeding

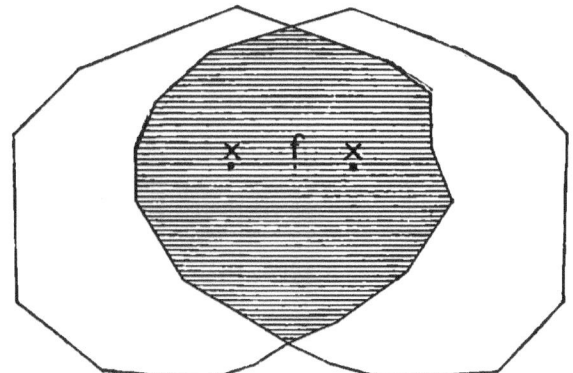

FIG. 153. THE VISUAL FIELDS (FIELDS OF SIGHT) OF THE TWO EYES WHEN THE EYES CONVERGE TO THE SAME FIXED POINT. (Aubert).

The shaded part is that common to the two eyes. f, the fixed point, corresponding to the fovea of each eye; x, the blind spots of the two eyes.

from the same object can fall on any part of the two retinas at the same time, and in these creatures vision is wholly monocular.

§ 787. *Corresponding or Identical Points.* Though when we use two eyes, we must receive from every object in the field of sight common to the two eyes two sets of visual impulses, indeed we may say two sets of sensations, our perception of the object is under ordinary circumstances a single one; we see one object, not two. By putting either eye into an unusual position, as by squinting, we can render the perception double; we see two objects where one only exists. This shews that certain parts of each retina are so related to each other that when an image of an object falls on these parts at the same time, the two sets of sensations excited in the two parts are blended into one; such parts are spoken of as *corresponding* parts; they have also been called *identical* parts. Since in the ordinary movements of the eyes we see objects single, and do not receive double impressions unless we move the eyes in an unusual manner, it is obvious that the movements of the eyeballs and these corresponding parts of the two retinas are so related, the one to the other, that the former bring the images of objects to fall on the latter.

We can easily determine which are the corresponding parts of the two retinas by tracing out the paths of the rays of light falling on the two retinas, § 706. As we have said, when we look at an object with one eye the visual axis of that eye is directed to the object, and when we use two eyes the visual axes of the two eyes converge at the object, the eyeballs moving accordingly. The corresponding points of the two retinas are those on which the two images of the object fall when the visual axes converge at the object. Thus in Fig. 154 if vl from X to x and X to x' be the two visual axes, x, x' being the centres of the foveæ centrales of the two eyes, then, the object $X\,Y\,Z$ being seen single, the point y on the one retina will 'correspond' to or be 'identical' with the point y' on the other, and the point z in the one to the point z' in the other.

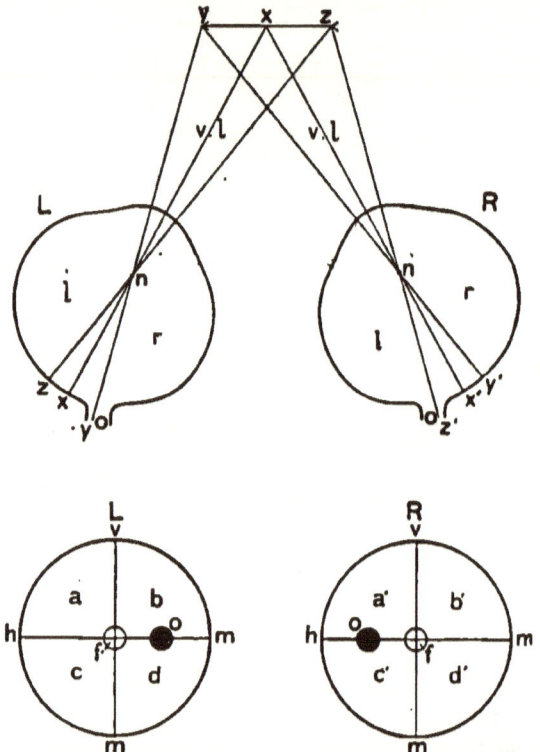

FIG. 154. DIAGRAM ILLUSTRATING CORRESPONDING POINTS.

L the left, R the right eye, $n.$ nodal point, $o.$ optic nerve, $x.$ fovea. $x'y'z'$ are points in the right eye corresponding to the points xyz in the left eye. $v.l.$ visual axis. The two figures below are projections of L the left and R the right retina. $f.$ fovea, $o.$ blind spot. It will be seen that a and c on the *temporal* side of L corresponds to a' and c' on the *nasal* side of R. $v.\,m.\,h.\,m.$ lines of separation.

When the whole area of the retina in each eye which we use for binocular vision is explored in this way we find, as follows geometrically from the paths of the rays of light, that the upper half of one retina corresponds to the upper half of the other, the lower half to the lower half, the right side to the right side, and the left side to the left side. But when we turn to the structure of the retina we find that the left or nasal side of the right eye, since it contains the entrance of the optic nerve, is comparable with, not the left, but the right or nasal side of the left eye, and in like manner the right or temporal side of the right eye is comparable with the left or temporal side of the left eye. Hence, considered in relation to the structure of the retina, the corresponding points appear to be reversed from side to side, though not from top to bottom. While the upper half of the retina of the left eye corresponds to the upper half of the retina of the right eye, and the lower to the lower, the nasal side of the left eye corresponds with the temporal side of the right, and the temporal of the left with the nasal side of the right.

It will be observed that in each eye a vertical plane through the visual axis ($v. l.$ in Fig. 154) cuts the retina in a vertical line $v. m.$, which divides the retina into two lateral, temporal and nasal, halves, each temporal and each nasal half corresponding with the nasal and temporal half respectively of the other eye. When the visual axes of the two eyes are parallel, the two vertical planes in question are parallel to the median plane and to each other. Further, a horizontal plane drawn through the visual axis at right angles to the above vertical plane cuts the retina in a horizontal line $h. m.$; and this also divides the retina into two halves, an upper and lower half, the upper and the lower halves of both retina being corresponding. These two lines, each of which may be considered as a series of corresponding points, are sometimes spoken of as *lines of separation*.

The blending of the two sensations into one occurs, we repeat, only when the two images of an object fall on corresponding points of the two retinas. Hence it is obvious that in single vision with two eyes the ordinary movements of the eyeballs must be such as to bring the visual axes to converge at the object looked at so that the two images may fall on corresponding points. When the visual axes do not so converge, and when therefore the images do not fall on corresponding points, the two sensations are not blended into one perception, and vision becomes double. It is therefore important to study in some detail the movements of the eyeballs, by means of which, in ordinary vision, the relative positions of the two retinas are so carefully adjusted that we habitually see objects single not double.

§ 788. *The movements of the Eyeball.* As we have said, the movements of the eyeball are movements of rotation round an immobile centre, the centre of rotation; but these movements are limited in a particular way, and it is necessary to pay attention to their characters and limitation.

One position of the eyeball, for reasons which we shall see presently, is called the *primary position*, and it will be desirable to start from this position. Though its exact determination requires special precautions it may be described as that which is assumed when, with the head erect and vertical, we look straight forwards to the distant horizon; the visual axes of the two eyes are then parallel to each other and to the median plane.

Let us now suppose three axes drawn through the centre of rotation, in the three planes of space: — one, the visual axis itself, which we may call the longitudinal axis; another at right angles to this and horizontal, the horizontal axis; and a third also at right angles, but vertical, the vertical axis. Corresponding to these three axes we have three main possible movements of rotation. The eyeball might be rotated round the vertical axis so that the visual axis moved from side to side. It might be rotated round the horizontal axis so that the visual axis moved up and down. And lastly, it might be rotated round the longitudinal axis, the visual axis itself remaining motionless and the pupil turning round like a wheel.

Now we can easily carry out by an exercise of the will the first and second of these movements. We can easily move the eyes up and down, rotating them on the horizontal axis, as when we look up to the heavens or down to the ground. We can also move the eyes from side to side, rotating them round the vertical axis, as when we look to the right or to the left. We can move the two eyes sideways together in the same direction keeping the visual axes parallel, or we may move them laterally in opposite directions, as when the visual axes being parallel we make them converge, or when convergent bring them back to or towards parallelism. And we can combine rotation round the horizontal axis with rotation round the vertical axis, and so give oblique movements to the eyeball. We can do all this by an exercise of the will, but we cannot by any voluntary effort carry out the third kind of movement, we cannot rotate the eyeball round the visual axis, we cannot twist the eye in a swivel movement round its longitudinal axis. There are certain movements of the eye in which such a swivel rotation, if we may so call it, does to a certain extent take place, and when we induce these movements we do bring about such a swivel rotation; but we cannot bring about swivel rotation by itself, we can only effect it as part of the particular movements in question.

And there is a reason why we are thus limited as to our power of moving the eyeball. In both rotation round the horizontal axis, and rotation round the vertical axis and in all the various combinations of these two movements which are possible, the two "lines of separation" (§ 787) on both the retinas keep their places; there is no dislocation of the corresponding regions of the two retinas. Obviously the two retinal circles in the lower part of Fig. 154 could be rotated round the vertical or round the horizontal axis or round any intermediate oblique axis without the two images of an external object ceasing to fall on corresponding parts. But if the retinal circles were twirled round their respective visual axes, the lines of separation, $v. m.$ and $h. m.$, would rotate in a clock-hand fashion and if the movements of the two eyes were unequal or in opposite directions, a dislocation of corresponding parts would ensue, and vision would become double. The limitation to the movements of the eyeball so as to avoid a swivel rotation is in the interests of binocular vision.

§ 789. Not only do we find ourselves thus limited in our power when we attempt by a direct effort of our will to execute particular movements of the eyeball, but a similar limitation obtains in the natural movements of the eye in vision. The various movements of the eyeballs which we carry out when we are looking at things conform to a general law, which is known as "Listing's Law," and which may be described as follows.

We stated a little while back that the "primary position" of the eyeball is one in which the visual axis lies parallel to the median plane and is directed to the distant horizon. When the eyeball is changed from this primary position into any other position, all of which may be called secondary positions, the change is effected without any swivel rotation round the visual axis itself; the visual axis may be directed up and down, or from side to side or in any intermediate oblique manner without any such swivel rotation taking place. In other words the movements by which the eyeball is brought from the primary position into any of the secondary positions are, in all cases, movements of rotation round the horizontal axis, or round the vertical axis, or round an axis, which though oblique, being neither horizontal nor vertical, lies in the same plane that they do; that is to say every movement from the primary to a secondary position is a movement of rotation round an axis lying in a plane which passing through the centre of rotation is vertical to the visual axis.

The experimental proof of "Listing's Law" may be obtained by the help of negative images (§ 782) in the following manner. Let the eye be directed to a grey wall or board which, otherwise of uniform appearance, is marked by parallel vertical and horizontal lines, placed at some little distance from each other so as to give a pattern of squares. At one of the intersections, which is to be used as the fixed point of vision, place two narrow strips of red paper in the form of a cross, one vertical coinciding with the vertical line and the other horizontal coinciding with the horizontal line. Having brought the eye carefully into the primary position stare at the red cross until on turning the eye away a green negative image is produced. If now the vision be directed from the fixed point either up or down along the vertical line of the pattern on the wall, or from side to side along the horizontal line, it will be found that the cross of the negative image coincides in turn with each of the crosses of the pattern on the wall, the horizontal limb coinciding with a horizontal line and the vertical limb with a vertical line. This shews that during the up and down and during the side to side movement, during the rotation of the eyeball round its horizontal or round its vertical axis, no swivel rotation has taken place, for otherwise the negative image would have been turned round, and its cross would make an angle with the image of the cross on the wall. If the pattern on the wall be changed so that the lines while still at right angles to each other are oblique, not vertical and horizontal (this is most conveniently done by using not a wall but a large board and turning the board round), and the observation be repeated except that the eye is turned not vertically or horizontally but obliquely so as to follow the lines of the pattern, it will still be found that the cross of the negative image coincides with the cross of the pattern, and that whatever be the angle round which the board has been turned. This shews that Listing's law holds good not only for up and down and side to side movements but also for oblique movements, for movements of rotation round an axis which whatever its obliquity lies in a plane at right angles to the visual axis.

The same result as regards oblique movements may be obtained in another way even while the lines of the wall or board are allowed to remain vertical and horizontal. If in this case the eye be directed not up and down, or from side to side, but diagonally from the fixed centre in the middle of the board to one of the corners of the board, the cross of the negative image will not coincide with the cross of the pattern at the corner but will appear to slant; it will appear to slant to the right in the right hand upper and left hand lower corner, to the left in the left hand upper and right hand lower corner. The slanting cannot be due to a swivel rotation, since in that case the slanting would be in the same direction at both right hand corners, and would be contrary to that occurring at both the left hand corners. The discrepancy between the cross of the negative image and the cross of the pattern at the corner is to be explained by the fact that a horizontal line in the extreme upper part or in the extreme lower part of the field of vision, appears to us curved, bent up or down at each right or left end of the line; and a vertical line in the extreme lateral part of the field of vision appears also curved, bent to the right or left at each, upper or lower, end of the line. Hence in the experiment in question we are comparing the cross of the negative image, not with a rectangular cross in the pattern, but with one, the arms of which seem to dip one way or the other and the two crosses necessarily slant towards each other. Our mind, however, corrects this dip, and regards the cross of the pattern as still rectangular, and in so doing judges the obliquity to belong to the negative image. The experiment therefore really proves that the cross of the negative image undergoes no twisting while the eye is being directed from the centre of the board to the corner, though at first sight it seems, and once was thought, to prove the contrary.

In the ordinary movements of the eye then, a swivel rotation round the visual axis does not take place; and this limitation, since it holds good for the two eyes used together, as well as for one eye used by itself, serves to secure single vision with two eyes inasmuch as it avoids changes which might cause the images of external objects to fall on the parts of the two retinas which were not "corresponding parts." In certain movements of the eyes, however, a certain amount of swivel rotation does take place. This is especially seen in somewhat unusual movements. For instance when the head is turned down to the shoulder, or again when in directing vision to any object, the head is moved from side to side, the eyes do not move with the head; they appear to remain stationary, very much as the needle of a ship's compass remains stationary when the head of the ship is turned. The change in the position of the visual axes to which the movement of the head would naturally give rise is met by compensating movements of the eyeballs; were it not so, steadiness of vision would be impossible; and these compensating movements are found, on careful examination, to include a certain amount of swivel rotation round the visual axes. In certain other more usual movements some amount of such a swivel rotation is also present; and indeed, though so long as the visual axes remain parallel, movement in any direction may take place without any such rotation, a slight amount does intervene during convergence of the visual axes, as when we turn our eyes from a distant to a near object. On careful examination, however, it appears that such an amount of swivel rotation as does take place is after all for the purpose of securing the end that corresponding parts of the two retinas should be affected by the same external object; and, though we cannot here enter more fully into the subject, we may say that not only the more general movements of the eye which obey Listing's law, but also those which form an exception to it, appear to be carried out in the interests of binocular vision. We may now turn to the study of the ocular muscles, by the carefully coordinated contractions of which the various movements, on which we have dwelt, are brought about.

§ 790. *The muscles of the eyeball* or *ocular muscles*. The eyeball is moved by six muscles, four of which are straight, *musculi recti, inferior, superior, internus* or *medialis* and *externus* or *lateralis*, and two oblique, *musculi obliqui, inferior* and *superior*. The four straight muscles, taking origin from the back of the orbit around the sphenoidal fissure and the entrance of the optic nerve, are directed, as their name indicates, straight forward, (the superior rectus however having a peculiar bend) and are inserted in positions corresponding to their several names into the sclerotic, behind the cornea, the bundles of fibres of the tendons being interwoven with those of the sclerotic. The tendon of the internal rectus on the median or nasal side of the eyeball is the broadest of the four; that of the superior rectus on the upper surface being somewhat narrower, and those of the inferior rectus on the under surface and of the external rectus on the lateral or temporal side, still narrower (Fig. 155). The insertion of the superior rectus lies nearer to that of the external rectus than to that of the internal rectus; its position therefore is not exactly median, indeed for two-thirds of its width it lies in the upper lateral quadrant of the sclerotic ring. The insertions of the external and of the internal rectus are both median. The insertion of the internal rectus is the one closest to, and that of the superior rectus the one farthest away from the cornea, and the latter slants so as to be nearer the cornea at its median than at its lateral end.

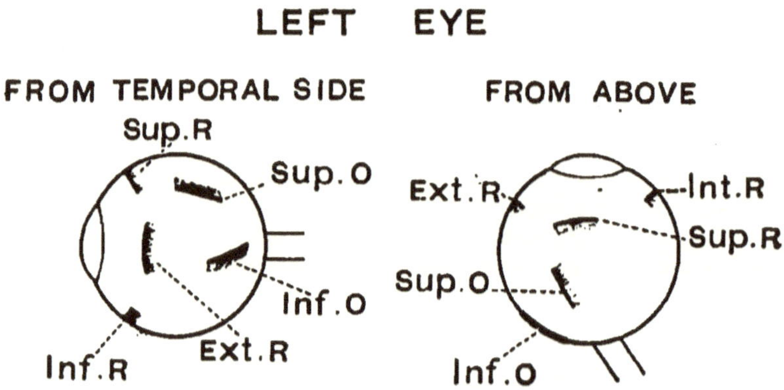

LEFT EYE

FROM TEMPORAL SIDE — Sup.R, Sup.O, Inf.O, Ext.R, Inf.R

FROM ABOVE — Ext.R, Int.R, Sup.R, Sup.O, Inf.O

FIG. 155. THE LEFT EYE SEEN FROM A, THE TEMPORAL SIDE. B, FROM ABOVE, SHEWING THE INSERTIONS OF THE OCULAR MUSCLES. (Jessop.)

The superior oblique muscle, or trochlear or pathetic muscle, taking origin from the back of the orbit near the origin of the straight muscles and running forward internal to the superior rectus, ends in a tendon, which changing its direction by means of a pulley (*trochlea*), and passing beneath the superior rectus is inserted into the sclerotic in the upper region of the bulb towards its hind part. The line of insertion of the tendon (Fig. 155) runs obliquely from the temporal towards the nasal side, its mid-point lying not far from the vertical meridian of the eyeball.

The inferior oblique muscle arises from the front of the floor of the orbit on the nasal side; it is directed at first backwards to the temporal side, underneath the inferior rectus, between that and the floor of the orbit, and then passing upwards and backwards is inserted into the sclerotic underneath the external rectus in the hind temporal part of the ball. The line of insertion (Fig. 155) is also an oblique one like that of the superior oblique but it is placed somewhat farther past it; its hind end lies not far from the entrance of the optic nerve and it runs thence forwards and downwards.

§ 791. The manner in which these muscles are thus severally attached to the eyeball suggests that in contracting they would move the eyeball in the following ways. Taking changes in the direction of the visual axis as indicating the nature of each movement we should expect that the superior rectus would turn the visual axis upwards, the inferior rectus downwards, the external rectus outwards towards the temporal side, and the internal rectus inwards towards the nasal side. The inferior oblique, its insertion being on the hind and lateral part of the eyeball, and the direction of the muscle being downwards, would in contracting turn the visual axis upwards, while the superior oblique having a somewhat similar insertion but acting in an opposite direction would turn the visual axis downwards. Both muscles however in thus raising or lowering the visual axis would, owing to the oblique direction of their insertions at the same time, turn it to the temporal side; the movement, as the names of the muscles suggest, would be an oblique one.

The six muscles would therefore seem to act as three pairs, the superior and inferior rectus, the internal and external rectus, and the inferior and superior oblique, each pair rotating the eyeball round a particular axis. Calculations based on a careful study of the attachments and directions of the several muscles, and the results of actual observations, shew that this is so, and that the movements carried out by the several pairs may be more accurately described as follows.

The superior rectus and the inferior rectus (see Fig. 156) rotate the eye round a horizontal axis, which may be described as one directed from the root of the nose to the temple; it is therefore not a line at right angles with the visual axis but one making an acute angle (20°) with such a line. The superior and inferior oblique rotate the eye round a horizontal axis which may be described as one directed from the centre of the eyeball to the occiput; it again is not a line at right angles to the visual axis, but makes an angle, with such a line, larger (60°) than the similar angle made by the inferior and superior rectus, and turned in a different direction. The internal rectus and external rectus rotate the eyeball round a vertical axis passing through the centre of rotation of the eyeball parallel to the median plane of the head when the head is vertical; this therefore is at right angles to the visual axis, and so differs from the other two.

When we compare the movements thus effected by these several pairs of muscles with the movements which we described above (§ 788) as the ordinary movements of the eye, namely movements of rotation round a vertical and round a horizontal

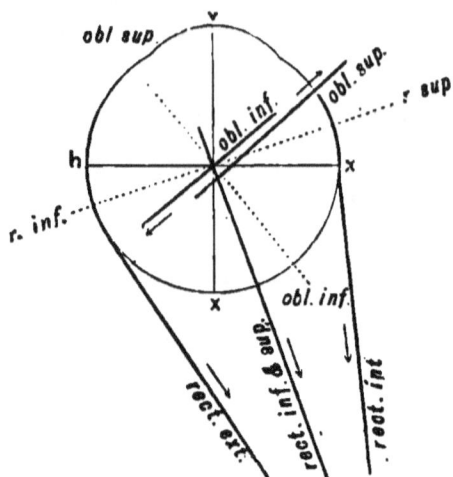

Fig. 156. Diagram to illustrate the actions of the muscles of the eye.

The eye represented is the left eye seen from above. The thick lines shew, by means of the arrows, the direction in which the several muscles pull, the beginning of each line also indicating the attachment of the muscle. The dotted lines indicate the axis of rotation of the superior and inferior rectus and of the oblique muscles. The axis of rotation of the internal and external rectus being perpendicular to the plane of the paper cannot be shewn. *v x* represents the visual axis and *h x* a line at right angles to it. (After Fick.)

axis both at right angles to the visual axis, we see that it is only the movements round the vertical axis which can be carried out by one pair of muscles acting alone, the particular pair being the internal and external rectus. Neither the horizontal axis of rotation of the inferior and the superior rectus, nor that of the oblique muscles, is placed exactly at right angles to the visual axis; each of them makes an oblique angle with that axis. Hence when in carrying out the ordinary movements of the eye we rotate the eyeball round the horizontal axis, we do not employ either of these pairs of muscles alone, but combine them, making use of one muscle of one pair with one of the other. The superior and inferior rectus in moving the visual axis up and down also turn it somewhat inwards, to the nasal side; but this is corrected if the oblique muscles act at the same time; and it is found that the rectus superior acting with the inferior oblique moves the visual axis directly upwards, and the rectus inferior acting with the superior oblique directly downwards in a vertical direction; that is to say the two combinations rotate the eyeball round a horizontal axis at right angles to the visual axis.

Hence there are only two movements of the eyeball which we

can carry out by the help of one muscle alone, namely that in which we simply turn the visual axis to the nasal side, employing the internal rectus, and that in which we turn it to the temporal side, employing the external rectus, the visual axis in both cases remaining in the same plane, the visual plane. In order to raise or lower the visual axis in the same vertical plane, without lateral movement, we must use two muscles; and if we wish to execute an oblique movement combining an up and down with a side to side movement of the visual axis we must employ three of the ocular muscles. These several movements, with the muscles concerned, may be stated as follows, the movement in each case being described with reference to changes in the direction of the visual axis.

Straight movements
- To nasal side. — Internal rectus.
- To temporal side. — External rectus.
- Upwards. — Superior rectus and inferior oblique.
- Downwards. — Inferior rectus and superior oblique.

Oblique movements
- Upwards and to nasal side. — Superior rectus, internal rectus and inferior oblique.
- Downwards and to nasal side. — Inferior rectus, internal rectus and superior oblique.
- Upwards and to temporal side. — Superior rectus, external rectus and inferior oblique.
- Downwards and to temporal side. — Inferior rectus, external rectus and superior oblique.

The fact that in our ordinary movements of the eye we do thus combine the actions of muscles, and the advantages of such a combination are further shewn in connection with that swivel rotation of the eye round the visual axis itself, which, as we have seen, is wholly avoided in many of our movements and which we cannot carry out by a direct effort of the will. The superior rectus acting by itself, owing to the position of its insertion in reference to the direction of the fibres, not only turns the visual axis inwards while directing it upwards, but also to a slight extent rotates the eye round the visual axis; and the inferior rectus as well as both the oblique muscles in like manner tend in contracting to give the eyeball such a swivel rotation. This tendency of the superior rectus like its tendency to turn the visual axis inwards is counteracted by the inferior oblique, the swivel rotation of the latter being contrary in direction to that of the former; and the like tendency of the inferior rectus is in like manner counteracted by the superior oblique. Thus the movements, in carrying out which these muscles are combined, are rendered free from the swivel rotation element. On the other hand this tendency of the muscles in question is utilized in the particular movements in which the swivel rotation does take place.

§ **792.** *The co-ordination of the movements of the eyes.* The external rectus is governed by the sixth nerve, nervus abducens, the nucleus of which, as we have seen (§ 620), lies in the floor of the fourth ventricle in a position indicated by the eminentia teres. The superior oblique muscle is governed by the fourth nerve, nervus trochlearis, the nucleus of which (§ 622) lies in the floor of the aqueduct, in the region of the posterior corpus quadrigeminum. All the other ocular muscles are governed by the third nerve, the nucleus of which lies in the floor of the aqueduct in the region of the anterior corpus quadrigeminum; as we have said (§ 726), the fibres of the third nerve going to these ocular muscles seem to be more especially connected with the hind part of the nucleus.

From what has been said above it is obvious that, even in the movements of one eye, a coordination of the motor nervous impulses must in most cases take place. When we turn the visual axis outwards the motor impulses are confined to the sixth nerve, reaching the external rectus, and when we turn it inwards are confined to the third nerve, reaching the internal rectus; but in all other movements motor impulses must descend to at least two muscles along different nerve-branches, and in many cases must start from two or even all three of the cranial nuclei just mentioned. Even in movements of one eye there must be, in most cases, more or less coordination of actual motor impulses, in order to secure due efficiency of the movement; by actual motor impulses we mean impulses leading to the contraction of muscular fibres, irrespective of any influences which may at the same time be brought to bear on antagonistic muscles, in order to facilitate or qualify the movement.

But if this is true in the case of one eye, much more is it true when we use both eyes in binocular vision.

Two facts about binocular vision strike our attention. The one is that, as may be seen by watching the movements of any person's eyes, the two eyes move together. If the right eye moves to the right, so does also the left, and, if the object looked at be a distant one, exactly to the same extent; if the right eye looks up, the left eye looks up also; and so with regard to other movements. Very few persons are able by a direct effort of the will to move one eye independently of the other; though by some the power has been acquired. We shall refer immediately to particular movements in which one eye only is moved, while the other remains motionless. The other salient fact is that the movements of the two eyes are limited in certain ways. As we have seen one of the simplest ocular movements is the side to side movement of the visual axis, and one of the commonest binocular movements is the convergence of the visual axes, as when we turn our eyes from something far off to something near, or conversely the change from considerable convergence to less convergence as when we turn our eyes from something near to something farther off. In a large number of instances this change to convergence from parallelism, or this increase or decrease of convergence takes place without any change in the visual plane, without any raising or lowering of the visual axes; in such instances the movement is carried out in convergence by the two internal rectus muscles, or in decrease of convergence by the two external rectus muscles; and the only coordination necessary is one which secures that the muscle of one eye should work in harmony with the muscle of the other eye. But even this relatively simple movement is limited in a very marked way. We can bring the visual axes of the two eyes from a condition of parallelism to one of almost any degree of convergence, but we cannot, without artificial assistance, bring them from a condition of parallelism to one of divergence. The stereoscope will enable us to create such a divergence. If in a stereoscope the distance between the pictures be increased very gradually so as carefully to maintain the impression of a single object, the visual axes may be brought to diverge; and the subject of the experiment may himself be made aware of the divergence, by the sudden removal of the instrument from his eyes; his vision of external objects is for a moment double, but for a moment only. This experiment shews the reason of the limitation of which we are speaking. So long as the visual axes are parallel or appropriately convergent the images of external objects fall on corresponding parts of the two retinas, and single vision results; when the visual axes are carried beyond parallelism, the images on the two retinas are not on corresponding parts and vision is double. Thus, as regards convergence or divergence of the visual axes, the movements of the two eyes are governed by the principle that the will can of itself only carry out those movements which are consistent with images of external objects falling on corresponding parts of the two retinas. There is an exception to this in the case of extreme convergence; we can as in squinting make the visual axes converge too much, and in consequence by a simple effort of the will can obtain double vision; but this is probably in order to leave a margin which shall secure our being able to use to the utmost our accommodation mechanism for near objects; otherwise the rule holds good. Not only so, but as the above experiment also shews, when by artificial assistance, which is in itself directed towards securing single vision with the two eyes, we obtain divergence of the visual axes, immediately that the assistance is done away with the axes return, by an involuntary movement, to parallelism; the double vision occurring at the moment of removal of the instrument rapidly gives way to normal single vision. Other illustrations of the same principle may be met with. For instance, if a distant object be looked at with both eyes, but with a prism held horizontally before one eye, and if the image of the object be kept carefully single while the prism is turned very slowly from the horizontal to the vertical position, then on suddenly removing the prism a double image is for a moment seen · this shews that the eye before which the prism was placed had moved in disaccordance with the other. The double image, however, immediately after the removal of the prism, becomes single on account of the eyes coming into accordance.

When we examine all the various movements of the eyes which we are capable of making by a direct effort of the will, we find that they are all of such a kind that through them the two images of an external object are brought upon corresponding parts of the two retinas; conversely the movements which could be effected by the contractions of this or that ocular muscle, but the effect of which would be to bring the two images on to parts of the retina which do not correspond, are the movements which our unassisted will cannot carry out.

In an earlier part of the work (§ 643) we insisted at some length on the important share taken by sensations, or at least by afferent impulses, in the coordination of motor impulses; and the movements of the eye illustrate this in a very marked degree. All the various movements of the eye are dependent on visual sensations. The issue of each efferent motor volitional impulse is dependent on afferent visual impulses. In order to move our eyes, we must either look at or for an object; when we wish to converge our axes, we look at some near object real or imaginary, and the convergence of the axes is usually accompanied by all the conditions of near vision, such as increased accommodation and constriction of the pupil. And so with other ocular movements. Above all, the careful selection of this or that ocular muscle, the extent to which it is to be thrown into contraction, its accompaniment by the contraction of other ocular muscles and the due coordination of all the several contractions — all these things are so determined by visual sensations that the two images of each object looked at fall on corresponding parts of the two retinas.

A little reflection will shew how large an amount of coordination must thus take place in daily life, how in the various movements of the eye there must be, so to speak, the most delicate picking and choosing of the muscular instruments. When we look at an object to the right, since we thereby turn the right eye to the temporal side, and the left eye to the nasal side, we throw into action the external rectus of the right eye and the internal rectus of the left; and similarly when we look to the left we use the external rectus of the left and the internal rectus of the right eye. On the other hand when we look at a near object, and therefore converge the visual axes, we use the internal rectus of both eyes; and when we look at a distant object, and bring the axes from convergence towards parallelism, we use the external rectus of both eyes. Or to take another instance. Suppose the eyes, to start with, directed for the far distance, and that it is desired to direct attention to a nearer point lying in the visual line of the right eye. In this case no movement of the right eye is required; all that is necessary is for the left eye to be turned to the right, that is, for the internal rectus of the left eye to be thrown into action. But in ordinary movements the contraction of this muscle is always associated with either the external rectus of the right eye, as when both eyes are turned to the right, or the internal rectus of that eye, as in convergence; the muscle is quite unaccustomed to act alone. This would lead us to suppose that in the case in question the contraction of the internal rectus of the left eye is accompanied by a contraction of both the external and the internal rectus of the right eye, keeping that eye in lateral equilibrium. And the peculiar oscillating movements seen in the right eye, as well as the sense of effort in the right eye which is felt by the person, support this idea. We need not multiply these instances; it must be sufficiently obvious that a very large amount of coordination takes place in the daily use of our eyes.

§ 793. Such a coordination involves the existence of what, to continue the use of a term which we have previously used, we may call a coordinating nervous mechanism. The coordinated efferent impulses issue from one or more of the nuclei of the three cranial nerves concerned, namely the sixth, the fourth, and the third. The afferent visual impulses taking part in the coordination, we have in an earlier part of this book (§ 669) traced to the primary visual centres, and thence to the occipital cortex. The volitional impulses themselves are we have seen (§ 655) connected in some way or other with an area of the cortex lying in the monkey in the frontal lobe, in the neighbourhood and in front of the precentral fissure (Figs. 125, 126) and probably in man occupying a corresponding position. How are these three factors of the whole nervous action brought to bear the one on the other? When it is remembered how complex and delicately balanced are the movements in question, probably the most intricate and the most delicately balanced of all the movements of the body, it will readily be understood how difficult is the answer to such a question. Stimulation of the cortical areas for movements of the eyes leads as might be expected to bilateral movements, to movements of both eyes; but, so far as results hitherto obtained shew, the movements are bilateral in a special manner. The most common effect of stimulating the cortical area is a lateral movement of both eyes in the same direction towards the opposite side, a conjugate lateral deviation of both visual axes towards the opposite side. For instance when the cortical area of the left hemisphere is stimulated, the visual axes of both eyes are turned to the right, the external rectus of the right eye and the internal rectus of the left eye being thrown into contraction by impulses passing down the right sixth nerve and left third nerve; the efferent impulses therefore cross in the case of one nerve but not in the case of the other. Similarly, when the right hemisphere is stimulated, impulses pass down the right third nerve and left sixth nerve. Stimulation of the occipital region (§ 671) also leads to a similar conjugate turning of both visual axes to the opposite side, accompanied in certain cases by a raising or lowering of the visual axes. So far artificial stimulation of the cortex has not been found to give rise to lateral deviation towards the side stimulated, impulses have not been excited so as to pass down to the external rectus of the same side; but, as we have already urged (§ 657), little weight can be given to the negative results of what is at best a very rough mode of stimulation.

We have already (§ 671) urged that the ocular movements which result from the stimulation of the occipital region cannot be due to an indirect stimulation of the 'motor' frontal area, and that probably they are carried out through some special ties between the occipital cortex and the primary visual centres; and in this relation, it is worthy of notice that ocular movements, similar to those obtained by stimulating the cortex, may be obtained by stimulating the anterior corpora quadrigemina. Stimulation of these structures on one side leads to conjugate lateral movement of the visual axes to the opposite side and it is stated that a more median stimulation leads to a downward movement of both sides with convergence, or in cases where convergence previously existed, to an upward movement with a return to parallelism. If on the other hand the stimulus is brought to bear directly on the nucleus of the third nerve the movements excited are said to be limited entirely to the eye of that side. From this we may infer perhaps that some, at least, of the coordination of which we are speaking, takes place in some part or other of the anterior corpora quadrigemina.

Lastly we may remark that the tract of fibres which we described (§ 634) as the posterior longitudinal bundle, uniting as it does the three ocular nuclei, has been supposed to be an instrument assisting in coordination by serving as a tie between the several nuclei; and it has been especially urged that some of the fibres of this tract cross over from the sixth nucleus of one side and joining the tract of the opposite side pass to the third nucleus of the opposite side, thus affording an anatomical basis for what we have seen to be one of the most frequent associations in ocular movements, that of the external rectus of one eye with the internal rectus of the other eye. In connection with these nuclei it is worthy of note that while all the fibres proceeding from the sixth nucleus of one side pass into the nerve of the same side, some of the fibres issuing from the third nucleus appear to decussate (§ 623) and the whole of the fourth nerve (§ 622) crosses over from its nucleus before it leaves the brain. Yet there appears to be nothing special about the behaviour of the superior oblique to account for this feature.

The Horopter.

§ 794. When we look at any object we direct to it the visual axes, so that when the retinal image of the object is small, the

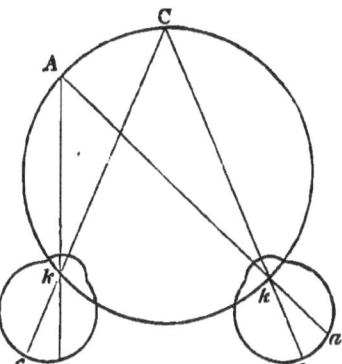

FIG. 157. DIAGRAM ILLUSTRATING A SIMPLE HOROPTER.

When the visual axes converge at C, the images a a of any point A on the circle drawn through C and the nodal points k k, will fall on corresponding points.

'corresponding' parts of the two retinas, on which the two images of the object fall, lie in their respective foveæ centrales. But while we are looking at the particular object the images of other objects surrounding it fall on the retina surrounding the fovea, and thus go to form what is called indirect vision. And it is obviously of advantage that other images, besides that of the object to which we are specially directing our attention, should fall on 'corresponding' parts in the two eyes. Were it not so, while our vision of the particular object would be single, our vision of all its surroundings would be double; and this, at least in certain cases, would be confusing. For, even when we are concentrating our attention on a particular object, we are still conscious of its surroundings, and besides, our appreciation of any image falling on the fovea is influenced by impressions which we are at the same time receiving from other parts of the retina.

Now for any given position of the eyes there exists in the field of sight a certain line or surface of such a kind that the images of the points in it all fall on corresponding points of the retina. A line or surface having this property is called a Horopter. The horopter is in fact the aggregate of all those points in space which, in any given position of the eyes, are projected on to corresponding points of the retina; hence its determination in any particular case is simply a matter of geometrical calculation. In some instances it becomes a very complicated figure. The case whose features are most easily grasped is that of a circle drawn in the plane of the two visual axes through the point of the convergence of the axes and the nodal points of the two eyes such as is shewn in Fig. 157. It is obvious from geometrical relations that the two images of any point in such circle, the rays from which can enter the two pupils and fall on the two retinas, will fall on corresponding points of the two retinas. When we study the various horopters of the several positions which the two eyes can take up, we find that the characters of the horopter are adapted to the needs of our daily life. Thus in the position assumed by the two eyes when we stand upright and look at the distant horizon the horopter is (approximately, for normal emmetropic eyes) a plane drawn through our feet, that is to say, is the ground on which we stand; the advantage of this is obvious.

Nevertheless, in most positions of the eyes a large number of the images which make up the binocular visual field, do not lie on any horopter, do not fall on corresponding points, and give rise not to one sensation only but to two sensations differing to a certain extent from each other. A great deal of what we see is seen double by us, we receive from many objects two unequal impressions; but the inequality chiefly serves to give an appearance of "solidity" to the objects, to assist in our judgment of solidity. To the consideration of these and other visual judgments as well as of some other psychological features of vision we must now turn.

SEC. 12. ON SOME FEATURES OF VISUAL PERCEPTIONS AND ON VISUAL JUDGMENTS.

§ 795. We may now turn our attention to some of those differences between the features of external objects and our perception of them which are more distinctly of psychological origin; but since the purpose of this work is physiological and not psychological we must be content to treat them very briefly.

Taking first of all the general features of the field of vision, we find psychical processes entering largely even into these. As we have incidentally seen, the sensations which an object excites are very different according as the object is in the central or in the peripheral region of the field of sight. Two parts of the object sufficiently far apart to give rise to two sensations in the former case may give rise to one sensation only in the latter case ; and the colour sensations excited by the same object may be widely different in the two cases. If we picture to ourselves the group of sensations excited by the image of an object, such as a flower, when the image falls on the fovea, and compare that group with the group of sensations excited by the same flower when the image of it falls on the periphery of the retina, supposing the comparison to be made before the sensations are moulded into psychical perceptions, the two groups would appear to belong to very unlike objects. Moreover, when we use both eyes, the images of some of the objects in the field of sight are falling on both retinas, while others are falling on one retina only, and of those which fall on both retinas, some lie on corresponding points, so that the sensations of the two eyes are blended, while others, not lying in the horopter, give rise to sensations in one eye different from those in the other. Could we become aware of the crude sensations which go to make up our field of vision, they would appear as a heterogeneous medley. But in the field of vision of which we are actually aware, that in which the crude sensations have by psychical operations been moulded into perceptions, we do not recognize the various discrepancies of which we are speaking; the field of vision is homogeneous. When we look at a landscape we are not aware that objects on the far left or far right hand are producing sensations in a way very different from that in which objects directly in the line of vision are producing sensations; it is only by special analysis that we become acquainted with the properties of the peripheral retina. In actual vision the activities of the central retina by virtue of psychical processes dominate those of the periphery. Conversely though, as we have said, when we wish to see anything very distinctly we habitually make use of the central retina; yet nevertheless in ordinary vision, at the same time that we are thus making use of the central retina we are also receiving impressions from the whole of the rest of the retina within the field of vision, and these more or less peripheral impressions influence to a certain extent the psychical effect of the central sensations. Our perception of an object, such as a flower, is not the same when we look at it as part of a landscape, making use of the whole field of vision, as when we look at it through a tube or otherwise in such a way as to exclude peripheral vision; the flower in the latter case seems much more brilliant, and more highly coloured. Some of the effect in this case may be physiological and due to retinal events, but the greater part is psychical. The influence of psychical processes is probably also illustrated by the experience that, if on turning our back on a landscape, we bend the body so as to get a view of the landscape backwards between the legs, all the objects seem to have an unusually brilliant colouring.

A striking difference between the objective field of sight and the subjective field of vision is illustrated by the fact that, though, as we have seen, that part of the retina which corresponds to the entrance of the optic nerve is quite insensible to light, we are conscious of no corresponding blank in the field of vision. When in looking at a page of print we so direct the visual axis that some of the print must fall on the blind spot, no gap in the print is perceived; we have to take special measures (§ 770) to discover the existence of the spot. We could not expect to see a black patch, because what we call black is the absence of the sensation of light from structures which are sensitive to light; we must have visual organs to see black. But there are no visual organs in the blind spot, and consequently we are in no way at all affected by the rays of light which fall on it. By psychical operations we "fill up," as it is said, the vacancy caused by the blind spot, so that there is in our subjective field of vision no gap corresponding to the gap in the retinal image; we treat the sensations coming from two points of the retina lying on opposite margins of the blind spot as if they were sensations excited in two points lying close together, thus preserving the continuity of the field of vision between them. Concerning the particular psychical actions by which this is carried out, and concerning the special effects which are produced when an object in the field of sight passes into the region of the blind spot there has been much discussion; but into this we cannot enter here.

In ordinary vision, the existence of the blind spot is of little moment. Since it lies outside the region of distinct vision, and since moreover in each movement of the eye the image of a fresh part of the external world falls upon it, the errors to which it may lead are not serious even when we use one eye only. The deficiency is further remedied by the use of two eyes, since, the two blind spots being each on the nasal side, the image of an object will not fall on both blind spots at the same time. Other smaller or accidental imperfections in one or both eyes are similarly remedied by the use of two eyes.

§ 796. Turning now to the psychical processes connected with the perception of particular objects, we find these to be very complex. Some of them relate to the very formation of the perception out of the sensations which the object excites, and are often of such a kind that the perceptions which they influence so distinctly fail to correspond with the actual objects that the lack of correspondence can in many cases be demonstrated: such erroneous perceptions are often spoken of as "illusions." In other cases the psychical processes relate to a further mental action by which we form judgments as to the features of external objects. It is not easy however always to draw a line between a 'visual judgment,' such as that involved in forming a conclusion as to the size of an external object, and what may be called a mere "modified perception," as when a line appears to us shorter or longer than it really is. We may be content here to treat them all together.

The complexity of the psychical processes in question comes about in various ways. On the one hand the characters of a perception are determined not alone by the sensations which actually give rise to it but also by the psychical conditions remaining as the effect of former like sensations. In the formation of perceptions and judgments, suggestions and associations play their part; so that each perception, while it adds to, is also in part the result of our 'experience.' A simple illustration of this is seen in some of the effects of colour. Blue colours as we have seen predominate in a dim light such as that of evening, of moonlight or of winter, whereas reds and yellows are marked in a bright light such as that of full sunshine, or of a summer's day. Hence, when a landscape is viewed through a yellow glass, the yellow hue suggests to the mind bright sunlight and summer weather, although the actual illumination which reaches the eye is diminished by the glass. Conversely when the same landscape is viewed through a blue glass the idea of moonlight or winter is suggested. And many other instances might be given in which the appreciation of the present is moulded by the experience of the past.

On the other hand the visual perception or visual judgment

is not formed exclusively out of the visual sensations which are excited by the image of an object falling on one or on both eyes in a given position. In looking at an object, a movement of one or both eyes often takes place, and the perception of the object or a judgment concerning the object is formed out of the two (or more) sensations excited by the same object in different positions of the eyes. And here other factors enter into the process, namely sensations other than visual sensations, sensations connected with the contractions of the muscles of the eye, affections of what is known as "the muscular sense." These come into play even when we use one eye only, but are especially potent when we use both eyes in binocular vision; a large number of our visual judgments are determined by the muscular sensations derived from the movements of the eyes through which we look at the object whose features we are judging.

Other influences also, such for instance as sensations of touch, take part in the psychical processes in question. The mere visual sensations which external objects excite, the immediate and direct effects of the visual impulses, form after all but a small part of what we call our vision. Such sensations and other like sensations derived through other senses are to us but symbols of things, upon which the mind puts its own interpretation. But into these matters we cannot enter here. We must confine ourselves to certain common facts concerning perceptions, illusions and visual judgments, and more especially to those which relate to the size and distance of external objects and to the characters of form which are indicated by the word "solidity."

§ 797. *Appreciation of Apparent Size.* The foundation of our judgment of the size of any object is the size of the retinal image of the object. We can distinguish a sensation involving a large retinal area from one involving a small area, and in the region of distinct vision can appreciate even small differences; this is of course only an exercise of the power of localization. We have seen however that, even in the case of a simple and single sensation such as that of a white patch on a black ground, the sensation does not correspond exactly to the objective stimulation of the retina; the white patch through irradiation § 780 appears larger than it really is. When we come to deal with more complex groups of sensations we find that over and above any such physiological modifications of the sensations, the psychical processes mentioned above affect our perceptions and judgments of size, often giving rise to illusions. If a line such as AC, Fig. 158, be divided into two equal parts AB, BC, and AB be divided by distinct marks

Fig. 158.

into several parts, as is shewn in the figure, while BC be left entire, the distance AB will always appear greater than CB.

The retinal images of the spaces from A to B and from B to C are equal and the corresponding primary visual sensations are also equal, but the mental appreciation of AB is interfered with by the concurrent sensations of the several intervening dots and intervals, and this leads to a mental exaggeration of the interval between A and B. So also, if two equal squares

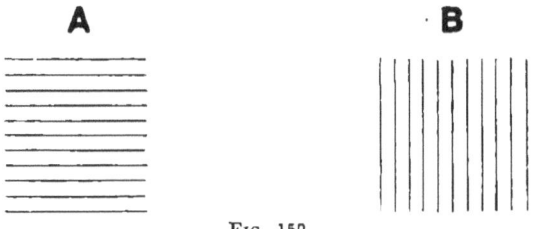

Fig. 159.

(Fig. 159) be marked, one with horizontal and the other with vertical alternate dark and light bands, the former will appear higher, and the latter broader, than it really is. Hence short persons often affect dresses horizontally striped in order to increase their apparent height, and very stout persons avoid longitudinal stripes. Again, when a short person is placed side by side with a tall person, the former appears shorter and the latter taller than each really is. By reason of somewhat similar psychical processes two perfectly parallel lines or bands, each of

Fig. 160.

which is crossed by slanting parallel short lines (Fig. 160), will appear not parallel, but diverging or converging according to the direction of the cross-lines; the direction of the cross-lines affects our perception of the distance between the parallel lines.

§ 798. *Judgment of Distance and Actual Size.* The size of the retinal image gives us by itself a measure not of the real size but only of the apparent size of the object. The size of the retinal image will depend on the distance of the object and on the dioptric arrangements of the eye; with the same dioptric arrangements it will depend on the angle subtended by the diameter of the object, and this may be the same for a small object near as for a large object far off. In order to form a judgment as to the actual size of an object, we must adjust our perception of the apparent size by means of a judgment of the distance at which the object is placed; and here the great use of two eyes comes in.

Even with one eye we can, to a certain extent, form a judgment not only as to the position of the object in a plane at right angles to our visual axis, but also as to its distance from us along the visual axis. If the object is near to us, we have to accommodate for near vision; if far from us, to relax our accommodation mechanism so that the eye becomes adjusted for distance. The muscular sense of this effort enables us to form a judgment whether the object is far or near. Seeing the narrow range of our accommodation, and the slight muscular effort which it entails, all monocular judgments of distance must be subject to much error. Everyone who has tried to thread a needle or to pour out a glass of wine without using both eyes, knows such errors.

When, on the other hand, we use two eyes, we have still the variations in accommodation, and in addition have all the assistance which arises from the muscular effort of so directing the two eyes on the object that single vision shall result. When the object is near, we converge our visual axes; when distant, we bring them back towards parallelism. This necessary contraction of the ocular muscles affords a muscular sense, by the help of which we form a judgment as to the distance of the object. We can judge of the distance of a vertical line more easily than of a horizontal line, because we can converge our vision more easily upon the former; this is seen in attempting a 'high jump' over a horizontal cord, the judgment of the distance of the cord is facilitated by hanging a vertical cord or tape to it. Conversely, when by any means the convergence which is necessary to bring the object into single vision is lessened, the object seems to become more distant; when the convergence is increased, the object seems to move towards us; this may be seen in the stereoscope.

The judgment of size is, as we said above, closely connected with that of distance. The real size of the object can be inferred from the apparent size, that is to say from the size of the retinal image, only when the distance of the object from the eye is known. Thus when an object gives rise to a retinal image of a certain size, that is to say has a certain apparent size, we estimate the distance from us of the object giving rise to the image, and upon that come to a conclusion as to its real size. Conversely, when we see an object, of whose real size we are otherwise aware, or are led to think we are aware, our judgment of its distance is influenced by its apparent size. Thus when part of our field of vision is occupied by the image of a man, knowing otherwise the ordinary size of a man, we infer, if the image be very small, that the man is far off. The reason of the image being small may be because the man is far off, in which case our judgment is correct; it may be, however, because the image has been lessened by artificial dioptric means, as when the man is looked at through an inverted telescope, in which case our judgment becomes an illusion. So also a picture on a magic lantern screen when gradually enlarged seems to come forward, when gradually diminished seems to recede. In these cases the influence which the absence of any muscular sense of binocular adjustment or monocular accommodation ought to bring to bear on our judgment, is thwarted by the more direct influence of the association between size and distance. An instructive illusion of a similar kind is produced by developing in the eye a strong negative image (§ 782) and projecting the image on to a screen which is made to move backwards and forwards, or is alternately inclined at various angles; the negative image appears to change in size and shape, although it is absolutely subjective in nature and wholly independent of the movements of the screen.

The complex reaction on each other of judgments as to distance and size is illustrated by the experience that an object such as a person looks unnaturally large when seen in a fog; being seen indistinctly, he is judged to be farther off than he really is, and so appears larger than he naturally would do at the distance at which he is supposed to be; and we are similarly influenced by the greater or less brightness or saturation of colours. Conversely, distant mountains when seen distinctly in a clear atmosphere appear small, because on account of their distinctness they are judged to be nearer than they really are. The indistinctness of the image of the moon or sun when seen on the horizon, similarly contributes to its appearing larger than when seen in the zenith; our judgment however is probably in this case also due to our being better able to compare the moon or sun with terrestrial objects. We seem moreover in this matter to be especially influenced by our conception (which is itself an illustration of the subject we have in hand) that the vault of the heavens is flatter than it really is; the zenith appears to be less distant than the horizon; a geometric construction will shew that a body of the same size placed at different parts of the real (spherical) vault will appear greater near the horizon than near the zenith of the flatter, apparent vault. An amusing illustration of visual judgments may be obtained by asking a number of persons in succession what they regard as the size of the moon in mid heavens. Even making allowance for dioptric differences in individual eyes the size of the retinal image of the moon must be about the same in all eyes. And yet while some persons will be found ready to compare the moon in mid heavens with a three-penny piece, others will liken it to a cart-wheel; and others will make intermediate comparisons.

§ 799. *Judgment of Solidity.* When we look at a small circle all parts of the circle are at the same distance from us, all parts are equally distinct at the same time, whether we look at it with one eye or with two eyes. When, on the other hand, we look at a sphere, the various parts of which are at different distances from us, a sense of the accommodation, but much more a sense of the binocular adjustment, of the greater or less convergence of the two eyes, required to make the various parts successively distinct, makes us aware that the various parts of the sphere are unequally distant; and from that we form a judgment of its solidity. As with distance of objects, so with solidity, which is at bottom a matter of distance of the parts of an object, we can form a judgment with one eye alone; but our ideas become much more exact and trustworthy when two eyes are used. We are further much assisted by the effects produced by the reflection of light from the various surfaces of a solid object, and the shadows cast by its raised parts; so much so that raised surfaces may be made to appear depressed, or *vice versa*, and flat surfaces either raised or depressed, by appropriate arrangements of shadings and shadow.

Binocular vision, moreover, affords us a means of judging of the solidity of objects, inasmuch as the image of any solid object which falls on to the right eye cannot be exactly like that which falls on the left, though both are combined in the single perception of the two eyes. Thus, when we look at a truncated pyramid placed in the middle line before us, the image which falls on the right eye is of the kind represented in Fig. 161 R, while that

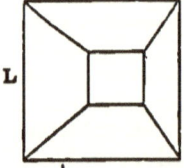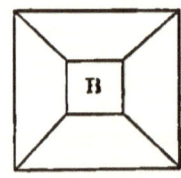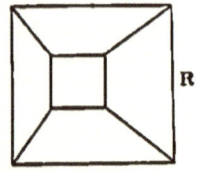

Fig. 161.

which falls on the left eye has the form of Fig. 161 L; yet the perception gained from the two images together corresponds to the form of which Fig. 161 B is the projection. Whenever we thus combine in one perception two dissimilar images, one of the one, and the other of the other eye, we judge that the object giving rise to the images is solid.

This is the simple principle of the stereoscope, in which two slightly dissimilar pictures, such as would correspond to the vision of each eye separately, are, by means of reflecting mirrors, as in Wheatstone's original instrument, or by prisms, as in the form introduced by Brewster, made to cast images on corresponding parts of the two retinas so as to produce a single perception. Though each picture is a surface of two dimensions only, the resulting perception is the same as if a single object, or group of objects, of three dimensions had been looked at.

It might be supposed that the judgment of solidity which arises when two dissimilar images are thus combined in one perception, was due to the fact that all parts of the two images cannot fall on corresponding parts of the two retinas at the same time, and that therefore the combination of the two needs some movement of the eyes. Thus, if we superimpose R on L (Fig. 161), it is evident that when the bases coincide the truncated apices will not, and *vice versa ;* hence, when the bases fall on corresponding parts, the apices will not be combined into one image, and *vice versa ;* in order that both may be combined, there must be a slight rapid movement of the eyes from the one to the other. That, however, no such movement is necessary for each particular case is shewn by the fact that solid objects appear as such when illuminated by an electric spark, the duration of which is too short to permit of any movements of the eyes. If the flash occurred at the moment that the eyes were binocularly adjusted for the bases of the pyramids, the two summits not falling on exactly corresponding parts would give rise to the perceptions of two summits, and the whole object ought to appear confused. That it does not, but, on the contrary, appears a single solid, must be the result of psychical operations, resulting in what we have called a judgment.

As we have seen, in any one position of the two eyes, only a small portion of the field of sight lies in the horopter and falls on corresponding points of the two retinas. Most of the objects in a scene on which we look give rise to dissimilar images in the two eyes; and we attribute solidity to them by reason on the one hand of the movements of the eyes, and on the other hand of the psychical processes just mentioned. Conversely the same processes which thus give rise to apparent solidity assist us in forming judgments of distance.

§ 800. If the images of two surfaces, one black and the other white, are made to fall on corresponding parts of the eye, so as to be united into a single perception, the result is not always a mixture of the two impressions, that is a grey, but, in many cases, a sensation similar to that produced when a polished surface, such as plumbago, is looked at: the surface appears brilliant, is said to have a "lustre." The reason probably is because when we look at a polished surface the amount of reflected light which falls upon the retina is generally different in the two eyes; and hence we associate an unequal stimulation of the two retinas with the idea of a polished lustrous surface.

We may in this connection refer to what is known as "the struggle of the two fields of vision," though the matter is one of sensations and not of judgments or intricate psychical processes. When the impressions of two colours are united in binocular vision, the result is in most cases not a mixture of the two colours, as when the same two impressions are brought to bear together at the same time on a single retina, but a struggle between the two colours, now one, and now the other, becoming prominent, intermediate tints however being frequently passed through. This may arise from the difficulty of accommodating at the same time for the two different colours (§ 735); both eyes will be accommodated at the same time and to the same degree, but if two eyes, one of which is looking at red, and the other at blue, be at one moment both accommodated for red rays, the red sensation will overpower the blue, while if at another moment they are both accommodated for blue, the blue will prevail. It may be however that the tendency to rhythmic action, so manifest in activity of other simpler forms of living matter makes its appearance also in the cerebral changes involved in binocular vision.

SEC. 13. THE NUTRITION OF THE EYE.

§ 801. The main blood-vessels of the eye are, as we have incidentally seen, the arteria centralis supplying the retina, and the (posterior) ciliary arteries supplying the choroid, ciliary processes and iris, the vessels going to the choroid being called the short ciliary arteries, and those reaching forward to the ciliary processes and iris, the long ciliary arteries. From the arteria centralis retinæ the blood is returned by the vena centralis, while the venae vorticosae of the (posterior) ciliary veins gather up the blood of both the long and the short (posterior) ciliary arteries. These two systems communicate to some extent with each other by anastomoses at the entrance of the optic nerve, but on the whole are independent.

In addition to the above, the anterior ciliary arteries pass to the eyeball with each of the four straight ocular muscles, supplying the front part of the sclerotic as well as the edge of the cornea, and sending through the sclerotic 'perforating' arteries to end in the iris, ciliary processes, and front part of the choroid, and so join the system of the posterior ciliary arteries. Corresponding to these anterior ciliary arteries are veins which make their way back to the ocular muscles, and the roots of which are especially connected with the circular canal of Schlemm (§ 717). Further, the edge of the cornea is in addition supplied by conjunctival blood-vessels.

The blood-supply of the various parts of the eye is therefore somewhat as follows. The inner layers of the retina are supplied in a direct manner by the arteria centralis retinæ, but the outer layers together with the pigment epithelium in an indirect manner by the close set choroidal network "choriocapillaris" of the posterior ciliary arteries. The choroid proper, that part which serves as an investment to the retina and specialized pigment epithelium, is supplied by the short (posterior) ciliary arteries; but the front ciliary part of the choroid, together with the ciliary processes and iris, receives blood from the long (posterior) ciliary arteries, and also from the anterior ciliary arteries. The cornea is supplied by the conjunctival as well as by the anterior ciliary arteries, the blood-vessels as we have said extending a short distance only within the circle of the corneal circumference, while the scanty supply of the sclerotic is furnished in the front part of the eyeball by the anterior and in the hind part by the posterior ciliary arteries.

The nutritive supply of the lens, with its capsule, and of the vitreous humour is an indirect one, by means of lymph; the anterior surface of the former is bathed by the aqueous humour; the lymph streams in the vitreous humour, of which we shall speak immediately, furnish that substance with the scanty nourishment it needs, and sweep by the posterior surface of the lens.

§ 802. In speaking of the movements of the pupil we referred to vaso-motor changes in the eye. So far as our present information goes, we have evidence chiefly of vaso-constrictor fibres which passing from the sympathetic to the ciliary ganglion (§ 725) reach the posterior ciliary arteries by the short ciliary nerves; but there are facts which seem to shew that the fifth nerve supplies vaso-dilator fibres through the ophthalmic branch.

The separate distribution of the short ciliary arteries to the hinder part of the choroid investment which is busy with the nourishment of the retina and which takes little or no share in the movements of accommodation, and of the long ciliary arteries to the front part of the investment which, as iris, ciliary processes and muscle, and front part of the choroid itself, is concerned in the movements of the pupil and of accommodation, suggests that a corresponding separate distribution of vaso-motor nerves also exists; but we have no exact experimental evidence of this.

We saw in speaking of the brain (§ 700) that clear evidence of the cerebral vessels being subject to vaso-motor influences was wanting; and in this respect once more the retina behaves like a part of the brain. Though by help of the ophthalmoscope changes of calibre in the retinal vessels can easily be observed, we have as yet no decisive proof that such changes can be brought about by vaso-motor nerves acting directly on the arteria centralis retinae. The changes which are observed seem to be determined not by the greater or less contraction of the muscular coat of the retinal vessels themselves, but by the pressure to which the blood in the vessels is subjected, and that may be varied by many extraneous causes.

The Lymphatics of the Eye.

§ 803. Though the lymph in the large serous cavities may be considered to play a mechanical part inasmuch as it facilitates the movements of the viscera, and though in such a tissue as the skin, the lymph in the cavities and vessels of the dermis may similarly perform a mechanical task in assisting to give at once firmness and suppleness to the skin, yet over the body at large the function of the lymph is preeminently a nutritive one, and its mechanical duties are insignificant. As regards the eye the case is different. The eyeball is broadly speaking a shell filled with fluid, the aqueous and vitreous humours; and for the various functions of the eye it is necessary that this shell should be filled to a certain extent, should be tense to a certain degree, not more and not less; and this fulness, this tension, "intraocular tension," which is considerable, probably much higher than the ordinary pressure in the lymph-spaces of the body at large, is provided by the lymphatic arrangements. If the retina were not adequately supported by the vitreous humour, if it could flap about or in any way alter its curvature, the dioptric arrangements of the eye would be upset; if the vitreous humour at one time shrank, at another expanded, the movements of accommodation could not be carried on; if the aqueous humour were now abundant, now scanty, the movements of the pupil would become irregular and uncertain; and if the whole globe were so flabby as to give way under the pull of each ocular muscle, the delicate movements of the eyeball on which we lately dwelt would become impossible. Hence the lymphatics of the eye have a double importance, inasmuch they not only, as elsewhere, assist in maintaining the due nutrition of the several tissues, but also in a mechanical way help to make the eye an adequate dioptric instrument. In accordance with this double duty we find a special lymph apparatus added to the more general lymphatic arrangements such as exist elsewhere.

As belonging to the more general arrangements we may note the following. The lymph-spaces of the cornea pass at the margin of the cornea into the lymphatic vessels of the conjunctiva. The scanty lymph-spaces of the sclerotic pass at the extreme front into the conjunctival lymphatics, but elsewhere are continuous either on the inner surface with the perichoroidal lymph-spaces, or on the outer surface and that more freely, with the large lymphatic *Tenonian cavity.* Tenon's capsule is a loose thin investment of connective tissue lying between the sclerotic and the ocular muscles and forming sheaths round the tendons of the latter. Between the looser capsule and the denser sclerotic is a large irregular lymphatic cavity bearing the above name. The perivascular and other lymph-spaces of the choroid join the perichoroidal spaces, which in turn communicate with the Tenonian cavity by lymph-spaces or lymphatics accompanying the ciliary veins and to some extent the ciliary arteries as these pierce the sclerotic. The Tenonian cavity itself joins a large lymphatic cavity surrounding the optic nerve, ' the supravaginal' cavity, whence the lymph is carried away by the ordinary lymphatics of the orbit.

The perivascular and other lymph-spaces of the retina are in connection with the lymph-spaces of the optic nerve, which in turn join the subarachnoid space of that nerve, and this is continuous with the corresponding space in the brain. There appear to be also paths uniting these lymphatics of the retina and optic nerves with the perichoroidal spaces and Tenonian cavity, and so with the external lymphatic system.

§ 804. In the special lymph apparatus the ciliary processes, the iris, the aqueous humour and the vitreous humour are concerned.

The aqueous humour. We have more than once spoken of the anterior chamber as a lymphatic cavity; nevertheless the aqueous humour contained in it differs greatly from ordinary lymph. Not only does it contain much more water, the total solids being not much more than 1 p.c. (1·3 p.c.) but also the relative proportion of the solids between themselves is different from that of lymph, and special substances are present in it. The proteids are particularly scanty, not more than about ·1 p.c.; these are serum-albumin, globulin, and apparently fibrinogen. Inorganic salts, are present in about the same proportion as in blood and lymph, viz. 8 p.c.; and these, chiefly sodium chloride, with an unusual proportion (·4 p.c.) of so-called extractives, furnish nearly all the solid matter. Among these extractives is a substance which reduces cupric solutions but which is not a sugar, though its exact nature is as yet unknown; urea and sarcolactic acid, (in some combination) are also said to be, at least often, present. The reaction is neutral or faintly alkaline.

Like the 'serous fluid' in the large serous cavities and the cerebro-spinal fluid in the cavities of the central nervous system, the aqueous humour comes and goes; the particular fluid which at any given moment is present in the eye has not always been there; some of the fluid is continually passing away and fresh fluid continually arriving. If fluid be withdrawn from the anterior chamber by puncture of the cornea, the chamber is soon refilled; indeed, under certain circumstances, a considerable quantity of fluid may be drained away from the chamber, fresh fluid taking the place of that which escapes. And, though under normal conditions the quantity of aqueous humour is fairly constant, the fluid may be in excess or may be deficient, and the one phase may pass into the other. The question therefore arises, Whence comes the fluid and whither does it go?

The characters of aqueous humour just given shew that in many respects it resembles cerebro-spinal fluid though differing in several features. That fluid, we have seen reason to believe (§ 694), is in part at all events furnished by the choroid plexuses, by a process which presents some analogies with the act of secretion. And the resemblance between the ciliary processes and the choroid plexuses, for both are vascular folds of pia mater covered with epithelium derived from the lining of the primitive medullary canal, suggests that the former furnish the aqueous humour in some such way as the latter furnish the cerebro-spinal fluid. There is a certain amount of experimental evidence in favour of this view, for when such a substance as fluorescin, which can be detected by the greenish tinge which it gives to the fluids and tissues, is injected into the body, into the subcutaneous connective tissue or peritoneal cavity for instance, not only does it speedily appear in the aqueous humour, but the ciliary processes are said to be the parts of the eye in which its presence may be first detected. It may be urged that, unlike the epithelium covering the choroid plexuses, the pars ciliaris retinæ bears no distinctive histological indications of secretory activity; but, as we shall presently have occasion to point out, a wholly analogous layer of epithelium, that lining the cavities of the internal ear, though possessing no marked secretory features, certainly furnishes, by an act very similar to secretion, a more or less lymph-like fluid, the so-called endo-lymph. The phrase 'secretion' however must not be strained. The somewhat specialized loose stroma of both the ciliary processes and iris undoubtedly contains in its meshes a large quantity of what we may suppose to be ordinary lymph; and what is intended by the above view is that while some of this lymph may pass by the perichoroidal spaces and so away as ordinary lymph, a much larger proportion passes on to the free surfaces abutting on the posterior and anterior chambers, and in so passing becomes modified in nature.

The fluid thus furnished by the ciliary processes makes its way, in the first place, into the posterior chamber; but though the iris, as we have seen (§ 722) lies close on the lens, there is undoubtedly a communication between the two chambers sufficiently free to allow fluid to pass readily from one to the other and so to fill the anterior chamber from the posterior. It is difficult to suppose that some of the lymph with which the sponge-like stroma of the iris is laden, does not find its way direct through the anterior surface of the iris into the anterior chamber; and such a transit would probably be assisted by the continual changes of the pupil. On the other hand the extent of surface furnished by the ciliary processes, which moreover also have the advantages of movement in each act of accommodation, is very large compared with that of the iris; hence we may probably with confidence conclude that the greater part of the aqueous humour is furnished by the ciliary processes, though the iris may contribute. We may add that probably the iridic contribution differs in nature from the rest, since the epithelium which the fluid has to traverse is a thin layer of flat epithelioid plates.

The answer to the question, How does the aqueous humour leave the anterior chamber? presents perhaps less difficulties. As we have seen (§ 717), the anterior chamber at the 'iridic angle' communicates freely with the spaces of Fontana, and these with the canal of Schlemm, which in turn is in direct connection with the radicles of the anterior ciliary veins. Since the ciliary muscle pulls on the tissue surrounding the canal of Schlemm it is possible, or even probable that the movements of accommodation help alternately to close and open the canal, and thus to pump its contents into the veins; by this means the exit of fluid from the anterior chamber is rendered less dependent on the relative pressures of the blood in the vein and of the fluid in the anterior chamber. By this channel the aqueous humour gains a ready, relatively direct, and short access to the blood-stream. And clinical experience shews that if this way be blocked an accumulation of aqueous humour results.

We may conclude then that the aqueous humour is a reservoir intercalated in a stream of a peculiar fluid which is passing from the ciliary processes through the small posterior and larger anterior chamber, the spaces of Fontana and the canal of Schlemm into the venous system. This reservoir on the one hand serves a mechanical purpose in preserving the natural form of the eye and in affording an adequate fluid bed for the movements of the iris, and on the other hand, by bringing new food material and carrying away waste products, enables the lens to carry out the slow and scanty metabolism necessary for its life.

§ 805. For mechanical purposes the due condition of the vitreous humour is perhaps even more important than that of the aqueous humour. We have already (§ 720) called attention to the fact that the vitreous humour in spite of its being originally a plug of mesoblastic tissue, in adult life closely resembles the aqueous humour in its chemical features; and indeed it is practically an attenuated mesoblastic sponge through which is continually streaming, though at a low rate, a fluid identical with or exceeding like to the aqueous humour. Through the optic disc the fluid of the vitreous humour has access to the lymph-spaces of the optic nerve; material injected into the pial sheath of the optic nerve finds its way through the optic disc into the vitreous humour passing along a 'central canal,' 'hyaloid canal' which remains after the disappearance of the prolongation of the arteria centralis retinæ (§ 703). And probably some of the fluid of the vitreous humour finds its way by this path into the subarachnoid space.

But the greater part of the fluid of the vitreous humour seems to belong to the same system as the aqueous humour. Fluids pass readily in some way or other through the suspensory ligament; fluid injected into the vitreous humour finds its way into the anterior chamber, and a block at the iridic angle leads to undue distension, not of the anterior and posterior chambers only, but of the whole globe of the eye; the pressures of the aqueous and vitreous humour are the same and vary similarly and concurrently. We have no satisfactory evidence that any large amount of fluid passes direct from the choroid through the retina, past the internal limiting and hyaloid membranes into the vitreous humour; as far as we know the whole of the lymph of the retina is carried away by the optic nerve in the manner mentioned above; and we must therefore conclude that the region of the zonule of Zinn serves as the door both for the entrance and exit of fluid, the circulation through the vitreous humour between its indistinct concentric lamellae being secured by diffusion assisted by the movements of the eyeball.

This important flow of what we may call modified lymph like that of the more ordinary lymph in other parts of the body, is determined in the first instance by the blood flow, and we may apply to the eye the remarks which were made when (§ 302) we treated generally of the relations of lymph to blood-supply. Broadly speaking the intraocular pressure rises and falls with the general blood-pressure; the dim cornea and sunk eye that betoken the approaching end are due to the fall of blood-pressure which accompanies death. A local fall, preceded by a transient rise, may be brought about by stimulation of the cervical sympathetic, and a local rise by stimulation of the ophthalmic branch of the fifth nerve, stimulation of the third nerve having apparently little effect in either direction. We may add that, tempting as the view may seem that the lymph arrangements of the eye are under the direct control of the nervous system, we have no evidence that such is the case.

Concerning the influence of the nervous system on the general nutrition of the eye, and the disorders which follow upon section or injury to the fifth nerve we have already, in an earlier part of the work (§ 549), said all that at present we have to say.

SEC. 14. THE PROTECTIVE MECHANISMS OF THE EYE.

§ 806. The eye is protected by the two eyelids, each of which is strengthened and rendered firm by a curved plate of dense connective tissue called the tarsus, (or incorrectly the tarsal cartilage), which is larger in the upper than in the lower eyelid. Elevation of the upper eyelid assisted by some depression of the lower eyelid is spoken of as "opening the eye"; depression of the upper eyelid assisted by elevation of the lower eyelid is spoken of as "shutting the eye." The latter movement is brought about by the contraction of the *orbicularis oculi*, a muscle of circularly disposed striated fibres placed beneath the skin of each eyelid and stretching also over the adjoining bony orbit. The muscle is governed by a branch of the seventh, facial nerve, and may be thrown into action as part of a reflex act or of a voluntary effort. When the facial nerve becomes incapable, through injury or disease, of carrying motor impulses, the eye can not be shut and remains widely open. There are some reasons however for thinking that the motor fibres for the orbicularis, though forming part of the facial nerve outside the brain, take origin within the brain, not from the facial nucleus but from the hind end of the third, oculo-motor nucleus. In the reflex contraction of the orbicularis, known as 'winking' or 'blinking,' which is so familiar as an almost typical reflex movement, but which in the waking hours is repeated so regularly, twice a minute or so, as to take on almost the characters of a rhythmic automatic act, the exciting afferent impulses are carried along the fibres of the fifth nerve distributed to the cornea and conjunctiva, and probably, but not certainly, pass some way down the ascending root (§ 621) of that nerve.

The eye is opened mainly by the raising of the upper eyelid through the contraction of the *levator palpebrae superioris*. This muscle, taking origin from the back of the orbit in company with the ocular muscles, is inserted into the upper surface of the tarsus of the upper eyelid, beneath the orbicularis. It is governed by a branch of the third nerve; hence injury or disease of this nerve is frequently the cause of a drooping of the upper eyelid and an inability to open the eye fully.

A portion of the tendon of the levator palpebrae closely united with an extension of the tendon of the superior rectus is inserted into the hinder part of the upper eyelid, where the conjunctiva lining it is about to be reflected over the eyeball; and a similar extension of the inferior rectus is similarly inserted into the lower eyelid. Hence a contraction of the superior rectus, while elevating the visual axis, at the same time raises somewhat the upper eyelid; and in like manner the inferior rectus, while depressing the visual axis, lowers the lower eyelid.

Between the main tendon of the levator palpebrae and the tendinous slip just mentioned lies a small bundle of plain, unstriated muscular fibres, which starting from the levator, ends in the hind border of the tarsus; it is sometimes spoken of as the middle insertion of the levator. A similar bundle of plain muscular fibres connects the insertion of the inferior rectus with the tarsus of the lower eyelid. These two small plain unstriated muscles appear to be governed by nervous filaments proceeding from the cervical sympathetic, stimulation of the cervical sympathetic leading to contraction of these muscles and so to a partial opening of the eye, and section of the same nerve preventing their being thrown into contraction and so contributing to closure of the eyelids. In some of the lower animals this closure of the eye upon section and opening upon stimulation of the cervical sympathetic is very distinct. In those animals which possess a third eyelid this is retracted by stimulation and comes forward upon section of the cervical sympathetic.

Stimulation of the cervical sympathetic also causes some protrusion and section causes recession of the whole eyeball; this is seen at times in man in disease.

§ 807. The conjunctiva which lines the ocular surface of the eyelids and is reflected from them over the eyeball, the line along which reflection takes place being spoken of as the *fornix conjunctivae*, consists like the skin of the body of which it is a continuation, of an epithelium or epidermis resting on a dermis of connective tissue. It differs from the skin in the dermis being delicate and in the epidermis being thin with a tendency for the constituent cells to become columnar; hence it is sometimes spoken of as a "mucous membrane." On the ocular surface of the eyelids the conjunctiva is thrown into irregular ridges or imperfect and fused papillae, giving rise to a satiny appearance, here the epithelium consists of several layers of cells, the uppermost of which are flattened. Over the fornix, the epithelium consists of two or three layers only, the cells in the uppermost layers being cubical or columnar; over the bulb the epithelium consists also of a few layers only, the upper cells being somewhat flattened and the dermis being thrown up into scattered papillae.

Imbedded in the tarsus, stretching from the hind border to the free edge of the lid lies, in each eyelid, a row of thirty or fewer largely developed sebaceous glands (§ 437) the Meibomian glands. Sebaceous glands are also attached to the follicles of the eyelashes, and into the ducts of some of these open the glands of Moll, which have the structure of a sweat gland (§ 436). Small mucous glands are moreover found in the conjunctiva especially in the neighbourhood of the fornix.

These several glands contribute to keep the surface of the eye and eyelids moist; but this is chiefly effected by the secretion of the lachrymal gland which is placed above the upper eyelid in the lateral region of the orbit, and which, imperfectly divided by an extension of the tendon of the levator palpebrarum into two masses, discharges its secretion by several ducts opening along the fornix conjunctivæ. Under ordinary circumstances the fluid thus secreted is carried away through the punctum lachrymale of the upper and of the lower eyelid, at the inner angle of the eye, into the lachrymal canaliculi, and so into the lachrymal sac, and finally into the cavity of the nose. When the secretion becomes too abundant to escape in this way it overflows on to the cheeks in the form of tears.

The structure of the lachrymal gland is in its main features identical with that of an albuminous salivary gland, or with that of the parotid, save that the epithelium of the·ducts is never striated; it will be unnecessary to describe it in detail. In some animals a somewhat peculiar gland, the Harderian gland, lies in the inner (median) region of the orbit; this varies in structure in different animals, being in some a sebaceous gland united with a gland similar in structure to the lachrymal gland.

If a quantity of tears be collected, they are found to form a clear faintly alkaline fluid, in many respects like saliva, containing about 1 p. c. of solids, of which a small part is proteid in nature. Among the salts present sodium chloride is conspicuous.

The nervous mechanism of the secretion of tears, in many respects, resembles that of the secretion of saliva. A flow is usually brought about either in a reflex manner by stimuli applied to the conjunctiva, the nasal mucous membrane, the tongue, and the interior of the mouth, or more directly by emotions. Powerful stimulation of the retina by light will also cause a flow, as will electrical or other stimulation of any of the cranial or upper spinal afferent nerves. Venous congestion of the head is also said to cause a flow. The efferent nerves are the lachrymal and orbital branches of the fifth nerve, especially the former, stimulation of these causing a copious flow. It is said that stimulation of the cervical sympathetic will also cause a somewhat scanty flow of turbid tears, but on this point all observers are not agreed.

The chief use of the act of blinking is to keep the surface of the cornea moist, and so transparent; if the cornea be kept uncovered for a few minutes its dried surface soon becomes dim. But besides this, blinking undoubtedly favours the passage of tears through the lachrymal canaliculi into the lachrymal sac, and hence when the orbicularis is paralysed tears do not pass so readily as usual into the nose; but the exact mechanism by which this is effected has been much disputed. According to some authors, the contraction of the orbicularis presses the fluid onwards out of the canals, which, upon the relaxation of the orbicularis, dilate and receive a fresh quantity. Others maintain that a special arrangement of muscular fibres keeps the canals open even during the closing of the lids, so that the pressure of the contraction of the orbicularis is able to have full effect in driving the tears through the canals.

CHAPTER IV.

HEARING.

SEC. 1. ON THE GENERAL STRUCTURE OF THE EAR, AND ON THE STRUCTURE AND FUNCTIONS OF THE SUBSIDIARY AUDITORY APPARATUS.

§ 808. WE have seen that the eye consists on the one hand of the special modified epithelium, the retina, so constituted that light falling upon it gives rise to visual impulses in the optic nerve and thus to visual sensations in the brain, and on the other hand of a special dioptric mechanism, into the construction of which several tissues enter and which is so arranged as to cause the rays of light to fall in a proper manner on the retina. In the ear we meet with a somewhat similar arrangement; we may recognize on the one hand a specially modified epithelium, which we may call the *auditory epithelium*, so constituted that the vibrations of matter, the rapidly alternating variations of pressure, which we call "waves of sound," generate in the auditory nerve connected with it, auditory impulses, developed in the brain into auditory sensations, and on the other hand an acoustic apparatus so arranged that waves of sound are conducted in a proper manner to the auditory epithelium. Just as visual impulses can be excited by light only through the mediation of the retina, so auditory impulses can be excited by sound only through the mediation of the auditory epithelium; but here the analogy between the optic auditory nerves seems to end, for while as we have seen the optic nerve conveys, so far as we know, visual impulses only, we have reason to think (§ 642) that some fibres at least of the auditory nerve convey impulses which do not give rise to auditory sensations, but enter in a peculiar manner into the mechanism of coordinated movements.

The retina as we have seen is developed out of the optic vesicle, and the subsidiary dioptric mechanism is built up around the optic vesicle; and in a somewhat similar way the auditory epithelium is developed into an *otic vesicle*, and the subsidiary acoustic apparatus is built up around the otic vesicle. The otic vesicle, like the optic vesicle, is lined by an epithelium of epiblastic origin, but is not like that vesicle budded off from the medullary canal. It takes origin in an involution of the skin covering the head; for a time the epithelium of the vesicle is continuous with the epidermis of the skin, and wholly unconnected with the developing brain; later on the epithelial involution separates from the skin, becomes a closed independent vesicle, and makes connections with the brain through the auditory nerve growing out to meet it. The otic vesicle therefore is not like the optic vesicle a part of the brain, and we find accordingly the structure of the auditory epithelium much more simple than that of the retina; it corresponds only to a part of the retina, to the more external layers of the retina, not to all of them.

We have seen that the optic fibres are connected with a part only of the optic vesicle, with the anterior wall only of the retinal cup and not with the whole of this; the part of the anterior wall which forms the pars ciliaris retinæ and the whole of the posterior wall make no connections with the optic fibres and remain in the form of a relatively simple epithelium. The connection of the auditory nerve with the walls of the otic vesicle is still more partial; the nerve fibres become connected with the epithelium in a few limited areas. It is only in these areas that the epithelium lining the otic vesicle becomes differentiated into the special auditory epithelium; elsewhere it possesses relatively simple characters.

The cavity of the optic vesicle is, as we have seen, soon obliterated by the coming together of the anterior and posterior walls. The cavity of the otic vesicle is permanent, growing with the growth of the organ and becoming filled with a peculiar fluid secreted by the walls, called *endolymph*. The vesicle as it grows soon loses its early simple, more or less spherical form and assumes a most complicated shape, becoming divided into the parts known as the utricle with the semicircular canals, the saccule, and the canalis cochlearis; of these we shall speak in detail later on.

§ 809. While the vesicle is assuming this complicated shape, the mesoblastic tissue investing it undergoes a differentiation. The tissue immediately in contact with the epithelium becomes connective tissue serving as a dermis to support the epithelium, so that the vesicle becomes a (complicated) sac with membranous walls lined with epithelium specially modified into auditory epithelium at particular places, at which places and at which places alone, the auditory nerve makes connections with the walls.

The outer portion of the mesoblastic tissue is converted into bone of a somewhat dense character, and thus furnishes a bony shell or envelope enclosing and to a large extent following the contour of the complicated membranous sac. Between the outer

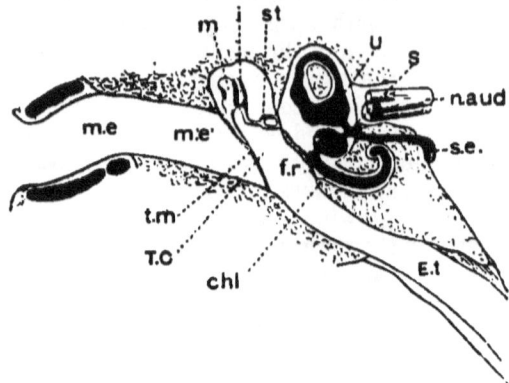

FIG. 162. DIAGRAM TO ILLUSTRATE THE GENERAL STRUCTURE OF THE EAR. (After Schwalbe.)

The figure is purely diagrammatic, intended only to shew in one view all the several important parts in relation to each other; such a view is in the actual ear impossible.

m. e. the external meatus or auditory passage, in the outer part where the walls are cartilaginous, *m'. e'.* the same in the inner part where the walls are osseous.

T. C. the tympanic cavity. *t. m.* the tympanic membrane. *m.* malleus, *i.* incus, *st.* stapes, attached to the fenestra ovalis. *f. r.* fenestra rotunda. *E. t.* Eustachian tube.

U. the utricle with the perilymph space around. One semicircular canal with its ampulla is shewn, with the bony core of the hoop. *S.* Saccule. *s. e.* sacculus endolymphaticus lying within the cranial cavity, and connected by the ductus endolymphaticus with both saccule and utricle. *chl.* the canalis cochlearis, connected with the saccule by the canalis reuniens, and surrounded by its perilymph space, scala vestibuli, and scala tympani, the latter ending at the fenestra rotunda, the former continuous with the perilymph space of the vestibule around the utricle and saccule, the cochlea is shewn diagrammatically as a simple curve, the scala vestibuli and scala tympani being continuous at the top.

N aud., the auditory nerve shewing the three main divisions of the trunk.

bony envelope and the inner membranous sac is developed a large irregular lymphatic space which (Fig. 162) follows to a great extent the contour of the sac, but is broken up by broad adhesions of the membranous sac to the periosteum lining the bony envelope or by narrower bridles of connective tissue crossing the space; some of these form beds for the branches of the auditory nerve on their way to the auditory epithelium. The fluid in this space, which is lymph and which has access to the lymphatics of neighbouring parts, receives the special name of *perilymph*. A portion of the sac, with its surrounding perilymph space and bony envelope, undergoes a development differing materially from that of the rest of the sac, and is known as the *cochlea*. The bony envelope surrounding the parts of the membranous sac known as the utricle and saccule does not follow closely the contour of those parts but remains as an undivided part called the *vestibule* (Fig. 163); the parts of the membranous sac called the semicircular canals are however followed somewhat closely by the bony envelope. The whole bony envelope may be dissected out from

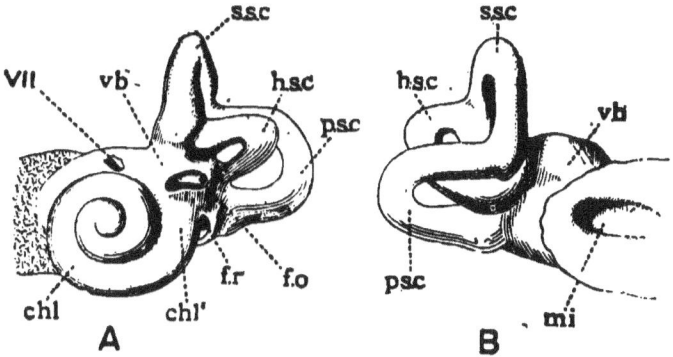

Fig. 163. The Bony Labyrinth. Left Ear. (Schwalbe.)

A. seen from the outside. B. seen from the median side. Both magnified twice.
Vb. vestibule. *Chl*. cochlea. *Chl'*. the beginning of the first turn of the cochlea. *F. o*. fenestra ovalis, *f. r*. fenestra rotunda, *s.s.c*. superior, *p.s.c*. posterior, *h.s.c* horizontal semicircular canals. *m.i*. meatus auditorius internus, canal for the auditory nerve. VII. opening of the canal containing the seventh nerve.

the spongy bone surrounding it, and may be obtained as a separate mass (Fig. 163), known by the name of the *labyrinth*, or *bony labyrinth* to distinguish it from the *membranous labyrinth* which lies within it, separated from it by the perilymph space. The bony labyrinth consists of cochlea, vestibule and semicircular canals, but the part of the membranous labyrinth corresponding to the vestibule is divided into utricle and saccule. The auditory nerve pierces the bony labyrinth at the so-called *meatus auditorius internus* (Fig. 163 *m. i*.) on its way to be distributed to the walls of the membranous sac.

All these structures, lying at first not far beneath the skin and forming together the 'internal ear,' as they grow come into close connection with a passage on the side of the head leading from the exterior into the pharynx and known as the "first" or "hyomandibular visceral cleft." By a series of changes, which we need not describe here, and indeed about which there is some divergence of opinion, this simple primitive passage is replaced in the adult by two passages separated from each other by a partition known as the *membrana tympani* (Fig. 162 *t. m*.) or tympanic membrane. On the outer side of the membrane lies a tubular channel, the *external auditory meatus* (*m.e.*, *m.'e.'*), lined by skin, and opening on to the exterior by an orifice guarded with the "pinna" or "auricle." On the inner side of the membrane lies the drum-shaped *tympanic cavity* (*T. C.*), often called the "middle ear," which through the tubular *Eustachian tube* (*E. t.*) opens into the pharynx, and which is lined throughout by mucous membrane continuous with that of the pharynx.

The 'internal ear' forms the mesial side of the more or less flattened and drum-shaped tympanic cavity opposite to the outer side which is to a large extent formed by the tympanic membrane; and at two places the osseous tissue of the bony envelope of the internal ear is wanting, the gaps giving rise to what in the dried skull appear as two foramina, but in the fresh state are two membranous fenestræ. One of these, oval in shape, called the *fenestra ovalis* (Figs. 162, 166, 168 *f. o.*), lies between the tympanic cavity on the outside and that part of the perilymph space which surrounds the division of the membranous labyrinth known as the utricle on the inside; in the dried bony labyrinth (Fig. 163 *F.o.*) it appears as a hole in the vestibule. The other, round in shape, called the *fenestra rotunda* (Figs. 162, 166 *f. r.*) lies between the tympanic cavity and a part of the perilymph space which enters into the construction of the cochlea; as we shall see, the perilymph space of the cochlea may be regarded as a peculiar tubular prolongation of that of the vestibule, and the membrane of the fenestra rotunda closes as it were the end of this prolongation.

Certain bones of the skull, converted by striking developmental changes into a jointed chain of minute bones, *the auditory ossicles* (Fig. 162 *m. i. st.*), are by processes of growth thrust into the tympanic cavity in such a way that they eventually seem to lie wholly in the cavity, and to form a bridge across the cavity between the tympanic membrane on the outer side, and the fenesta ovalis on the mesial side. The ossicles are three in number; to the tympanic membrane is attached the *malleus;* this is joined to the *incus,* which in turn is joined to the *stapes,* the end of which is attached to the fenestra ovalis. Into the details of these ossicles we shall enter presently.

§ 810. The affections of consciousness, which we call sensations of sound, are the result of auditory impulses reaching certain parts of the brain along the auditory nerve; and these auditory impulses are generated through vibrations, or rhythmically repeated variations of pressure which we call, 'waves of sound,' in some way or other acting upon the terminations of the auditory fibres in the auditory epithelium. The waves of sound gain access to the epithelium by means of the perilymph, passing probably in some parts directly through the dermis of the membranous sac to the overlying epithelium, and being in other parts transmitted to the endolymph from the perilymph across the membranous walls, and acting on the epithelium through the endolymph.

Waves of sound may be and to a certain extent are conducted in a direct manner to the perilymph, through the tissues, especially the harder bony tissues, of the head, reaching the perilymph across its bony envelope. The vast majority however of the waves of sound which fall upon the head travel through the medium of the air, and in order to reach the perilymph have to pass from a gaseous medium, the air, into the solid and liquid media of the head. Now the vibrations of particles constituting waves of sound are not readily communicated from a gaseous to a liquid or solid medium; special conditions are required to effect this. The transference of sound from the air to the perilymph is attended with considerable difficulty; and the parts of the ear which we have spoken of above as constituting the middle and outer ear, serve as an acoustic apparatus for facilitating this transference and thus bringing the aerial waves to act on the auditory epithelium, the action of the apparatus being somewhat as follows.

Waves of sound falling on the side of the head reach the tympanic membrane by the external meatus, and throw that membrane into vibrations. These vibrations are transmitted through the chain of ossicles to the membrane of the fenestra ovalis and so to the perilymph lying on its far side; sweeping over the perilymph in its continuous cavity the waves eventually break upon the fenestra rotunda, having on their way affected the auditory epithelium. We have first to inquire how this subsidiary acoustic apparatus performs its work.

The Conducting Apparatus of the Tympanum.

§ 811. *The auditory ossicles.* The *malleus*, or hammer bone (Fig. 164 A and D), has a rounded head (*cp.*) bearing a peculiar saddle-shaped surface for articulation with the incus, and ends below in a tapering process, the *manubrium*, or handle (*mbr.*) by which it is attached, in a manner to be described presently, to the inner surface of the tympanic membrane. To the handle at its upper part is attached on the inner side the tendon of the tensor tympani muscle (Figs. 167, 170, 173 *T.T.*); and at the top of the handle, on the outer side, is a short blunt process, *processus brevis*, (Fig. 164 A *p.b.*) which as we shall see abuts on a particular part of the tympanic membrane. Still higher up is a thinner and generally much longer process (Fig. 164 A *p.f.* and Fig. 173 *p.f.*), *processus gracilis* or *Folianus*, the base of which with part of the thick neck of the malleus above serves for the attachment of ligaments, and the end of which is inserted into a fissure in the bony wall of the cavity.

The *incus*, *ambos* or anvil bone (Fig. 164 B and D) has a less well defined head, bearing a surface for the articulation with the malleus, and a short body which immediately divides into two processes at right angles to each other. One of these, the "short process" ($p'.\ b'.$), takes up a horizontal position (Fig. 173 $p'.\ b'.$) and is attached to the wall of the tympanum by a ligament (Fig. 171 $lg.\ inc.$). The other "long process" (Figs. 164, 170, 173 $p'.\ l'.$) or

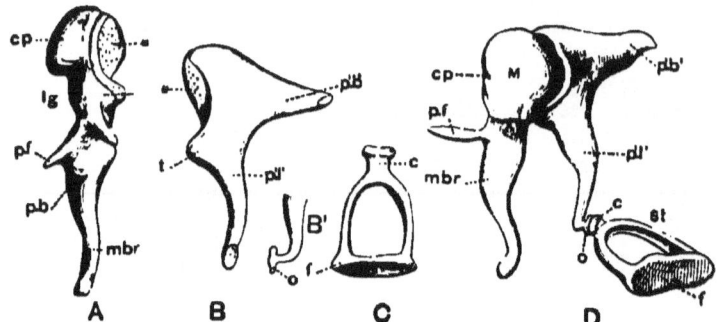

Fig. 164. THE AUDITORY OSSICLES. (After Schwalbe and Helmholtz.) Magnified four times.

A. *The malleus.* $cp.$ the head (caput). *the articulating surface for the incus, $t.$ Tooth locking with tooth of incus. $lg.$ is placed opposite the attachment of the ligaments. $p.f.$ processus gracilis or Folianus, represented as short. $p.\ b.$ processus brevis. $m.\ br.$ handle (manubrium.)

B. *The incus.* * surface articulating with malleus. $t.$ tooth locking with tooth of malleus. $p'.\ b'.$ processus brevis. $p'.\ l'.$ processus longus.

B'. The lower end of the processus longus seen sideway; $o,$ its expanded termination or os orbiculare.

C. *The stapes.* $c.$ the head. $f.$ the foot-plate.

D. The three ossicles in connection. $M.$ malleus, $I,$ incus, $st.$ stapes; the other letters as above.

'shaft,' tapering and somewhat curved, takes up a vertical position parallel to the handle of the malleus, but at its end makes a sudden bend mesially towards the internal ear (Fig. 162) and terminates in a flattened knob, with which it is articulated to the stapes (Fig. 164 D). This knob, having frequently at least an independent ossification, is sometimes spoken of as the *os orbiculare* or *lenticulare*.

The *stapes* or stirrup bone (Fig. 164 C and D), which is placed horizontally at right angles to the shaft of the incus, consists of a head (c) articulating with the extremity of the shaft of the incus, and an oval foot-plate (f), attached in a manner, which we shall presently describe, to the fenestra ovalis, the two being united by curved limbs after the fashion of a stirrup.

§ 812. The tympanum, in which these ossicles lie, may be compared to a low drum placed obliquely at the end of the external meatus. The outer lateral side or surface of the drum is furnished by the tympanic membrane, the inner mesial side or surface by the bony labyrinth; but the two sides are not exactly parallel,

HEARING.

and the mesial side is much broken up by irregularities. The cavity of the drum is not completely circumscribed by a circular

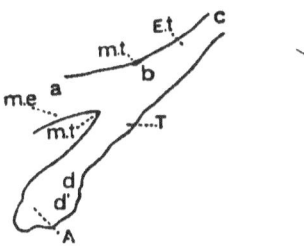

FIG. 165. DIAGRAM TO ILLUSTRATE THE RELATIONS OF AUDITORY PASSAGE, TYMPANUM AND EUSTACHIAN TUBE. (After Schwalbe.)

The figure represents a section not quite horizontal, being inclined downwards in front; the right-hand edge of the page may be taken to represent the median plane of the head.

m. e. external meatus, *T.* the tympanic cavity. *E.t.* the Eustachian tube. *A.* is placed in the antrum mastoideum. *m.t.* indicates the attachment of the tympanic membrane.

a. b. the axis of the external meatus, *c.b.d.* that of the Eustachian tube. *dd'.* shews the curved axis of the antrum.

wall like that of a drum proper. In about the lower half of the cavity (Figs. 166, 167) the wall though irregular is complete; but the upper half of the cavity is continuous along an axis oblique to the axis of the meatus (Fig. 165) with two more or less tubular cavities, one stretching upwards and backwards (Fig. 165, *dd.'*), the other downwards and forwards (Figs. 165, 166, 167 *E.t.*). The latter, distinctly tubular, is the Eustachian tube leading from

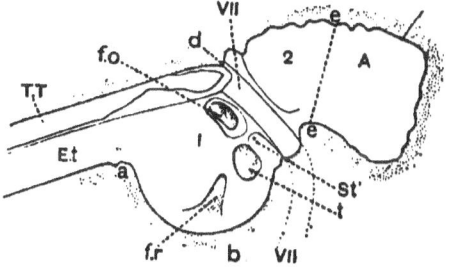

FIG. 166. DIAGRAM OF THE MEDIAN WALL OF THE TYMPANUM OF THE LEFT EAR Magnified twice. (After Schwalbe.)

1 The tympanic, 2. the epitympanic region; below the reference figure is seen the gentle prominence due to the ampullæ. *A*, the antrum mastoideum, the line *ee* marking its limits. *E. t.* the Eustachian tube. *T.T.* the groove for the tensor tympani. *f. o* the depression of the fenestra ovalis, the fenestra itself being shaded. *f. r* the depression leading to the fenestra rotunda; above, and obliquely to the left of this, lies the projection caused by the base of the cochlea. *St.* the prominence for the stapedius, with the orifice for the exit of the tendon. VII, the course of the facial nerve. The tympanum proper lies within the letters *a. b. d. e.*

the upper front part of the tympanum obliquely forwards, downwards and towards the median plane of the head into the pharynx. The former continues the upper hind part of the tympanum upwards, backwards and away from the median line of the head, first as an irregular space, sometimes called the "epitympanic region," and then farther backwards as a larger space, the *antrum mastoideum* (Fig. 166 A), which in turn communicates with the labyrinth of spaces or "air cells" of the mastoid bone.

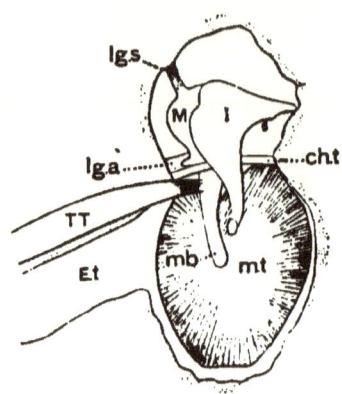

Fig. 167. Diagram of the outer wall of the Tympanum (Right Ear) as seen from the mesial side. Magnified twice. (After Schwalbe.)

m.t. membrana tympani. *mb.* handle of *M* the malleus. *I.* the incus. *E. t.* Eustachian tube. *T. T.* tensor tympani, the tendon of which is seen attached to the upper part of the handle of the malleus. *lg. a.* the anterior and *lg. s.* the superior ligament of the malleus. *ch. t.* the chorda tympani nerve traversing the tympanic cavity.

The ossicles are placed more or less vertically but yet obliquely in the tympanic cavity in such a way that the heads and bodies of the malleus and incus lie above the tympanum proper in the epitympanic region (Fig. 167), but the handle of the malleus and the shaft of the incus descend to the centre of the tympanum. Opposite but rather above this centre is seen in the median wall of the tympanum a funnel-shaped depression (Figs. 166, 168 *f. o.*) at the bottom of which lies the fenestra ovalis; and to this the stapes, horizontal but slightly inclined upwards, passes from the end of the shaft of the incus.

In the lower part of the median wall, some distance below the fenestra ovalis, is seen an irregular depression (Fig. 166 *f.r.*) which leads to the fenestra rotunda; and by the side and above this, occupying the central portion of the median wall of the tympanum proper is a projection marking the position of the first whorl or base of the cochlea. The fenestra ovalis itself marks the position of the junction of the utricle and saccule, and in the epitympanic region, above a rounded ridge caused by the projection of the Fallopian canal carrying the seventh or facial nerve (Fig. 166 VII.) is a protuberance marking the position of the swollen ends

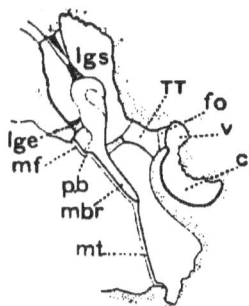

FIG. 168. FRONTAL (TRANSVERSE VERTICAL) SECTION THROUGH THE TYMPANUM. (LEFT EAR.) (Schwalbe.)

The figure, partly diagrammatic, is magnified twice, and shews the front part of the tympanum as seen from behind; the incus has been removed, the articular surface on the head of the malleus being indicated.

mt. The membrana tympani. *mf.* membrana flaccida. *mbr.* handle of the malleus. *p. b.* short process of the malleus.

lg. e. external ligament, *lg.s.* the superior ligament of the malleus.

TT. The bony projection from which the tendon of the tensor tympani passes to the malleus. *f. o.* the fenestra ovalis. *v.* the front part of the vestibule. *c.* the beginning of the first (basal) turn of the cochlea.

or ampullae of two of the semicircular canals, namely those known as the horizontal and superior.

The distance across the tympanum between the beginning of the chain of ossicles at the point of the handle of the malleus and its end at the fenestra ovalis is very short, much shorter than the length of the chain itself; the greater part of the chain, including the bodies and processes of, that is to say nearly the whole of, the malleus and incus, owing to the peculiar form and articulation of the ossicles, lies above a line drawn from the point of the handle of the malleus to the fenestra ovalis. Cf. Figs. 167, 168. We must now turn to the details of the manner in which the ossicles are attached to the membrana tympani, to each other, and to the fenestra ovalis.

§ 813. The *membrana tympani* (Fig. 169) irregularly elliptical in form with the long axis vertical is placed obliquely (Figs. 162, 168 *m. t.*), at the inner end of the external meatus. Nearly the whole of the circumference of the membrane is fixed in a groove of the ring-shaped bone, the annulus tympanicus. At the extreme top the ring is wanting; and the portion of the membrane thus attached, not to the ring but to the bone above, being less tense than the rest of the membrane, and indeed thrown into folds, is distinguished as the *membrana flaccida* (Fig. 169 *m. f.*). The short process of the malleus abuts against this part of the membrane (Fig. 168 *p. b.*), and when the membrane is viewed from the outside, as in looking down the meatus, is seen shining through it (Fig. 169 *p. b.*).

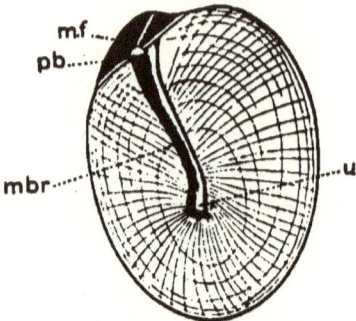

Fig. 169   The Membrana Tympani. (After Schwalbe.) (Magnified four times.)

The membrane is seen from the external meatus and the handle of the malleus, *mbr*, is represented as shining through. *m. f.* the membrana flaccida, the folds of which are represented radiating from *p. b.*, the projection outwards caused by the end of the short process of the malleus. *u* the umbo of the membrane, to which is attached the end of the handle of the malleus. The figure shews diagrammatically, the radial and circular fibres of the membrane.

This larger tenser part of the membrane forms a shallow funnel, the apex of which, called the *umbo* (Fig. 169, 170 *u*), projects into the tympanic cavity; and the handle of the malleus is attached to this part of the membrane on its inner side in such a way that the umbo is supported by the tip of the handle. The umbo is somewhat eccentric in position, lying nearer the bottom than the top; and the sides of the funnel are not flat but slightly convex towards the meatus, though not equally so in all parts.

The membrane consists of a basis of connective tissue, *membrana propria*, covered on the outside by a continuation of the skin of the meatus, and on the inside by the mucous membrane lining the tympanum. The connective tissue basis, which is absent from the small flaccid part of the membrane, consists of bundles of connective tissue, somewhat peculiar in appearance, but yet ordinary fibrillated, inelastic gelatiniferous connective tissue, arranged in an outer layer of radiating bundles, and a thinner less complete inner layer of circularly disposed bundles; both layers, especially the circular, are thinner towards the centre than at the circumference. The handle of the malleus is imbedded in, and wrapped round by the bundles of this connective tissue, the radial bundles radiating from the umbo, or in the upper part of the membrane diverging by the sides of the handle.

The skin covering the outer side of the membrane is ordinary

skin consisting of dermis and a rather thin epidermis, in which the
distinction between the malpighian and corneous layers is not

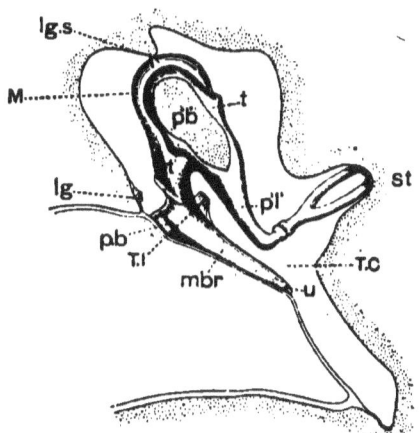

FIG. 170. THE OSSICLES IN POSITION. Magnified four times. (After Hensen.)
The figure represents a section through tympanum in the line of the long axis of the
malleus and incus; the short process of the incus $p'b'$ has been cut through.

T.C. The tympanic cavity, *mbr.* handle of malleus. *u.* umbo. *p.b.* short process
of the malleus shewn in dotted outline as pushing outwards the membrana
flaccida. *T.T.* the attachment of the tendon of the tensor tympani. *lg.* the
attachment of the external ligament of the malleus. *lg.s.* the superior ligament
of the malleus. *t.t.* the teeth of the incus. *p'l'.* the long process, shaft, of the
incus. *St.* the stapes.

very sharp. Blood vessels and a nerve (nervus membranæ tympani, a branch of the auriculo-temporal) run in the dermis. The mucous membrane, lining the inner surface of the membrane, consists of a single layer of flattened non-ciliated epithelium cells lying on a dermis in which is much reticular connective tissue. It will be understood that this mucous membrane is continued over the handle of the malleus, as indeed over the rest of the ossicles.

§ 814. The handle of the malleus being thus firmly imbedded in the substance of the tympanic membrane moves with every movement of it; the attachment of the short process to the flaccid part of the membrane is of a looser character. Besides this attachment to the tympanic membrane the malleus is further bound to the wall of the tympanum by three ligaments of connective tissue. One, the superior ligament (Figs. 167, 168, 170, *lg.s.*) descends from the upper wall of the tympanum to the head of the malleus, whose movements it thus steadies. More important however than this are the other two ligaments which pass more or less horizontally from the outer wall of the tympanum above the tympanic membrane, to the neck of the malleus. One placed anteriorly (Figs. 167, 171 *lg.a.*), the anterior ligament, embraces the long or Folian process, the other, the exterior ligament

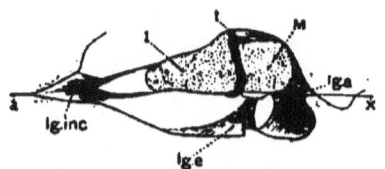

FIG. 171. THE LIGAMENTS OF THE OSSICLES. (After Hensen.)

The figure represents a nearly horizontal section of the tympanum, carried through the heads of the malleus and incus.

M. malleus. *I.* incus. *t.* articular tooth of incus. *lg. a.* anterior and *lg. e.* external ligament of the malleus. *lg. inc.* ligament of the incus.
The line *ax* represents the axis of rotation of the two ossicles.

(Fig. 171 *lg.e.*), is placed more externally; the two are nearly continuous, leaving however a distinct gap (Fig. 171) between them. They serve to limit the movements of and so to keep in place the head of the malleus, which it is said still remains in position even after the incus has been removed.

The joint between the malleus and incus, which like other joints has articular cartilages, synovial membrane, a capsule and ligaments, the latter being very slender, is of a peculiar shape, the lower part of the articular surface of each bone projecting in the form of blunt teeth (Figs. 164, 170, 171, 173, *t.*). These teeth lock into each other in such a way that when by an inward movement of the tympanic membrane the malleus is carried inwards, the incus is necessarily carried inwards also, but that when the malleus is moved outwards, that is towards the external passage, the incus does not necessarily follow it. Hence while every inward movement of the tympanic membrane leads to an inward movement of the malleus, incus and stapes, in succession, the three falling back into their previous positions when the movement ceases, should the tympanic membrane for any reason be pushed unduly outwards into the meatus, the joint between the malleus and incus gapes and so prevents the stapes being pulled out of the fenestra ovalis.

A ligament, ligament of the incus (Fig. 171 *lg. inc.*), more or less divisible into two parts, passing from the median wall of the tympanum to the end of the short process of the incus, firmly secures that part of the chain of ossicles. The long process of the incus hangs nearly vertically downwards but its end turning sharply round at right angles expands into a flattened knob, which, covered with cartilages, forms a joint with the cartilage covered head of the stapes (Figs. 164 B′ and D, 172).

The foot of the stapes (Fig. 172), an irregularly oval plate of bone, covered on the side towards the internal ear with cartilage, has its long diameter, about 3 mm., placed horizontally, the vertical diameter being about 1·5 mm. It corresponds in form, but is

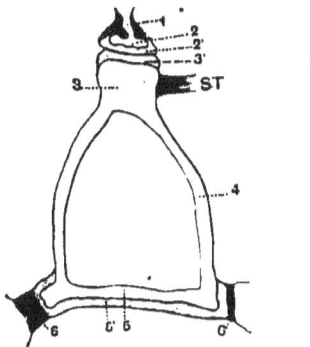

FIG. 172. THE STAPES IN POSITION. Much magnified. (Schwalbe.)
1. The end of the shaft of the incus. 2. Its expansion or os orbiculare. 2'. The articular cartilage of the same. 3. The capitulum of the stapes; 3'. Its articular cartilage. 4. The hoops of the stapes. 5. The foot-plate of the stapes. 5'. Its articular cartilage. 6. The membrane of the fenestra ovalis. *ST*. The tendon of the stapedius muscle attached to the capitulum of the stapes.

rather smaller than the fenestra, and between its cartilaginous rim and the cartilage-lined rim of the fenestra is attached a ring-shaped membrane (Fig. 172, 6), consisting of radially disposed bundles of connective tissue with which many elastic fibres are mixed. The ring though slightly broader in the front part than in the hind part of the oval is very narrow, at most 100 μ; hence the movements of the stapes within the fenestra are very limited in extent, and probably do not exceed a small fraction ($\frac{1}{18}$ to $\frac{1}{14}$) of a millimeter. The tympanic surface of the fenestra and included stapes is covered with a continuation of the mucous membrane which covers the tympanic membrane, and which not only lines the whole tympanic cavity but is also reflected over the whole chain of ossicles; the other surface of the stapes, that which forms part of the perilymph space of the labyrinth, is lined like the rest of the cavity with a lymphatic epithelium.

§ 815. The chain of ossicles, thus jointed together, attached to the tympanic membrane at one end, and to the fenestra ovalis at the other, and secured by ligaments, may be regarded as a lever. Observations and experiments shew that the end of the short process of the incus serves as the fulcrum, the power being applied at the umbo in which the handle of the malleus ends, and the effect being brought to bear on the end of the long process of the incus attached to the stapes. In thus acting as a lever the heads of the malleus and incus rotate round a horizontal line drawn through them in the direction of the line ax in Fig. 171. Such a lever may be represented by the line xx' in Fig. 173.

Careful measurements shew that the whole length of the line from F the fulcrum to P, where the power is applied, is about

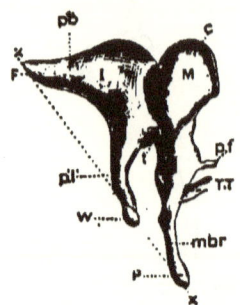

FIG. 173. THE MALLEUS AND INCUS, IN POSITION. (Helmholtz.)

M. The malleus. *c*, the head, *mbr*, the handle, *p.f*, processes Folianus. *T. T.* the tendon of the tensor tympani.

I. The Incus. *p'b'*, the short process, *p'l'*, the long process, *t.* tooth locking with the malleus

The line XX represents the lever formed by the two ossicles, with, F, the fulcrum at the attachment of the short process of the incus, P, the point where the power is applied at the end of the handle of the malleus, W, the point where the effect is produced at the os orbiculare of the incus.

9·5 mm., while the length from F to W, where the effect is brought to bear, is about 6·3 mm. Hence when the tympanic membrane is driven inwards, the corresponding inward movement of the stapes in the fenestra is as far as extent is concerned only about two-thirds of that of the tympanic membrane. By the principle of the lever however the amount of pressure exerted by the movement of the stapes, the force of the movement, is one and a half times greater than the force expended in producing the movement of the tympanic membrane. The arrangement of the lever of ossicles therefore is such as to convert a relatively large movement into a smaller movement of greater intensity; the benefit of such a conversion is obvious.

The conduction of sound through the Tympanum.

§ 816. The conduction of sound from the external air to the labyrinth takes place by means of the tympanic membrane and the chain of ossicles acting as a lever in the manner just described.

Stretched membranes have the property of being readily thrown into vibrations by aerial waves of sound, and of transmitting the vibrations to bodies in contact with themselves. The tympanic membrane is a stretched membrane which, by its size, nature and conformation is specially adapted to take up and transmit a great variety of vibrations. Sound is a vibration of the particles of matter, a series of movements of the particles from and to a fixed point. In air and other gases the movements of the particles lead to alternating condensation and rarefaction of the medium, the sound is propagated as waves of alternating condensation and rarefaction, which since the to-and-fro movement of the particles is in the same direction as that in which the undulations are travelling, are spoken of as 'longitudinal' waves. In liquids the transmission of sound also takes place by longitudinal waves of alternating condensation and rarefaction, and sound may travel through solids in the same way. But solids in the form of membranes or plates, strings, and rods may also give rise to sounds by being thrown into bodily vibrations, a rod for instance bending alternately to-and-fro in rapid succession. In such a case the particles of the rod move sensibly in a direction transverse to the long axis of the rod; and the vibrations of this kind, thus giving rise to sounds, are spoken of as "transversal" vibrations. It will be understood that a rod, membrane, plate or string, may also be the subject of longitudinal vibrations; but the sound given out by such longitudinal vibrations differs from that given out by transversal vibrations of the same body. A rod, string, or membrane thrown into sufficiently rapid and strong transversal vibrations, will communicate its vibrations to the surrounding air, and so give forth a sound, which will travel through the air in the form of waves of longitudinal vibrations. Conversely, sound travelling through the air in waves of longitudinal vibrations, and striking upon a rod, string or membrane, may throw it into transversal vibrations. And this is what takes places in the ear. Aerial waves of sound, in the form of longitudinal vibrations, alternating condensations and rarefaction, of the air, travelling along the meatus, fall upon the tympanic membrane, and throw it into transversal vibrations; the membrane bends bodily inwards and outwards in time with the condensations and rarefactions of the air in the meatus on its outer surface.

The vibrations of a rod, a tuning-fork for example, are comparatively simple in character; and we find, correspondingly, that a tuning-fork is very limited in its power of 'taking up' sounds from the air, of being thrown into vibrations by sounds falling upon it; it will only take up from the air the particular sounds, the particular tones as we shall presently call them, which it itself gives forth when thrown into vibrations by being struck. The vibrations of a membrane are much more complex; and for this reason a membrane takes up much more readily a variety of different sounds reaching it through the air. Still every membrane has its fundamental tone or tones, as they are called, those which it naturally gives forth when thrown into vibrations; and it takes up these from the air much more readily than any other sounds. It is a feature of the tympanic membrane that it takes up, without any marked distinction, a very great variety of sounds within a very large range. It probably has a fundamental tone of its own, but this is kept in the background; it is not prominent, and does not materially influence our hearing. Were it otherwise, were the tympanic membrane thrown into vibration much more readily by a particular sound than by any others, that sound would be dominant in all our hearing; and unless, as in vision, psychical experience intervened to correct the mere sensation, we should be misled in our judgments as to what was taking place around us.

This general usefulness of the tympanic membrane is secured partly by features proper to itself, partly by the fact that it is 'damped' by the attachment to it of the chain of ossicles. Without attempting to enter into a discussion of a matter which is in many ways complex, we may say that the following features contribute to make the tympanic membrane sensitive to a large variety of sounds. In the first place its dimensions are relatively small. In the second place the material of which it is composed is peculiar, so that it is in a special way unyielding and rigid; it retains its form when cut away from its bony attachments by a circular incision, and the malleus, including its handle, may be removed from it without distorting it. In the third place, its remarkable form, that of a shallow funnel with sides gently convex towards the meatus, has a marked effect upon its capabilities of vibration. The chain of ossicles, attached at its far end, to the membrane of the fenestra ovalis has a 'damping' effect similar to that, familiar to every one, of lessening or stopping the sound of a vibrating empty wine-glass or tumbler by pressing the finger on it; and this 'damping' while it diminishes to a certain extent all the vibrations of the membrane is especially effective in the case of excessive vibrations, such as those which might be produced by the sound which is the fundamental note of the membrane.

§ 817. The vibrations thus set going in the tympanic membrane are transmitted from it to the chain of ossicles. The transmission might take place in two ways. In the first place the vibrations, the alternate bendings inwards and outwards of the membrane, might, by carrying with it the attached handle of the malleus, work the chain of ossicles as a lever, in the manner described in § 815, so that each inward flexion of the tympanic membrane led to the membrane of the fenestra ovalis pushing the perilymph of the labyrinth inwards, while the return outwards again of the one led to a like return of the other. In the second place the transversal vibrations of the tympanic membrane might set up longitudinal vibrations in the substance of the malleus, which would travel as longitudinal vibrations through the chain, and so reach the perilymph. In the one case the whole chain of ossicles swings to and fro, in the other case the sound is propagated by molecular movement. That the ossicles do move *en masse* has been proved by recording their movements in the usual graphic method. A very light style attached to the end of the incus or to the stapes is made to write on a travelling surface; when the tympanic membrane is thrown into vibrations by a sound, the curves described by the style indicate that the chain of bones moves with every vibration of the membrane. On the other hand, the comparatively loose attachments of the several ossicles is an obstacle to the molecular transmission of sonorous vibrations through them. Moreover, sonorous vibrations can only be transmitted to or pass along such bodies as either are very long compared to the length of the sound-waves, or, as in the case of membranes and strings, have one dimension very much smaller than the others. Now the bones in question are not only not especially thin in any one dimension, but are in all their dimensions exceedingly small compared with the wave lengths of the vibrations of even the shrillest sounds we are capable of hearing; hence they must be useless for the molecular propagation of vibrations. We may conclude then that when waves of sound throw the tympanic membrane into vibrations, each inward excursion of the membrane is followed by a corresponding impulse given by the foot of the stapes to the perilymph. As we have seen the space through which the end of the incus moves is less than that through which the handle of the malleus moves, and the movements of the stapes are in addition restricted by the manner of its attachment to the rim of the fenestra ovalis; but the energy with which the end of the incus and hence the stapes moves is proportionately increased, so that we might almost speak of the gentle swingings of the tympanic membrane being converted into smart taps on the perilymph of the labyrinth.

The impulses thus given to the perilymph at the fenestra ovalis travel along the intricate passages of the perilymph spaces, the details of which we shall presently study, and finally break upon the fenestra rotunda; if the membrane which closes this orifice be watched it may be observed to pulsate in sequence with the pulsations of the fenestra ovalis. During their passage these impulses are communicated to the endolymph and in some way or other affect the endings of the auditory nerve. How they do this we shall presently study; but we may here call attention to the fact that the waves of sound which fall on the tympanic membrane are for the most part not simple in character but complex, and in many cases exceedingly so. This complexity is carried on into the vibrations of the tympanic membrane and so into the impulses given to the perilymph; the waves which sweep past the endings of the auditory nerve are, so to speak, reproductions of the complex aerial waves passing down the meatus.

§ 818. By far the greater number of the sounds which we hear reach the tympanic membrane by passing through the air down the meatus. One great use of the long external passage is probably to protect the delicate tympanic membrane from the accidents to which it would be subject were it freely exposed on the surface of the body; but it has also a use in transmitting to the tympanic membrane sounds travelling to the ear in certain directions more readily than those coming in other directions. The constriction of the meatus at the junction of the outer and middle third serves as a sort of diaphragm by which waves of sound travelling too much out of the line of the meatus are turned back. The external ear, auricle, or pinna has also probably a similar effect, reflecting into the meatus waves which fall upon it in a particular direction or waves of a particular kind. But of these uses, which are of more importance in some animals than in man, we shall speak again in considering the manner in which we recognize the directions of sounds.

Sounds however may reach the ear by paths other than the meatus. If a tuning-fork be struck and then held near the ear it will after a while cease to be heard, the sound dies away; but the sound is heard again if the handle of the fork be placed between the teeth; and when the sound again dies away, it may be revived by gently closing the external meatus, care being taken not to cause compression of the air within. When the tuning-fork is held between the teeth its vibrations are transmitted, through the bones of the head to the tympanic membrane, which thus set in motion acts in the same way as when it is set in motion through the air of the meatus. That the vibrations which thus reach the internal ear are, for the most part at least, conducted through the tympanum, and not brought to bear on the perilymph directly through the bony walls of the labyrinth is not only indicated by the effect just mentioned of closing the meatus, for this could have no influence on the labyrinth itself, but may be also proved by experiment. If a style be attached to the stapes laid bare in the skull, the vibrations of a tuning-fork brought into contact with the skull, will lead to corresponding movements of the style.

Not only may vibrations be transmitted from the skull to the tympanic membrane, but also conversely the vibrations of the membrane, brought about in the usual way through the meatus, may be transmitted to the bones of the skull. If a long tube introduced into one meatus be spoken or sung into, the sounds may be heard by help of a stethoscope placed over various parts of the head. They are heard best perhaps at the opposite meatus; the vibrations of the bones of the skull set going by one tympanic membrane throw the other tympanic membrane also into vibrations.

§ 819. Two muscles act upon the auditory apparatus of the tympanum; one, the *tensor tympani*, acts upon the malleus and hence upon the tympanic membrane, the other, the *stapedius*, acts upon the stapes.

The tensor tympani (Fig. 174) is a slender muscle, lying in a groove above the bony canal of the Eustachian tube, and having very much the direction of that tube. The tendon in which it ends, turns round, almost at right angles to the line of the muscle,

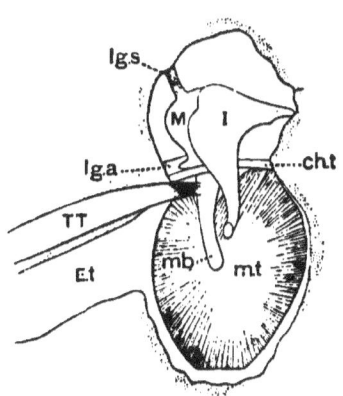

Fig. 174. DIAGRAM OF THE OUTER WALL OF THE TYMPANUM AS SEEN FROM THE MESIAL SIDE. Magnified twice. (After Schwalbe.)

m.t. membrana tympani. *mb.* handle of *M* the Malleus. *I.* the incus. *E.t.* Eustachian tube. *T. T.* tensor tympani, the tendon of which is seen attached to the upper part of the handle of the malleus. *lg. a.* the anterior and *lg. s.* the superior ligament of the malleus. *ch. t.* the chorda tympani nerve traversing the tympanic cavity.

over a bony prominence at the end of the groove, and passing athwart the cavity of the tympanum from the median side outwards (Fig. 168 *T. T.*) is attached to the upper part of the handle of the malleus.

The effect of the contraction of the muscle is to pull the handle of the malleus and so the tympanic membrane inwards towards the median side. Even in a quiescent state it may be of use in keeping up a certain amount of tension and in preventing the tympanic membrane being pushed out too far. When it contracts it certainly renders the tympanic membrane more tense; hence it has been supposed on the one hand to act as a damper lessening the amount of vibration of the membrane in the case of too powerful sounds, and on the other hand to accommodate the apparatus to the sounds falling upon it since the more tense membrane is more readily thrown into vibrations by higher notes and is less sensitive to lower notes. It has been urged that it is readily thrown into contraction at the commencement of a sound, especially of a noise, and returns to rest during the continuance of a prolonged musical note, the contraction being a simple contraction or twitch, rather than a continued tetanic contraction; it is suggested that this may serve to tune the membrane as it were for the sound which follows. Efferent impulses reach it through fibres of the fifth nerve from the otic ganglion, and its activity is regulated by reflex action, vibrations of the tympanic membrane starting the afferent impulses. In some persons the muscle seems to be partly under the dominion of the will, since a peculiar crackling noise which these persons can produce at pleasure appears to be caused by contraction of the tensor tympani.

The stapedius is a small muscle imbedded in the bone of the median wall of the tympanum, the tendon issuing by a hole close to the fenestra ovalis (Fig. 166 *St.*) and being inserted into the head of the stapes (Fig. 172 *ST*). It is supposed to regulate the movements of the stapes, and especially to prevent the foot plate being driven too far into the fenestra ovalis during large or sudden movements of the tympanic membrane. Contractions of the muscle pull the front part of the stapedial foot plate towards the tympanum, the hind part being thereby pressed somewhat into the labyrinth and the whole membranous ring round the foot being rendered more tense; but the total effect is to diminish the pressure in the labyrinth. It perhaps may be regarded as the antagonist of the tensor tympani. It is governed by fibres from the facial nerve.

§ **820.** The cavity of the tympanum is, as we have seen, continuous with the Eustachian tube. The walls of the tube in the first third of its length adjoining the tympanum are osseous, but in the remaining two-thirds are cartilaginous and membranous. The tube, whose lumen is of varying diameter and special shape, passes obliquely forwards, downwards, and towards the median line (Figs. 166, 167 *E.t.*) to open at the side of the upper part of the pharynx. The mucous membrane lining the tube consists of a ciliated epithelium resting on a dermis rich in reticular and adenoid tissue, and bearing glands. The action of the cilia is such that the movement which they effect is directed from the tympanum to the pharynx. The mucous membrane lining the tympanum is a continuation of that lining the tube and, like that, ciliated except over the tympanic membrane, the chain of ossicles, and probably some other parts; in these situations the epithelium consists of a single layer of flat non-ciliated cells, and a similar epithelium lines the antrum and mastoid cells which continue the cavity of the tympanum backwards and upwards.

One use of the Eustachian tube is to carry down to the pharynx the fluid, normally very small in amount, which is secreted by the mucous lining of the tympanum, but a far more important use is that of placing the air in the tympanum in communication with that in the pharynx and so with the external air, by which means the pressure on the two sides of the tympanic membrane is equalized. If as sometimes happens the tube is definitely closed, the absorption of the gases in the air at first present in the tympanum diminishes the pressure on the inner side of the tympanic membrane, and so interferes with the vibrations of the membrane. Moreover it is desirable that general changes of pressure in the external atmosphere should be rapidly followed by corresponding changes in the pressure within the tympanum, since the tympanic membrane would not vibrate normally if any marked difference of pressure on the two sides were brought about; and this would result if the way from the tympanum to the external air through the tube were blocked.

The lumen of the tube has in its lower part the form of an obliquely vertical slit, the sides touching or nearly so; and much dispute has taken place as to whether the tube is normally closed or open. It is undoubtedly opened during the act of swallowing, and during the act, by the action of certain muscles of the palate, air is forced up into the tympanum. It may be opened also by a forced inspiration or a forced expiration when the nose and mouth are kept closed; in the former case the pressure of the air in the tympanum is diminished, in the latter case increased. Although under normal circumstances the lumen is so far patent as to allow the escape of the fluid driven by the cilia, the evidence goes to shew that it is practically closed; sounds for instance generated in the pharnyx do not throw the tympanic membrane into vibrations in such a way as they would do if the tube were thoroughly open. Apparently the occasional opening, such as that effected by swallowing, is sufficient to keep the pressure within the tympanum at its proper level. When the general pressure of the external atmosphere is rapidly increased or diminished, temporary deafness, especially to low notes, frequently ensues, in consequence of the pressure within the tympanum not following the changes of the pressure without. This however is soon remedied by the act of swallowing, which opens the tube and thus equalises the pressure.

An abnormal permanency in the closure of the tube is recognized as a cause of deafness, and may be remedied by catheterism of the tube, that is to say, opening up the tube by passing an instrument into it from the pharnyx.

SEC. 2. THE STRUCTURE OF THE LABYRINTH.

§ 821. The membranous labyrinth, into which the primitive otic vesicle is developed, though very complicated in form, is virtually a sac the cavity of which filled with endolymph is continuous throughout. We may in the first place consider it as consisting of two divisions, which differ from each other both as to

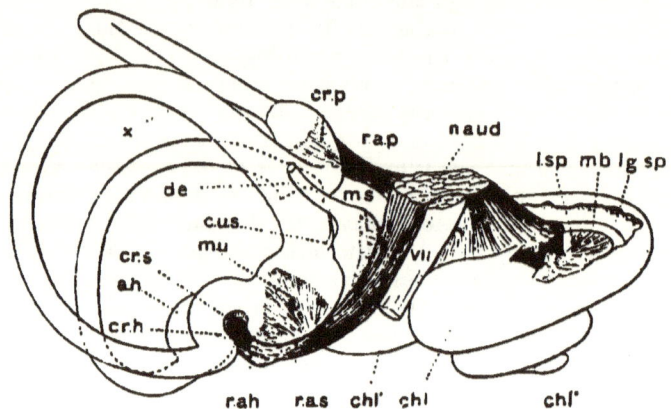

Fig. 175. The Membranous Labyrinth (of the Right Ear) as seen from above, magnified six times. (After Retzius.)

The bony envelope has been wholly removed from the vestibular division, but only in part broken through in the cochlear division.

chl the cochlea, *chl'* the first part of the basal whorl, *chl''* the summit. To the right, where the bony wall has been broken through, are seen: *l.sp* the spiral lamina, *m.b* the basilar membrane, *lg.sp* the spiral ligament.

n aud the auditory nerve, lying along side of which is seen: VII, the seventh, facial nerve.

m.s macula of the saccule. *m.u* macula of the utricle. *cr.p* the crista of the posterior semicircular canal, with *r.a.p* the branch of the auditory nerve distributed to it, *cr.s* crista of the superior canal with *r.a.s* its nerve *a.h* ampulla and *cr.h* crista of the horizontal canal, with *r.ah* its nerve.

x the conjoined posterior and superior canals. *d.e* ductus endolymphaticus, with *c.u.s* its junction with the utricle.

their relations to the perilymph space and as to the manner in which the auditory nerve ends in them. One division is that part of the sac which enters into the construction of the cochlea, and is called the *canalis cochlearis* (Fig. 176, *Coch.*). The other division comprises the rest of the sac. The two correspond respectively to the cochlea and the vestibule of the bony labyrinth, including with the vestibule the semicircular canals; we may speak of them as the cochlear and vestibular portions of the sac. As we saw in studying the cranial nerves the auditory nerve (§ 618), though usually spoken of as one nerve, really consists of two nerves, different in origin, in ending, and to a certain extent in structure; one of these two, distributed to the cochlear division of the membranous labyrinth, we called the cochlear nerve, the other, distributed to the vestibular division, we called the vestibular nerve.

The vestibular division of the membranous labyrinth consists of an oval sac, about 6 mm. long, the *utricle* (utriculus) (Fig. 176, *U*), and lying below this a smaller, 3 mm. in diameter, more spherical, though somewhat oval flattened sac, the *saccule* (sacculus) (Fig. 176, *S*). Into the utricle open both ends of each of the

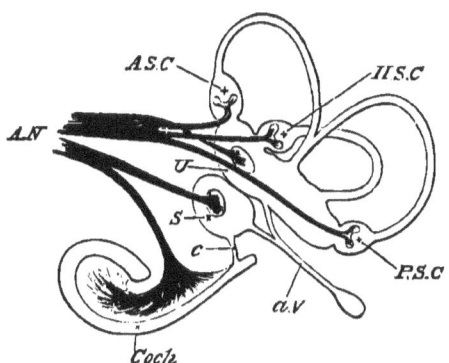

Fig. 176. THE MEMBRANOUS LABYRINTH AND THE ENDINGS OF THE AUDITORY NERVE.

The figure is wholly diagrammatic, and is introduced as giving a simpler view of the essential parts of Fig. 175; it should be used only in conjunction with that.

U. utricle. *S*. saccule. *A.S.C.* Superior (or anterior), *P.S.C.* posterior, *H.S.C.* horizontal, semicircular canals.

Coch. The canalis cochlearis represented as a tube partially unrolled. *c*. canalis reuniens, joining the saccule with the canalis cochlearis. *a v*. ductus endolymphaticus, shewing its origin from both saccule and utricle, and its dilated blind end, the saccus endolymphaticus.

A.N. The auditory nerve ending in the cristæ of the ampullæ, in the maculæ of the utricle and saccule, and along the whole length of the canalis cochlearis. The branch of the vestibular division of the nerve ending in the saccule remains in close contact with the cochlear division, longer than does the rest of the vestibular division ending in the utricle and ampullæ (the branch to the posterior canal should have been represented as lying in contact with that to the saccule.)

three semicircular canals. These are disposed in the three planes of space. One (Fig. 176, *H.S.C.*) lies in a horizontal plane, and is called the *horizontal* or, since its hoop is directed outwards, the *external* semicircular canal. The other two lie in two vertical planes at right angles to the above. One lying in a vertical plane more or less parallel to the median plane of the head has its hoop directed backwards and hence is called the *posterior* canal (*P.S.C.*); the other lying in a vertical place more or less parallel to the transverse plane of the head has its hoop directed upwards, and hence is called the *superior* canal (*A.S.C.*). The three planes in which the three canals lie are placed with great exactitude at right angles to each other; they do not however coincide with the three planes of the head (sagittal or horizontal, median, and frontal or transverse) but make angles with these.

Each canal, at one of the ends by which it opens into the utricle, is dilated into a flask-shaped swelling, the ampulla, (Figs. 175, 176), but at the other end does not shew any such marked swelling. The two ends of the two vertical canals, superior and posterior, which do not bear ampullæ, join together into a common canal (Fig. 175 *X.*) before they open into the utricle, but both ends of the horizontal canal are separate. Hence the canals, taken together, open into the utricle by five openings, three of which are marked by ampullæ, two are not.

The saccule, though lying close to the utricle, is wholly distinct from it and indeed is separated from it to a certain extent by a bony partition stretching inwards from the bony envelope; the cavity of the one has no direct communication with that of the other. An indirect communication is however supplied by the *ductus endolymphaticus* (Figs. 175 *d.e.* 176 *a.x.*) which formed by the union of a narrow tube, springing from the utricle with a wider one leading from the saccule, runs inwards towards the median line, in a canal hollowed out of the petrosal bone, and ends in a flattened sac, *saccus endolymphaticus* (Fig. 162 *s.e.*), placed in the cavity of the skull and supported between two layers of the dura mater. Through this hollow tube the cavity of the utricle is made continuous with that of the saccule.

From the saccule there starts also another narrow tube, the *canalis reuniens* (Fig. 176 *c.*), which opens into the canalis cochlearis, or cochlear division of the membranous labyrinth; by this the cavity of the vestibular division of the sac is made continuous with that of the cochlear division.

§ 822. The wall of the membranous labyrinth consists throughout of an epithelium, modified in certain places into auditory epithelium (§ 808) and of a connective tissue or dermis, on which this epithelium rests. Around this dermis is developed the lymphatic cavity, lined with lymphatic epithelium (§ 809) and filled with perilymph, the outer wall of the cavity being furnished by connective tissue continuous with the periosteum of the bony envelope. The contour of the bony labyrinth follows in a general way only, not closely the contour of the membranous labyrinth; hence the perilymph space is not uniform but irregular. In some places, as for instance in the convexities of the semicircular canals, and where the nerves join it, the membranous labyrinth is fixed to the bony envelope, the periosteum of the latter being continuous with the dermis of the former or broken only by small lymph spaces And where in other situations the perilymph cavity is large, it is much subdivided, in some places more than in others, by bridles and bands of connective tissue. Where the vestibule abuts on the median wall of the tympanum, in the situation of the fenestra ovalis, which is placed over against the line of division of utricle and saccule (Fig. 162) the space contains few such bridles, and here a considerable portion of the perilymph is gathered into what is sometimes spoken of as the 'cisterna.' In the hoops of the semicircular canals the membranous canal, much smaller in section than the bony canal, seems imbedded in the connective tissue of the latter, leaving a considerable space, broken by some few bridles, for the perilymph. In other places the bands of connective tissue, passing from the inner lining of the bony envelope to the walls of the membranous sac, are so abundant that the perilymph space becomes a labyrinth of small irregular passages. Nevertheless, however broken up, the whole perilymph space of the vestibular division of the ear is a continuous space, and the pulses given to it by the movements of the stapes sweep over the whole of it

§ 823. The auditory nerve, both the vestibular and the cochlear division, plunging into the floor of the cranium together with the facial nerve and the nervus intermedius by the canal known as the *meatus auditorius internus* (Fig. 163 *m.i.*), and traversing some compact bone continuous with the compact shell of the bony labyrinth, reaches the labyrinth at the open angle between the base of the cochlea and the vestibule. Here the cochlear nerve passes to the cochlea in a way which we shall presently describe. The vestibular nerve consists of two branches; one (ramus superior), lying above the other, is distributed to the utricle, and to the superior and horizontal semicircular canals, the other lying beneath the former ends in the saccule and in the posterior semicircular canal. In the utricle the nerve comes into connection, in a manner which we shall study in detail, with an area of modified auditory epithelium in the form of an oval low swelling, the *macula acustica* (Fig. 175 *m.u.*), and the connection of the nerve with the saccule is likewise in the form of a *macula* (Fig. 175 *m.s.*); the maculæ of the utricle and of the saccule are the only parts of these two structures in which the epithelium has any connections with the auditory nerve. In the case of each of the semicircular canals the nerve is in connection with a part and a part only of each ampulla. The area of modified auditory epithelium has in each ampulla the form of a forked or horse-shoe shaped ridge placed athwart the long axis of the ampulla and projecting into the interior; it is called a *crista acustica* (Figs. 175 *cr.p., cr.s., cr.h.*, 176). Hence the vestibular nerve ends exclusively in the macula acustica of the utricle, the macula acustica of the saccule and the crista acustica of each of the three ampullae. The superior branch before it ends in the utricle and superior canal bears a ganglion of nerve cells, and the division which the same branch gives off to the horizontal canal also bears a group of nerve cells just before it joins its crista. The median branch to the saccule and posterior canal likewise bears a ganglion which is more or less continuous with the ganglion of the superior branch.

§ 824. The cochlea may be considered as a prolongation of the vestibule in the form of an elongated cone; and indeed in some of the lower animals, in birds for instance, it is a short blunt cone. But it differs very widely from the rest of the labyrinth; and its special features may be broadly considered under three heads.

In the first place, the elongated, almost tubular, bony cone is not straight, but twisted closely on itself in two and a half whorls (Fig. 163), and the whorls grow together so as to form a short cone, the markings of the whorls being visible on the outside after the fashion of a gasteropod shell; hence the name. In the natural position in the head the cochlea is nearly horizontal, with the beginning of the first whorl at the base abutting on the median wall of the tympanum and with the apex directed forwards, and towards the median line; but when we are dealing with it by itself it will be convenient to consider it as if it were vertical in position with the apex above and the base below. The axis, or 'modiolus' as it has been called, round which the whorls are coiled differs from the walls of the whorls themselves in being formed of spongy, not compact bone, and is traversed by canals for the passage of the cochlear nerve, which entering it at the base from the meatus internus, ascends along it to the apex giving off fibres as we shall see all the way along.

In the second place, in the vestibule and semicircular canals the membranous labyrinth hangs, for the most part, loose within its bony shell, supported by irregular bridles, or is so attached that in any case there is no very definite arrangement of the perilymph space in relation to the sac which it surrounds; in the cochlea on the contrary a very definite arrangement obtains. This is best seen in a vertical section of the cochlea, in which the whorls are cut transversely in succession. The whole lumen of the coiled bony tube (Fig. 177) is seen to be roughly circular in section. Within this lies the canalis cochlearis (*C. Chl.*), the tubular continuation of the membranous labyrinth. This however is not, as in the semicircular canals round or oval, but triangular in section. It is moreover so placed that while the base of the triangle is firmly attached to the outer wall of the bony tube, no perilymph space lying between the two, the apex of the triangle is also attached to the end of a thin sheet of bone (*Lam. sp.*) which projects outwards into the

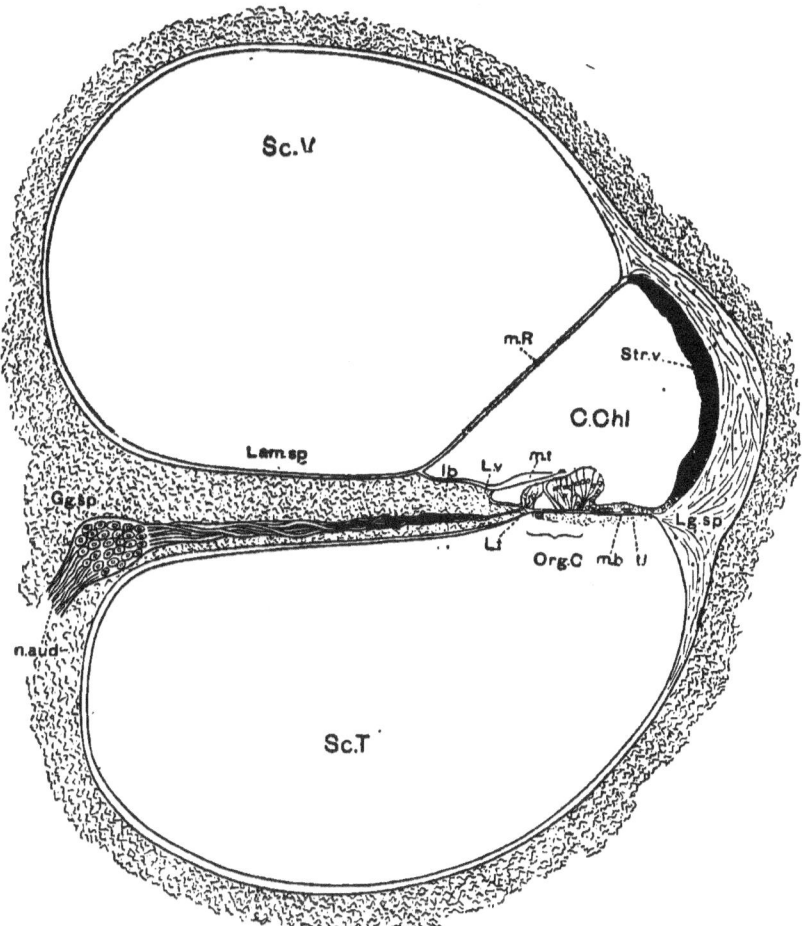

FIG. 177. DIAGRAM OF A TRANSVERSE SECTION OF A WHORL OF THE COCHLEA.

Sc.v. Scala Vestibuli. *Sc.T.* Scala Tympani. *C.chl.* Canalis cochlearis.
n.aud. auditory nerve. *Gg.sp.* Spiral ganglion. *Lam.sp.* Lamina spiralis. *lb.* limbus. *L.v.* labium vestibulare. *Lt.* labium tympani. *m.R.* membrane of Reissner. *Lg.sp.* spiral ligament. *Str.v.* stria vascularis. *Org.C.* organ of Corti. *m.b.* basilar membrane *t.l.* lymphatic epithelioid lining of the basilar membrane on the tympanic side. *m.t.* tectorial membrane.

tube from the axis for some considerable distance. The triangular canalis cochlearis and the projecting sheet of bone thus completely divide the perilymph space of the tube into two spaces, one above and one below, (the cochlea being supposed to be placed vertical).

These two spaces moreover are each entire, not being broken up or subdivided in any way by bridles or bands of connective tissue. Such an arrangement obtains all the way along the successive whorls except at the extreme top and extreme bottom. The projecting sheet of bone, as it is traced from the bottom to the top, describes a spiral, and hence is called the *lamina spiralis*. The canalis cochlearis also describes a spiral, winds in fact like a turret staircase, as do as well the two perilymph spaces, and the latter are hence called *scalae*. The one lying above the canalis cochlea, when followed to the bottom of the cochlea is found to be continuous with, to open freely into, the perilymph space of the vestibule; hence it is called *scala vestibuli* (*Sc. V.*) The one lying below the canalis cochlearis ends blindly at the bottom of the cochlea, but in the bony wall of its blind end is an orifice, which we have already spoken of as the fenestra rotunda, the membrane covering which shuts off the scala in question from the cavity of the tympanum; hence this scala is called the *scala tympani* (*Sc. T.*). The canalis cochlearis thus lying between these two scalæ is sometimes called the *scala media*; but this name is undesirable since the canalis cochlearis being a part of the membranous labyrinth, a derivative of the otic vesicle, differs essentially in nature from the two scalæ, which are merely lymphatic, perilymph spaces.

The whole tube of the cochlea diminishes in size from the bottom of the lowermost whorl to the top of the highest; but the diminution affects the two scalæ alone, and the scala tympani more rapidly than the scala vestibuli; the canalis cochlearis so far from growing less, increases, except at the very top, from below upwards and especially, as we shall see in the dimensions of one of its sides. At the top the lamina spiralis comes to an end, finishing off in the form of a hook, *hamulus*, and the canalis cochlearis suddenly diminishing ends blindly. Beyond the tip of the canalis cochlearis, the scala vestibuli which formed its roof, becomes, by a round orifice, *helicotrema*, continuous with the end of the scala tympani which formed its floor; here, and here alone does the one scala open into the other; and by this connection only has the fluid in the scala tympani access to the scala vestibuli and so to the perilymph space surrounding the vestibular portion of the labyrinth. The pulse which each thrust of the stapes at the fenestra ovalis imparts to the perilymph of the vestibule at the cisterna, passes into the scala vestibuli, and must either be transmitted to the scala tympani across the canalis cochlearis or must travel along the scala vestibuli to the apex of the cochlea, and down the whole length of the scala tympani before it breaks on the membrane of the fenestra rotunda.

If then the bony tube of the cochlea were unrolled and made straight it would appear as a tube diminishing to a pointed end; the lamina spiralis would appear as a longitudinal plate running along the whole length of the tube, and partially dividing it lengthways into two, while the canalis cochlearis would appear as a smaller tube, of triangular section, slid into the larger tube in such a way that the apex of the triangle all the way along touched the edge of the lamina spiralis, and the base all the way along was adherent to the opposite side of the main tube, thus separating completely the scala vestibuli on the one side from the scala tympani on the other. Only at the pointed end of the main tube would the canalis cochlearis be wanting, and here the two scalae would run into each other.

The third great feature of the cochlea is that the auditory nerve is connected with the canalis cochlearis, not in a circumscribed patch, macula, or ridge, crista, but along its whole length; and there is accordingly an area of modified auditory epithelium along the whole length of the canal from close to the bottom of the undermost whorl to the tip, or nearly to the very tip of the topmost whorl.

The three sides of the canalis cochlearis differ markedly in structure and appearance. We shall study the details of these later on; meanwhile we may say that the wall which separates the canal from the scala vestibuli is a thin membrane (Fig. 177 *m. R.*), known as the *membrane of Reissner*, the epithelium on which lining the canal is of a simple character; that the epithelium which lines the base of the triangle, firmly attached to the bony wall, is also simple in character: but that a part of the epithelium lining the wall which separates the canalis cochlearis from the scala tympani is, along the whole length of the spiral, modified auditory epithelium and is known as the *organ of Corti* (Fig. 177, *Org. C.*).

The auditory, cochlear nerve, leaving the meatus internus, passes up the axis of the cochlea. As it ascends it gives off fibres passing outwards in a spiral manner into the lamina spiralis (cf. Fig. 175), which, thick towards the central axis, thins out towards its attachment to the canalis cochlearis. As these fibres traverse the lamina on their way outwards, numerous bipolar nerve-cells appear on their course, thus forming along the whole length of the spiral a continuous spiral ganglion, *the ganglion spirale* (Fig. 177, *Gg. sp.*). Having passed this ganglion and having reached the edge of the lamina spiralis, the fibres pass into and become connected, in a manner presently to be described, with the auditory epithelium of the organ of Corti.

We may now turn to the minute structure of the membranous labyrinth and its auditory epithelium, but before doing so it will be well to recall the position in relation to the tympanic cavity of the several parts of the internal ear which we have just described. The whole internal ear lies to the median side of and forms in part the median wall of the tympanic cavity. Nearly opposite the middle of the tympanic ring and membrane, the median wall of the tympanum is marked by an elevation (the promontorium), (cf. Fig. 166) which corresponds to the base of the cochlea, and which overhangs the depression leading to the fenestra rotunda. From this base the cochlea placed nearly horizontally, and thus nearly at right angles to the median wall of the tympanum, runs forwards inclining to the median side, the apex abutting on the bony wall of the Eustachian tube. Above the promontory the fenestra ovalis marks on the tympanic wall the position of the projecting portion of the vestibule; from this the semicircular canals project backwards and laterally, in their several planes, the ampulla of the superior and external canal forming a projection on the wall of the epitympanic cavity above the fenestra ovalis. The auditory nerve entering, in company with the facial nerve, on the median side in the open angle between the base of the cochlea and the vestibule, is distributed, as we have seen, to both these structures.

The Vestibular Labyrinth.

§ 825. Little need be said concerning the minute structure of those parts of the vestibular division of the labyrinth with which the auditory nerve makes no connections. The epithelium consists of a single layer of cells which are for the most part flat and polyhedral, though they differ somewhat in form and in other features in different regions. The epithelium rests on a thin connective tissue basis known as the "tunica propria;" this is hyaline, and except for some fibrillation or striation more obvious in some parts than in others, appears to be structureless; nuclei are absent from it and blood vessels do not pass into it. This tunic, which in the semicircular canals, in man, is with its epithelium frequently raised into irregular papillæ or warts, rests on an outer coat of vascular connective tissue, in some places continuous with, in others connected by bridles with the periosteum of the bony envelope. The spaces between the bundles of connective tissue of this coat gradually open out into the general perilymph cavity; it and they are lined with lymphatic epithelioid plates. The fluid (perilymph) contained in them, though it is not ordinary lymph, being viscid through the presence of mucin, finds access along the sheath of the auditory nerve into the subdural and subarachnoid spaces of the brain.

§ 826. The three cristæ acusticæ are as we have said ridges projecting crosswise into the cavities of the ampullæ to which they respectively belong, each ridge being formed partly by a development of the tunica propria and underlying connective tissue into a thick cushion of somewhat peculiar nature, and partly by an increase in the epithelium which, thick on the top of the ridge, gradually thins away at the sides.

Immediately below the epithelium the connective tissue cushion, especially at the top of the ridge, consists of a hyaline or faintly fibrillated ground substance, traversed by blood vessels, but free from nuclei. Lower down scattered nuclei or rather connective tissue corpuscles make their appearance, but the ground substance in which they lie remains for the most part hyaline, the tissue having an aspect not unlike that of cartilage. Still lower down the groundwork is broken up, in the ordinary way, into bundles of connective tissue. The branchlet of the auditory nerve, reaching the ridge at its base, not far from its middle, spreads out fanwise into nerve-fibres and bundles of nerve-fibres, which in a more or less plexiform manner run vertically upwards towards the summit of the ridge along its whole length. Hence in a transverse section of the ridge the nerve-fibres are seen ascending, in a vertical direction, through the connective tissue cushion to the epithelial cap, in which they are lost to view. So long as it remains in the connective tissue cushion each fibre retains all its constituents, neurilemma, medulla, and axis cylinder; upon entering the epithelium it loses as we shall see its neurilemma, and in most instances its medulla.

In a vertical section of a ridge, the thickened epithelium forming a cap to the ridge is seen to have special characters

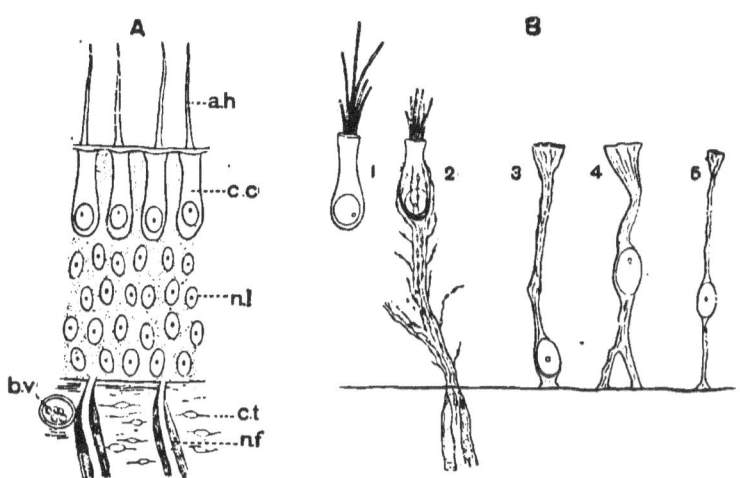

FIG. 178. DIAGRAM TO ILLUSTRATE THE STRUCTURE OF A CRISTA OR A MACULA

A. a portion of a crista seen under a low power, shewing c.c. the cylinder cells, with a.h. auditory hairs, and n.l. the nuclear layer, the nuclei being diagrammatically shewn as if imbedded in a uniform ground substance. ct. the connective tissue corpuscles of the dermis with b.v. a blood vessel, and n.f. nerve fibre, the latter being shewn entering into the epithelium as a simple axis cylinder.

B. cylinder and rod cells. 1, an isolated cylinder cell. 2, the same surrounded with a nest of nerve-fibrillæ proceeding from a nerve-fibre. 3, 4, 5, various forms of rod-cells.

which are maintained from some little distance down the sides, and then almost suddenly cease. This is the *auditory epithelium* and with this alone do the nerve-fibres make connections. Seen *in situ* (Fig. 178 A) this epithelium appears to consist of an outer row of cylindrical or columnar cells (*c.c.*), forming the free surface, and between this and the tunica propria, of a part calling to mind the nuclear layers of the retina, since it appears to be composed of nuclei (*n.l.*) closely packed together; we may speak of it as "the nuclear layer." From the free surface of this auditory epithelium a number of elongated, rigid, spoke-like processes (*a.h.*), of great length in some animals such as fishes, but shorter in man, project into the cavity of the ampulla; these are the *auditory hairs*. According to some authors at all events the free surface of the epithelium is guarded by a cuticular membrane, pierced for the passage of the auditory hairs.

At some little distance down the sides of the ridge, the nuclear layer disappears as do also the auditory hairs; the auditory epithelium is almost suddenly transformed into a single layer of epithelial cells, which differ chiefly from the epithelial cells forming the general lining of the labyrinth, in that they are tall, cylindrical and large, with the cell substance rich in granules. These cylindrical cells, which in no way enter into connection with the fibres of the auditory nerve, gradually change at some distance from the auditory epithelium. into the flat polyhedral cells of the general lining.

In hardened and prepared specimens the auditory hairs appeared to be imbedded in a cap of mucous or fibrinous material. This, which has been called the "cupula," is supposed to be an artificial product, the result of a coagulation of the endolymph, and not to exist during life.

Contrary to what occurs in an epithelium elsewhere, the blood vessels instead of being absolutely confined to the connective tissue basis or dermis pass occasionally into the epithelium itself, and form loops among the cells; as we shall see this also occurs in the cochlea.

§ 827. The features which we have just described may be recognized without any great difficulty in sections of ampullæ prepared in various ways; but considerable difference of opinion obtains as to the exact nature of the constituent elements of the epithelium and especially of their relations to the nerve fibres; nor is the existence of a difference of opinion to be wondered at when the difficulties of investigation, greater perhaps than in any other histological subject, are born in mind.

According to one view, the auditory epithelium consists of two kinds of cells. The one kind (Fig. 178 B, 1, 2) is a cell cylindrical in form or rather flask-shaped, with a flat top forming part of the free surface of the epithelium, and a conical but rounded base reaching less than half-way down the thickness of the epithelium.

The cell substance, very delicate in nature, contains a number of granules, and bears near the base a large, conspicuous, spheroidal nucleus. From the free surface there projects a bundle of long stiff hairs, the auditory hairs, which often stick together in the form of an attenuated cone. Cells of this kind may be called *hair-cells*, or for reasons which we shall see directly, *cylinder cells*. The other kind of cell (Fig. 178 B 3, 4, 5) possesses a nucleus smaller than that of the cylinder cell and having the form of a short ellipsoid, placed vertically; around this nucleus lies a relatively small quantity of cell substance, delicate like that of the cylinder cells but probably of a different nature. This scanty cell body is prolonged upwards between the cylinder cells as a rod-shaped process terminating abruptly at the surface, and stretches in the opposite direction as a process which, frequently but not always branched and irregular, reaches to and ends at the surface of the dermis. These *rod cells* or *spindle cells* are much more numerous than the cylinder cells, and their nuclei are placed at different levels, some close upon the dermis, others at different distances from it up to the level of the bases of the cylinder cells. The nuclei of these rod cells thus occupy the space between the bases of the cylinder cells and the dermis; they form in fact the nuclear layer spoken of above. It should be added that the nuclei which form a row immediately above the dermis are regarded by some authors as belonging to cells differing from the rod cells, their cell substance being said to be confined to the neighbourhood of the nucleus, and not to extend to the surface of the epithelium; these are spoken of as *basal cells*.

According to the view which we are relating, a nerve fibre of the auditory nerve after traversing the auditory cushion (Fig. 178 A) passes into the epithelium and losing both neurilemma and medulla, though sometimes retaining the latter for a short distance, makes its way as a naked axis cylinder between the rod cells, taking sometimes a vertical, but often a more horizontal direction. In its course it gives off fine lateral irregular branches, (Fig. 178 B 2) often divides, is frequently very distinctly fibrillated, and eventually ends in a nest or brush of fibrillæ, into which the conical base of a cylinder cell fits, or with which the cell substance of the cell is continuous; though appearances support this latter view, it cannot be regarded as certain. The nerve fibres appear to make no connections with the rod cells, which are hence regarded as of the nature of supporting or subsidiary structures; the cylinder cells alone, and according to this view it is these which bear the auditory hairs, are to be looked upon as the functional terminal organs of the fibres of the auditory nerve.

Other observers, on the other hand, maintain that the auditory hairs are borne not by the cylindrical cells but by the rod cells (hence it is perhaps better to call the former cylinder cells rather than hair cells), and that the nerve fibres, forming a plexus of fibrillae in the lower layers of the epithelium, become connected with the rod cells and not at all with the cylinder cells. According to this view, the rod cells are the functional terminal organs and the cylinder cells subsidiary structures. A third class of observers maintain that both cylinder cells and rod cells bear hairs, that both are connected with the fibres of the auditory nerve, and that both serve as terminal organs. Probably however the view which we first related is the more correct one, though the matter cannot as yet be regarded as definitely settled.

§ 828. The maculæ acusticæ, both that of the utricle and that of the saccule, resemble the cristæ so far as essential features of structure are concerned. In them as in the cristæ both the dermis and the epithelium are thickened, but the elevation thus caused is in the form of a low swelling, not a steep ridge. The epithelium, which bears auditory hairs, consists of cylinder cells and rod cells, and the fibres of the auditory nerve enter into and are lost among the cells in the same way as in the cristæ. Perhaps the most conspicuous difference is that in the maculæ the auditory hairs are distinctly shorter than in the cristæ. Above each macula lies a fibrinous mass, not unlike the cupula, but having the form of a flat membrane; it is stated however to be a natural structure and not, like the cupula, an artificial product. There are no essential differences in structure between the macula of the utricle and that of the saccule. So far as we may permit ourselves to draw from structural features inferences concerning function, we may conclude that both the semicircular canals and the utricle with the saccule perform very much the same functions; there is, so far as structure is concerned, nothing to indicate that the afferent impulses started in the cristæ differ from those started in the maculæ.

§ 829. The endolymph, which we may look upon as secreted by the epithelium lining the membranous labyrinth and so far differing in origin from perilymph, resembles nevertheless that fluid very closely, containing however rather less solid matter. It is to a certain extent viscid, apparently owing to the presence of mucin.

In both the utricle and saccule are suspended in the endolymph crystalline bodies composed of calcium carbonate, with traces of other salts and with 25 p.c. or less of organic matter. In man and the higher animals, though varying in size, they are small crystals, generally rhombic or octohedral; and are then generally called *otoconia*. In some of the lower animals, for instance fishes, they form large masses, and are then generally called *otoliths*. The otoliths, and for the most part the otoconia, are confined to the utricle and saccule, and are usually found imbedded in the membrane spoken of above, which is hence sometimes called the "otolith" or "otoconial membrane." Otoconia may however occasionally be found in the ampullæ, or even in the perilymph chambers of the cochlea.

The Cochlea.

§ 830. As we have seen, the canalis cochlearis is a long spiral tube triangular in section, the apex of the triangle being attached to the edge of the spiral lamina, and the base to the opposite wall of the bony canal. In a dried specimen the bony spiral lamina ends in a thin edge, but in the fresh state the edge is thickened by connective tissue into a projection of peculiar form called the *limbus* (Fig. 177 *lb.*), which presents two edges, placed one above the other and seen in vertical section as two lips separated by a groove. The upper lip, which when looked at lengthways is seen to end in a number of projections or teeth, "auditory teeth," is called the *vestibular lip, labium vestibulare* (Fig. 177 *Lv.*, 179 *l.v.*), the lower lip is called the *tympanic lip, labium tympanicum* (177 *Lt.*, 179 *l.t.*), and the groove between them is called the *spiral groove, sulcus spiralis*. The vestibular lip and upper portion of the limbus is composed of a somewhat peculiar connective tissue, consisting of a homogeneous matrix in which are imbedded corpuscles; this is covered, except over the auditory teeth themselves, by a thin layer of flat epithelial cells, and deeper down passes into the bony tissue of the lamina. The vestibular lip serves for the attachment of the structure known as the *tectorial membrane, membrana tectoria* (Figs. 177, 179 *m.t.*). The tympanic lip, jutting farther outwards than does the upper lip, is the more direct continuation of the bony spiral lamina, and ends in an even edge composed of connective tissue which serves for the attachment of the *basilar membrane* (Figs. 177, 179 *m.b*).

The membrane of Reissner, stretching across from the limbus of the spiral lamina to the opposite wall and so forming the vestibular wall of the canalis cochlearis (Fig. 177 *m.R.*), is a thin membrane, the basis of which is a sheet of homogeneous or obscurely fibrillated connective tissue, continuous on the inner, median side with the connective tissue of the limbus and on the opposite side with the periosteum of the bony shell of the cochlea. On the side looking towards the scala vestibuli this basis is covered with a single layer of lymphatic epithelioid plates; on the opposite side, in the cavity of the canalis cochlearis, it is covered with a single layer of flat polygonal cells, similar to those lining the non-auditory part of the vestibule, and like them presenting minor differences between themselves, some cells being more granular than others.

The periosteum which lines the whole of the bony canal of the cochlea, and which over the limbus may be supposed to be represented by the peculiar connective tissue spoken of above, is on the outer side, where the base of the triangle of the canalis cochlearis is attached, developed into a thick cushion (Fig. 177), consisting of interwoven bundles of connective tissue, among which are interspersed numerous branched cells imbedded in a clear matrix. Opposite the tympanic lip of the spiral lamina, the bundles of fibres of this tissue converge to form a projection, the *spiral ligament, ligamentum spirale* (Figs. 177, 179 *Lg. sp.*), which is attached to, or rather which passes into the outer edge of the basilar membrane.

This cushion of connective tissue extends above for a short distance into the region of the scala vestibuli, and for a greater distance below into the region of the scala tympani; it is however thickest and best developed opposite the canalis cochlearis. Here it is lined by the epithelium of that canal, but the characters of the epithelium in this region are somewhat special. The cells are cubical or even columnar, and frequently irregular in form; they are also granular and have the aspect of cells in which metabolism is active. The special characteristic however is that blood vessels which are abundant in the underlying connective tissue cushion traverse the line of demarcation between dermis and epithelium, and pass between the epithelial cells themselves, so that, in this region, a confusion between connective tissue and epithelial elements takes place. Owing to the peculiar prominence of the blood vessels this portion of the lining of the canalis cochlearis has been called the *vascular band, stria vascularis* (Fig. 177 *Str. v.*). We may probably regard it as secretory in function, analogous to the choroid plexus of the brain and the ciliary processes of the eye, and as taking at least a large part in furnishing the endolymph, which it must be remembered is useful not only for mechanical acoustic purposes, but also for the nourishment of the delicate auditory epithelium.

§ 831. The remaining tympanic wall of the canalis cochlearis consists, like the membrane of Reissner, of a connective tissue basis with an epithelium derived from the epithelium of the otic vesicle on the one side, and with lymphatic epithelioid plates on the other; but part of the epithelium, namely a portion lying midway between the spiral lamina on the inside and the spiral ligament on the outside, is along the whole length of the spiral, except at the extreme ends, diffentiated into auditory epithelium of remarkable characters; and the connective tissue basis possesses corresponding special features, as indeed does also the lymphatic epithelium.

At the extreme edge of the tympanic lip of the spiral lamina the ordinary bundles of fibres of connective tissue are gathered up into a thin sheet which stretches radially across to the spiral ligament, and there fuses again with the more ordinary connective tissue of that ligament. It is this sheet which is called the *basilar membrane* (Figs. 177, 179 *m.b.*). It may be regarded as consisting of two parts, one reaching radially from the tympanic lip to a point marked by the attachment of what we shall presently speak of as the feet of the outer rods of Corti, the other continued on from this point to the spiral ligament. In its first more median part the basilar membrane is a thin rigid sheet which, though distinctly fibrillated radially, cannot be said to be composed of definite fibres. In its second, more lateral part, the membrane becomes somewhat thicker, thinning however again as it approaches the spiral ligament, and is obviously composed of fibres, lying side by side and cemented by or imbedded in a homogeneous ground-substance differing in nature from the fibres themselves. The fibres, when isolated are stiff, bending at a sharp angle, not curling, and are easily broken.

On its tympanic side, the basilar membrane bears, resting on a thin layer of homogeneous ground substance, a lymphatic epithelium (Figs. 177, 179 *t.l.*), the cells of which, often more

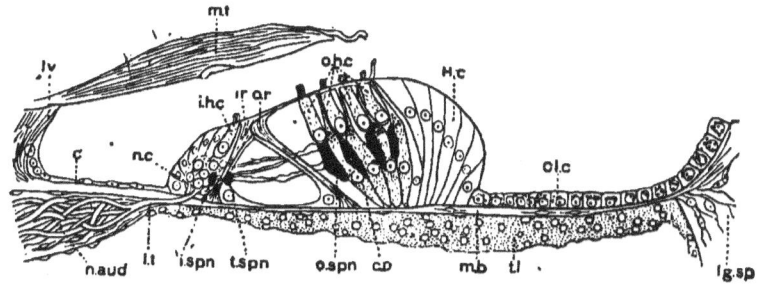

FIG. 179. DIAGRAM OF THE ORGAN OF CORTI. (After Retzius.)

i.r. inner rod of Corti, *o.r.* outer rod of Corti.

i.h.c. inner hair-cells, *n.c.* the group of nuclei beneath it. *o.h.c.* outer hair-cell, or cell of Corti, of the first row, *c.D.* its twin cell of Deiters; four rows of these twin cells are shewn.

n.aud. the auditory nerve perforating the tympanic lip *l.t*, and lost to view among the nuclei beneath the inner hair-cell. *i.sp.n.* the inner spiral strand of nerve-fibrillæ. *t.sp.n.* the spiral strand of the tunnel. *o.sp.n.* the outer spiral strand belonging to the first row of outer hair-cells; the three succeeding spiral strands belonging to the three other rows are also shewn. Nerve fibrillæ are shewn stretching radially across the tunnel.

H.c. Hensen's cells, *Cl.c.* Claudius cells, *m.b.* basilar membrane, *t.l.* lymphatic epithelioid lining of the basilar membrane on the side towards the scala tympani. *lg. sp.* spiral ligament. *c'.* cells lining the spiral groove, overhung by *l.v.* the vestibular lip. *m.t.* the tectorial membrane; a fragment of it is seen torn from the rest and adherent to the organ of Corti just outside the outermost row of outer hair-cells.

than one layer deep, are spindle shaped, the cell substance being prolonged into filamentous processes taking a longitudinal, that is to say spiral, direction along the length of the canal. Near the tympanic lip, **beneath** what we shall speak of as the "tunnel" of the organ of **Corti**, a blood vessel, lying apparently in the midst of the epithelioid cells, may be traced for some distance up the spiral; this *vas spirale* as it is called, serves to secure the due nourishment of the important structures lying over it.

§ **832.** On the side of the basilar membrane which looks towards the canalis cochlearis, the epithelium of the canal lies immediately on the membrane or is separated from it by a thin layer of homogeneous ground substance in which here and there a corpuscle may be seen; and the part of this epithelium which is modified into auditory epithelium is called, as we have said, the *organ of Corti*.

At about the middle of the organ of Corti and forming as it were the key-stone of its structure, are the bodies known as the *rods of Corti* (Figs. 179 *i.r.*, *o.r.*, 180 B, B'), peculiarly modified epithelial cells, arranged along the length of the spiral in two rows, an inner row and an outer row. Each cell in each row has become, in large part, converted into a curved rod of peculiar shape, inclined at an angle to the basilar membrane in such a way that the inner rods and the outer rods lean against each other, their upper parts or "heads" being in contact, but their lower parts, or "feet," which rest on the basilar membrane, being wide apart; hence they with a strip of the basilar membrane lying between their feet, enclose a space, triangular in vertical section, forming along the length of the spiral the "tunnel" spoken of above (Fig. 179).

On the inner or median side of the row of inner rods, lies a single row of epithelial cells bearing hairs, the *inner hair-cells*, seen in vertical section as a single cell (Fig. 179 *i.h.c.*). At the base of the inner hair-cells lies a group of nuclei (*n.c.*) not unlike those forming the nuclear layer of a crista or macula; and just below these, the fibres of the auditory nerve pierce, as we shall see **in** detail presently, the tympanic lip in order to make connections with the epithelial organ of Corti. Next to the inner hair-cells, to the inner side, comes a row of tall columnar epithelium cells (seen in vertical section of course as a single cell); and this row is succeeded in the direction of the spiral groove, by other rows of cells diminishing in altitude until in the groove itself they thin away altogether leaving bare of epithelium the auditory teeth which, as we have said, overhang the groove.

To the outside of the row of outer rods come four (or in certain parts of the spiral three or five) rows of complicated cells, which we may for the present speak of as *outer hair-cells* (Fig. 179 *o.h.c.*). These are succeeded outwards by a group **of** tall columnar or conical cells, of simple character, massed together in a group forming a hump to the outside of the outer hair-cells. These, which are called "Hensen's cells" (*H.c.*), are in turn succeeded outwards by shorter cubical or low columnar cells, called "Claudius cells" (*Cl.c.*), which at the spiral ligament pass into the epithelium of the stria vascularis. In both the cells of Hensen and those of Claudius, the cell substance is granular and is very often loaded with pigment or with material staining deeply with osmic acid or other reagents.

As we shall see, the row of inner and the rows of outer hair-cells are the only cells with which the fibres of the auditory nerve make any connections, and these with the rods of Corti are alone to be regarded as the functional terminal organs of the nerve; it is by means of these structures that the waves of sound are enabled

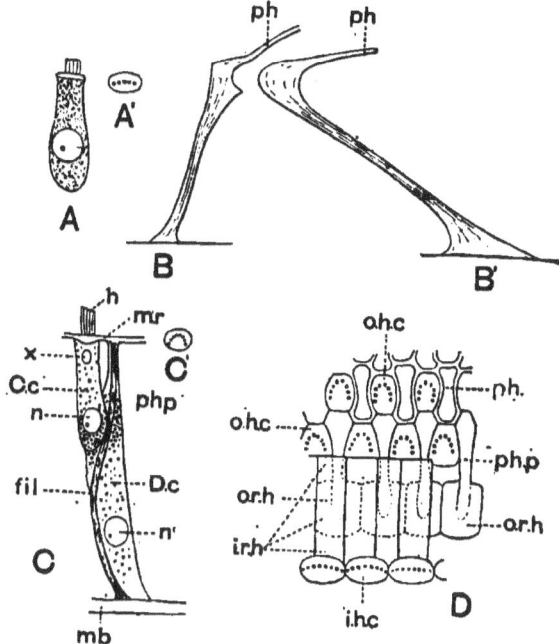

FIG. 180. DIAGRAM OF THE CONSTITUENTS OF THE ORGAN OF CORTI. (After Retzius.)

A. Inner hair-cells. A'. the head seen from above.

B. Inner, B'. outer rod of Corti, *ph*. (in each) phalangar process.

C. The twin outer hair-cell. *C.c.* cell of Corti, *h.* its auditory hairs, *n.* its nucleus, *x*, Hensen's body. *D.c.* cell of Deiters, *n'*. its nucleus, *ph.p.* its phalangar process, *fil.* the cuticular filament, *m.b.* basilar membrane, *m.r.* reticulate membrane.

C'. The head of the cell of Corti as seen from above.

D. The organ of Corti seen from above. *i.h.c.* the heads of the inner hair-cells. *ir.h.* the head and phalangar process of the inner rod. *o.r.h.* the head of the outer rod, with *ph.p.* its phalangar process, covered to the left hand by the inner rods, but uncovered to the right. *o.h.c.* the heads of the cells of Corti supported by the rings of the reticulate membrane. *ph.* one of the phalangæ of the reticulate membrane.

to give rise to auditory impulses in the auditory nerve. The cells of Hensen and Claudius on the outside of the outer hair-cells, and the cells in the spiral groove on the inside of the inner hair-cells, have doubtless parts to play; but their function is probably in some way or other nutritive only, they are not immediately concerned in the production of auditory impulses.

§ **833.** The *inner rod of Corti* (Fig. 180 B) consists of a head, more or less round but with flattened sides, and of a thinner cylindrical body, or limb, which sloping with a gentle curve downwards and inwards ends in an enlarged foot cemented to rather than fused with the basilar membrane just at its beginning outside the tympanic lip of the spiral lamina. From the head a thin flat plate is continued outwards over the outer rods (Fig. 180 B *ph.*, D *i.r.h.*).

The *outer rod of Corti* (Fig. 180 B′) also consists of a head, the rounded surface of which directed inwards fits into a hollow supplied by the head of the inner rod, while the upper surface is prolonged outwards in the direction of the outer hair-cells as a long plate, the "phalangar process" (B′ *ph.*). To the head succeeds a slender cylindrical gently curved body or limb, which sloping downwards and outwards ends in an expanded foot cemented to the basilar membrane at some distance to the outer side of the attachment of the foot of the inner rod.

The substance of which the rods are composed is peculiar. In a perfectly fresh state the rods seem homogeneous or, especially in the region of the limb, obscurely striated; they are somewhat easily decomposed and are readily acted upon by reagents; under the influence of hardening reagents they become rigid, and the the limb is then distinctly striated longitudinally. At each angle, formed by the limb and foot of the inner rod and of the outer rod with the basilar membrane, is seen a nucleus surrounded by ordinary protoplasmic cell substance, and a thin layer of the same cell substance is continued as a delicate lining up the limbs. The rods may be considered as portions of the substance of two cells, represented by the two nuclei just spoken of, which have become specially differentiated in nature, and in being differentiated have assumed a special form. They have been spoken of as 'cuticular' formations, and indeed the phalangar process of the outer rod is the beginning of a structure which may be considered cuticular and which, consisting of rings joined together by flat bars, or "phalangae," stretches outwards so as to form a covering over the whole region of the outer hair-cells (Fig. 180 D). Through the holes of the rings of this *reticulate membrane, membrana reticulata* as it is called, the heads of the outer hair-cells project, and processes from cells which we shall presently describe as connected with or forming part of the outer hair-cells are attached to the bars between the rings. The word "phalangae" means the poles on which a burden is slung between two men's shoulders; and the cells of the organ of Corti may be regarded as slung from the trellis work of the reticulate membrane, or perhaps more exactly slung between it and the basilar membrane. The whole reticulate membrane thus seeming to serve as a support to the outer hair-cells, may justly be regarded as cuticular; and if so we may regard the rods of Corti as cuticular also. It must however be remembered that the rods are very peculiar in nature; one might be inclined to compare them on the one hand with the hyaline border of a ciliated cell, and on the other hand with the outer limbs of the rods and cones of the retina.

The rods thus form along the length of the spiral of the cochlea a double row, inner and outer. In each row the heads are in contact, the adjoining sides being as we said flat; the phalangar processes of the outer row are also in contact with each other, side by side; and the outer heads fit into the inner heads. Hence when the organ of Corti is viewed from above (Fig. 180 D) along a portion of its length, this part of the organ very strikingly resembles the keys of a piano. The inner rods however are more numerous than the outer rods, in the proportion of 5·6 to 3·8, so that each outer rod is not exactly opposite an inner rod, but fits into more than one inner rod.

While, in the case of both inner and outer rods the heads are in contact sideways, that is to say along the length of the spiral, and the same is true of the expanded feet also, the more slender limbs are not in contact but leave spaces or clefts between every two rods in each row. Through these clefts nerve filaments as we shall see make their way from the region of the inner hair cells into the spiral tunnel formed by the rows of inner and outer rods on each side and by the basilar membrane at the base, and beyond this, from the tunnel into the region of the outer hair-cells.

§ 834. The *inner hair-cells* form a single row to the inner side of the inner rods. Each hair-cell (Fig. 180 A) bears much resemblance to, and may be regarded as analogous to a cylinder cell of a crista or macula. It is flask-shaped, ending below in a blunt cone, and bears in its lower part a large spherical nucleus. Its upper end, circular or oval in outline, has a hyaline border like that of a ciliated cell, and from this project a number, a dozen or more, not of long hairs but of short (5 μ) rods, definitely arranged in a gentle curve (Fig. 180 A'), lying at about the middle of the free surface of the cell with the hollow of the curve looking inwards towards the spiral lamina. The substance of the cell is granular, but very watery and very delicate, readily shrinking and becoming deformed under the influence of reagents.

The pointed base of the inner hair-cell dips down into a group of nuclei, which form, as we have said, something like the nuclear layer of the crista or macula. There has been much difference of opinion about these nuclei, but they appear to belong to cells very like the rod cells of the crista and macula. The scanty cell substance round the nucleus is prolonged downwards as a thin process to or towards the tympanic lip and beginning of the basilar membrane and upwards also as a thin process, which running by the side of the inner hair-cells, ends apparently in a cuticular expansion. The nuclei may therefore be considered as belonging to supporting or subsidiary structures. The row of inner hair-cells abuts, on the inner side, on the heads of the inner rods, which thus afford a support to them on this side. On the other side, towards the spiral lamina, the hair-cells are supported by the elongated epithelial cells mentioned above as continuous with those lining the spiral groove and also by the supporting cells of the nuclear layer, the processes of the latter apparently passing also in between the hair-cells in the row, for the hair-cells though near together in the row do not absolutely touch.

§ 835. The *outer hair-cells* are arranged as we have said in a series of rows between the outer rods of Corti and Hensen's cells, the number of rows along the greater part of the spiral being, in man, four. Each row is almost exactly like the others, and the description of any one row will apply to all the others. We have hitherto spoken of the cells as simply outer hair-cells, but each row is composed not of single cells in a file, but of twin cells or of pairs of cells. In each pair we may recognize a cell which bears hairs (or rather rods), the hair-cell proper, or *cell of Corti* (Fig. 180 C. *C.c.*) and a cell which does not bear hairs (or rods) the *cell of Deiters* (*D.c*), the two cells in each pair being in close apposition or according to some observers actually united.

The cell of Corti very closely resembles an inner hair-cell. The body is flask-shaped, and ends in a blunt cone at some distance below the reticulate membrane, between it and the basilar membrane; near its end is placed a large spherical nucleus. The cell-body appears granular, especially at its lower part, but the granules seem to be superficial in position, and the greater part of the interior of the body appears to be of a fluid nature: hence the cell readily shrinks and becomes deformed under the influence of reagents. The upper end, circular in form, projects through, and is as it were grasped by a ring of the reticulate membrane (Fig. 180 D), and the free surface bears, like the inner hair-cell, a row of short rods, but these are arranged as a distinct horse-shoe or a semicircle, with the hollow of the curve looking inwards. The top portion of the cell supplies a hyaline border, and immediately below this is placed a peculiar nuclear looking body, called "Hensen's body" (Fig. 180 C *x.*).

The cell of Deiters (Fig. 180 C. *Dc.*) consists of a cell-body, the median portion of which is placed at the level of the lower end of the cell of Corti, so that this seems to rest on or according to some to be fused with it. From this body there stretches upwards a tapering process (Fig. 180 C. *php*), which joins the overlying reticulate membrane, and becomes attached to the phalangar bar lying to the outside of the ring encircling the head of its twin cell of Corti; it may be called the 'phalangar process.' Downwards the body is prolonged, slanting outwards, as a cylindrical process reaching as far as and becoming attached to the basilar membrane. This part of the cell bears a rounded nucleus (n'), and is, in the greater part of its extent, of a delicate nature and easily destroyed; but on its inside, looking towards the rod of Corti, the cell-substance is differentiated into a cuticular band or thread (fil), which below is cemented to the basilar membrane, and above may be traced into the upward phalangar process and so to the reticulate membrane. There seems good reason for regarding this cell of Deiters as a structure analogous to the rod cells of the crista and macula, but a structure more specially modified than are they. We may probably consider it, like them, to be essentially supporting or at least subsidiary in function, that is to say, not in itself giving rise to auditory nervous impulses but in some subsidiary way assisting the hair-cell proper, that is the cell of Corti, to do so. It will be observed that the cell of Deiters serves as a brace or tie between the reticulate membrane above and the basilar membrane below, while at the same time it is in such complete apposition to if not in continuity with the cell of Corti, that we may justly suppose molecular processes started in it to be readily communicated to that.

The first row of outer hair-cells is placed at some little distance from the outer rod of Corti, being separated from it by a space corresponding to the length of the phalangar process of the rod-head; but the succeeding rows follow close on each other, and the series is closed on the outside towards the spiral ligament by the group of Hensen's cells. The phalangar process of the inner rod is inclined somewhat upwards, and the same inclination is maintained by the reticulate membrane and the whole row of outer hair-cells, there being a gradual ascent from the inner hair-cells to the cells of Hensen.

§ 836. *The cochlear nerve* reaches the organ of Corti through the spiral lamina. The centre of the base of the cochlea round which the first whorl winds is scooped out into a hollow, and from this a central canal, gradually narrowing, runs up the axis of the whorls. The cochlear nerve lies in the hollow and is continued up into the central canal; in its course it gives off from the hollow and from the central canal a series of bundles of nerve fibres which pass radially into the spiral lamina, being like it arranged in a spiral. At the bottom the nerve is thick; it diminishes in bulk as each bundle is given off, and ends by giving off its last bundle near the top of the spiral.

On their way through the spiral lamina in a radiate direction all the bundles become connected with a collection of nerve cells, arranged in a spiral band, the *ganglion spirale*, lying in the spiral lamina (Fig. 177 *Gg.sp.*). The cells of this ganglion closely resemble those of a ganglion on the posterior root of a spinal nerve (§ 97); the nerve fibres are however not connected with the cells by **T** pieces, since the cells are typically bipolar, the fibre entering the cell at one pole and issuing at the opposite pole. The issuing fibre as well as the entering fibre is medullated.

The fibres, issuing from the ganglion in bundles, break up into a loose plexus, and ascend obliquely towards the tympanic lip. As they approach the membranous, connective tissue end of the lip, the fibres are gathered into a-series of more close set plexuses, which pierce the lip through a series of slit-like orifices, *foramina nervina*, and thus, as a series of bundles, enter the overlying epithelium in the region of the inner hair-cells. As they issue from the connective tissue of the lip the fibres lose neurilemma and medulla, and enter the epithelium as naked axis-cylinders.

Concerning the farther course and ultimate endings of the nerve there is much diversity and uncertainty of opinion; but the following account is probably the one deserving the greater amount of confidence.

The axis-cylinders, passing into the epithelium in the region of the nuclear layer beneath the inner hair-cells, split up into fibrillae and bundles of fibrillae. Some of these changing their course from a radiate to a longitudinal, spiral one, contribute to form a strand of fibrillae which runs in a spiral along the length of the cochlea, in the nuclear layer at the base of the inner hair-cells, and may be seen in vertical sections as a group of dots in this position (Fig. 179 *i.sp.n.*); it is known as the *inner spiral strand*. Fibrillae from this strand, or coming directly from the entering axis cylinders, invest the bases of the inner hair-cells with nests of fibrillae, very similar to those which invest the hair-cells of the crista and macula.

Other fibrillae passing between the limbs of the inner rods of Corti, give rise to a second spiral strand lying within the tunnel close to the inner rods (Fig. 179 *t.sp.n.*). This is known as the *spiral strand of the tunnel.*

Other fibrillae again, possibly connected with or joined by fibrillae from the tunnel strand just mentioned, traverse the tunnel in a radiate direction, pass between the limbs of the outer rods, and form beneath the bases of the outer hair-cells proper, that is the cells of Corti, at the level where the bodies of the cells of Corti and of Deiters join each other, four *outer spiral strands* (Fig. 179 *O.sp.n.*), one for each row of hair-cells. From these spiral strands or in connection with these spiral strands, fibrillae invest the bases of the cells of Corti with nervous nests similar to those belonging to the inner hair-cells.

So far therefore as can be at present ascertained the fibres of the auditory nerve end in fibrillae which form nests around both the inner and outer hair-cells; and if we accept the view laid down in § 827 we may say that the mode of ending of the cochlear nerve is fundamentally the same as that of the vestibular nerve. Whether the fibrillae are actually continuous with the substance of the hair-cells, or whether the material of the one is only in close juxtaposition with the material of the other must be left for the present uncertain. We may however conclude that it is in the hair-cells, inner and outer, that auditory impulses are originated, and that the other parts of the organ of Corti are subsidiary in function, helping or guiding in some way the development of the nervous impulses, but not actually giving rise to them.

The space of the tunnel between the rods of Corti appears to be occupied by fluid, which gives support to the bundles of fibrillae stretching across the tunnel. This space is continuous, through the clefts between the limbs of the outer rods, with the space between the outer rods and the first row of hair-cells, and so with the narrow spaces between the several rows of hair-cells. But the whole of this space is completely shut off from the endolymph space of the canalis cochlearis by the reticulate membrane above, by the closely packed epithelial cells on the inner side of the inner hair-cells to the inside, and to the outside by the closely packed cells of Hensen lying on the outside of the outermost row of outer hair-cells; it is also completely shut off from the scala tympani by the basilar membrane.

§ 837. The organ of Corti is overhung by a peculiar structure projecting from the vestibular lip of the spiral lamina, and known as the *tectorial membrane, membrana tectoria* (Fig. 179 *m.t.*). It is in a fresh state soft and elastic, is fibrillated in a radial direction, and indeed appears to be largely composed of fine fibres or fibrils; these resist the action of acetic acid. It begins on the surface of the limbus of the spiral lamina not far from the attachment of Reissner's membrane; it is thin over the vestibular lip, but beyond the free edge of the lip, overhanging the inner hair-cells and rods of Corti, it becomes thick, ending gradually in a thin and often ragged edge at about the zone of the outermost row of outer hair-cells. It is frequently pitted or otherwise sculptured on its under surface.

§ 838. The features and the structure of the canalis cochlearis and especially of the organ of Corti differ in details in different mammals; the description given above applies to man. We said above that the number of rows of outer hair-cells was in man four; but in the lowermost turn of the spiral three only are present, and even in the upper turns, the number four is not constant; in places five rows are sometimes seen. In the majority of mammals there are three rows of outer hair-cells, but a fourth row is sometimes present.

In the same cochlea, the features differ along the length of the spiral, so that a vertical section from an upper whorl presents many differences when compared with a section from a lower whorl. We must not dwell on these differences; but we may call attention to what seems an important fact, namely that the basilar membrane especially that outer part of it which reaches from the foot of the outer rods to the spiral ligament, increases in length from below upwards, except at the very top.

At the top of the spiral the organ of Corti becoming rapidly less conspicuous comes suddenly to an end; the inner and outer hair-cells and rods of Corti suddenly stop, and with them the fibres of the cochlear nerve stop also; the blind end of the canalis cochlearis is lined merely with epithelial cells of a simple character, the continuation of the cells of the spiral groove, of the cells of Claudius, and of the other cells lining the canal; the vascular band is continued a little way beyond the organ of Corti and then comes to an end too, and by thus ending indicates its functional connection with that organ.

At the other, lower extremity or beginning of the spiral, the organ of Corti similarly ceases, and the blind end or "cup" beyond the canalis reuniens, is, like the extreme top, lined by a simple epithelium.

Vertebrates below mammals do not possess an organ of Corti strictly so called; in the rudimentary cochlea of birds and reptiles a basilar membrane and hair-cells are present; but the rods of Corti are absent. As we pass in review the features of the membranous labyrinth in various vertebrates from the lowest upwards, we find the several parts becoming more and more distinct from each other; the saccule becomes more and more separate from the utricle and the semicircular canals, and the cochlea, which is at first a process of the saccule, becomes more and more distinct from it.

§ 839. It may be worth while to call attention to the following data concerning the cochlea of man which have been obtained by careful partial measurements and calculations.

Length of the Canalis Cochlearis	35 mm.
Length of the organ of Corti	33·5 mm.
Radial width of the Basilar Membrane (measured from the entrance of the nerve fibres to the spiral ligament)	
In the Basal whorl of spiral	·21 mm.
„ Middle „ „	·34 mm.
„ Topmost „ „	·36 mm.
Number of perforations for nerve fibres	4000.
„ „ Inner Hair-cells	3500.
„ „ Inner Rods of Corti	5600.
„ „ Outer „ „ „	3850.
„ „ Outer Hair-cells (in 4 rows)	12000.
„ „ Fibres of the Basilar Membrane	24000.

SEC 3. ON AUDITORY SENSATIONS.

§ 840. The vibrations which we call sound are transmitted as we have seen to the perilymph through the fenestra ovalis, by means of the tympanic membrane and chain of ossicles. The vibrations of the perilymph in some way or other, by help of the auditory epithelium, give rise in the fibres of the auditory nerve to auditory impulses, and these reaching the brain are developed into auditory sensations. Before we attempt to consider how the vibrations of the perilymph thus give rise to auditory impulses it will be convenient to adopt the plan which we pursued in the case of vision and to deal first with some of the leading characters of auditory sensations such as can be ascertained by psychological methods.

We readily recognize two classes of sensations; the objective causes of the one class we speak of as *noises*, those of the other class as *musical sounds*. When we inquire into the physical features of the two classes we find that the vibrations which constitute a musical sound are repeated at regular intervals, and thus possess a marked periodicity or rhythm. When no marked periodicity is present in the vibrations, when the repetition of the several vibrations is irregular, the sensation produced is that of a noise. There is however no abrupt line between the two. Between a pure and simple musical sound produced by a series of vibrations each of which has exactly the same period, and a harsh noise in which no consecutive vibrations are alike, there are numerous intermediate stages. Much irregularity may present itself in a series of sounds called music, and in some of the roughest noises the regular repetition of one or more vibrations may be easily recognized. Still it will be desirable to consider the two classes as distinct, and it will be convenient to deal first with musical sounds.

§ 841. The sensations which are produced by musical sounds possess three marked characters. In the first place our auditory sensations like our other sensations, may be more or less intense; and the character in a musical sound which corresponds to the intensity of the sensation we call *loudness*. This is determined by the amplitude of the vibrations, by the amount of energy which is expended in producing the vibratory movements; the greater the disturbance of the air (or other medium) the louder the sound. Using the term 'wave' to denote the characters of the vibrations, the loudness of a sound is indicated by the *height* of the wave.

In the second place we recognize a character which we call *pitch*. This is determined by the frequency of repetition of the vibrations, by the time taken up by each vibration; the greater the number of consecutive vibrations which fall upon the ear in a second, the shorter the time of each vibration, the higher the pitch. Hence the pitch of a sound is indicated by the *length* of the wave, a low note having a long, a high note a short wave-length. We are able to distinguish a whole series of musical sounds of different pitch, from the lowest to the highest audible note.

In the third place, we distinguish musical sounds by what is usually called their *quality* (*timbre*); the same note sounded on a piano and on a violin produces very different sensations, even though the two instruments give rise to vibrations having the same period of repetition. This arises from the fact that the musical sounds generated by most musical instruments are not simple but compound vibrations; the instrument sets going in the surrounding air not one series only of vibrations of one wave-length, but several series of different wave-lengths; as we shall see however, the several vibrations travel through the air, not as a group of waves but as one compound wave. When the note C in the bass clef is struck on the piano, and we analyse the total sound, we find that it can be resolved partly into a series of vibrations with a period characteristic of the pure tone of C of the bass clef, and partly into other series of vibrations with periods characteristic of the C in the octave above (middle C), of the G above that, of the C of the next octave, and of the E above that. And the sensation which we associate with the sound of the C in the bass clef on the piano is determined by the characters of the complex vibration rising out of these several constituent simple vibrations. Almost all musical sounds are thus composed of what is called a *fundamental tone* accompanied by a number of *partial tones*. When a violin string gives out a musical note, the fundamental tone is produced by the string vibrating along its whole length, the partial tones by the string vibrating at the same time in segments or definite parts of the whole length; and so with other instruments; hence the name 'partial.' Since these partial tones have a higher pitch than the fundamental tone they are frequently spoken of as 'partial upper-tones or overtones' or simply as 'overtones.' The partial tones vary in number and relative prominence in different instruments and thus give rise to a difference in the sensation caused by the whole sound. Hence while a 'tone' is a single series of simple vibrations, a 'note' may be and generally is a number of series of different vibrations occurring together. While the fundamental tone determines the pitch of a note, the quality of the note is determined by the number and relative prominence of the partial tones.

If we compare auditory with visual sensations it is obvious that loudness of sound corresponds to brightness or luminosity of light; in both cases the terms denote the intensity of the sensation. We may perhaps compare the character of pitch, dependent on the wave-length of the sound vibration, to the character of colour dependent on the wave-length of the ray of light; and we may, in a general way, liken the auditory effect produced by a sound of a particular quality to the visual effect produced by an object which excites mixed colour sensations, owing to rays of several different wave-lengths falling at the same time on the retina; on examination however it will be found that the differences in these respects between the two sets of sensations are more striking than the resemblances.

§ 842. In much the same way that rays of light of more than or of less than a certain wave-length are incapable of exciting the retina, our vision being limited to the range of the visible spectrum, waves of sound of more than or of less than a certain wave-length are unable to affect the ear so as to produce a sensation of sound. Vibrations having a recurrence below about 30 a second are unable to produce a sensation of sound; the note of the 16-feet organ pipe, 33 vibrations a second, gives us the sensation of a droning sound; a tone of 40 vibrations is quite distinct. Some authors however place the limit at 24 or even 15 a second. If waves of long wave-length are powerful enough we may feel them by the sense of touch, though not by that of hearing. What we have just said applies to vibrations which are simple, such as give rise to a pure tone; if the fundamental tone is accompanied by partial tones we may hear one or other of these, and are thus apt to say we hear the former when in reality we only hear the latter. As regards the limit of high notes, it is possible to hear a note caused by vibrations as rapid as 40,000 a second; at least some persons have this power, though the limit for most persons is far lower, about 16,000. Some persons hear low sounds more easily than high ones, and *vice versa*. This may be so pronounced as to justify the subjects being spoken of as deaf to low or high tones respectively, a condition which may be compared in a general way to colour blindness. The range in different animals differs very widely, the high notes of the instrument known as Galton's whistle, though inaudible to man, are distinctly heard by some other animals, for instance cats.

The limitations which are thus imposed on our hearing do not wholly correspond to the limitations of our vision. In the latter case the limits are fixed wholly by the capacities of the retina and cerebral centres; radiant rays of longer wave-length than the extreme visible red are able to get access to the retina through the dioptric apparatus though they are unable to excite visual impulses, or at least such visual impulses as can affect consciousness. In the case of hearing, though the auditory epithelium is probably, like the retinal structures, limited in its powers, narrower limits are fixed by the subsidiary acoustic apparatus; the tympanic membrane, extensive as is its range compared with that of most artificial membranes, cannot respond to all vibrations; and hence its powers fix the limits of hearing. The reason why we appreciate high notes more readily than low ones is probably to be referred to the tympanum rather than to the auditory epithelium. And the condition of the tympanal apparatus as affected by disease will determine the relative appreciation of low or high tones; in certain states of the tympanum the ear becomes unusually sensitive to high notes; an instance of this is seen in the paralysis of the stapedius muscle due to injury or disease of the seventh nerve.

§ 843. We dwelt, in speaking of vision, on our power of appreciating differences of brightness or luminosity; we have a similar power of appreciating differences in loudness; and that relation between differences in the intensity of the stimulus and differences in the intensity of the sensation, which we spoke of as Weber's law (§ 747), holds good for hearing as well as for vision.

The power of distinguishing difference of pitch, the power of recognizing the difference between two notes of different pitch, and the appreciation of the qualities of various musical sounds which is built up on this, may in a general way be compared to acuteness of colour vision. It is however, as we have said, very different from that in many respects, and varies much more widely than does that. As is well known the difference in this power between different individuals, according as they have or have not a 'musical ear,' is very great. Some persons even though fairly sensitive to differences of loudness, are unable to distinguish two notes differing considerably in wave-length. On the other hand a well-trained ear can distinguish the difference of a single or even of a half vibration a second, and that through a long range of notes. As might be expected the power of appreciating difference of pitch is not the same for all audible notes. The range of an ordinary appreciation of tones lies between 40 and 4000 vibrations a second, *i.e.* between the lowest bass C (C_1 33 vibrations) and the highest treble C (C^5 4224 vibrations) of the piano; tones above and below these, even though audible, are distinguished from each other with great difficulty. The power of recognizing, and being able to name, a note when heard, is an extension of and based upon this power of recognizing differences of pitch, though not by itself exactly the same thing.

§ 844. We said, in speaking of vision (§ 748) that, probably, several undulations falling in succession upon the retina were necessary for the development of a visual sensation. In like manner, in order that a distinct sensation of a musical sound may be developed, several, or at least more than one wave of sound must fall on the ear. The various observers are not agreed as to the lower limit of the number of vibrations necessary in order that the affection of consciousness may take the form of a definite musical sound; some place it at five, others higher, while it has been asserted that two vibrations are sufficient. When the vibrations are thus limited in number, the sound even though it is recognized as a musical sound, is not clearly appreciated; its pitch is not distinctly recognized. In such a case the recognition may be made more full and certain by increasing the number of vibrations; in order that we may appreciate the pitch of a sound the ear must receive a larger number of vibrations than are necessary merely to enable us to recognize that the sound is a definite one. Conversely even when the vibrations are too few to give rise to a sensation of a definite tone, consciousness is not wholly unaffected, an auditory sensation is produced, though it cannot be called one of tone. These facts indicate the complex nature of the nervous processes which form the basis of auditory sensations; we might say this of sensations in general, for similar results are observed in the case of all sensations.

§ **845.** As we said above (§ 840) noises are not sharply defined from musical sounds, they differ only in being more complex and less regular; and what has just been said in respect to musical sounds, holds good to a large extent for noises. We readily distinguish, in noises, difference of loudness; we may also in many cases recognize a dominance of pitch, due to the fact that among the multifarious vibrations certain groups of vibrations are repeated periodically; we distinguish a rumbling noise in which vibrations of slow recurrence are prominent from a harsh shrill noise in which rapid vibrations are similarly prominent; we also recognize qualities in noises, we distinguish one noise from the other by the characters of the predominant constituent vibrations. Owing to the fact to which we just now referred that in a musical sound the effect on consciousness is a summation of the individual effects of the several vibrations we are more sensitive to a musical sound of not too short duration, than to a noise involving an equal expenditure of energy. On the other hand the limit of the number of movements necessary to give rise to a sensation of noise is less than that required for a musical sound; a few vibrations insufficient in number to give rise to the sensation of a tone are able to give rise to an auditory sensation which we may call a noise, and probably one movement of the tympanic membrane might if ample enough give rise to such an auditory sensation. Moreover owing to the very irregularity of a noise, to the varied character of the constituent molecular movements, we have a very great range in distinguishing various noises; persons who have great difficulty in detecting different notes can often readily recognize differences in noises.

§ **846.** In treating of vision we dwelt at some length on the phenomena of exhaustion which make their appearance when the stimulus is continued. These occur in hearing also, and indeed are indicated by such common phrases as "a deafening noise;" but they are not so prominent as in vision, and do not so distinctly serve as the basis for theoretical discussions. They are best studied by means of musical sounds, since with these owing to their very nature the stimulation is more uniform than with noises. With almost any note, the sensation diminishes and finally disappears if the sound be maintained long enough; but the exhaustion comes on more rapidly with high than with low notes, especially with very high ones. If a sounding tuning-fork be held up to one ear, and then, just as the sound becomes inaudible be transferred to the other ear, the sound may be distinctly heard; the fresh untired sensory apparatus of the one side is sensitive to the vibrations which the tired apparatus of the other side can no longer feel. Or, if the tuning-fork which the tired ear can no longer hear, be replaced by one vibrating at the moment as far as can be arranged with the same intensity as it, but of distinctly different pitch, this will be heard; the first tuning-fork only tired certain parts of the sensory apparatus, those affected by vibrations of a certain period characteristic of the pitch of that tuning-fork, but left untired the parts of the sensory apparatus responding to the vibration of other periods, such as those of the second tuning-fork.

Again, the quality of a note struck on a musical instrument depends as we have seen on the presence of partial tones, having certain relations to the fundamental tone. Now, if immediately before striking a note on an instrument, choosing especially an instrument whose notes are 'rich' by virtue of the number or prominence of the partial tones, we cause one of the partial tones of the note to be sounded powerfully in the ear, the note when subsequently struck does not possess its full quality; it appears 'thin' or 'poor.' This is because the previous sounding of the partial tone has tired the particular part of the auditory apparatus with which we hear the partial tone, and in the whole sensation of the subsequent full note the constituent sensation corresponding to that particular partial tone is absent or at least is below its normal intensity. Thus we have in auditory sensations something analogous to the "negative image" of visual sensations.

We do not in hearing experience a sensation analogous to the visual sensation of white light, a simultaneous stimulation of the apparatus by vibrations of all kinds, and cannot therefore experience an auditory sensation corresponding to the visual sensation of black; the nearest approach perhaps to such a psychological condition is that in which we are placed upon the sudden cessation of powerful and varied music; at such times we seem to be the subject of a "silence which can be heard."

§ 847. As in the case of visual sensations, so likewise in the case of auditory sensations the duration of the sensation is longer than that of the action of the stimulus, the auditory sensation lasts after the waves of sound have ceased to fall upon the ear. Hence when two sensations follow each other within a sufficiently short interval, they are fused into one. Since a membrane, thrown into vibrations by a passing sound may continue to vibrate after the sound has ceased, we might perhaps expect that this would be the case with the tympanic membrane, and that hence the interval of fusion would be longer in the case of hearing than in that of vision, for in the latter case we have no corresponding behaviour of any part of the dioptric apparatus. But we have seen (§ 816) that the acoustic arrangements of the tympanum very rapidly damp the tympanic membrane; and, as a matter of fact, the interval in question is decidedly shorter in hearing than in vision. Visual sensations separated by less than $\frac{1}{10}$ sec. become fused (§ 749); but auditory sensations separated by not more than $\frac{1}{100}$ sec. may remain distinct; if two seconds pendulums be set swinging not quite in accord with each other and made to tick, the tick of the one can be distinguished from that of the other even when they differ in time by not more than $\frac{1}{100}$ sec.

§ 848. When two notes are sounded at the same time the two sound waves (we may suppose the notes to be pure ones, consisting of a fundamental tone only without partial tones) do not travel as two separate waves, but are compounded as we have already said, into a single wave, the characters of which will depend on the relative characters of the two constituents. If the two notes have the same period, that is to say are identical, the effect will be simply an increase in amplitude; the compound wave will have its crests higher, and its troughs deeper than those of either of the single waves, but will otherwise be like both of them. If two tuning-forks of exactly the same pitch be struck, the sensation which we experience is the same as that which we experience from either of them alone, only more intense; the sound is louder.

If however the two tuning-forks are not of the same pitch, but so related that the period of vibration of the one is not an exact multiple of that of the other, the sensation which we experience when the two sound together has certain marked features. We hear a sound which is the effect on our ear of the compound wave formed out of the two waves; but the sound is not uniform in intensity. As we listen the sound is heard now to grow louder and then to grow fainter or even to die away, but soon to revive again, and once more to fall away, thus rising and falling at regular intervals, the rhythmic change being either from sound to actual silence or from a louder sound to a fainter one. Such variations of intensity are due to the fact that, owing to the difference of pitch, the vibratory impulses of the two sounds do not exactly correspond in time. Since the vibration period, the time during which a particle is making an excursion, moving a certain distance in one direction and then returning, is shorter in one sound than in the other, it is obvious that the vibrations belonging to one sound will so to speak get ahead of those belonging to the other; hence a time will come when, while the impulse of one sound is tending to drive a particle in one direction, say forwards, the impulse of the other sound is tending to drive the same particle in the other direction, backwards. The result is that the particle will not move, or will not move so much as if it were subject to one impulse only, still less to both impulses acting in the same direction; the vibrations of the particle will be stopped or lessened, and the sensation of sound to which its vibrations are giving rise will be wanting or diminished; the one sound has more or less completely neutralized or "interfered" with the other, the crest of the wave of one sound has more or less coincided with the trough of the wave of the other sound. Conversely at another time, the two impulses will be acting in the same direction on the same particle, the movements of the particle will be intensified, and the sound will be augmented. And the one condition will pass gradually into the other. The repetitions of increased intensity thus brought about are spoken of as *beats*.

The length of the interval at which the beats recur will depend on the difference of period of the two sounds in relation to the actual period or pitch of each. It may be stated generally that the number of beats in a second is equal to the difference between the number of vibrations per second of the two sounds; thus two very low pitched tuning-forks, vibrating respectively at 64 and 72 a second, will give 8 beats a second, and two very high pitched tuning-forks, vibrating respectively at 4224 and 4752 a second will give 528 beats a second; but in this respect there are complications which we cannot consider here.

Beats are produced when the periods of the coincident sounds are not exact multiples of each other. When the periods are exact multiples no beats occur; two tuning-forks, for instance, the period of one of which is exactly double that of the other, give rise to no beats when sounding together; and so in other instances.

By beats then a continuous musical sound may be broken up into a series of discontinuous sounds. When the beats are repeated a few times only in a second the discontinuous sounds give rise to discontinuous sensations; we hear the separate beats. But if the beats are repeated sufficiently rapidly the successive sensations are fused in one, we cease to hear the beats as such, though we have other evidence that the beats continue to be produced. Just as a series of simple vibrations when repeated sufficiently rapidly, say 40 times a second, gives rise, by summation, to a single musical sound, to a tone, so a series of groups of vibrations, each group corresponding to the interval between two beats, gives rise when the groups follow each other rapidly, by a similar summation, to a continuous sensation. And, though the matter is one which has been much disputed, the evidence seems to shew that the continuous sensation thus produced is a musical sound, a tone, which has been called a "beat-tone," whose pitch is determined by the number of beats repeated in a second.

The rapidity however with which beats must be repeated in order to give rise to a continuous sensation, is different from that with which single vibrations must be repeated in order to give rise to a musical sound. Beats repeated 30 or 40 times a second are readily distinguished as such; it is not until they reach a rapidity of repetition of about 132 a second that they cease to be distinctly recognized. Before they disappear or as they disappear, at the time when they can no longer be recognized as separate beats, but have not as yet become fused into a completely continuous sensation, they give to the sound which they accompany a peculiar quality, a particular roughness and harshness. This quality if excessive is disagreeable to the ear; we speak of it as dissonance.

From what has been said it is obvious that when a piece of music is played on an instrument and still more when it is played, as in a concert, on several instruments of different kinds, the disturbance in the air, and the consequent vibrations of the tympanic membrane and of the perilymph, are in the highest degree complex. If the disturbance has certain characters, the sound gives us pleasure, if other characters, we regard the sound as disagreeable; and it is found that the disagreeable features of music are associated with the presence of beats, and still more with the presence of that ill-defined roughness which, as we said just now, is the characteristic of beats when, through rapidity of repetition, they are about to disappear. At the same time there are reasons for thinking that it is the prominence rather than the mere presence of this element which offends the ear, that the element is a necessary ingredient of effective music, and that even the very quality of a musical sound is dependent in part on a certain minute admixture of vibrations disagreeing in period with the fundamental tone and with the regular partial tones. But this is a matter into which we cannot enter here; we have referred to it because it illustrates the extreme complexity of the processes which underlie our sensations of sound.

SEC. 4. ON THE DEVELOPMENT OF AUDITORY IMPULSES.

§ **849.** We may now turn for a little while to the obscure question, How the vibrations of the perilymph give rise to auditory impulses and so to auditory sensations.

In speaking of the ossicles (§ 817) we gave reasons for thinking that the vibrations of the tympanic membrane are carried onward by the chain of ossicles swinging as a whole, and not conveyed through the chain from molecule to molecule. A similar argument may be applied to the perilymph. The dimensions of the whole labyrinth compared with the length of the waves of sound are so minute that molecular vibrations may be neglected. Moreover the walls of the labyrinth may, as a whole, be regarded as absolutely rigid so that, the perilymph being incompressible, each blow given at the fenestra ovalis is transmitted instantaneously through the whole mass of perilymph; the fluid driven in by the inward thrust of the stapes has to find room for itself elsewhere, and that room is furnished by the outward bulge of the membrane of the fenestra rotunda, for we may neglect other means of escape such as the lymph spaces around the endolymphatic duct, the nerves and the blood vessels. Hence at each movement of the stapes the whole mass of the perilymph swings bodily, the membrane of fenestra rotunda moving outwards and inwards at the same instant that the stapes moves inwards and outwards; and each such mass-vibration of the perilymph repeats the characters of the vibration of the ossicles and tympanic membrane, of which it is the continuation.

As they sweep over the vestibule, these vibrations are communicated through the walls of the enclosed membranous labyrinth to the endolymph. The vibrations of the endolymph, or of the walls themselves, affect in some way or other the auditory epithelium of the three cristae and the two maculae.

The vibrations also travel from the vestibule into the scala vestibuli of the cochlea, ascending the spiral from below upwards. As they ascend they are transmitted across the membrane of

Reissner, the endolymph of the canalis cochlearis, and the basilar membrane to the scala tympani, and so reach the fenestra rotunda. The bulk of the vibrations ascending the scala vestibuli thus reach the scala tympani by crossing the canalis cochlearis, and in so crossing affect in some way or other the auditory epithelium of the organ of Corti, it is probably only a remnant which at the summit of the spiral passes directly from the one scala to the other.

The features of the basilar membrane point to its being readily thrown into vibrations, and we may conclude that the vibrations started at the fenestra ovalis and transmitted from the scala vestibuli to the scala tympani throw the basilar membrane into corresponding vibrations.

In the cochlea the connection of the fibres of the auditory nerve seems to be exclusively limited to the hair-cells, inner and outer; and we may conclude that these hair-cells are in some way or other concerned in the development of auditory impulses. This view is supported by the analogy of vision; for we have seen reason to think that visual impulses begin in the rods and cones, which like the hair-cells are modified epithelium cells, and we shall presently find that modified epithelium cells also play an essential or important part in the development of sensations other than those of vision and hearing.

Accepting provisionally the view (we will return to the point later on), that the vestibular nerve also conveys auditory impulses started in the cristae and maculae, and leaning on the analogy of the cochlea in which as we have just said it seems clear that the hair-cells alone are connected with the auditory fibres, we may conclude, in spite of the divergence of opinion as to the results of histological inquiry, that in the cristae and maculae it is the hair-cells also which in some way or other originate auditory impulses. The relatively long hairs of the cristae, exceedingly long in some of the lower animals, seem specially fitted to take up the vibrations of the endolymph, that is to say, to be thrown into bodily (transversal) vibrations and to communicate those vibrations to the substance of the cell which bears them.

It has been observed that, among the auditory hairs of the crustacea, some will vibrate to particular notes; and as we shall see presently, the vibration of particular hairs in response to particular tones might play an important part in the development of auditory sensations. But the auditory hairs of the mammal even in the cristae are far too much of the same length to permit the supposition that they can act satisfactorily in this way. Moreover even such fitness as may seem to be present in the cristae is less conspicuous in the shorter hairs of the maculae, and vanishes wholly in the case of the peculiar short rods of the hair-cells of the cochlea; we can hardly imagine that the latter vibrate bodily, and even in the maculae and cristae, the hairs consisting as they do of fine filaments tending to cohere together, seem imperfect instruments for vibratory purposes. On the other hand in the case both of the cochlea, the maculae and the cristae, the hairs of the hair-cells have close relations with structures which have the aspect of being 'damping' organs, namely the tectorial membrane over the organ of Corti, and the otolith membrane over the macula, to which we may add the cupula over the crista. It has indeed been suggested that the otoliths may serve as exciting adjuvants of the vibrations of the fluid endolymph, may act as it were as minute hammers enforcing the blows of the fluid; but this is negatived by observations on the auditory organs of certain molluscs. In these animals the organ of hearing is a closed spherical vesicle, part of the lining of which is furnished by an auditory epithelium of hair-cells with short rod-like hairs, and the rest by ciliated epithelium, the former being alone in connection with a nerve which seems to function as an auditory nerve. Within the vesicle lies a large otolith, and it has been observed that while under ordinary circumstances when sounds of not too great intensity are directed towards the object the otolith is kept away from the auditory hairs, but that when too intense a sound is brought to bear upon the organ, the otolith is driven by the cilia down upon the auditory hairs, and obviously behaves as a damper. We may probably conclude therefore that the membrana tectoria, and the otolith membrane act as dampers; but if so the view seems natural that the so-called auditory hairs are useful for conveying the damping action from the damper to the hair-cell, and do not serve to communicate the vibrations of sound from the endolymph to the hair-cell. This would lead us to the conception that both in the organ of Corti and in the macula and crista, the vibrations reach the bodies of the hair-cells in some other manner, not by means of the hairs, but by means of the other structures which go to make up the auditory epithelium; and indeed it is much more in respect to the general arrangements of the auditory epithelium than to the special characters of the hair-cells that the organ of Corti, which we must certainly regard as the chief auditory mechanism, is so much more highly differentiated than the macula or crista. The hair-cells of the cochlea and of the vestibule are exceedingly alike; but very different is the organ of Corti, with its basilar membrane, reticulate membrane, rods of Corti, and rod cells specialized into cells of Deiters, from the simple macula or crista, which contain besides the hair-cells nothing except the relatively plain rod cells. We can hardly resist the conclusion that the structures just mentioned play an important part in bringing the vibrations to bear on the hair-cells. But what that part is remains for the present mere subject of speculation.

§ 850. A complex sound, consisting of vibrations of more than one period, travels as we have said not as a group of discrete waves, each corresponding to a vibration of a particular period, but as a complex wave in which the simple waves are compounded into one ; and the vibrations of the tympanic membrane, followed by the vibrations of the perilymph, have the same composite character. When for instance a note is sung, or sounded on a musical instrument, the air in the external auditory passage is not the subject of one set of waves corresponding to the fundamental tone, and of other sets corresponding to the several partial tones, but vibrates in the pattern of one composite wave; the tympanic membrane executes one complex vibration, and a corresponding single complex vibration excites the auditory epithelium. And this holds good not for a single sound only but for a mixture of sounds. We can in a clumsy way take a graphic record of the vibrations of a dead tympanic membrane, by attaching a marker to the stapes ; could we take an adequate record of the movements of the living tympanum of one of the audience at a concert, we should obtain a curve, a phonogram, which though a single curve only would be on the one hand a record of the multitudinous vibrations of the concert, and on the other hand a picture of the actual blows with which the perilymph had struck the auditory epithelium.

Now, whatever be the exact nature of the process by which the vibrations of the perilymph give rise to auditory impulses, we may consider it as probable that, in giving rise to those impulses, the complex vibration is analysed again into its constituent simple vibrations, that the vibrations start afresh so to speak in the auditory epithelium, marshalled in the same array as that in which they started from the sounding instruments, as if the auditory epithelium itself constituted the band playing the music. And indeed that something of this kind does take place is indicated by the fact that an adequately sensitive ear can in a musical sound detect one or more of the partial tones as distinct from the fundamental tone, or still more easily can in a mixed concert detect the several notes of the several instruments, though as we have just said in the movements of the tympanic membrane, all the constituent factors are merged into one complex sweep. We may conclude then that we possess some means of analysing the composite waves of sound which sweep through the perilymph, and of sorting out their constituent vibrations.

There is at hand a simple and easy physical method of analysing composite sounds. If a person standing before an open pianoforte, the loud pedal being held down, sings out any note, it will be observed that a number of the strings of the pianoforte will be thrown into vibration, and on examination it will be found that those strings which are thus set going correspond in pitch to the fundamental tone and to the several partial tones of the note sung. The note sung reaches the strings as a complex wave, but the strings are able to analyse the wave into its constituent vibrations, each string taking up those vibrations and those vibrations only which belong to the tone given forth by itself when struck. If we suppose that each terminal fibril or each group of fibrils of the auditory nerve is connected with a terminal organ so far like a pianoforte-string that it will readily vibrate in response to a series of vibrating impulses of a given period and to none other, and that we possess a number of such terminal organs sufficient for the analysis of all the sounds which we can analyse, and that each terminal organ so affected by particular vibrations gives rise to a sensory impulse and thus supplies the basis for a sensation of a distinct character — if we suppose these organs to exist, our appreciation of sounds is in part explained.

When the rods of Corti were first discovered, it was thought that they were specially connected with the nerve fibres, and served mechanically to stimulate the fibrils passing along their limbs, by striking them after the fashion of minute hammers. Since these rods, to whose striking resemblance to the keys of a pianoforte we have already called attention, are arranged in a long series the members of which vary regularly in the length and in the span of their arch, from the bottom to the top of the spiral, it was supposed that each pair would vibrate in response to a particular tone, and hence that the whole series served for the analysis of sound.

But this view proved untenable. Whatever purpose they serve, the rods of Corti produce their effect, not by acting directly on nerve fibrils, but by contributing in some way or other to the play of the hair-cells; and, whatever be the way in which they intervene, they do not vary in length and arrangement along the spiral to such an extent as the above view demands. Moreover, they are wholly absent from the rudimentary cochlea of birds, though these creatures very clearly can appreciate musical sounds. This last fact proves indubitably that the rods in question are not absolutely essential for the recognition of tones, since it is in the highest degree improbable that birds are able to recognize tones in some manner absolutely different from that employed by mammals.

In the face of these difficulties it has been suggested that the basilar membrane, which is present in birds as well as in mammals, and which, being tense radially but loose longitudinally, *i.e.* along the spiral of the cochlea, may be considered as consisting of a number of parallel radial strings, each capable of independent vibrations, is the sought-for organ of analysis; for it may be shewn mathematically that a membrane so stretched in one direction only is capable of vibrating in such a manner. And the radial dimensions of the basilar membrane increasing as they do upwards from the bottom of the spiral to near the top give a much greater range of difference than do the rods of Corti. According to this view, when a composite vibration sweeps along the cochlea it throws into sympathetic vibrations those small portions and those portions only of the basilar membrane, the vibrations of which correspond to the single vibrations of which the composite vibration is made up, and the vibrations in turn so affect the overlying structures, that auditory impulses are generated in particular groups of fibrils of the auditory nerve. These auditory impulses reaching the brain give rise to a corresponding sensation of a particular sound.

But the dimensions of even the basilar membrane do not seem wholly adequate for the purpose; since the latest measurements shew that in man its range is very limited. If we take the whole width of the membrane, the range is from ·21 mm. at the base to ·36 mm. at the top, though if we take the specially modified part reaching (§ 831) from the outer feet of the rods to the spiral ligament, we get a wider range, namely, from ·075 mm. at the base to ·126 mm. at the top. On the other hand the estimated number of radial fibres of the membrane is very large, 24,000; and even if we suppose that several fibres always vibrate together, this would still leave some thousands of groups of strings, each group acting as an analyser.

In the present state of our knowledge the whole matter must be left as uncertain. Even if the basilar membrane acts in some such way as suggested, the other structures in the auditory epithelium present problems as yet insoluble. The true function of the rods of Corti and of the reticulate membrane of which these form a part, of the cells of Deiters, of the inner hair-cells as distinguished from the outer hair-cells, as well as the reason there are four rows of the latter (whereby probably the effect of the vibrations of a group of the basilar fibres is increased) and only one of the former, all these are as yet merely questions which cannot be answered.

§ 851. Even admitting that, in some way or another, sets of vibrations or, to use a more general term, sets of molecular movements are started in the auditory epithelium, in more or less complete correspondence with the sets of vibrations which originate from the musical instrument or other sounding body, and admitting further that each set of such molecular movements in the auditory epithelium starts a particular nervous impulse in a fibril or in a set of fibrils of the auditory nerve, we are very far from having solved the problem of hearing.

It must be borne in mind that making the fullest allowance for the assistance afforded us by the organ of Corti, the appreciation of any sound is ultimately a psychical act. The analysis of the vibrations by help of the basilar membrane or otherwise is simply preliminary to a synthesis of the auditory impulses so generated into a complex sensation. We do not receive a distinct series of specific auditory impulses resulting in a specific sensation for every possible variation in the wave-length of sonorous vibrations any more than we receive a distinct series of specific visual impulses for every possible wave-length of luminous vibrations. In each case we have probably a number of primary sensations, from the various mingling of which, in different proportions, our varied complex sensations arise; but there is this difference between the eye and the ear that whereas in the former the number of primary sensations appears to be limited to three or at least to six, in the latter the number is probably very large; what the exact number is has not at present been even suggested. Our appreciation of a sound is at bottom an appreciation of the combined effect produced by the relative intensities to which the primary auditory sensations are, with the help of the organ of Corti, excited by the sound. The appreciation and the subjective analysis of sounds is ultimately a psychical process; and though there are individual differences in the structural finish and physical capabilities of the auditory epithelium as of other parts of the ear, the differences in the psychical or at least cerebral powers of individuals are far greater; and when we speak of a musical ear we really mean a musical mind or a musical brain.

§ **852.** If the organ of Corti, as appears from the above, affords the means by which we appreciate tones, it is evident that by it also we must be able to estimate loudness, for the quality of a musical sound is dependent on the intensity, as well as on the pitch, of the partial tones in relation to the fundamental tone and to each other. And since noise is but confused music, and music more or less orderly noise, the cochlea must be a means of appreciating noises as well as sounds. But if this be so, what functions are fulfilled by the vestibular division of the labyrinth?

We have seen (§ 643) reason to think that the fibres of the vestibular nerve convey afferent impulses which are not auditory in nature in the sense of serving as the basis of sensations of sound, but which affect in a special way the movements of the body. We then discussed the matter exclusively in reference to the cristæ of the semicircular canals; but the view has been extended to the maculæ of the utricle and saccule. It is contended that while the former are affected by movements involving a rotation of the head, the latter are affected by movements in which the head is carried in a straight line. either vertical as when the body rises or falls, or horizontal as when the body progresses forwards or backwards. It is maintained that the utricle and saccule thus share with the ampullæ in providing impulses by which we become aware of the direction and amount of the movements we execute. Indeed it is urged that the true function of the otoliths and otoconia has no connection with hearing, that these act neither as adjuvants to nor as dampers of an auditory mechanism, but by impinging on the so-called auditory hairs in the movements of the head give rise to the impulses of which we are speaking; and arguments in favour of this view are drawn not only from the arrangement of the otoliths in vertebrates, but also from the distribution and behaviour of similar structures in invertebrates. We have further seen (§ 618) that the vestibular nerve, the division of the auditory nerve distributed to the maculæ and cristæ, differs in its characters, in its mode of development and especially in its central tracts and central endings from the cochlear nerve, the division of the auditory nerve distributed exclusively to the organ of Corti in the cochlea; and it is stated that the vestibule may be injured without any marked effect on hearing. But we are not thereby justified in concluding, as some have done, that the vestibular nerve does not in any way serve as the channel of auditory impulses. In the first place, the evidence concerning the ampullar afferent impulses, as (§ 643) we called them, is by no means complete; it is still less complete as regards the utricle and saccule. It is certainly not strong enough to justify the conclusion that auditory impulses are not generated in the vestibule, whatever else may happen there. In the second place, vertebrates, lower in the scale than birds and reptiles, namely, fishes, though they have a well-developed vestibular labyrinth, possess either no cochlea at all or the merest trace of one, and yet undoubtedly are the subject of auditory sensations, in some cases of acute sensations. The evidence that fishes hear seems irresistible, they are said to respond to musical sounds; and yet those who hold the views just explained are driven to maintain either that fishes do not hear in the true sense of the word but only feel vibrations, or that they hear by means of an insignificant fragment of their relatively large vestibule. The structure of the piscine and amphibian vestibular auditory epithelium is in the main, putting aside smaller matters, such as the length of the auditory hairs, the size and abundance of otoliths and otoconia and the like, so identical with that of birds and reptiles and of mammals, that it is impossible to resist the conclusion that it serves the same purpose in all the several classes. In birds and reptiles the short rudimentary nearly straight tubular cochlea possesses a short basilar membrane, an auditory epithelium in which a distinction of outer and inner hair-cells is foreshadowed, and a tectorial membrane. But if we are to suppose that these creatures receive auditory impulses exclusively from the cochlea, and none at all from the vestibule, it is a matter of wonder that the cochlea of the, for the most part, dumb crocodile should appear almost as highly developed as that of the vocal bird. Or again, if the bird and reptile already possessing a cochlea still derive auditory sensations by means of the vestibule, we may conclude that mammals also do the same.

The whole evidence then goes to shew that even in man the vestibule plays an important part in hearing. As we said above the distinction between noise and music is a quantitative and fluctuating one; indeed the tendancy of inquiry seems to shew that the quality of timbre of a sound, and it is this which so largely contributes to the value of a sound as an element of music, is in part dependent on vibrations, which being irregular, that is, having no exact arithmetical relation to the fundamental tone, may be spoken of as noise. We are thus driven nearer and nearer to the conclusion, that when we are listening to sounds we do not hear this sound with the cochlea and that with the vestibule, but that we hear each sound, whether it be music or noise, with so to speak the whole ear. But if this be so, then the origin and nature of auditory impulses must be still more complex and difficult than appears from the study of the cochlea alone, perplexing as they even then seem.

SEC. 5. ON AUDITORY PERCEPTIONS AND JUDGMENTS.

§ 853. In spite of the many and striking differences between the two senses, it is possible to draw several parallels between auditory and visual sensations. When we are the subject of a visual sensation we refer the cause not to changes taking place in the retina, but to some luminous object in the external world. So also, when we are the subject of an auditory sensation we refer the cause not to changes taking place in the internal ear, but to some sounding body outside the ear and in the vast majority of cases to some sounding body outside ourselves. We do not simply feel auditory sensations, we perceive sounds, cf. § 779.

We have seen that in the case of the eye, visual sensations, excited by events taking place in the visual apparatus itself, may be confounded with sensations excited by objects in the external world; and much the same happens with the ear also. The tympanic membrane for instance may be thrown into vibrations not by waves of sound, but by objects coming mechanically into contact with it; particles of the dried secretion of the external auditory passage, the 'wax of the ear,' playing on the tympanic membrane, may give rise to auditory sensations, a 'buzzing' or 'singing in the ears,' which we cannot by the mere psychological examination of the sensations themselves distinguish from auditory sensations excited in the ordinary way by sonorous vibrations reaching us from some sounding body at a distance. And in a general way, we may speak of *entotic* phenomena, corresponding to the entoptic phenomena on which we dwelt (§ 736) in speaking of vision.

Auditory sensations moreover may arise, in the complete quiescence of the tympanic apparatus and perilymph, as the result of changes either in the auditory epithelium or in the central auditory nervous apparatus. We may be subject to *auditory phantoms* or *hallucinations*, corresponding to ocular phantoms or hallucinations, and like them often misleading or distressing. Few persons, moreover, can listen to exciting music or can hear impressive cries without experiencing "recurrent" auditory sensations.

§ **854.** In one important respect the parallel between hearing and sight fails. When we see an object, the rays of light coming from the object excite a particular part of the retinal expansion; and our appreciation of the position which that object holds in space is based on our power of "localizing" retinal changes. The terminal expansion of the auditory nerve however has no such definite relations to the positions in space of objects from whence sounds are proceeding; we have no evidence that any particular part either of the organ of Corti or of the maculae is alone or specially affected by sounds coming from a particular quarter; and the evidence that sounds affect the three cristae differently according to the direction of the sound is at least doubtful. Hence we possess no "auditory field" which can be directly compared with the "visual field;" and our conclusions as to the direction in which the sounds which reach our ears have travelled, our judgments as to the position in space of bodies exciting auditory sensations are formed in an indirect manner.

The vast majority of the sounds which we hear reach the auditory epithelium by way of the tympanic membrane and chain of ossicles; even the sounds which are conducted to the ear through the bones and hard parts of the head pass to a large extent by this way (§ 818); in normal hearing the auditory sensations which are generated by vibrations transmitted directly through the bony walls of the labyrinth to the perilymph are probably insignificant. Now it is only in relation to these latter that the disposition in space of the three semicircular canals can possibly have any meaning; the vibrations reaching the perilymph by way of the tympanic membrane, whatever their original direction, have all the same direction when they enter at the fenestra ovalis, and fall in the same way upon the three semicircular canals. We may therefore conclude that the position in space of the three canals in question has nothing to do with our ordinary judgments as to the direction of sounds. In forming those judgments we are assisted mainly by two things.

In the first place a peculiar character of the outwardness which we attribute to our usual auditory sensations, that by which we judge the sound to arise not only outside the internal ear but outside our whole body, seems, in some way, largely dependent on the vibrations which cause the sensation having travelled along the external auditory passage. If the two passages be filled with fluid the hearer refers the sounds which he hears, in spite of their starting at some distance off, not to the external world outside himself, but to the inside of his own head; the sounds appear to him to come, not it may be remarked from the internal ear or any part of it, but from the roof of the mouth, or the top of the skull or the back of the head. So also if the ear-pieces of a binaural stethoscope be pushed well up into the auditory passages, the sounds heard through the instrument seem to come from the roof of the observer's own mouth.

The difference between such an abnormal mode of hearing and ordinary hearing does not lie in the fact that in the former case the tympanic membrane is not used at all; for even when the external passage is filled with fluid, a layer of air which always adheres to the tympanic membrane permits at least a certain amount of vibration of that membrane; and on the other hand when the sound is actually generated in the roof of the mouth, and rightly judged to be generated there, the tympanic membrane by its vibrations conducts the greater part of the sound to the internal ear. How it is that the passage of the vibrations through the external passage imparts to the sensation this attribute of outwardness is not clear. Indeed certain sounds may be made to lose this particular outwardness, though the external passage be still employed. If two musical sounds of the same pitch be listened to with the two ears separately by means of two telephones, the sound will, under certain conditions, appear to originate somewhere in the head of the observer.

§ 855. In the second place our appreciation of the particular quarter from which a sound, recognized by help of the external passage to be of outward origin, has travelled is dependent on our using two ears. As our ordinary vision is largely binocular, so our ordinary hearing is, to a still larger extent, *binaural*. In the case of the ear there are no sharp limitations to the range of the organ of either side; through the medium of the air and external auditory passage or of the hard parts of the head a sound which affects one ear affects to a certain extent the other ear also; hence all our hearing is, under ordinary circumstances, binaural. And in some such way as two visual sensations excited in "corresponding parts" of the two retinas are fused into one, so every sound which reaches us is heard not as two sounds, one by one ear and the other by the other, but as one sound by the two ears together.

When the sounding body is on one side of the head, say the right side, the sensations excited through the right internal ear are more powerful than those excited through the left internal ear; we are not distinctly conscious of the difference between the two sensations, the combined effect is a single sensation; but the difference does affect our consciousness in a certain way, and that affection of consciousness serves as a basis for the judgment that the sounding body is somewhere on our right hand. Hence we are able to judge the lateral much more readily than the fore and aft position of a sounding body. If a tuning-fork be held in the median vertical plane over the head, the eyes being shut, though it is easy to recognize it as being in the median plane, it is very difficult to say what is its position in that plane, *i.e.* whether it is more towards the front or back of the head. Hence also a man who is absolutely deaf of one ear has great difficulty in recognizing the direction of sounds.

Further, when we desire to judge particularly as to the direction of a sound, we listen to it, and in the act move the head into the position in which we hear the sound most distinctly. In this way the movements of the head in hearing play a part somewhat analogous to the movements of the eyes in vision.

Even in the case of ourselves, and still more in the case of some animals, the form of the external ear favours the entrance into the meatus, and hence the access to the tympanic membrane, of sounds travelling in a particular direction ; this also assists our judgment of the direction of sounds. Hence, by tying back the ears and affixing artificial ears, differing in shape or position from the natural ones, we may make false judgments in this matter.

Moreover, in forming a judgment as to the direction of sounds we appear to be guided by something more than the mere relative intensity of the sounds falling on the two ears. When a complex sound emanates from a body on one side of us, the constituent vibrations do not travel equally and uniformly over and around the head; some are refracted more than others, so that they do not reach the two ears equally; and besides when they reach them are not equally reflected by the two pinnæ. In this way partial tones of different pitch, and this applies especially to high tones, reach the two tympanic membranes in unequal intensities, and the sound of which they form part appears as heard by the one ear of a quality slightly different from that heard by the other ear; this difference of quality, like the difference in mere intensity of the sound as a whole, serves as a basis for recognizing the direction of the sound. Such a difference will be more marked in the complex sounds which we call noises than in purer and more simple musical sounds; and, as a matter of fact, our appreciation of direction is more accurate in the case of noises than of musical sounds. An exception to this rule is met with in the case of the human voice, the direction of which, though it is as a whole a musical sound, can be judged better than even that of a noise; but noises enter largely into the human voice, and besides we are much more practised in relation to it than in relation to any other kind of sound. All our judgments of the direction of sounds are however at the best imperfect.

§ **856.** Our judgment of the *distance* of sounds is even still more limited. A sound whose characters we know appears to us near when it is loud, and far off when it is faint. A blindfold person will be unable to distinguish between the difference of intensity produced on the one hand by a tuning-fork being held before him, first with the broad edge of the fork toward him and then with the narrow edge, and the difference on the other hand caused by the removal of the tuning-fork to a distance. And our judgments in this respect may be false, as is seen in the effects produced by the ventriloquist. We can on the whole better appreciate the distance of noises than of musical sounds, differences of quality as well as of intensity playing the same part in the judgment of distance as of direction; when a sound becomes distant the intensity of the fundamental tone diminishes more rapidly than do those of the higher partial tones, and hence the quality of the sound is affected.

CHAPTER V.

TASTE AND SMELL.

SEC. 1. THE STRUCTURE OF THE OLFACTORY MUCOUS MEMBRANE.

§ **857**. THE organ of smell resembles the organs of sight and hearing in that it consists of a special nerve, ending in a specialized epithelium. The nerve is the olfactory nerve (§ 674) and the epithelium forms part of the mucous membrane which lines the upper region, comprising the upper and middle turbinal bones and the upper third of the septum, of each nasal chamber. This region of the nasal mucous membrane is called the *olfactory mucous membrane*, the remaining lower region being called the *respiratory* mucous membrane, or sometimes the *Schneiderian* membrane. In ordinary breathing the currents of inspired and expired air are mainly limited to the lower respiratory region, the upper region being reached by means of diffusion or of eddies.

The respiratory nasal mucous membrane closely resembles that of the trachea, and consists of a ciliated epithelium, of several layers, resting on a vascular dermis fairly rich in lymphatic elements. Goblet cells are abundant in the epithelium, and numerous small branched mucous or albuminous glands are found in the deeper layers of the dermis. Filaments from the fifth nerve are distributed to this region, but none from the olfactory nerve.

The olfactory mucous membrane is thicker than the respiratory portion, and has to the naked eye a yellowish colour. The thickness is chiefly due to the development in the epithelium of a nuclear layer similar in general appearance to that of a crista or macula acustica; indeed the resemblance between the olfactory epithelium and the auditory epithelium is in many ways striking. A vertical section discloses, very much as in a macula acustica, a layer of cylindrical or columnar cells, the *cylinder cells*, resting on a thick nuclear layer, below which is seen the dermis traversed in various directions by bundles of non-medullated olfactory fibres. And very much as in a macula, the nuclei of the nuclear layer belong to numerous more or less rod shaped or spindle shaped cells, the *rod cells.*

A cylinder cell consists of a cylindrical cell body of granular cell-substance, the upper border of which forms part of the free surface of the epithelium, and the lower part of which, after bearing a more or less ovoid nucleus, becomes narrow and ends in an irregular branched less granular process lost to view in the nuclear layer. Near the confines of the respiratory mucous membrane, these cylinder cells bear, in some cases, cilia; and indeed at the margin of the olfactory membrane, pass into ordinary columnar ciliated cells; but, as a rule, they bear no cilia elsewhere.

A rod cell consists of a spherical nucleus, surrounded with an exceedingly thin layer of cell-substance, which is prolonged peripherally, as a rod-shaped process directed between the cylinder cells and ending between them at the free surface of the epithelium. In the lower animals, amphibia, the free end bears a fine bundle of delicate hairs, which like cilia are capable of movement, though moving in a peculiarly slow, gentle fashion; whether these cilia-like processes are present in mammals is not as yet settled. Besides this peripheral process the cell bears at the other pole of the nucleus a central process directed towards the dermis. This is more delicate than the peripheral process, often appears varicose, and has much the appearance of a nerve filament; it is lost to view in the nuclear layer. In some rod cells the peripheral process is longer than others, and the nucleus accordingly placed deeper down; some of the nuclei lie close to those of the cylinder cells, others close to or nearly close to the dermis, and some at intermediate levels; in fact the nuclei which make up the thick nuclear layer are the nuclei of rod cells, these greatly exceeding in number the cylinder cells.

An exception to this last statement should be made in favour of the layer of nuclei lying immediately above the dermis. These are oval not spherical nuclei, and the cell bodies are not in the form of rods but more or less irregularly branched. They have been distinguished as *basal cells.*

Among these cylinder, rod, and basal cells which make up the mass of the epithelium, numerous cylinders of flattened cubical cells, surrounding a narrow lumen, pass vertically down from the surface of the epithelium to the dermis. These are the intra-epithelial ducts of small tubular twisted glands, generally albuminous but sometimes mucous in character, which lie in the deeper parts of the dermis, and which more simple and smaller than the corresponding glands in the respiratory nasal membrane, have been called the "glands of Bowman."

The nuclei of the rod cells not only differ in form from those of the cylinder and basal cells, being spherical, while the latter are oval, but also behave in different ways towards various staining and other reagents; with hæmatoxylin and similar staining reagents they stain less deeply than do the nuclei of the cylinder cells. They are obviously of a very different nature; and since the nuclei of the cylinder and basal cells behave towards reagents very much in the same way as do the nuclei of the cells of the gland ducts, and indeed of the cells in the respiratory portion of the membrane, we may speak of the nuclei of the rod cells as possessing specialized characters. The cell-substance of the rod cells is also of a different nature, more specialized than that of the cylinder and basal cells. The yellowish tint of the olfactory membrane is due to pigment deposited chiefly in the epithelium.

§ 858. The connective tissue, vascular and lymphatic constituents of the dermis present no special characters; no cushion of modified connective tissue like that seen in the auditory crista or macula is present. The special feature of the dermis is the presence of bundles of fibres of the olfactory nerve arranged in more or less plexiform manner.

An olfactory nerve fibre is a non-medullated nerve fibre consisting of an axis-cylinder, whose fibrillation is at times conspicuous, surrounded by a sheath or neurilemma, and bearing oval nuclei at intervals. Fibres of this kind are bound up by delicate connective tissue into small bundles and these by coarser connective tissue into larger bundles. From the olfactory bulb (§ 674) numerous large bundles are given off which, piercing the cribriform plate of the ethmoid bone, run in a plexiform manner in the walls of the nasal chamber in the olfactory region, giving rise to bundles which become successively finer.

From the bundles lying in the dermis immediately below the epithelium, fibres pass into the epithelium itself, the demarcation between epithelium and dermis being in the olfactory region not very sharply defined. As they pass into the epithelium the fibres break up into fibrils, which running in various directions in a plexiform manner are lost to view in the nuclear layer. Hence the ground substance in which the nuclei of the nuclear layer appear imbedded consists partly of the more or less fibrillar processes of the numerous rod cells themselves, partly of the more branched processes of the cylinder cells and of the basal-cells, but also and indeed largely of a plexus or at least a tangle of nerve fibrils proceeding from the olfactory nerve fibres.

Filaments from the fifth nerve are also distributed to the olfactory as well as to the non-olfactory region of the membrane.

It is contended by many that the fibrils of the olfactory nerve are continuous with the central processes of the rod cells and make no connections with the cylinder cells. From this it is concluded that the rod cells are primarily concerned in the development of olfactory nervous impulses, the cylinder cells acting only as supporting or at most subsidiary structures; and the rod cells have hence been called "olfactory cells." On the whole the evidence is in favour of this view, which as we shall see is supported by the analogy of the organ of taste; but it is maintained by others that the cylinder cells and not the rod cells are connected with the olfactory fibres, and that they are the functional endings of the olfactory nerve. And it may be borne in mind that in the auditory macula and crista, on the resemblance of which to the olfactory epithelium we have already remarked, the evidence is strongly in favour of the cylinder cells being the functional organs. Others again maintain that both cylinder cells and rod cells are connected with the olfactory fibres, both serving as terminal organs, and urge that there is no fundamental difference between the two. To determine beyond dispute the actual terminations of delicate nerve fibrils and to place beyond doubt the continuity of a nerve fibril with other fibrils like those which form the processes of the rod cells is a task of extreme difficulty; and an assertion that the continuity has been successfully traced may well be received with caution. Moreover it must be borne in mind that, as we urged in speaking of the central nervous system, we are not justified in assuming that anatomical continuity is essential to physiological conduction. We may add that even admitting one kind of cell, for instance the rod cell, to be the chief agent in the processes whereby odoriferous particles give rise to nervous impulses, we are not thereby compelled to conclude that the other kind of cell, the cylinder cell, is a 'supporting' cell in the sense that it simply affords a mechanical support to adjoining rod cells. It is possible that the cylinder cell has some intimate relations, not of a purely mechanical kind, to the well being and activity of the rod cell. In any case it would at present be dangerous to found any important physiological conclusion on either the view that the rod cells or the view that the cylinder cells are exclusively the terminal functional organs.

SEC. 2. OLFACTORY SENSATIONS.

§ 859. Particles of odoriferous matters present in the inspired air, passing through the lower nasal chambers, diffuse into the upper nasal chambers, and falling on the olfactory epithelium produce sensory impulses which, ascending to the brain, give rise to sensations of smell. We may assume that the sensory impulses are originated by the contact of the odoriferous particles with the free endings of the rod cells; but we are wholly in the dark as to the manner in which the contact of the particles with the cells brings about the molecular changes constituting a nervous impulse. We cannot even say whether we ought to speak of the first step by which the contact of the particle begins the series of changes as a chemical or as a physical process.

In nearly all cases the odoriferous particles are conveyed to the membrane in a gaseous medium, namely, the atmosphere; but before they can gain access to the cells they must become dissolved or at least suspended in fluid; for the whole olfactory membrane is kept moist by a layer of fluid, the secretion of the glands described above, and the odoriferous particles must pass into this layer of fluid before they can gain access to the cells. Indeed, the proper condition of this layer of fluid is one of the essential conditions of the exercise of the sense. If on the one hand the membrane be too dry, or if on the other hand the secretion be too abundant or altered in quality, the power of smelling is diminished or even wholly suspended. It is a matter of common experience that a nasal catarrh interferes with smell. When the nostril is filled with rose-water the odour of roses is not perceived; and simply filling the nostrils with distilled water suspends for a time all smell, the sense gradually returning after the water has been removed; the water apparently acts injuriously on the delicate olfactory cells. If instead of using rose water, the rose scent be dissolved in "normal saline solution" which (§ 14) more closely resembles the natural secretion, the cells can perform their function, and the scent is perceived. The glands of the olfactory membrane form an important subsidiary apparatus for the development of olfactory sensations.

The other subsidiary apparatus of smell is exceedingly meagre. By the forced nasal inspiration, called sniffing, we draw air so forcibly through the nostrils that currents pass up into the upper as well as the lower nasal chambers; and thus a more complete contact of the odoriferous particles with the olfactory membrane than that supplied by mere diffusion is provided for.

§ 860. We have every reason to think that any stimulus applied to the olfactory cells will produce the sensation of smell; but the proof of this is not absolutely clear; and we have no definite evidence as to what is the result of directly stimulating the fibres of the olfactory nerve. The olfactory membrane however is certainly the only part of the body in which odours as such can give rise to any sensations : and the sensations to which they give rise are always those of smell. The mucous membrane of the nose is however also an instrument for the development of afferent impulses other than the specific olfactory ones. Chemical stimulation of the nasal mucous membrane by pungent substances such as ammonia gives rise to a sensation distinct from that of smell, a sensation which does not afford us the same imformation concerning the chemical nature of the stimulus, as does a real olfactory sensation, and which is much more allied to the sensations produced by chemical stimulation of other surfaces sensitive to chemical action. This sensation moreover seems to be developed both in the non-olfactory and in the olfactory regions of the nasal mucous membrane; and it is probable that these two kinds of sensations, the one produced by odours, the other by pungent substances, thus arising in the olfactory membrane are conveyed by different nerves, the former by the olfactory, the latter by the fifth nerve.

Each substance that we smell causes a specific sensation, and we are not only able to recognize a multitude of distinct odours, but also in certain cases to distinguish individual odours in a mixed smell. And though we may recognize certain odours as more like to each other than to other odours, or can even make a rough classification of odours, we cannot, as we can in the case of visual colour sensations, reduce our multifarious olfactory sensations to a smaller number of primary sensations mixed in various proportions. Nor have we at present any satisfactory guide to connect the characters of an olfactory sensation with the chemical constitution of the body giving rise to it.

The sensation takes some time to develope after the contact of the stimulus with the olfactory membrane, and may last very long. When the stimulus is repeated the sensation very soon dies out: the sensory terminal organs speedily become exhausted. The larger, apparently, the surface of olfactory membrane employed, the more intense the sensation; animals with acute scent have a proportionately large area of olfactory membrane. ⸜The greater the quantity of odoriferous material brought to the membrane, the more intense the sensation up to a certain limit; and an olfactometer for measuring olfactory sensations has been constructed, the measurements being given by the size of the superficial area, impregnated with an odoriferous substance, over which the air must pass in order to give rise to a distinct sensation. The limit of increase of sensation however is soon reached, a minute quantity producing the maximum of sensation and further increase giving rise to exhaustion. The minimum quantity of material required to produce an olfactory sensation may be in some cases, as in that of musk, almost immeasurably small.

In ordinary circumstances odoriferous particles reach both nostrils, and we receive two sets of olfactory nervous impulses, one along each olfactory bulb. These however are fused into one sensation; our olfactory sensations are almost exclusively binasal. When two different odours are presented separately to the two nostrils, by means of two tubes for instance, the effect is not always the same. Sometimes an oscillation of sensation similar to that spoken of in binocular vision (§ 800) takes place. At other times, the particular result depending on the nature of the odours, one sensation only is felt, the one sensation wholly destroys the other. And we may infer from this that when, as frequently happens, in a mixture of odours we can only recognize one dominant odour, the suppression of the missing sensations is not due to the chemical action of one odour upon another, or to the one odour preventing the other from acting on the olfactory cells; but from a central cerebral obliteration of all the sensations but one.

§ 861. As in the cases of the previous senses, we project our olfactory sensations into the external world; the smell appears to be not in our nose, but somewhere outside us. We can judge of the position of the odour however even less definitely than we can of that of a sound. Our chief guide seems to be that we by turning the head ascertain in which direction we experience the strongest sensations.

The sense of smell seems to play a far more important part in the lives of the lower animals than it does in our own life; and what we now possess is probably the mere remnant of a once powerful mechanism. We may perhaps connect with this on the one hand the fact that, even in ourselves, the olfactory fibres have allotted to them what is virtually a whole segment of the brain, namely the olfactory lobe, and on the other hand the fact that olfactory sensations seem to have an unusually direct path to the inner working of the central nervous system. Mental associations cluster more strongly round sensations of smell than round almost any other impressions we receive from without. And powerful reflex effects are very frequent, many people fainting in consequence of the contact of a few odorous particles with their olfactory cells.

The assertion that the olfactory nerve is the nerve of smell has been disputed. Cases have been recorded of persons who appeared to have possessed the sense of smell, and yet in whom the olfactory lobes were found after death to be absent. Direct experiments on animals however shew that loss of the olfactory lobes entails loss of smell. On the other hand, it is stated that section or injury of the fifth nerve causes a loss of smell though the olfactory nerve remains intact; but in these cases it has not been shewn that the olfactory membrane remains intact, and it is quite possible that, as in the case of the eye, changes may take place in the nasal membrane as the result of the injury to the fifth nerve, sufficient to prevent its performing its usual functions.

SEC. 3. THE STRUCTURE OF THE ORGANS OF TASTE.

§ 862. Sensations of taste, like those of sight, hearing and smell, are brought about by means of special nerve fibres whose endings are connected with a modified epithelium; but these special nerve fibres do not form a nerve devoted exclusively to taste as the optic nerve is to sight, the olfactory nerve to smell, and though less distinctly, the auditory nerve to hearing; they run in company with fibres of other kinds in one or other or both of two cranial nerves, in the glossopharyngeal nerve and in the lingual branch of the fifth nerve. In this respect the sense of taste shews a tendency to a more general distribution similar to that of touch; and in the lower animals this tendency is still more strongly marked. Moreover, though as we shall see, a modified epithelium, subserving taste, does exist in the lining of the mouth, it is not so clear, as in the case of sight, hearing and smell, that the development of gustatory sensations is exclusively dependent on this modified epithelium; there are reasons for thinking that taste may be exercised by parts of the mouth where no such distinctly modified epithelium exists.

Although the lining membrane of the mouth is generally spoken of as a "mucous membrane," its histological plan is identical with that of the skin. The epithelium overlying the dermis is a stratified epithelium, consisting of many tiers of cells, in which we may distinguish a lower malpighian layer and an upper corneous layer. One of the chief differences between the epithelium of the mouth and the epidermis of the skin is, that in the former the stratum granulosum and stratum lucidum (§ 434) so characteristic of the latter, are absent, or present in a few places only, being replaced by a layer in which the polygonal prickle cells of the malpighian layer, losing their prickles and becoming flatter, are gradually transformed into the cells of the corneous layer.

Another difference is that the uppermost cells of the corneous layer, though transformed into flat scales, still retain a conspicuous nucleus, and are not so wholly corneous as the corresponding scales of the skin. The differences appear to be connected with the fact that the epithelium of the mouth is always soaked in moisture, whereas the epidermis of the skin is as a rule dry.

The dermis, thrown up into numerous conspicuous papillæ, contains a considerable quantity of reticular, or at places, even adenoid tissue, is extremely vascular, plexiform arrangements of the veins being very frequent, and in it in many situations, for instance in the tongue, striated muscular fibres end with branched terminations; numerous small chiefly albuminous glands are also present, especially in certain regions.

In the tongue, the papillæ, which are for the most part compound, have been divided into three classes; filiform, fungiform, and circumvallate. The two first, which are much the most numerous, differ chiefly in form, the one being more or less pointed, the other having a broad head; both generally bear secondary papillæ. The circumvallate papillæ, few in number, arranged in the form of a V with the apex at the root of the tongue, are distinguished by the fact that the papilla is as it were sunk into the substance of the tongue, so that its base is surrounded by a circular groove like a fort surrounded by its fossa and vallum.

§ 863. Appearing irregularly and scantily on the fungiform papillæ and then seated sometimes at the sides, sometimes on the top of the papilla, more abundant on the circumvallate papillæ, in which they are seated generally on the sides of the papilla, looking into the groove, but sometimes on the opposite wall or vallum, are the structures known as *taste-buds* or "taste-bulbs." These are especially abundant and most easily studied in the aggregations of papillæ, which in some animals, for instance the rabbit, are found at the back of the tongue and are called *papillæ foliatæ*. In the rabbit these are seen by the naked eye as two oval patches, one on each side, placed obliquely at the root of the tongue. Upon examination each patch is found to consist of a number, twenty or so, of folds of the mucous membrane placed side by side like the leaves of a book. In each fold the dermis is raised up into three narrow parallel ridges which in a transverse section appear as three pointed papillæ. The epithelium, which here as elsewhere consists of a malpighian and a corneous layer, is somewhat thin at the sides of the fold, but thicker at the top, where it is even, not following in contour the ridges of the dermis beneath. Hence, in each fold, the two outside ridges of the dermis, which look towards the deep narrow furrows separating each fold from adjacent folds, are covered by a somewhat thin but complete epithelium. And in a transverse section this epithelium is seen to contain a vertical row of oval structures, four or five in number, which reach from the dermis to the surface of the epithelium, each furrow having a row of such structures on each side. These are the taste-buds, having essentially the same structure as those seen in man less regularly distributed on the sides of the circumvallate papillae, and still less regularly on the fungiform papillae.

Each taste-bud consists on the outside of an oval hollow structure, which may be compared to an oval nest, the walls of which are formed of epithelial cells arranged thatch-wise. The base of the nest lies bare on the dermis, the top of the nest comes to the level of the surface of the epithelium, and here bears a circumscribed circular hole, or mouth by means of which the fluids of the mouth can gain access to the inside of the nest. The cells forming the walls are elongated, flattened, fusiform, appropriately curved, nucleated cells, arranged in an imbricate fashion, the outer ones passing somewhat abruptly into the ordinary cells of the malpighian and corneous layers. Those ends of the cells which converge to form the top of the nest are scooped out, or bear distinct perforations, and thus in the one way or the other, form the margin of the mouth or pore leading to the inside of the nest. The cell-substance forming the scooped margin or surrounding the perforation seems to be cuticularized so as to secure the patency of the pore.

The cells thus forming the walls of the nest differ from the surrounding ordinary cells chiefly in form ; at least they cannot be spoken of as distinctly modified in respect to their actual nature. The interior of the nest however is occupied by cells, which we must regard as specialized sensory cells. These are of two kinds. The one kind, very similar to the rod cells of the olfactory organ, are called *rod cells;* the other kind, analogous to the cylinder cells of the olfactory organ, we may call *subsidiary cells;* they have also been called "cover-cells." The rod cell consists of an elongated oval nucleus placed vertical, at about the level of the middle of the nest. The scanty cell-substance surrounding the nucleus is prolonged peripherally as a rod shaped nearly hyaline, somewhat refractive process which reaches up, or projects through the pore of the nest; the cell-substance is also prolonged centrally as a delicate thread-like process, which either branching or unbranched, stretches towards and is lost to view at the dermis.

The subsidiary cell possesses a nucleus which is larger and more rounded than that of the rod cell, and its cell-substance larger in amount and more granular in appearance than that of the rod cell, has a more or less fusiform shape, the part peripheral to the nucleus ending in a tapering fashion, and the part on the other, central, side of the nucleus becoming branched and lost to view near the dermis.

The axis of the nest is occupied by a group of rod cells, the peripheral processes of which converge in a bunch at the pore of the nest. With these however are interspersed a few subsidiary cells, and the core thus formed is surrounded by a wrapping consisting of subsidiary cells only. Both the rod cells and subsidiary cells differ from the cells forming the wall of the nest not only in form but in chemical nature, and behave very differently from them towards various reagents; they stain with gold chloride for instance very readily, and may with this reagent be prepared deeply stained, while the walls of the nest and the rest of the epithelium are hardly stained at all.

In the dermis underlying the taste-buds, nerve fibres, derived from the glosso-pharyngeal nerve, are abundant; they may very readily be seen in preparations of the papillæ foliatæ. Some of these are medullated and others sooner or later lose their medulla; some are distributed to parts other than the taste-buds, but some form at the base of the taste-bud a plexus of non-medullated nerve filaments. From this plexus, fibrils ascend into the substance of the bud and are there lost to view. Some observers state that they have been able to assure themselves of the continuity of the fibrils with the central processes of the rod cells. Be this as it may, the existence of the taste-bud is dependent on its connection, in some manner or other, with the nerve fibres. If the glosso-pharyngeal nerve be divided, and regeneration of the nerve prevented, the rod cells and subsidiary cells degenerate and vanish, and finally the whole taste-bud disappears. There can be no doubt that the specialized cells are in some way the functional endings of the nerve.

The small albuminous glands occurring as we have said in the mucous membrane of the mouth are especially abundant in the neighbourhood of the taste-buds; they serve to supply fluid for the solution of substances to be tasted, and as we shall presently see such a solution seems to be a necessary condition for the activity of the sensory organs. The dermis in the neighbourhood of the taste-bud is always very vascular, and venous plexuses, assuming almost the form of venous sinuses, are not infrequent; such a venous sinus is conspicuous in the central ridge of a fold of a papilla foliata.

Taste-buds are found in other parts of the mouth besides the hind part of the tongue, where as we have said they occur on the fungiform and circumvallate papillæ. They are found in the soft palate, and in not inconsiderable numbers on the hinder surface of the epiglottis. They are absent or extremely rare in the front part and sides of the tongue. Their distribution in fact appears not to coincide exactly with that of the sense of taste itself; and we may therefore conclude that gustatory nerve fibres have other terminations besides taste-buds. The only other mode of termination however of which we are at present aware, is that mode by which the nerve fibres, breaking up into fibrillæ, are lost among the cells of the epithelium without ending distinctly in any specially modified cell; but this mode of termination we shall study more closely in dealing with the sense of touch.

It is obvious that a very complete analogy exists between the taste-buds and the endings of the olfactory nerve. The rod cells of the one closely resemble the rod cells of the other, and the subsidiary cells of the taste-bud are clearly analogous to the cylinder cells of the olfactory mucous membrane; and much that we urged as to the relative values of the rod cells and cylinder cells of the olfactory membrane may be applied to the rod cells and subsidiary cells of the taste-buds. The resemblance is carried still further in the case of the olfactory organ of some of the lower animals. In these the olfactory membrane is not a continuous sheet, but is broken up into patches by the intervention of non-olfactory epithelium, and the isolated patches often take on the form of nests very similar indeed to taste-buds. It is obvious that the senses of taste and smell are closely akin, and in the peculiar organs distributed over the skin of some of the lower animals such as those along the lateral line in fishes, we probably have to do with structures which subserve a sense not exactly that of smell or of taste as we ourselves experience these sensations, but of something intermediate between the two, or possibly between both of them and the more general sense of touch.

SEC. 4. GUSTATORY SENSATIONS.

§ 864. The word taste is frequently used when the word smell ought to be employed. We speak of 'tasting' odoriferous substances, such as an onion, a wine, a savoury dish, and the like, when in reality we only smell them as we hold them in our mouth; this is proved by the fact that the so-called taste of these things is lost when the nose is held, or the nasal membrane rendered inert by a catarrh. If the nose be held and the eyes shut, it is very difficult to distinguish in eating between an apple, an onion and a potato; the three may be recognised by their texture, but not by their "taste." Most of what we call 'flavours' appeal in reality to the sense of smell not to that of taste.

We also experience by means of the surfaces with which we taste sensations other than those of taste. We feel by means of the mucous membrane of the mouth sensations of the same kind as those which we feel by means of the skin, and which we shall study presently as tactile sensations or sensations of pressure, sensations of heat and of cold; indeed the tactile sensations of the tip of the tongue are remarkably acute. We also experience by means of the mouth sensations of pain and other more or less indefinite sensations which we shall presently speak of as phases of "general" or "common sensibility;" and in this respect the mucous membrane of the mouth is much more sensitive than the skin towards chemical substances; an acid for instance or other corrosive liquid, in such a concentration as when applied to the skin produces a sensation not essentially different from that of mere contact with an innocuous liquid, may when applied to the mouth produce a very painful sensation. Again, when the interrupted current is applied to the tongue we not only feel the contact of the electrodes but experience a peculiar sensation which is probably due to the contractions excited by the current in the muscular fibres of the tongue; we say we "feel the current."

§ 865. There are however certain sensations quite distinct from those just mentioned and quite independent of smell which we experience when various substances are placed in the mouth; and these, which are the gustatory sensations proper, may be broadly classified into 'bitter,' 'sweet,' 'acid' or 'sour,' and 'salt,' to which perhaps should be added 'metallic' and 'alkaline.' The sensation of bitterness, such as that produced by quinine, and the sensation of sweetness, such as that produced by sugar, are very definite and specific sensations; they appear to be of an order different from those of acidity or sourness and of saltness; indeed an acid 'taste' is apt to merge into an affection of general sensibility mentioned above. The characters 'metallic' and 'alkaline' should perhaps be regarded as qualifying one or other of the other sensations rather than as being independent sensations.

In the ordinary course of things these sensations are excited by the contact of specific sapid substances with the mucous membrane of the mouth, the substances acting in some way or other, by virtue of their chemical constitution, on the endings of the gustatory fibres. When we taste quinine, the particles of the quinine, we must suppose, set up chemical changes in the cells of the taste-buds or in other parts of the epithelium, and by means of those changes gustatory impulses are started. But mechanical or electrical stimuli, in the absence of sapid substances, will give rise to gustatory sensations. When the tongue is smartly tapped, in addition to the sensation of touch or the more or less painful sensation which may be produced, a sensation, which we must call a sensation of taste, is developed and often lasts for some considerable time. If a constant current be applied to the tongue, sensations of taste are developed at the two electrodes, that at the anode differing from that at the kathode, and the exact nature of each being dependent upon the region of the mouth stimulated. It is probable that in this case electrolysis either of the fluids covering the epithelium or of the substance of the epithelial cells themselves generates bodies which act as chemical stimuli; and it is possible that the mechanical disturbance of the cells, when the tongue is tapped, also sets free chemical stimuli. But sensations of taste may be provoked by an interrupted induced current, so feeble as not to be felt as an electric current, and so arranged that the make and break shocks are equalized; in this case there can be little or no electrolysis, and we may infer that the current acts in some way or another on the specific nerve endings. It is somewhat singular that heat when applied to the tongue appears not to produce any sensations of taste.

As we shall presently see, the nerve fibres concerned in taste belong either to the fifth nerve or to the glosso-pharyngeal nerve or to both nerves. We saw in dealing with vision that the evidence as to whether direct stimulation of the optic fibres without the intervention of the retinal structures could produce visual sensations was uncertain. We have no satisfactory evidence whatever that direct stimulation of the gustatory fibres along their course in either the above two nerves will produce sensations of taste. As far as the sense of taste is concerned we have no adequate evidence that specific gustatory impulses can be developed in the gustatory fibres apart from changes in the nerve endings. But the evidence is negative only; and the case is one- not suited for experiment, since both nerves along their whole course are mixed nerves containing other afferent fibres than those of taste.

§ 866. It is essential for the development of taste, that the substance to be tasted should be dissolved; hence, the value of the glands, which as we have seen are especially abundant in the neighbourhood of the taste-buds. The effect is also increased by friction; and the tongue and lips may be regarded as a subsidiary apparatus which by their movements assist in bringing the sapid substances into contact with the mucous membrane of the mouth. A substance may give rise to hardly any sensation of taste when simply placed on the extended tongue, and yet excite very distinct sensations when rubbed between the tongue and the soft palate; indeed we generally make use of this movement known as "smacking the lips," when we desire to obtain strong taste sensations. In this act however we not only make use of the most sensitive surfaces and call in the aid of friction, but we also increase the sensation by employing a large area of sensitive surface; for the larger the surface the more intense is the sensation.

The sensation takes some time to develope, and endures for a long time, though this may be in part due to the stimulus remaining in contact with the terminal organs.

A temperature of about 40° is the one most favourable for the production of the sensation. At temperatures much above or below this, taste is much impaired. A weak solution of quinine readily tasted at the normal temperature of the mouth is not tasted if, immediately before, very cold or very hot water be held in the mouth for a little while.

We may experience at the same time coincident taste sensations of different kinds, such for instance as one of bitterness with one of saltness; but in some cases one sensation interferes with the other, as for instance bitterness and sweetness. A taste sensation following upon a previous sensation of a different kind, is frequently influenced by its predecessor, being sometimes augmented, sometimes inhibited.

Though we can hardly be said to project our sensations of taste into the external world, as we do those of sight, hearing and smell, we assign to them no subjective localisation. When we place quinine in our mouth, the resulting sensation of taste gives us no information as to where the quinine is, though we may learn that by concomitant general sensations arising in the buccal mucous membrane. And it must be remembered that all our gustatory sensations are always accompanied by tactile or other sensations. We do not, as in the case of smell, experience the specific sensation alone and apart by itself. And not infrequently, as when substances at once sapid and pungent are placed in the mouth, the general sensation of pungency overcomes and hides the specific gustatory sensation. In the case of acids, it is often difficult to distinguish between the acid taste and the more general effect of the acid on the common sensibility of the buccal membrane of which we spoke above § 864.

Though we possess a gustatory apparatus with separate nerves on each side of the mouth all our sensations are single. Nor can we distinguish a pure gustatory sensation developed on one side only from one developed on both sides, if the two are equally intense.

As in the case of the senses previously dealt with, we may experience subjective gustatory sensations, sensations of central origin due to changes in the central sensory organs (§ 676); and these, though originated not by gustatory impulses but by other events, may seem to us identical with those set up in an ordinary way by gustatory impulses reaching the centre along the gustatory fibres.

§ 867. Sensations of taste are not originated, either by sapid substances or otherwise, equally in all parts of the lining membrane of the mouth. The part in which they are best developed, and always developed if developed at all, is the back of the tongue, in the neighbourhood of the circumvallate papillæ. They are also developed at the tip and along the sides of the tongue, but to a variable extent in different individuals; some persons have very acute and distinct taste sensations in these parts, others little or none at all. On the dorsal surface of the middle of the tongue very feeble taste sensations, if any at all, are developed; they are always wholly absent from the under-surface of the tongue. Some taste sensations are also developed in the soft palate and front surface of the palatine arches; but these again vary much in distinctness in different individuals. In the cases recorded in which taste remained after the entire extirpation of the tongue including the circumvallate papillæ, the sensations seem to have been chiefly developed in the soft palate. There is also some evidence that taste sensations may be developed on the hinder surface of the epiglottis.

In individuals who receive sensations from all or several of the various parts above mentioned, it commonly happens that bitter things are most readily appreciated at the back of the tongue and sweet things at the tip; and this distribution may perhaps be considered as the normal one; but individual variations in this respect are met with; many persons taste both bitter and sweet things best at the back of the tongue; and some persons taste bitter things quite distinctly at the tip. The salt taste is said to prevail at the tip and the acid taste at the sides of the tongue; but many persons experience acid and salt tastes in those regions and those regions only in which they experience bitter and sweet tastes.

We have already said that bitter and sweet tastes seem to be on a different footing from acid and salt tastes ; and we have a certain amount of evidence that the two former sensations are brought about by means of terminal organs different from those by means of which the two latter are brought about. If some of the leaves of a plant which grows in India and is called *Gymnema sylvestre*, be chewed, or if the mouth be washed with a decoction of the leaves, for some little time afterwards bitter and sweet tastes are lost, neither quinine nor sugar exciting the usual sensations, though acid and salt tastes remained unaffected. We may interpret this result as indicating that the drug in some way or other 'paralyses,' that is to say, suspends the action of, the terminal organs, whatever they may be, by means of which bitter and sweet tastes are developed, but leaves untouched those by which other gustatory sensations are developed. The action of the same drug supports the further conclusion that the terminal organs of bitter tastes are different from those of sweet tastes ; since by using an adequately weak dose of the drug the sweet taste may be abolished while the bitter taste remains distinct.

Indeed it is probable that the distribution of the several kinds of tastes over different regions of the mouth, which we mentioned above, is dependent on the distribution of different kinds of terminal organs; it is probable that we experience bitter tastes by means of the back of the tongue because the terminal organs of the bitter taste are limited to, or at least most abundant in, the back of the tongue, those of the sweet taste by the front of the tongue because the terminal organs of the sweet taste are more abundant there ; and so on. If a small quantity of a particular bromine derivative of the substance which from its remarkably sweet taste has been called 'saccharine,' be placed carefully on the tip of the tongue, a sweet taste is developed; but if the same substance be carefully placed on the back of the tongue the result is not a sweet but a bitter taste. At least this is the result in the case of those individuals who taste bitter at the back of, and sweet at the tip of, the tongue. From this we may infer that, in such tongues, the specific terminal organs of the sweet taste are more or less completely limited to the front, and those of the bitter taste to the back of the tongue, both sets of terminal organs being of such a nature that while quinine affects the one only and sugar the other only, the substance of which we are speaking is able to affect both of them. In a somewhat similar way certain salts, magnesium sulphate for instance, when applied to the back of the tongue excite a bitterish taste, but when applied to the tip of the tongue excite an acid or a sweetish acid taste.

We said a little while back that a weak interrupted current, so applied as to produce little or no electrolytic effect, was able to develope sensations of taste, varying in kind according to the region of the tongue stimulated. When the electrodes are applied to the tip of the tongue, the more usual result is that though an acid taste is the most prominent, a mixed gustatory sensation is developed, in which a sweet taste may be often recognised as a constituent. In like manner a bitter constituent may be recognised in the sensation developed when the electrodes are placed at the back of the tongue. If the tongue be previously subjected to the influence of *Gymnema*, the taste at the tip is free from all sweetness and that at the back free from all bitterness, the sensations which are then experienced being variously described as simply "metallic," or "salt" or "acid." From this result we may draw the important inference that the interrupted current developes a bitter and a sweet taste by acting in some way or other directly on the specific terminal organs of the two respective tastes, very much in the same manner as do bitter and sweet things.

We have already said that when an acid, especially a somewhat strong acid, is placed on the tongue or in the mouth, the pure gustatory acid sensation is apt to be confused with the sensation of pungency, the affection of general sensibility which the acid also brings about and which speedily merges into pain. These two sensations may be differentiated by means of cocaine. If the tongue be painted with a weak solution of cocaine, the general sensibility, the groundwork so to speak of pain, is abolished, while the pure gustatory sensations are at first hardly affected at all; a relatively strong acid which previously made the tongue "smart" so that real gustatory sensations were obscured, now developes a pure 'rich' acid taste alone. It is moreover said that cocaine applied to the tongue in increasing strength of solution abolishes the several classes of sensations in the following order: general sensibility and pain, bitter taste, sweet taste, salt taste, acid taste, tactile sensations.

Taking all these facts, and others which we might bring forward, into consideration, we are led to the conclusion that the development of the several kinds of gustatory sensations depends on the presence of specific terminal organs in the surfaces by means of which we taste. There appear to be distinct terminal organs for bitter tastes, for sweet tastes, for acid tastes, for salt tastes, and possibly for other tastes, all differing from the terminal organs for tactile sensations, and from the structures whatever they may be which are concerned in general sensibility. But beyond this it is very difficult to say anything decisive. From the fact that sensations of taste are, as a whole, most constant and most acute at the back of the tongue, in the neighbourhood of the circumvallate papillæ in which taste-buds are present, we may infer that the name taste-bud has been wisely chosen. But the development of taste sensations, including bitter sensations, at the tip of the tongue, from which taste-buds are said to be absent, presents a difficulty. Unless we suppose that taste-buds, though often absent from the tip of the tongue are present in those cases in which sensations are developed, we must conclude that gustatory sensations may originate by the help of some kind of nerve ending other than that of taste-buds. It might be suggested that bitter and sweet tastes are developed by means of taste-buds and acid and salt tastes by means of other endings; but there is no satisfactory evidence of this.

§ **868.** The question which nerve or nerves subserve taste and what is the course of the gustatory fibres is one which presents great difficulties. The front surface of the tongue is supplied by the lingual or gustatory branch of the fifth nerve, the hind surface by the glossopharyngeal nerve, which nerve also supplies the soft palate, though a branch (palatine) of the fifth nerve goes there also. The nerves traced to the taste-buds in the papillæ foliatæ and circumvallatæ belong to the glossopharyngeal nerve, and it can hardly be doubted that gustatory fibres run in the branches of that nerve which go to the back of the tongue. On the other hand in the cases in which sensations are distinctly developed in the tip of the tongue we must infer that gustatory fibres run in the lingual branch of the fifth, since no glossopharyngeal fibres are distributed to this part of the tongue.

But it by no means follows from this that gustatory fibres pass straight both up the trunk of the glossopharyngeal nerve, and up the trunk of the fifth nerve to their respective nuclei in the bulb.

On the one hand there is a good deal of evidence to shew a connection between sensations of taste and the chorda tympani nerve. Cases have occurred in which disease of the ear, involving destruction of the chorda tympani within the tympanum, has been followed by loss of taste in the tongue on the same side; and stimulation of the chorda tympani within the tympanum has been known to give rise to sensations of taste. Neither of these results is conclusive. The chorda tympani contains afferent fibres which have a remarkable effect on the nutritive processes of the tongue, and the loss of taste due to destruction of the chorda might be due to disordered nutrition of the tongue, and so be analogous to the loss of smell which may follow injury of the fifth nerve. Again, where stimulation of the chorda within the tympanum produces sensations of taste these may be due to efferent impulses producing changes in the tongue, which in turn give rise to sensations of taste reaching the brain by other channels than the chorda itself; we have no satisfactory evidence that direct stimulation of the central stump of a divided chorda will give rise to sensations of taste. The connection between the chorda and taste, however, may be of a more real kind.

On the other hand we must bear in mind how varied and complex are the junctions in the skull between the fifth nerve, the seventh nerve, and the glossopharyngeal nerve, by way of the Vidian nerve, the petrosal nerves, the tympanic plexus, Jacobson's nerve, and the otic and sphenopalatine ganglia. And it seems possible to suppose that fibres leaving the brain by the fifth nerve might find their way not directly to the lingual branch but by a roundabout way through the chorda tympani, and that at the same time other fibres from the same fifth nerve might ultimately join the glossopharyngeal nerve. There are no cases on record in which disease of the glossopharyngeal nerve within the cranial cavity has led to distinct loss of taste; but cases have been recorded in which disease of the fifth nerve within the cranial cavity, and as far as could be ascertained limited to the fifth nerve, has led to an entire loss of taste over the whole of one side of the tongue, both back and tip. Such cases lead to the at least provisional conclusion that the gustatory fibres are fibres belonging to the fifth, though they may reach the tongue partly by way of the glossopharyngeal, partly by way of the chorda tympani.

CHAPTER VI.

ON CUTANEOUS AND SOME OTHER SENSATIONS.

SEC. 1. THE NERVE ENDINGS OF THE SKIN.

§ 869. WE may speak of all the sensations which we derive from the skin as " cutaneous sensatious ; " we shall later on discuss their various kinds. The afferent nerves of the skin, those which appear to serve as the channels for impulses giving rise to cutaneous sensations, appear to end in two main ways. In the first place nerve fibres given off from a plexus of medullated fibres lying in the dermis immediately below the epidermis pass upward with loss of their medulla into the epidermis, and there, dividing into delicate nerve fibrils, come into connection with, or are lost to view between the epidermic cells. In the second place, nerve fibres still retaining their medulla and for the most part running singly, end in the dermis at a variable distance below the epidermis in special structures made up of modifications of the neurilemma, or other wrappings of the nerve fibre, associated more or less distinctly with cellular elements. We will consider the latter kind of ending first. In man the endings of this kind are two : the *end-bulbs* of limited, and the *touch-corpuscles* of much more general distribution. To these we must add the *Pacinian corpuscles*, which however are found not in the dermis proper, but· in the subcutaneous connective tissue, and which cannot be considered as distinctly cutaneous organs, since they also occur far away from the surface of the body, in connection with joints and the periosteum of bones and even with internal organs. They can hardly be considered as cutaneous organs, but they cannot be left unnoticed. In various animals we meet with other forms of nerve endings, the structure of which throws light on that of the kinds just mentioned.

The *end-bulbs* are found not in the skin generally but in certain situations only, namely the conjunctiva and the lips, and a special form occurs in the sensitive surfaces of the genital organs. They are not confined to the outer surface of the body, but are found also in mucous membranes, in the tongue and palate, in the rectum and elsewhere.

An end-bulb is a small rounded or oval body, 30 μ to 100 μ in diameter, consisting of a capsule of connective tissue enclosing a core of peculiar material. The capsule is indistinctly fibrillated, and bears nuclei which are sometimes so arranged as to indicate an outer capsule with nuclei placed lengthwise to the longer axis of the bulb, and an inner capsule with nuclei placed crosswise. The core consists of an homogeneous ground substance in which are imbedded granules, or in some cases nuclei; in some instances the core appears to be made up of small nucleated cells. A medullated fibre, with its neurilemma and sheath of Henle, approaches the end-bulb, and frequently becoming much coiled close to the bulb, plunges into the bulb at its base or side. The sheath of Henle, and perhaps the neurilemma also, become continuous with the capsule; the axis cylinder covered with medulla passes into the core, and here, frequently coiling about and at times dividing, loses the medulla, and ends in the midst of the core in a blunt slightly swollen end, or when it divides, in more than one such knob. Sometimes more fibres than one end in the same end-bulb. The marked feature of an end-bulb is a special development of the connective tissue wrappings of a nerve fibre, in the midst of which the axis-cylinder, with or without previous division, ends abruptly. The nature of the "core" is at present uncertain; it has been regarded as a modification of connective tissue but presents some analogies with the medulla of a nerve fibre.

§ **870.** *Pacinian corpuscles.* Though these are relatively large bodies, often more than a millimeter in length and easily seen by the naked eye, their general plan of structure is similar to that of an end-bulb; they may be regarded as very large highly developed end-bulbs.

The axis of the Pacinian corpuscle is furnished by a core, cylindrical or rather an elongated oval in form, resembling that of an end-bulb in that it consists of a homogeneous ground substance to which occur granules or small nuclei; in structures very similar in Pacinian bodies occurring as in the beaks of some birds (duck), called "Herbst's corpuscles," the core is distinctly composed of nucleated cells. This core is surrounded not by a single capsule only, but by a large number, twenty to sixty, of concentric capsules which are crowded together towards the axis but wider apart in the more superficial regions. A longitudinal or transverse section of a

Pacinian corpuscle accordingly exhibits a series of concentric lines separating spaces, the lines being especially close together near the axis. Each space represents a capsule or coat of connective tissue consisting of hardly more than fluid, traversed by transverse and longitudinal fibres, and defined by a membrane both on its inner and outer side. The lines in the section of a corpuscle represent the junctions of the inner membrane of one capsule with the outer membrane of the one lying to its inside. They are really the expression of lymphatic spaces, since each membrane of two adjoining capsules is covered with a layer of flat lymphatic epithelioid plates, the two layers leaving a linear space between them. Frequently a blood-vessel entering into the corpuscle at the side is distributed to the capsules, or at least to the outer ones.

A single nerve-fibre approaches the base of the corpuscle and as it draws near undergoes a great development of its sheath of Henle, which becomes transformed into a series of concentric, or in section parallel, sheets of connective tissue separated by lymph spaces. After reaching the corpuscle, the several sheets of the enlarged sheath of Henle leave the nerve fibre in succession to become the capsules of the corpuscle; and when the nerve fibre penetrates to the bottom of the core it consists only of axis, medulla, and neurilemma. The neurilemma is lost, passing off either to the innermost capsule or to the core itself; the medulla also disappears; but the axis-cylinder is continued undivided up the axis of the core to the far end where it ends in a blunt somewhat swollen point, or more frequently dividing ends in a number of such terminal knobs. Sometimes the nerve fibre passes straight through the corpuscle without ending; at the far end of the core it regains its medulla and neurilemma, takes up in turn the various capsules to form a thickened sheath of Henle, and leaves the corpuscle much as it entered it.

It is obvious that in the Pacinian corpuscle as in the more minute end-bulb, we have to do with a special development of connective tissue around an abruptly ending axis-cylinder. The whole arrangement has the air of a mechanical device; but it would be hazardous to insist too much on this.

Pacinian corpuscles are found in abundance clustered round the nerve branches running in the subcutaneous tissue of the palmar surface of the hand and sole of the foot, especially in the regions of the digits. It has been calculated that about 600 are present in the under surface of each hand, and about as many in each foot. They are much more sparse on, and often wholly absent from, the nerves of the skin in the back of the hand and foot, the upper arm and the neck. They occur on the nerves of the mamma and of the genital organs. They are very numerous along the nerves of the joints, especially in the flexures, about a hundred being counted at the elbow; they are also found along the periosteal nerves of the shafts of some bones, and occur occasionally in the inside of large nerve trunks. Lastly, they are scattered over the sympathetic nerves of the abdomen. It is obvious that their functional connection with cutaneous sensations is a very indirect one; but to this point we shall return.

§ 871. *Grandry's Corpuscles.* Although these are not found in connection with the skin of man, or indeed with true skin anywhere, their occurrence being limited to the dermis of the mucous membrane of the beak, of the tongue and of the palate of certain birds, it will be desirable to say a few words about them.

In its simplest, typical form a corpuscle of Grandry, about 50 μ in diameter, consists of two nearly hemispherical or dome-shaped cells placed with their flat surfaces face to face. Each cell contains a conspicuous round nucleus surrounded by granular cell substance; it has all the appearance of a large well-nourished epithelium cell. The two cells are surrounded by a capsule of connective tissue bearing nuclei. A medullated nerve fibre with its neurilemma and sheath of Henle, approaching the corpuscle in a more or less coiled course, joins it at the side at the level of the junction of the two cells. Here the sheath of Henle and probably also the neurilemma become continuous with the capsule, the medulla ceases, and the naked axis cylinder passing in between the opposed flat surfaces of the cells ends in the centre between them in a round flattened biconvex disc or "plate," forming as it were a "buffer" between the two cells in their central region. As far as can be ascertained there is no continuity between the axis cylinder plate and the substance of either of the cells; the former simply lies in contact with each of the latter. These epithelium cells, if we may venture so to call them, appear to be "subsidiary" cells, assisting in some unknown way the equally unknown functions of the terminal plate of the axis cylinder; and we may perhaps draw an analogy between them and the subsidiary cells of the organs of smell and taste. As we shall see, the nerves of the skin enter in the epidermis into similar relations with special cells distributed among the ordinary cells of the Malpighian layer in certain situations.

Sometimes there are three such cells, the uppermost and undermost being hemispherical or dome-shaped, and the middle one having the form of a disc or short cylinder, all being surrounded by the same capsule. In such a case the nerve fibre divides into two branches, one forming a terminal disc between the upper and middle cell, the other between the middle and lower cell. Where there are more than three cells the divisions of the nerve fibre are correspondingly increased. Sometimes a compound corpuscle is formed by the aggregation of several simple corpuscles, each with its epithelium cells and one or more terminal plates of an axis cylinder.

§ 872. *Touch Corpuscles.* These are minute bodies of an

oval form, 60 to 100 μ in the long diameter, which, in certain situations, are found in the papillæ of the dermis lying immediately beneath the epidermis. It is easy to recognize that each touch corpuscle contains a number of oval nuclei placed transversely, and that between the nuclei run irregular, but on the whole transverse lines, having the appearance of transverse fibres or septa, the whole being surrounded by an indistinct capsule. This arrangement gives the whole body an appearance not unlike that of a miniature fir cone. A medullated nerve fibre, generally with a more or less coiled course, reaches the corpuscle, most commonly at the side, and after twisting round the body to a variable extent, and frequently dividing, is finally lost to view in the midst of the corpuscle. Sometimes the corpuscles appear compound, as if made up of more corpuscles than one joined together, the whole body appearing lobed.

It is easy to recognize this much; but the further details of the structure of a touch corpuscle are matters of great difficulty, about which much divergence of opinion obtains. Perhaps the view which at present most commends itself is to regard them as formed on the same plan as a compound corpuscle of Grandry, but with the constituent elements less distinct. According to this view the transverse nuclei are the nuclei of subsidiary cells, in connection with which the divisions of the axis cylinder of the nerve fibre end in expansions comparable to the terminal plates of the corpuscle of Grandry, the characteristic transverse striæ being the expression of the divisions of the nerve fibre. But in the touch corpuscle the subsidiary cells are not so well formed as in the corpuscle of Grandry, and moreover seem especially placed towards the surface of the corpuscle, leaving the interior free at least from nuclei; the division of the nerve fibre is also more complete, the fibre often breaking up into clusters or bunches of branches; the medulla accompanies the dividing axis cylinder for a greater distance into the interior of the body; and the conspicuous terminal plates of the corpuscle of Grandry are replaced by mere slightly swollen knobs. Without insisting too much on the value of the analogy, we may probably conclude that in the touch corpuscle we have to do with an axis cylinder, which dividing frequently, ends abruptly in contact with, or in connection with, but not in continuity with certain cells of a peculiar nature.

Touch corpuscles are found in man most abundantly on the palmar surface of the hand, especially of the fingers. It has been calculated that the palmar surface of the tip of the forefinger contains about 100 touch corpuscles for each two square millimeters, that of the second joint 40, and that of the third joint 15 in a like area. They are also found, though somewhat less plentifully, on the sole of the foot and toes, about 30 being present in two square millimeters of the last joint of the great toe, and seven or eight in the like area of the middle of the sole. They occur in the nipple of the breast, on the under, volar, surface of the fore arm, being here however exceedingly scanty, at the edge of the eyelids and lips, and on the genital organs. From the greater part of the surface of the body they appear to be wholly absent.

§ 873. We may now turn to the second kind of ending of nerve fibres in the skin, that in which fibrils pass into the very epidermis. It will be useful to begin with the nerves of the cornea, the examination of which is relatively easy.

The medullated nerve fibres, which enter into the body of the cornea at its circumference, soon lose their medulla and subsequently their neurilemma. The axis cylinders dividing form towards the front of the cornea a plexus called the 'primary plexus,' at the nodal points of which, as also occasionally on the bars of which, nuclei are seen. From this primary plexus bundles of fibrillæ form other plexuses in the substance of the cornea, especially in front, where beneath the anterior elastic lamina (§ 716) a plexus called the "sub-basal plexus" may sometimes be found. From this plexus, or directly from the primary plexus, bundles of fibrillæ, or divisions of axis cylinders, of some little thickness, run straight outwards to the epithelium, piercing the anterior elastic lamina, and hence called *rami perforantes*. Reaching the base of the lowermost, vertical tier of cells, each ramus breaks up into a pencil of delicate fibrils, which taking a direction at right angles to that of the ramus itself, and converging towards the centre of the cornea, form between the upper surface of the elastic lamina and the base of the lowermost tier of cells, a close-set horizontal plexus of delicate fibrils, the *subepithelial plexus*. From this subepithelial plexus still more delicate fibrils run upwards into the epithelium, forming between the cells an *intraepithelial plexus;* and from this or directly from the subepithelial plexus, extremely fine fibrillæ pass upwards towards the surface of the epithelium. It has been asserted that these project beyond the surface, the free ends swollen into tiny knobs waving freely in the fluid which always covers the surface of the cornea. This however has been doubted; but in any case the nerves of the cornea send off branches which end in the manner described, quite close to the surface of the cornea as free fibrils, not directly connected with any cells. It need hardly be said that the detection of these fine endings requires special preparation; they can be seen only with the gold chloride method.

§ 874. A similar mode of ending also occurs in the skin. In all parts of the skin medullated nerve fibres are present, assuming more or less a plexiform arrangement in the dermis close beneath the epidermis, the fibres being more numerous and their arrangement more intricate in some regions of the body than in others. These medullated fibres give off fibres which, losing their medulla, penetrate into the epidermis, and there forming a more or less distinct plexus analogous to the subepithelial plexus of the cornea, give off numerous delicate fibrillæ which, forming a network among the cells of the malpighian layer, end in a manner similar to that obtaining in the cornea, namely, in free ends, between the cells of the upper region of the malpighian layer beneath the stratum granulosum ; this latter layer apparently they never penetrate.

The penetrating fibres giving rise to these intræpithelial fibrillæ, pass into the epidermis at the tops and at the sides of the papillæ, as well as in the valleys between the papillæ. As far as can be ascertained, they are present in all regions of the skin, being probably more abundant in some regions than in others. Their relative abundance cannot be quantitatively determined with exactness owing to the inconstancy of the gold chloride method; in the same region the method will disclose at one time a very large number of fibrillæ, at another time very few. We are probably justified in asserting that this method of ending, by means of intraepithelial fibrillæ, is the general mode of ending of the afferent nerve fibres distributed to the skin itself.

§ 875. In the epidermis of the pig's snout there are found in the deeper regions of the malpighian layer between the papillæ and elsewhere, oval nucleated cells which differ from the ordinary cells of the layer, both in their form, being oval not polygonal, in the absence of prickles, and in their behaviour towards reagents, especially gold chloride, with which they stain somewhat deeply. Non-medullated nerve fibres penetrating the epidermis from the dermis, and dividing into branches, pass to these cells, which frequently occur in clusters, and each branch terminates in a sort of plate or disc, which is applied closely to the under surface of one of the cells, but apparently is not continuous with the substance of the cell. The cells are not so well developed as are the cells of a corpuscle of Grandry, and the terminal plate has not such a definite form as in that body ; but the resemblance between the two is very striking. Moreover cells of this kind with the belonging nerve filaments have been observed lying partly within the epidermis and partly in the dermis beneath. It would seem as if these cells with their nerve filaments formed within the epidermis an organ of the same nature as and probably playing the same part as the corpuscle of Grandry in the dermis.

Similar cells, which have been called " touch-cells," have been observed in other regions of the skin of various animals, and are well developed in the outer root sheath or malpighian layer of the tactile hairs, such as those in the " whisker " of the cat. They have been found in man in regions where the touch corpuscles are sparse or absent, in the skin of the back, belly, arms, legs and neck. It seems probable that further investigation will disclose that they have a very wide distribution ; if so they will have to be regarded as a significant mode of ending of the cutaneous nerves.

SEC. 2. THE GENERAL FEATURES OF CUTANEOUS SENSATIONS.

§ 876. The sensations which we experience by means of the skin and cutaneous nerves appear, in the first instance, to be of at least three kinds. In the first place, all bodies, whatever their chemical or physical nature, be they gaseous, liquid or solid, when brought into contact with the skin, when made to exert mechanical pressure on the skin, produce sensations of a certain kind; these sensations, whose characters depend mainly on the amount of pressure exerted and on the region and area of the skin pressed upon, may be conveniently spoken of as *tactile sensations* or *sensations of touch proper*. In the second place, when either by actual contact with, or by the mere proximity of hot or cold bodies, of whatever nature, the temperature of an area of the skin is changed with sufficient rapidity, we experience sensations of a kind different from the tactile sensations just mentioned; these we may speak of as *sensations of temperature, sensations of heat and cold*. In the third place, when too violent a pressure is exerted on the skin, or when the changes of temperature are excessive, or when certain changes giving rise neither to tactile nor to temperature sensations are produced, or take place in the skin, we experience sensations which we call *sensations of pain*. This third kind of sensation stands, in many respects, apart from the other two, and it will be convenient to study sensations of pain by themselves. Sensations of touch proper and of heat and cold are much more akin and may be treated of together.

Tactile Sensations or Sensations of Pressure.

§ 877. Many of the characters of tactile sensations are of the same order as those of visual sensations, which we studied somewhat fully, and indeed similar characters may be more or less distinctly recognized in all sensations. The amount, that is to say the intensity of the sensation, varies with the amount of the stimulus, with the amount of pressure brought to bear on a given area of the skin. Taking the same spot of skin, the tip of the forefinger for instance, we can experimentally ascertain the minimum of pressure of which we can become conscious, such for example as that exerted by a minute fragment of some light body, pith or wool, falling through a certain small height. Starting from this minimum and increasing the pressure, we find the sensation also to increase up to a certain limit; and Weber's law (§ 747) holds good for tactile sensations, indeed may be more easily verified in their case than perhaps in the case of other sensations.

When two sensations follow each other in the same spot of skin at a sufficiently short interval they are fused into one; thus, if the finger be brought to bear lightly on the edge of a rotating card cut into a series of teeth, the teeth cease to be felt as such when they follow each other at a rapidity of about 1500 in a second. The vibrations of a cord cease to be appreciable by touch when they reach the same rapidity.

When two sensations are generated at the same time at two points of the skin too close together they become fused into one; but to this feature, which is of a different nature from the preceding, we shall return presently.

The sensation caused by pressure is at its maximum soon after its beginning, and thenceforward diminishes. The more suddenly the pressure is increased, the greater the sensation; and if the increase be sufficiently gradual, even very great pressure may be applied without giving rise to any sensation. A sensation in any spot is increased when the surrounding areas of skin are not subject to pressure at the same time. Thus if the finger be dipped into mercury the pressure of the mercury will be felt more at the surface of the fluid adjoining the skin which is not in contact with the mercury, than in the parts of the skin wholly covered with the mercury; and if the finger be drawn up and down, the sensation caused will be that of a ring moving along the finger. This effect may be compared with those of 'contrast' in visual sensations (§ 781).

All parts of the skin are not equally sensitive to pressure; the minimum of pressure which can be felt or the smallest difference of pressure which can be appreciated differs very much at different parts of the skin. Measured in this way, tactile sensations are much more acute on the palmar surface of the finger, or on the forehead, than on the arm or on the sole of the foot or on the back. In making these determinations all muscular movements should be avoided in order to eliminate the muscular sense of which we shall speak later on; and the area stimulated should be as small and the contact as uniform as possible.

In a similar manner small consecutive variations of pressure, as in counting a pulse, are more readily appreciated by certain parts of the skin, such as the tip of the finger, than by others. In all cases variations of pressure are more easily distinguished when they are successive than when they are simultaneous.

§ **878.** *The localization of tactile sensations.* When anything touches a spot of our skin, we not only experience a 'pressure sensation' of greater or less intensity according to the amount of pressure exerted and the particular region of skin pressed upon, we are also at the same time aware that the sensation has been started in that spot, that the spot in question and not another has been touched. When we are touched on the finger or on the back we refer the sensations to the finger or to the back respectively, and when we are touched at two places on the same finger at the same time we refer the sensations to two parts of the finger. We localize our touch sensations with reference to the surface of our body after the same fashion that we localize our visual sensations with reference to the external world. Our whole skin serves us as a 'field of touch' analogous to the 'visual field' of the eye; and as when experiencing a visual sensation, we refer it to its presumed cause and say we perceive a light in some part or other of the field of sight, so when we experience a tactile sensation we say we perceive that something has touched this or that part of our skin; the tactile sensation has become a tactile perception. As the accuracy of our visual perceptions is largely dependent on the smallness of the retinal interval which must separate two simultaneous retinal stimulations in order that these shall give rise to two separate sensations, vision being most distinct in the fovea centralis where this interval is smallest, so also the accuracy of our tactile perceptions is dependent on the smallness of a like cutaneous interval. Where, as in the tip of the finger, the interval is small, contact with even a small area of surface may give rise to several simultaneous but distinct sensations, each of which we localize; and we thus obtain by means of one contact several perceptions affording a considerable amount of information concerning the nature of the surface. Where, as in the skin of the back, the interval is great, contact with even a large area of surface may give rise to one sensation, which we do not resolve into its components, all the several sensory impulses from the skin fusing into one common sensation; we only localize this one sensation, we have only one perception of something touching that part of our back, and the information which we thus acquire concerning the nature of the surface in contact with the skin is limited.

As the above remark indicates, the interval in question varies very widely in different parts of the surface of the body; our power of localization is much finer in certain parts than in others. Moreover the distribution of the fineness of localization is not identical with that of the mere appreciation of pressure; some parts may be very sensitive and yet possess imperfect localization. The magnitude of the interval of space which must separate two simultaneous stimulations of the skin in order that the two consequent sets of impulses may give rise to two distinct sensations may be conveniently determined for different regions of the skin by measuring the distance at which two points (preferably blunted) of a pair of compasses must be held apart, so that when the two points are in contact with the skin, the two consequent sensations can be localized with sufficient accuracy to be referred to two points of the body, and not confounded together as one.

The following tabular statement of results thus obtained may be taken as shewing in a general way the relative power of localization in the more important regions of the surface of the skin.

Tip of tongue	1·1 mm.
Palm of terminal phalanx of finger	2·2 „
Palm of second „ „	4·4 „
Tip of nose	6·6 „
White part of lips	8·8 „
Back of second phalanx of finger	11·1 „
Skin over malar bone	15·4 „
Back of hand	29.8 „
Forearm	39·6 „
Sternum	44·0 „
Back	66·0 „

As a general rule it may be said that the more mobile parts, or those which execute the widest movements, or execute movements most easily and frequently, such as the hands and lips, are those by which we can thus discriminate sensations most readily. The lighter the pressure used to give rise to the sensations, provided that the impulses generated are adequate to excite distinctly appreciable sensations, the more easily are two sensation distinguished; thus two compass points which, when touching the skin lightly, appear as two, may, when firmly pressed, give rise to one sensation only. The distinction between the sensations is obscured by neighbouring sensations arising at the same time. Thus two points readily distinguished as two when the skin is under ordinary conditions, are confused into one when brought to bear inside a ring of heavy metal pressing on the skin.

It need hardly be said that these tactile perceptions, like all other perceptions, are increased by exercise. We may speak of our 'field of touch,' as being composed of tactile areas or units, in the same way that we spoke (§ 754) of our field of vision as being composed of visual areas or units; but all that was there said concerning the subjective nature of the limits of visual areas, applies equally well, *mutatis mutandis*, to tactile areas. –When two points of the compasses are felt as two distinct sensations, it is not necessary that two, and only two, nerve-fibres should be stimulated, or, putting the matter more generally, that two or only two discrete sets of sensory impulses, should travel along separate paths to separate cerebral centres. All that is necessary is that the two cerebral sensation-areas should not be too completely fused together. The improvement by exercise of the sense of touch must be explained not by an increased development of the terminal organs, not by a growth of new nerve-fibres in the skin, but by a more exact limitation of the sensational areas in the brain, as for example by the development of a resistance which limits the radiation taking place from the centres of the several areas.

Sensations of Heat and Cold.

§ 879. When we bring into contact with, or even into the immediate neighbourhood of a spot of skin, a body distinctly hotter than is the skin at the spot for the time being, we experience a special sensation; we feel something in the skin that was not there before, but that something is wholly unlike the effect of pressure, and we call the sensation a sensation of heat. The sensation is obviously due to the rise in the temperature of skin which is the direct effect of the contact with or the nearness of the hot body. Our skin has a certain temperature which varies from time to time, according to circumstances, and is not the same in all regions of the skin at the same time. A given spot of skin at a given time will have a certain temperature; that temperature does not give rise to a distinct sensation though its effects may enter into what we may call general sensibility; we may not be directly conscious, for instance, that the forehead has one temperature and the hand another, though the two temperatures may differ widely. It appears then that we are only conscious of a cutaneous sensation of heat when the temperature of a region of the skin which has previously been fairly constant is raised; we may add suddenly raised, for in sensations of heat as of pressure the stimulus must act with a certain rapidity in order to produce a distinct effect on consciousness.

If the body brought into contact with or near to the skin, instead of being distinctly hotter is distinctly colder than the skin we also experience a special sensation, a sensation of cold; and this sensation differs in kind not only from that of pressure, but also from that of heat. We might expect perhaps that since cold only differs from heat in degree, both being degrees of temperature, that the sensations of heat and cold would also be alike, differing only in degree; but when we appeal to our consciousness we recognize that they differ in kind. So long as sensations of heat and cold remain sensations of heat and cold, they appear to us not as merely different phases of the same thing but as quite unlike; when the exciting heat or cold is excessive we perhaps may fail to distinguish between the two, but that is because both are lost in the sensation of pain. It appears then that we are conscious of a specific sensation of cold when the temperature of a region of the skin which has previously been fairly constant is with sufficient rapidity lowered. To how large an extent we are, under ordinary circumstances, unconscious of the actual temperature of the skin and how sensitive we are to even slight changes of temperature may be illustrated by using one region of the skin as a stimulus of heat or cold for another. At a time, for instance, when we are not directly conscious of the hand being either colder or hotter than the forehead, by putting the one up to the other we may experience a distinct sensation telling us that the hand decidedly differs in temperature from the forehead; we feel at once that one is warmer or colder than the other, though it may take some little time to recognize which is the warmer or the colder.

§ 880. These sensations of heat and cold behave very much in the same way as sensations of pressure. We have already said that the change of temperature like the change of pressure must be effected with a certain rapidity in order to produce a distinct sensation, and in general the more gradual the change the less intense is the sensation.

As might be expected from the fact that it takes a longer time to produce a change of temperature than to exert pressure, the sensation of either heat or cold is somewhat slowly developed and lasts some considerable time; hence consecutive sensations readily fuse into one.

Since it is the changed temperature and not the particular temperature arrived at which is the basis of the sensation, a hot body or a cold body gives rise to a sensation only at the first contact or approach and for some little time afterwards, the effect diminishing from the very moment that the change has been established. Hence a hot body or a cold body, even when kept itself at a constant temperature and not cooled or heated by contact with the cooler or warmer skin, ceases after a while to be felt as hot or cold. For this reason the repeated dipping of the hand into hot or cold water produces a greater sensation than when the hand is allowed to remain all the time in the water, though in the latter case the temperature of the skin is most affected.

The effects of contrast are obvious in sensations of heat and cold as in those of pressure; when the hand is dipped in hot water the sensation is most intense at the ring where the hand emerges from the surface of the water.

We can with some accuracy distinguish small differences of temperature, especially those lying near the normal temperature of the skin. In this respect these sensations follow Weber's law, though apparently slight differences of cold are more easily recognized than those of slight heat. The range of the greatest sensitiveness seems to lie between 27^0 and 33^0.

The regions of the skin most sensitive to variations in temperature are not identical with those most sensitive to variations in pressure. Thus the cheeks, eyelids, temples and lips, are more sensitive than the hands. The least sensitive parts are the legs, and front and back of the trunk; to this matter however we shall return.

As with pressure sensations so also with sensations of heat and cold, two sensations excited at a certain distance apart may or may not be fused into one, the distance necessary for the separation of the sensations varying in different regions of the body, and being, as might be expected from the ease with which heat and cold are conducted, much greater than in the case of pressure sensations. We also 'localize' the sensations of heat and cold; we can recognize which region of the skin is being heated or cooled; and thus these sensations also enter into our perceptions of the external world.

§ 881. We have treated of the sensations of touch and of heat and cold as cutaneous sensations; but they are not confined to the skin commonly so called. We experience the same sensations in varying degree by help of the lining of the mouth and pharynx, which is called a mucous membrane; and they may also be traced for a short distance up the rectum beyond the margin of the skin proper. But in both these situations, the lining membrane is by origin and in structure epiblastic, that is to say cutaneous, and in possessing cutaneous functions shews its real nature. These functions are most marked at the beginning of the passages, the tip of the tongue being very sensitive to touch and heat and cold, with a well-developed power of localization; they are very rapidly lost in the rectum, and more gradually disappear at the lower part of the pharynx and in the œsophagus; a fluid which in the mouth is felt distinctly as hot gives rise to a sensation of pain not of heat when it is swallowed, and a cold or warm drink is only felt as cold or warm when swallowed in quantity sufficient to affect by conduction the abdominal skin. The maintenance of these cutaneous functions in the initial parts of the alimentary canal, which are under the dominion of the will, is, like the sense of taste, a safeguard against the introduction into the canal of noxious substances; in the subsequent parts, no longer subject to the will, any warning which such sensations might give would be too late and useless.

SEC. 3. ON PAINFUL AND SOME OTHER KINDS OF SENSATION.

§ 882. When excessive pressure is exerted on the skin, or when the change of temperature passes certain limits, the sensation which is excited ceases to be recognized as one either of touch or of temperature and takes on characters of its own; we then call it a sensation of pain. In this respect the skin as a sensory organ appears at first sight to differ from the other organs of sense which we have studied. We have no evidence that simple stimulation of the retina, however excessive, will give rise to pain, meaning by pain the kind of sensation we feel when the skin is cut or burnt. We often speak it is true, especially in cases of disease of the eye, of exposure to light causing pain, but the pain in such cases is felt through the eyeball, not through the retina and optic nerve; the pain is not an excessive development of visual sensations, it is a phase of that sensibility which the subsidiary structures of the eye share, in common as we shall see presently, with not only the skin but nearly all other structures of the body. In like manner we have no evidence that an auditory or an olfactory or a gustatory sensation can, through mere intensity, become converted into a sensation of pain, though we may, in the act of hearing, smelling or tasting, receive sensations of pain from the ear, nose or mouth. We often of course apply the word 'painful' to a sound, or a group of visual sensations, or a smell or a taste; but that is in the sense of being exceedingly disagreeable, and has reference to our classification of the complex psychical effects of all our sensations into those which are pleasurable and those which are painful. Without entering into any psychological analysis, we may assume that the pain which we feel when the finger is cut is a wholly different thing from the pain which is given to a most delicately musical ear by even the most horrible discord; and in what follows we shall use the word pain in the first of these two meanings.

§ 883. The above considerations suggest that in the case of the skin as in the cases of the other organs of special sense, a sensation of pain is not simply an exaggeration of a sensation of pressure or of a sensation of temperature, but is a separate sensation, developed in a different way in the skin, a sensation which may override and so seem to replace the sensation of pressure or temperature developed at the same time, but which must not be confounded with it. And this view derives support from the fact that events taking place in many other parts of the body, from which we experience sensations neither of touch nor of temperature, may under favourable circumstances give rise to pain in varying degree. When, for instance, a tendon is laid bare contact with a body will not produce tactile sensations, heating or cooling will not produce temperature sensations; one cannot by means of the tendon as one can by means of the skin perceive that a rough or smooth body, that a hot or cold body, has been brought to act upon it. Indeed in respect to all structures other than the skin and nerves, to such structures namely as muscles, tendons, ligaments, bones, and the viscera generally, there is a large amount of experimental and clinical evidence shewing that, so long as these are in a normal condition, experimental stimulation of them does not give rise to any distinct change of consciousness; a muscle or a tendon, the intestine, the liver or the heart may be handled, pinched, cut or cauterized without any pain or indeed any sensation at all being felt or any signs given of consciousness being affected. Nevertheless when the parts are in an abnormal condition even slight stimulation may produce a very marked effect on consciousness. If, for instance, a tendon becomes inflamed, any movement causing a change in the tendon, especially one putting the tendon on the stretch, will affect consciousness and give rise to a sensation. But the sensation is one of pain and not of any other kind. Moreover we simply 'feel' the pain, we do not 'perceive' the cause of it; because we feel the pain we infer that something has caused it, but we cannot from the nature of the pain itself decide whether that something is a stretching of the tendon, the contact of a hard or soft body, the approach of some hot or cold body, the application of some chemical substance, the passage of an electric current, or intrinsic events taking place in the tendon itself as the result of physiological changes. And so in other instances; there is hardly a part of the body changes in which may not, under certain circumstances, give rise to sensations of pain. We can to a variable extent, in a more or less ill-defined manner, localize the sensation; we can distinguish a pain in the foot from one in the leg, a pain in the thumb from one in a finger; we may occasionally fix the pain in a very small limited area, though especially if the sensation be intense, the pain radiates and its localization becomes obscure. And we may here remark that when we thus localize a pain arising in the structures of which we are speaking, we refer the pain not to the structures themselves but to neighbouring parts and especially to the skin; the intense pain, for instance, of "renal colic," caused by the impact of a calculus in the ureteris referred by us not to the ureter itself but to adjoining parts, to the corresponding somatic segment; and so in other instances. We can also recognize certain characters in different pains, beyond that of the mere degree of intensity; we speak of pains as being burning, aching, gnawing, cutting, throbbing and the like. But in all cases the pain remains a mere sensation; when it comes, all we can say is that we feel in a particular region of the body a pain of a certain intensity and having a certain character. We infer that something is wrong, but the pain in no way tells us what the wrong is; we may call the pain a burning one because it is more or less like the pain which we feel when the skin is burnt; but in the vast majority of cases heat has nothing whatever to do with pains of a burning character; and so with other kinds of pain, the character of the pain does not in itself tell us anything about its cause.

Are we then to regard pain as a sensation of a kind by itself, the very threshold of which, the very least amount of which that can in any way affect our consciousness, must be regarded as already pain? In attempting to answer this question the following considerations deserve attention.

We are in a certain obscure way aware of what we may call the general condition of our body. To put an extreme case, if the whole of our abdominal viscera were removed we should be aware of the loss. We should be aware of this through more ways than one. The tactile sensations from the abdominal skin would be in such a case different from the normal, and moreover the muscular sense of the abdominal walls and of all the muscles whose actions bear on the abdomen, would make us aware of the void. But beyond all these indirect ways, it is probable that we should in a more or less obscure manner be directly conscious of the loss. It is probable that sensory impulses, not of the character of pain, are continually, or from time to time, passing upwards from the abdominal viscera to the central nervous system. These do not affect our consciousness in such a distinct manner as to enable us to examine them psychologically in the same way that we are able to examine special sensations such as those of sight, or even sensations of pain; they are even less well defined than those of the muscular sense; nevertheless they do enter, though obscurely, into our consciousness, so that we become aware of any great change in them, and they have been spoken of under the title of "common" or "general sensibility." In discussing (§ 643) the manner in which the manifold coordinate movements of the body were carried out we saw reasons for thinking that the central processes of the nervous system were largely determined by varied afferent impulses which produce their effects without giving rise to any sharp and decided change of consciousness; many of these are probably afferent impulses of the common sensibility of which we are now speaking.

If we suppose that the skin in common with the other tissues of the body possesses this common sensibility, and if we further suppose that in the skin as elsewhere, these afferent impulses when developed, as is the case under normal circumstances, to a slight extent only are not distinctly recognized by consciousness, and that when they do assume such a magnitude or intensity as to break in upon consciousness the change of consciousness which they produce is of the kind which we call pain, we reach a conclusion which is also supported by other considerations. On the one hand such a view is in accord with the conclusion that cutaneous sensations of pain are wholly distinct from and developed in a wholly different way from sensations of touch and temperature; and, as we shall see, to this conclusion we are led by several different arguments. On the other hand it relieves us from the following difficulty. It may happen to a man to suffer pain in a particular region or tissue of the body, once only in the course of his lifetime or possibly not even once; nay, we may suppose that in this or that region or tissue pain is felt once only in one individual among a large number of persons. If we suppose that pain is not as suggested above an excessive phase of something which is continually going on in a lower phase, but a something by itself quite distinct from all other sensations, we are driven to conclude, since such a sensation must have a special mechanism, including special afferent nerve fibres to carry it out, that in the case in question such a mechanism of pain has been preserved intact but unused through whole generations in order that it may once in a while come into use; which is in the highest degree improbable. This difficulty disappears if we suppose that the constantly smouldering embers of common sensibility may be at any moment fanned into the flame of pain.

We may conclude then that the skin in common with other tissues possesses common sensibility, and that when this is excited in excess, so as to distinctly affect consciousness, we call it pain. We thus experience through the skin three kinds of sensations, those of touch, of temperature, and of common sensibility, but the two former only are developed by further psychical processes into perceptions; it is by them alone that we obtain through the skin knowledge of external objects.

§ 884. There is another consideration to be taken into view. The agents which applied to the skin produce pain, act violently on the skin, in many cases injuring the epidermis and affecting the dermis. Moreover if the epidermis be removed, and the stimulus, mechanical, thermal or chemical, be applied to the dermis or to the nerves running in it, we still experience sensations of pain, though no longer those of touch and temperature when a sharp or hot body is made to touch, not the intact skin but a wound, we suffer pain, but do not recognize the sharpness or the heat which is causing the pain. This suggests that the special sensations of touch and temperature are brought about by special, epithelial structures serving as the differentiated ends of nerve fibres, but that common sensibility and pain need no such special endings; this however opens up questions which we must consider separately by themselves.

§ 885. *Hunger and thirst.* We may introduce here the few words that we have to say concerning two affections of consciousness, which may perhaps be considered as kinds of sensation, namely, hunger and thirst.

We refer our feelings of thirst to, or at least we associate them with, a particular condition of the mucous membrane of the mouth, especially of the soft palate. When the mucous membrane of this region becomes drier than normal, as for instance by being exposed to too great an evaporation, we feel 'thirsty,' and the feeling is at once removed by adequately moistening the membrane. Under ordinary circumstances however the condition of thirst is brought about, not by anything bearing specially or exclusively on the mucous membrane of the soft palate or even of the whole mouth, but by the diminution of the water present in the body either through restriction of the intake, or through excess of the output in the secretions, such as that of sweat, or through both together. This is often spoken of as diminution of the water of the blood; but most probably the specific gravity of the blood is kept constant by the withdrawal of water from the lymph, so that the loss falls on the latter fluid. Such a diminution of the water of the body may be brought about by circumstances such as excessive sweating which in themselves do not cause special dryness of the mucous membrane of the soft palate; this part then undergoes a loss of water in common with the other tissues, but not in a special degree. Nevertheless thirst thus brought about may be temporarily assuaged by simple moistening of the soft palate. From this we may infer that the sensation of thirst is brought about by afferent sensory impulses started in the mucous membrane of the soft palate by a deficiency of water in that membrane, perhaps by a drain on the lymph spaces of that membrane.

We are in the habit of assuaging thirst by drinking water, or watery fluids, and in doing so produce both a direct local effect on the palate and a general indirect effect on the body. In the absence of the local effect, the indirect effect is slow in coming and needs a large quantity of fluid; when in cases of gastric fistula water is introduced into the stomach through the fistulous opening, large quantities may be given before thirst is assuaged.

The sensation of hunger is in a somewhat similar manner referred to, or associated with, the condition of the gastric mucous membrane. We feel hungry when the stomach is empty. But even more distinctly than in the case of thirst the main cause of the sensation seems to be a general condition of the body, namely, that produced by the products of digestion ceasing to be thrown into the blood. The sensation is not due to the mere emptiness of the stomach, though the emptiness of the stomach is one of the results of the abstinence from food; for the feeling of hunger may disappear though the stomach may remain empty, if adequate nourishment be conveyed in other ways, as by injection into the bowels; conversely even we ourselves may under abnormal conditions feel hungry on a full stomach, and in some animals, herbivora, the stomach is always more or less full. The sensation however does seem to be in some way specially connected with the condition of the gastric walls, much in the same way that thirst is specially connected with the palate; the products of digestion have a much greater power in appeasing hunger when they act locally and directly on the gastric membrane than when they are simply brought to bear on the body at large, and a small quantity of food will immediately satisfy hunger when introduced into the stomach, though it will have no effect when introduced otherwise. Moreover our own consciousness clearly connects the sensation in some way or other with the stomach.

As to what is the particular change in the gastric membrane which thus gives rise or assists in giving rise to the sensation we know little or nothing; indigestible substances such as cannot be properly called food when taken into the stomach at least temporarily remove the sensation. And we have little or no knowledge as to the particular nerves which serve as the paths for the afferent impulses which we may suppose to be generated in the gastric membrane. Division of the vagus nerve on both sides is said to have no effect on hunger; from this we may conclude that the impulses do not pass up this nerve, though it appears to be the sensory nerve of the stomach. But we have no evidence that the impulses pass along the sympathetic nerves.

Allied somewhat to hunger is the peculiar feeling which we may perhaps also speak of as a sensation, known as nausea, the precursor of vomiting (see § 272) and brought about like vomiting by a variety of events. We have little or no knowledge of it viewed as a sensation.

The affection of consciousness which is produced by the form of cutaneous stimulation known as "tickling" is of a peculiar character, differing from tactile sensations. Indeed it is probably undesirable to speak of it or of other psychical effects of cutaneous stimulation as a sensation, since it seems to be not the direct effect of the sensory cutaneous impulses, which are probably ordinary tactile impulses, but rather the effect on consciousness of changes in the central nervous system brought about by those sensory impulses.

SEC. 4. ON THE MODE OF DEVELOPMENT OF CUTANEOUS SENSATIONS.

§ 886. Our studies so far point to the conclusion that sensations of touch and temperature stand on the same footing as visual, auditory and other special sensations; and it will be profitable now to compare in some detail the former with the latter. In doing so we may, in order to make the matter more simple, confine ourselves in the first instance to sensations of touch proper, that is to sensations of mere contact and pressure, discussing later on the relations of these to sensations of heat and cold.

In studying vision we came to the conclusion that the undulations of the ether so affect the rods and cones and other retinal structures as to give rise to visual impulses, and that these visual impulses, travelling up the fibres of the optic nerve to the visual centres, gave rise by means of those centres to the affections of consciousness which we call visual sensations; we may leave aside in the present instance all reference to the complexity of the visual centres.

We obtained absolute proof that the only way in which light can give rise to visual impulses in the optic fibres is by acting on the retinal structures. Since the optic fibres are the only nerve fibres in direct connection with the retinal structures visual impulses can be carried by them alone. As we pointed out we know absolutely nothing about the nature of visual impulses themselves; our conclusions concerning the various characters and kinds of visual impulses are simply deductions from the psychological examination of our sensations; our objective knowledge of them is limited to the fact that when light falls on a functionally active retina an electric change is developed in the optic fibres. As we mentioned in § 750 the statement that stimulation of the optic fibres themselves, as when the optic nerve is cut with a knife, gives rise to visual sensations, has led to the adoption of the view that any impulse passing along the optic fibres, however started, whether by the action of light on the retina, or by direct stimulation of the fibres themselves, gives rise to a visual sensation and must therefore be regarded as a visual impulse.

This view, under the title of the doctrine of "the specific energy of nerves," has been extended to the nerves of the other special senses and indeed to nerves in general. This doctrine teaches that, owing either to the constitution of the central ending of a sensory fibre or to that combined with the nature of the fibre itself (the view may also be adapted to motor fibres), whatever impulses are generated in the fibre can give rise to those events only which are specific to that central ending, impulses of all kinds along an optic fibre giving rise to visual sensations, impulses of all kinds along an auditory fibre giving rise to auditory sensations, and so on. Hence under this view the purpose of the specific terminal organ is simply to allow the specific stimulus of the sense, light in the case of the retina, to develop impulses in the specific nerve, a result which, in the absence of the terminal organ, it is powerless to achieve.

We saw however (§ 750) that according to some observers direct stimulation of the optic fibres, as when the nerve is cut, does not produce visual sensations, and therefore does not give rise to visual impulses; so far as can be ascertained such a stimulation of the fibres appears to produce no effect at all on the central nervous system. If we accept this result as true, we must modify the doctrine of the specific energy of nerves in the following way. We must suppose that the visual centres are so constituted that they are stirred up to the development of visual sensations by the advent only of those kind of impulses which are started by means of the terminal organ. Since electric changes are developed in the optic fibres as in other nerve fibres when the optic fibres are directly stimulated, we may infer that direct stimulation does lead to nervous impulses; and we may further infer that these reach the visual centres but are unable to develop visual sensations because they are not true visual impulses such as are generated by help of the terminal organs.

This modified view is supported, though the support is of a negative kind only, by the behaviour of the other organs of special sense. We have no satisfactory experimental or other evidence that stimulation of the auditory fibres or of the olfactory fibres otherwise than through the terminal organs will give rise to auditory or olfactory sensations. We have evidence that stimulation of the centres by various means will give rise to the specific sensations, but not that stimulation of the fibres of the nerves themselves will. The branches of the glosso-pharyngeal and fifth nerves distributed to the organs of taste are, unlike the above, mixed nerves, and when they are stimulated sensations other than specific taste sensations are also developed, and the former might obscure the latter; still the evidence so far as it goes supports the view that stimulation of gustatory fibres otherwise than through their terminal organs does not lead to the development of gustatory sensations.

In the case of touch the evidence is perhaps still stronger. We must in any case suppose that each cutaneous nerve distributed to a given area of skin contains fibres which subserve the sense of touch exercised by that area, and which pass from the terminal organs in that area, whatever their nature, to the parts of the central nervous system, whatever they may be (§ 679), which act as centres of touch sensations. If these fibres when directly stimulated, apart from their terminal organs, necessarily give rise to touch sensations, stimulation of the nerve itself while running in the subcutaneous tissue should give rise to touch sensations. But experience shews, as we said a little while ago, that this is not the case. Whenever the nerve fibres themselves are directly stimulated, as for instance when the epidermis is removed from the skin or when a nerve is laid bare, then however they be stimulated, be the stimulus weak or strong, if consciousness be affected at all, the affection takes on the form of pain; psychological examination of the subjective result discloses nothing that can be called a sensation of touch. A familiar instance of the difference between the effects of stimulating a nerve trunk, and those of stimulating the cutaneous terminal organs of special sense, is seen in the effect of dipping the elbow into a freezing mixture. The cold affects the skin of the elbow and gives rise to sensations of cold in that part; but the cold, if intense enough, also affects the underlying trunk of the ulnar nerve, and by direct stimulation of the fibres in the trunk develops sensory impulses; these impulses however are those not of sensations of cold, but of pain; and the pain, in accordance with a principle to which we shall presently call attention, is referred to the terminal distribution of the ulnar nerve on the ulnar side of the hand and arm. In speaking above (§ 883) of pain we said that excessive pressure or excessive heat or excessive cold applied to the skin, overrides or annuls pressure and temperature sensations and gives rise to mere sensations of pain; and it might be urged that when a nerve is directly stimulated the specific sensations of touch and temperature are similarly annulled. But in the case of the skin an excessive or violent stimulation is necessary to produce this effect, whereas a nerve may be directly stimulated by so slight a stimulus as to give rise to hardly more than discomfort without distinct pressure or temperature sensations being felt, and we can hardly suppose that in such a case these are present but are annulled by an amount of pain so slight as that which is produced. Thus making every allowance for the suggestion that sensations of pain may override and obscure concomitant sensations of touch and temperature, we seem driven to the conclusion that the latter sensations can only be developed by help of special terminal organs, and that a stimulation of the nerve fibres themselves if it produces any effect at all on consciousness gives rise to pain, and to pain alone.

· We are in this way led to conceive of the skin as provided on the one hand with specific fibres ending in specific terminal organs and serving for sensations of touch and temperature, and on the other hand with fibres of common sensibility having no such specific terminal organs, the two kinds of fibres being mixed together in the common cutaneous nerve. These fibres moreover have not only different peripheral but also different central endings, and during at least some part of their course run in different tracts or in a different manner in the central nervous system; for as we saw in treating of the central nervous system (§ 683) cases of disease of the central nervous system have been recorded in which over certain cutaneous areas sensations of touch had been lost, while common sensibility and sensations of pain remained, or vice versa. We may add that the difference between the central paths or endings of the nerves of touch and those of pain is further shewn by the fact that in certain nervous diseases (tabes) when the skin is pricked with a pin, the contact of the pin may be felt as mere touch for so long a time as one or two seconds before pain is felt; the diseased condition enormously delays the transmission of the impulses of pain but has not so much effect on those of touch.

§ 887. We may go a step further; there is a certain amount of evidence that the terminal organs and fibres concerned in touch proper, in sensations of pressure, are different and separate from those concerned in sensations of heat and cold. In the first place the general topographical distribution over the surface of the body of sensitiveness to pressure is different from that of sensitiveness to temperature. A familiar instance of this is seen in bringing the palm of the hand to touch the forehead. In the former the sense of touch is highly developed, in the latter the sense of temperature; hence with the forehead we feel that the hand is warm or cold, with the hand we feel that the forehead is rough or smooth; at least these two feelings respectively preponderate, the one in the one part, the other in the other. In the second place cases of disease of the central nervous system of the spinal cord have been recorded in which, over certain cutaneous areas, sensations of pressure were lost but sensations of temperature remained, and vice versa. In the third place, if the stimulation of the skin be confined to extremely minute areas, if the pressure, or the change of temperature be brought to bear as much as possible on a mere point of the skin, it is found that some points of the skin are sensitive to pressure but not to change of temperature, while others again are sensitive to change of temperature but not to pressure. If a blunt pointed but otherwise fine needle be used to exert pressure, a little exploration will ascertain that at some points the amount of pressure can readily be recognized, the sense of touch is acute, while at other points, and these may be quite near the others, the amount of pressure cannot be recognized, and indeed no sensation is experienced until the pressure is excessive and then the sensation felt is not one of touch proper but of pain. Similarly if heat or cold be applied by means of a metal tube or rod narrowed to a point, it will be found that some points of the skin are very sensitive to changes of temperature, while other points are insensitive to temperature, the application of heat or cold giving rise to pain only and not to specific sensations of heat or cold. Further, the points of the skin which are sensitive to pressure are those which are not sensitive to heat or cold, and vice versa. Such results as these are only intelligible on the supposition that the terminal organs for pressure are different from those for heat and cold and differently distributed over the surface of the skin.

§ 888. The punctiform method of exploring the sensitiveness of the skin has further led to a result which is unexpected and indeed presents difficulties. Heat and cold in themselves differ only in degree; they are positive and negative phases of the same thing. We should therefore naturally expect that the same terminal organs would be employed for sensations both of heat and of cold, and that the same points of the skin would be alike sensitive both to heat and to cold. But the results of experimentation by the method in question contradict this expectation. It is found that some points are sensitive to heat, that is to say a sensation is developed when the temperature of the point of the skin is raised above what it happens to be at the time of experimenting, but are not sensitive to cold, that is to say no sensations are developed when the temperature of the point of the skin is lowered below what it happens to be at the time of experimenting; and other points may similarly be found to be sensitive to cold but not to heat. Moreover this result is in accord with results gained otherwise. If the arm or leg be "sent to sleep" by pressure on the brachial or sciatic nerves the skin will be found at a certain stage to be little sensitive to warmth though distinctly sensitive to cold. So also the whole surface of the glans penis, in contrast to the prepuce, is very slightly sensitive to cold, but distinctly sensitive to warmth. Moreover cases of disease of the central nervous system have been recorded in which the skin of a limb was sensitive to warmth, that is to degrees of temperature above that of the limb, but insensitive to cold. It may be remarked that in these cases, as in that of the limb "gone to sleep," the sensations of touch proper and of cold seem to run together and sensations of pain and of heat also to run together.

It seems probable then from these considerations that we possess three sets of terminal organs and three sets of fibres, one for pressure, a second for heat and a third for cold. It must be borne in mind however that the three sensations are not wholly independent, since sensations of pressure are modified if changes in temperature be taking place at the same time in the same spot of skin. Thus a penny cooled down nearly to zero and placed on the forehead will be judged by most people to be as heavy or even heavier than two pennies of the temperature of the forehead itself, that is to say the sensation of pressure is increased by a concomitant sensation of cold; and a similar modification of the sensation of pressure is also often observed when the object pressing is not colder but warmer than the skin pressed on. A similar effect seems to be shewn in certain cases of disease of the central nervous system in which it has been recorded that a hot body such as a heated spoon was felt when brought in contact with the skin, though the same spoon applied at the temperature of the skin itself produced no sensation at all, and the heated spoon was recognized not as a hot body, but simply as something touching the skin. The exact explanation of these facts is not very clear, but it may perhaps be argued that the effect is brought about amid the central processes through which the sensations are developed and does not shew that the sensations have common terminal organs.

§ 889. In attempting to understand the nature of the peripheral events through which the sensory impulses giving rise to sensations of pressure of heat and of cold are developed two or three matters must be borne in mind. In the first place, as we have already said, though the skin has a temperature of its own, we are not directly conscious of that, or at all events are not distinctly conscious of it in the same way that we become conscious of any sudden change in that temperature; nor indeed are we, except in extreme cases, distinctly conscious that the temperature of one region differs from that of another, or that the temperature of the same region gradually varies from time to time. It would seem as if the development of a clear and distinct sensation was largely dependent on the contrast as to temperature between an area of the skin and surrounding areas; and indeed we have already pointed out the marked effects of contrast. The same applies to pressure; we are not, at least distinctly and directly, conscious of the uniform pressure of the atmosphere over the whole surface of the body, when we stand naked in still air. We are not however justified in assuming that under the above circumstances nothing whatever is taking place in the sensory nerves of the skin, that when we feel a sensation the change in the sensory apparatus (using that phrase in its widest sense to include both peripheral and central parts) is one from absolute quiescence to activity; it is not impossible, and some facts indeed seem to suggest, that even when we feel no distinct cutaneous sensations, afferent impulses still continue to stream onwards from the periphery to the central nervous system, supplying as it were a groundwork of nervous events which enter largely in various ways into the conduct of the whole body, but which do not distinctly affect consciousness. If this be so, we may infer that the affection of consciousness which we call a sensation is the immediate effect of an adequately large change in this groundwork, rather than of a set of quite new isolated impulses passing straight up from the peripheral organ to the "seat of consciousness."

In the second place when we do experience sensations of temperature the sensation is caused not by the mere change of temperature but by the altered condition of the skin which results from that change. When an area of the skin having a normal temperature is brought in contact with a cold body, the skin undergoes a change from a normal to a lower temperature, and we experience a sensation of cold. Now, if it were only the change from a normal to a lower temperature which gave rise to the sensation, though the sensation might and probably would last much longer than the change itself, it could not be prolonged by the mere maintenance of the lower temperature when once the change had been established. But experience shows that it is; we still feel a sensation of cold, at a time when the contact of the cold body is not producing any further lowering of temperature and at most is only maintaining the lower temperature already brought about. Nay, more, the sensation of cold continues after the cooling body has been removed, at the time when the skin is returning to its normal temperature, that is to say is undergoing the very opposite change of temperature, namely, one from cold to heat. And the same considerations apply to sensations of heat.

§ 890. We may conclude then that when the application of cold or of heat to the skin causes a sensation of cold, the cold or heat produces a condition in the material of the skin, which condition starts nervous impulses in the afferent nerves of cold and heat sensations. Since the application of cold or of heat to the nerve fibres underlying the skin does not produce a sensation of cold or heat, but only a sensation of pain, we may further conclude that the material whose condition starts the sensation is placed in the skin itself, in the epidermis or in the immediately underlying dermis. Since we experience sensations of cold and heat in regions of the skin, not only free from touch corpuscles but also free from any dermic terminal organs as yet known, the "points" of the skin determined experimentally to be points of cold and heat sensations, having been repeatedly found when extirpated to be free from all such dermic organs, we may, though with less certainty, still further infer that the material exists somewhere in the epidermis. We may add that sensations of temperature may be felt in the cornea, from which all dermic terminal organs seem certainly to be absent. And our knowledge that the nerve fibres end as fine fibrillæ between and among the cells of the malpighian layer (§ 874) brings us to the final conclusion that the material of which we are speaking is to be sought for either in the fine nerve fibrillæ themselves, or, as seems more likely, in some or other of the cells of the malpighian layer specially connected with those fibrillæ.

Beyond this we cannot go; and even admitting thus much, it

is difficult to understand how, if the change be one from a higher to a lower temperature, the lower temperature, whatever may have been the exact degree of the higher temperature, should in giving rise to sensations of cold affect one set of fibres only, or how the higher temperature should similarly affect another set of fibres only; but we must leave the matter here.

The considerations which have just been brought forward in relation to sensations of heat and cold, may be also applied to sensations of pressure; with regard to them also we are driven to the conclusion that they take origin in the lower layer of the epidermis through some condition brought about by the pressure. We can appreciate pressure by the cornea, from which as we have said dermic organs are absent. If the 'points of skin' in various parts of the body, determined experimentally to be points of pressure sensation, be extirpated and examined it is found that dermic organs are not necessarily present; indeed such points of pressure sensations do not differ essentially in structure from points of heat or cold sensations, though some slight difference in the manner of distribution of the dermic nerve filaments has been described.

We are thus brought to the conclusion that the so called touch corpuscles are in no way essential to touch. At the same time their remarkable prominence in those parts of the skin in which touch is most sensitive would seem to shew that, even if not necessary, they are in some way adjuvant to pressure sensations. But what that aid may be is at present a mere matter of speculation; and we are perhaps still more in the dark as to the functions of the end-bulbs and of the Pacinian bodies.

SEC. 5. THE MUSCULAR SENSE.

§ 891. Before we go on to deal with some of the psychical aspects of cutaneous sensations it will be desirable to speak of certain sensations accompanying and belonging to the movements of the body which are carried out by means of the skeletal muscles; for these sensations, often spoken of as constituting a "muscular sense," are in many ways related to or mixed up with cutaneous sensations.

When we examine our own consciousness we find that we are aware of the position of the several parts of our body. In this we are under ordinary circumstances assisted by sight; but sight is not necessary. If for instance, with the eyes shut, we place the arm in any attitude, we are aware of the attitude, and can describe, or by movements of the other arm imitate with considerable accuracy the details of the attitude, the relative positions of the upper arm, forearm, hand, fingers and the like. If we change the attitude by moving the arm or part of the arm we can, though the eyes be still shut, tell the amount and characters of the change.

Again, when we examine our own consciousness we find that we possess a measure of the amount of resistance to our movements which we from time to time meet with. When we come into contact with an external object we are conscious not only of the pressure exerted by the object on our skin, but also of the pressure which we exert on the object; we can appreciate the amount of effort which we make to produce by pressure an effect upon the object. A similar appreciation of our own efforts assists us largely in forming a judgment as to the weight of an object. If we place the hand and arm flat on a table, we can estimate the pressure exerted by a body resting on the palm of the hand, and so come to a conclusion as to its weight; in this case we are conscious only of the pressure exerted by the body on our skin. If however we hold the body in the hand, we not only feel the pressure of the body, but we are also aware of the exertion required to support and lift it. And we find by experience that when we trust to this appreciation of the amount of effort needed to lift an object as well as to sensations of pressure, we can form much more accurate judgments concerning the weight of the object than when we rely on sensations of pressure alone. When we want to tell how heavy a thing is, we are not in the habit of allowing it simply to press on the hand laid flat on a table or otherwise at rest; we hold it in our hand and lift it up and down.

The above instances deal with three things which it might be desirable to keep separate, namely, 'position,' 'movement' and 'effort;' it might seem desirable to speak of "a sense of position," "a sense of movement," and "a sense of effort." But, if we leave out of consideration the problems connected with our appreciation of the position of the head, which as we have seen seems especially dependent on afferent impulses passing up the auditory (vestibular) nerve, we may say that the position of the various parts of our body is so closely dependent on movement, that is on the contraction of skeletal muscles, some muscle or other playing its part in almost every position and every change of position, that in the discussion on which we are now entering it will be hardly profitable to distinguish between the two; and we may use the term "muscular sense" to denote our appreciation both of movement and of position resulting from movement.

§ 892. There are more valid reasons for distinguishing between our appreciation of an effort and our appreciation of the movement which is the result of that effort. For the view has been put forward and supported by argument that when we make a muscular effort, we are directly conscious of the nervous processes of the central nervous system underlying the effort, that the changes in the central nervous system involved in initiating and executing a movement of the body so affect our consciousness that we have a sense of the nervous effort itself, of the innervation as it has been called; and it is urged that the condition of the central nervous system through which we appreciate the nature and magnitude of the effort is thus the direct effect of central changes, and not the outcome of afferent impulses proceeding from the part moved.

Whether it be the case or not that consciousness is thus directly affected by changes in the central nervous system, such for instance as those taking place in the motor cortical area or in the pyramidal tract, the evidence goes to shew that any such affection has, at most, very little share in that appreciation of our movements which is generally called "the muscular sense." Not only is our appreciation of passive movements very similar to our appreciation of active movements (we are as well aware of an attitude in which our arm has been placed by others as of one in which we have placed it ourselves), but also if a muscular contraction be brought about not by any action at all of the central nervous system, but by the direct electric or other stimulation of the muscles or motor nerves, the muscular sense of the movement which results differs little from that of a like voluntary movement. If for instance, while our eyes are shut, the wrist be bent by direct stimulation of the flexor muscles, we are aware of the movement and can appreciate its character and amount; we can even use such an artificial movement to judge of weight and resistance. It is indeed urged that our judgment under such conditions is less secure than when the movement is a voluntary one; and from this it is argued that our judgment is at least assisted by our appreciation of the central changes by a "sense of the effort" as distinguished from a muscular sense of peripheral origin; but even this is disputed. We may at least conclude that our appreciation of our movements and muscular efforts is largely, if not wholly, dependent on what may be called a muscular sense which is the outcome of afferent impulses proceeding from the periphery and started in the parts concerned in the movement.

§ 893. Coming next to the questions, What is the exact nature of these afferent impulses? In what tissues are they started, and along what paths do they travel? we find the answers beset with considerable difficulties. Every movement of the body, even a simple one, is in reality a complex affair, and the carrying it out involves changes in several tissues. In the first place there are changes in one or more muscles, changes of contraction in active movements, of extension and relaxation in passive movements. In the second place there are changes in the skin which during a movement is in one spot stretched, in another relaxed or folded; and in movements of locomotion the pressure of the foot on the ground is continually changing. In the third place, by far the majority of movements affect a joint, and hence involve changes in the relations of the articular surface, in the capsule and ligaments and in the tendons. All these are possible sources of afferent impulses.

Now we know that the skin is a source of afferent impulses and so of sensations, namely, the sensations of pressure, of temperature and of pain; and we may fairly suppose that stretching or slackening the skin gives rise to impulses either analogous to those caused by the pressure of an external object or, it may be, of a nature more akin to those which belong to general sensibility. Hence it is possible that these do at least contribute, under normal circumstances, to what as a whole we call the muscular sense.

Indeed it is maintained by some that these cutaneous impulses furnish the whole basis of what is called the muscular sense, the name on this view being of course erroneous. In attempting to judge of such a view we may appeal on the one hand to our own consciousness, and on the other hand to the phenomena of incoordinate movements. In a previous part of this work, § 643, we dwelt upon the importance of afferent impulses as factors in the coordination of movements. We have had occasion repeatedly to insist that all the movements of the body, a large number of those which are involuntary as well as all those which are voluntary, are guided by afferent impulses, and that in the absence of these afferent impulses the movements are apt to become uncertain and imperfect, or even to fail altogether. We need not here repeat what we have previously urged; it is sufficient for our present purpose to say that conspicuous among these afferent impulses are those which form the groundwork of the muscular sense; at times they may do their work without directly affecting consciousness but at other times they bring about a distinct affection of consciousness, and it is this affection of consciousness which is more properly called the muscular sense.

Now, on the one hand, we find upon examination that coordination of movements is not distinctly affected by the diminution of cutaneous sensations, but may be maintained in the absence of cutaneous sensations and indeed in the absence of the skin. Thus frogs are said to be able to execute their ordinary movements without signs of incoordination after the whole skin has been removed. Cases of nervous diseases have been recorded in which, if not complete absence of, at least great failure in, cutaneous sensations has not been accompanied by any loss of coordination. And if we appeal to our own consciousness we do not find the muscular sense notably diminished by temporary anæsthesia of the skin; if, for instance, the skin of the arm be rendered for a while anæsthetic, we do not find any marked change in our power of judging weights or resistance, or in appreciating, with the eyes shut, the position of the limb.

On the other hand we find recorded cases of nervous diseases in which loss of coordination, and loss of the muscular sense, as indicated by the difficulty or inability to judge weights and resistance and to recognize with the eyes shut the position of the limbs or other parts of the body, have occurred without notable loss of cutaneous sensations. This is often strikingly shewn in cases of the disease or group of diseases known as "tabes dorsalis," often spoken of from one of its prominent symptoms as, "locomotor ataxy," the conspicuous pathological condition of which is a structural change in the posterior columns of the lower part of the cord. In certain stages of this disease the patient may retain good cutaneous sensations, he may experience tactile, temperature and painful sensations in the skin of his legs, for instance, and possess adequate muscular strength in his legs, and yet, from want of coordination, be unable to move them properly unless he be assisted by sight. So long as his eyes are open he may be able to stand and walk, but if his eyes are shut he often falls, and when he moves, moves with a staggering uncertain gait; he fears, in the dark, to go up or down stairs even though he knows them well. When a direct appeal is made to his consciousness he appears to possess little or no muscular sense; he is unaware, so long as his eyes are shut, of the position of the limbs affected by the disease, and if the arms are affected is unable properly to judge weights.

These cases of "tabes" are very varied in their symptoms, which indeed alter as the disease advances. Concerning them and similar phenomena presented by other allied nervous diseases there has been much discussion; but the evidence afforded by them, supported as it is to a certain extent by experimental results, is strongly in favour of the view that the afferent impulses which determine coordination and which go to make up what we are now calling the muscular sense are other than those started in the skin.

We may therefore dismiss cutaneous sensations as not being essential factors of the sense.

§ 894. There remain on the one hand the muscles, on the other the joints with their belonging ligaments and tendons; the afferent impulses under discussion must come from one or other or both of these.

Against the view that the afferent impulses of the muscular sense come from the muscles themselves has been urged the fact that, tested experimentally, muscular fibres in a normal condition possess a very feeble general sensibility; when a muscle is cut or pinched comparatively little or, according to some observers, no pain is felt; it is only under abnormal circumstances, as when a muscle is inflamed, that direct stimulation of this kind causes pain; and the pain which we feel in cramp is similarly the product of an abnormal condition, for even an extremely violent muscular effort does not cause us actual pain.

This argument however is not valid, for not only may it equally well be applied to the other set of tissues, tendons, ligaments and the like, which in a normal condition possess a similarly feeble general sensibility, but it supposes that the muscular sense is merely a development of general sensibility not a special sense, like that of touch. We have no positive reasons for this supposition, and arguments based on the analogy of the skin oppose it. We have seen reason to regard the cutaneous sensations of pressure and temperature as wholly distinct from those of general sensibility, that is to say of pain; and we may conclude that the muscular sense is similarly a special sense, similarly distinct from affections of common sensibility in either muscular fibres or their connective tissue appendages.

On the other hand afferent impulses may proceed from muscles, for when a nerve twig going to a muscle is stimulated centripetally, after division, reflex movements result; if the stimulus is weak the movement is confined to the muscle itself (we are supposing that other nerve twigs going to the muscle are left intact); if the stimulus is strong, the movement spreads to neighbouring muscles. And we know that a muscle is supplied by nerve fibres which do not end in end-plates in the muscular fibres; some of these are vaso-motor, but others are probably afferent, more especially those described as ending in fine fibrils around and among the muscular fibres, that is in the perimysial and other connective tissue; and of these while some may serve for the impulses of pain, others may serve as impulses for the muscular sense.

Tendons and ligaments are also provided with afferent nerve fibres, and a special mode of ending, a plexiform arrangement of fibrils terminating in minute end-bulbs, has been described in tendons under the name of the "organ of Golgi."

Both muscles on the one hand and tendons and ligaments on the other may furnish the afferent impulses of which we are speaking. We must seek therefore other arguments to decide whether the muscular sense is derived from the muscles or from other parts. We cannot by an appeal to our own consciousness localize the sensation so as to lodge it either in one tissue or another, and must trust to indirect indications. On the one hand there seems to be a close connection between the muscular sense and the 'sense of fatigue;' and the latter appears to be determined by the condition of the muscles. Again, in many of our movements we only employ a part of a muscle, and it is difficult to suppose that the afferent impulses which guide us in using that part only, depend only on the effect which the partial use of the muscle produces in the tendons and the like. On the other hand, when we have a muscular sense of the movements of the fingers, we can hardly suppose that the sense is afforded by impulses coming exclusively from the muscles moving the fingers, distant as these often are from the joints which they move. And, again, the movements of which we are most distinctly sensible, are especially the movements affecting joints; indeed we have some difficulty in appreciating the amount and character of a movement not necessarily involving a joint such as one caused by contractions of the orbicular muscle of the mouth or of the eye, even though in these cases we are assisted by cutaneous sensations.

We ought therefore probably to conclude that the muscular sense though based in part on impulses derived from the muscular fibres is also, and possibly to a large extent, based on impulses derived from the tendons and other passive instruments of the muscles, though we cannot as yet assign accurately the relative share. If this be so the 'muscular' sense is not a wholly appropriate term; but it would be undesirable, at present at least, to attempt to replace it by a new one.

This muscular sense, using the term in its broad meaning, enters largely into our life. By it we are not only enabled to coordinate and execute adequately the various movements which we make, but through it we derive much of our knowledge of the external world. Through it we are also conscious of the varying condition of the several parts of our body even when the muscles are at rest; the tired and especially the paralysed limb is said to 'feel heavy.' In this way the state of our muscles and other tissues largely determines our general feeling of health and vigour, of weariness, ill health and feebleness.

The fact that the Pacinian bodies are found around joints has led to the suggestion that these serve as the terminal organs of the muscular sense; but especially bearing in mind what has just been said, the argument which we used against considering the touch corpuscles as the terminal organs of touch may, with perhaps still greater force be applied against regarding the Pacinian bodies as the terminal organs of the muscular sense.

SEC. 6. ON TACTILE PERCEPTIONS AND JUDGMENTS.

§ 895. As a means of gaining knowledge of external things the sense of touch ranks next in importance to that of sight. Auditory sensations enter largely and in several ways into our life; they serve as an important means of communication; together with smell and taste they afford pleasure and guide our acts; but, as regards our direct knowledge of the external world, we learn by means of them very little compared with what we learn by sight and touch. To a certain extent we make use of touch by itself; we bring the surface of a body into contact with some region of the skin such as the finger, and by the several sensations which we receive either from several points of that region at the same time, or from one or more points in succession, we learn certain characters of the surface, whether for instance it is rough or smooth. We thus also ascertain whether the body be hot or cold; and we may, within certain limits, form a judgment of the size of the surface by simply estimating the size of the area of our skin with which the body can be in contact at the same time.

But though we may and do thus base conclusions on tactile perceptions alone, we most frequently employ touch in association with sight on the one hand and with the muscular sense on the other.

The ties indeed between touch and the muscular sense are many and close. When we explore the nature of a body by touch we press the skin, of the finger for instance, on the body; and we do that not merely in order to determine to what extent the tactile sensation is increased by the increase of pressure, but also and indeed chiefly to ascertain the amount of resistance to pressure which is offered by the body. But that resistance, through which chiefly we judge whether the body be soft or hard, is appreciated not by the tactile but by the muscular sense.

Or again, placing the finger on the surface of a body, and moving the finger over the surface in such a way that the contact, as judged by the pure tactile sensation, remains the same, we find that in one case the movement has been continued in the same plane, whereupon we judge the surface to be flat, that in another case the finger has been gradually carried out of the plane, whereupon we judge the surface to be curved, and that in the third case the movement of the finger has been irregular, whereupon we judge that the surface is irregular; and so on. In each case we estimate the movement by the muscular sense, and thus by a combination of muscular sense and of touch we form a judgment of the conformation of external bodies. In the same way, and indeed as part of the same process, by a combination of the muscular sense and of touch we estimate the size of external objects. By a like double act we estimate the position in space in relation to our body of such objects as are within our reach, such as can be touched either directly by one of our limbs or indirectly by help of a stick or otherwise. So closely bound together are the muscular sense and the sense of touch proper, that in common language we speak of learning this or that by touch, when we really employ both senses.

§ 896. No less close are the ties between sight and touch; indeed a very large part of our psychical life is built up on the association of visual and tactile sensations. There is no part of the external world, including our own bodies, which we can explore by touch, which we cannot, either directly or by optical aids such as mirrors, also explore by vision; and our conceptions of the nature of all such things is the outcome of a combination of the two senses, or rather bearing in mind what has just been said, of the three senses, sight, touch and the muscular sense. It is relatively easy to recognize blindfold by touch alone, the characters of objects with which we are already previously familiar by help of vision; but it is very difficult to form by touch alone an accurate judgment of the form and size of objects which we have never seen. Were we limited to sight alone, we should form one set of conceptions of the world, were we limited to touch we should form another; and the two sets would be different.

In the conceptions which we form in actual life the two are combined. The congenitally blind are limited to one set only; and, when, as has happened in cases of congenital cataract, those who have been blind from birth are restored to vision after they have grown up and have accumulated a store of tactile conceptions, they fail at first to connect their new visual sensations with their old tactile experience. The stories of the first experiences in vision of such persons, as that for instance of the man who had to feel a cat in order to connect the visual image with his previous tactile image, and having carefully felt it all over said "Now, Puss! I shall know you again," illustrate the close dependence on each other of visual and tactile normal perceptions. This is also indicated by the zeal with which in former days the question was discussed whether a man who had been born blind and restored to sight in adult life, could recognize at first sight and by sight alone a cube, a square, and a sphere. It is perhaps especially in relation to size and space, that the two senses work together.

There are no converse cases of persons who, born without touch, and trusting to sight alone have, in later life, had touch restored to them; but there are many things within our vision, which are beyond our touch at the moment and some which we can never touch at any time; our conceptions of these latter are more or less uncertain, and the direct visual sensations have to be strengthened or corrected not by mere sensations but by intellectual efforts and reasoning. A group of visual sensations, constituting a visual image, may have an ordinary objective cause, but may be an ocular illusion; and the test which we at once apply to determine this is that of touch; the ordinary idea of a 'ghost' is that of a something which we can see but cannot touch, which excites visual sensations but affords no tactile sensations. Conversely a touch by something invisible, a touch as of a body which we ought to be able to see but cannot, we also recognize as unreal. The concordance of touch and vision affords in fact to a large extent the standard by which we judge of the reality of things.

§ 897. The last remark naturally leads to the statement that as in the case of the other sensations, so in the case of the several cutaneous sensations, we may have sensations which are not due to their ordinary objective causes.

We have seen that visual sensations may arise from changes in the retina started not by light but by other agents, mechanical and others; and the question presents itself, Can touch proper, the sensation of pressure, be excited otherwise than by pressure and sensations of temperature by changes in the skin other than those of temperature? No very definite answer can be given to this question, though the case quoted above (§ 888) in which a heated spoon applied to the skin produced a sensation not of heat but of contact, points perhaps to the affirmative, as does also the fact that electric currents applied to the skin may produce sensations, pricking sensations, which if not identical with, may at least be confused with those of pressure.

Cutaneous sensations of all kinds may however be of central origin, may be due to changes in the central nervous system quite independent of all events in the skin, and may yet be referred to this or that region of the skin and to the objective cause which ordinarily gives rise to the sensation. Painful sensations indeed may rise from changes not only in the central organs but at any part of the whole length of the nerve, all being referred to the cutaneous terminations of the nerves on which the cause of pain is usually brought to bear. Tactile and temperature sensations as we have said cannot originate in changes in the nerves themselves, but they may arise through changes in the central organs; we may be subject to tactile phantoms comparable to ocular phantoms. Compared with visual sensations however our tactile sensations are so to speak fragmentary. A momentary exposure of the retina may fill the mind with a complex visual image, full of the most varied incident; but the tactile impressions which we can receive at any one moment are few and simple. Hence our tactile phantoms are also simple; we may fancy that some invisible garment has swept past us, or that a scorching flame has passed near us, we may feel that the hand or that the head is swollen and large, and we may experience an imaginary pain in every region of the skin in turn; but the most that we can thus feel is simple compared with the possible complexity of an ocular or even an auditory phantom.

§ 898. Like other sensations our tactile sensations while they sometimes give us trustworthy information of the external world at other times may give rise to illusions. This is well illustrated by the so-called experiment of Aristotle. It is impossible in an ordinary position of the fingers to bring the radial side of the middle finger and the ulnar side of the ring finger to bear at the same time on a small object such as a marble. Hence when with the eyes shut we cross one finger over the other, and place a marble between them so that it touches the radial side of the one and the ulnar side of the other, we recognize that the object is such as could not under ordinary conditions be touched at the same time by these two portions of our skin, and therefore judge that we are touching not one but two marbles. Upon repetition however we are able to correct our judgment and the illusion disappears.

CHAPTER VII.

ON SOME SPECIAL MUSCULAR MECHANISMS.

SEC. 1. THE VOICE.

§ **899.** IN the trachea the respiratory passage is of nearly uniform bore; in the larynx it is less regular, and at one part is narrowed into a slit of variable width, stretching from front to back, forming as it were a throat (*glottis*) to the passage below. The mucous membrane lining the larynx is so modified at the edges of this slit as to form two more or less parallel elastic membranes capable of being thrown into vibrations when an adequate blast of air is driven through the slit. These vibrations communicated to the air give rise to the sound which we call the voice, and the two membranous edges, which are thus the essential means of producing the voice, are called the *vocal cords, chordae vocales*. The blast of air is supplied by the respiratory mechanism, the expiratory current being almost exclusively used because it is more manageable, and more favourable for the conduction of the sound outwards than is the inspiratory current.

When a sound is produced, as is the voice, by the vibrations of the membranous edges of a slit, placed in the course of a tubular passage, the characters of the sound are determined in part by the nature, arrangement and behaviour of the slit and its membranous edges, but they are also determined by the length and shape of the tubular passage, especially of that part through which the sound is conveyed after its formation at the slit, and which forms what is often called a "resonance tube" or "resonance chamber." Hence in studying the voice we have to consider on the one hand the essential parts of the apparatus, namely,

CHAP. VII.] SPECIAL MUSCULAR MECHANISMS. 307

the vocal cords, and on the other hand the subsidiary parts, namely, the larynx above the glottis, the pharynx, the nasal

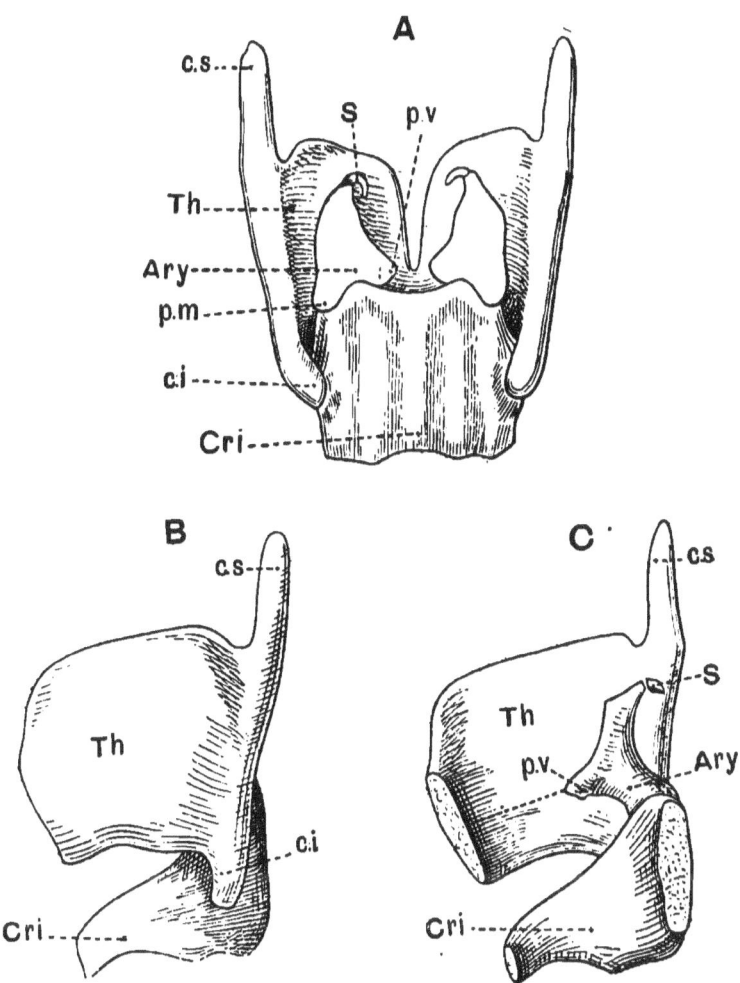

FIG. 181. THE CARTILAGES OF THE LARYNX.

A. Seen from the back. *B*. Seen from the left side. *C*. The right half seen from the inside after removal of the left half.

Th. Thyroid. *Cri*. Cricoid. *Ary*. Arytenoid. *c. s.* upper cornu, *c. i.* lower cornu of thyroid. *p. m.* processus muscularis, *p. v.* processus vocalis of the arytenoid. *S*. Cartilage of Santorini. The dotted line in *C* indicates the position of the vocal cords.

The parts are represented of natural size, as they are also in the succeeding figures unless otherwise indicated.

passages and the mouth, as well as to a certain extent the larynx below the glottis, the trachea and the lungs.

§ 900. The framework of the larynx consists chiefly of two cartilages, the thyroid and the cricoid cartilages.

The *cricoid* cartilage (Fig. 181 *Cri.*) is, practically, a largely developed tracheal ring (§ 319). Forming a complete thick hoop, high behind and low in front, the upper edge sloping downwards, it is attached all round by membrane to the tracheal ring below, and may be regarded as moving upon that ring as well as moving with the rest of the trachea.

The *thyroid* cartilage (Fig. 181 *Th.*) is also a hoop of cartilage but is largely opened behind so as to present in horizontal section the figure of a horse-shoe (Fig. 188), is larger in its vertical dimensions than is the cricoid, and is of peculiar form, both its upper and lower border ending behind in the form of a horn, the superior and the inferior cornu (Fig. 181 *c.s.*, *c.i.*). The end of each lower horn is definitely articulated to a portion of the hinder lateral surface of the cricoid; but otherwise the attachments of the two cartilages are membranous only, so that the thyroid can to a certain extent move downwards and forwards upon the cricoid or the cricoid, especially in its front part, can be pulled up towards the thyroid.

The thyroid cartilage is connected with the hyoid bone above partly by muscles but especially by the thyro-hyoid membrane, which stretches from the upper border of the thyroid along its whole length to the lower border of the hyoid, and the hind edges of which, passing from the ends of the upper cornua of the thyroid to the ends of the great cornua of the hyoid, are strengthened by elastic tissue into rounded cords, the lateral thyro-hyoid ligaments. This membrane permits the whole larynx to be drawn up within the sweep of the hyoid bone.

Placed behind the thyro-hyoid membrane and imbedded in the mucous membrane lining the pharynx and upper part of the larynx is the rhomboidal plate of yellow elastic cartilage which forms the body of the epiglottis. The greater part of the epiglottis projects into the cavity of the pharynx as a tongue-shaped process (Figs. 182, 183) presenting an anterior and posterior surface and placed generally at an angle of about 45° with the horizon; in children it is often more erect and in adults is sometimes more horizontal. The base forms part of the margin of an aperture which we shall presently speak of as the "superior aperture of the larynx" and which at times, as in swallowing, is covered over and so closed by the folding back of the free epiglottis.

§ 901. Seated on the upper border of the back part of the cricoid on each side of and at some little distance from the middle line are two smaller cartilages deserving especial attention, the *arytenoid* cartilages (Fig. 181 *Ary.*).

Each arytenoid cartilage is in form an irregular pyramid with the base seated on the cricoid, with the apex directed vertically, but somewhat obliquely, upwards, and with the three surfaces looking one towards the middle line and its fellow, one backwards, and one outwards and forwards. The median surface is not so tall as the other two so that the top of the pyramid appears as it were pinched into a plate, which is irregularly curved, and to the summit of which is attached a nodule of cartilage in the shape of a minute horn, the *cartilage of Santorini*, or *corniculum laryngis* (Fig. 181 *S.*). All the three surfaces are somewhat concave but more or less irregular.

The hind part of the irregularly triangular base is hollowed out into a small elliptical articular surface for articulation with the cricoid, the long axis of the ellipse being placed transversely. The rest of the base projects beyond the cricoid, and at the front angle, between the median and outer sides, forms a process which is important since it serves for the attachment of the vocal cord, and hence is called the *processus vocalis* (Fig. 181 *p.v.*); the two vocal cords stretch from the two processus vocales, across the larynx to the thyroid, to the re-entering rounded angle of which they are attached at a level which is somewhat nearer the lower than the upper border of the cartilage. The outer angle of the base of the arytenoid between the outer and hind surfaces also forms a process which, since it serves for the attachment of muscles, is called the *processus muscularis* (Fig. 181 *p. m.*); but the remaining angle of the base, that between the median and hind surfaces, is rounded off and does not form any projection. The greater part of the body of the arytenoid is above the level of the vocal cord, since the processus vocalis to which this is attached, though tilted somewhat upwards, is part of the base of the pyramidal cartilage.

While the movements of the cricoid and the thyroid on each other are on the whole simple in character, the articular surface of the arytenoid permits that cartilage to execute very varied movements. Of these the most important, as we shall see later on in detail, are on the one hand a movement of rotation by which the processus vocales converge towards or diverge from each other and the middle line, carrying with them in each case the vocal cords, and on the other hand a movement by which the bodies of the two cartilages are drawn close together in the middle line or dragged far apart. It is by means of these movements and by means of the movement of the cricoid on the thyroid that the changes in the shape and condition of the rima glottidis, upon which the formation of the voice so largely depends, are in the main brought about.

§ 902. The hind and outer surfaces of the arytenoid pyramid are imbedded in muscular and connective tissue, but the only covering of the median surface is a mucous membrane continuous with that of the vocal cords and the rest of the lining of the larynx. The two vocal cords form we have said the edges of the laryngeal slit called the glottis or rima glottidis. But this is true

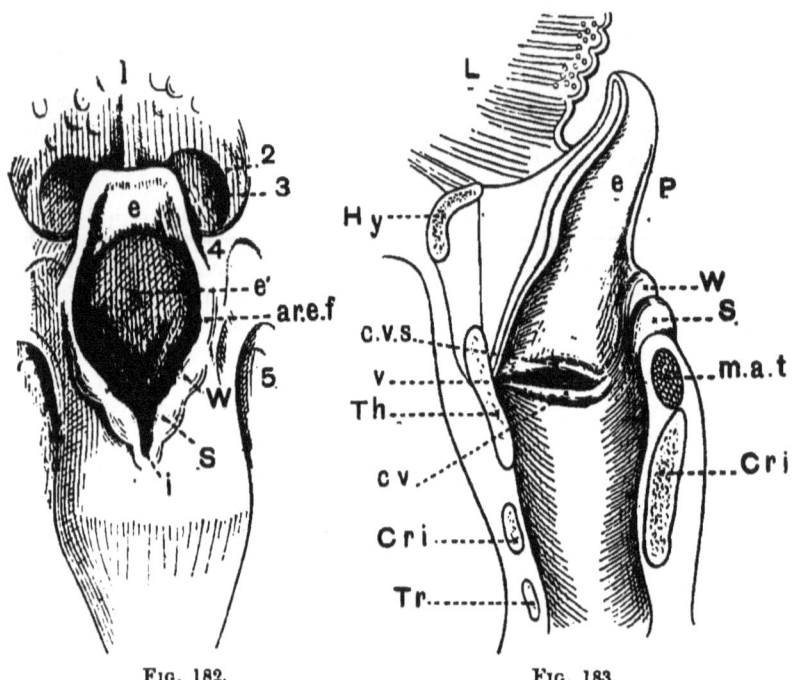

Fig. 182.   Fig. 183.

Fig. 182. Diagram of the Superior Aperture of the Larynx.

The œsophagus and pharynx are supposed to be laid open from behind.

e. the epiglottis with e' its cushion. ar. e. f. the ary-epiglottic fold, on which are seen the swellings or "capitula" caused (W) by the cartilage of Wrisberg and (S) by the cartilage of Santorini. i. the notch or incisura in the mucous fold uniting transversely the two arytenoid cartilages.

1. (placed in the middle line of the base of the tongue) the median and (2) the lateral glosso-epiglottic folds, the latter forming the boundary of the depression (3) called the vallecula. 4. the pharyngo-epiglottic fold. 5. the pharyngo-laryngeal or pyriform recess.

Fig. 183. Diagram of the Larynx in vertical section.

e. The epiglottis. l. The base of the tongue. Hy. Hyoid bone. Th. Thyroid cartilage; Cri. Cricoid cartilage; Tr. Tracheal cartilage; all cut across.

W. the swelling due to the cartilage of Wrisberg and S. that due to the cartilage of Santorini; from these eminences folds descend towards the processus vocalis of the arytenoid. c.v. the true, and c.v.s. the false vocal cord or "ventricular band," with the mouth of the ventricle of the larynx v. between them. m.a.t. the transverse arytenoid muscle cut across. P is placed in the cavity of the pharynx.

only of the front part of the slit from the thyroid to the processus vocales; from this point backwards, to the muscular and other tissue which braces together the hind surfaces of the arytenoid pyramids the slit is continued by the median edges of the bases of the pyramids. Hence the whole slit consists of two parts: of a front part from the thyroid (Fig. 185) to the processus vocales, along which the edges are furnished by the membranous vocal cords, and of a hind part from the end of the vocal cords backwards, the edges of which are not membranous but are furnished by the bases of the arytenoids covered with mucous membrane. The front part about 15 mm. long in the adult human male, is sometimes called the "rima vocalis," and the hind part, about 8 mm. long, the "rima respiratoria;" but these names are not free from objection, and it is better to speak of the former as the membranous or ligamentous and the latter as the cartilaginous or inter-cartilaginous rima or glottis.

The tubular passage of the trachea as it ascends within the cricoid ring is narrowed funnelwise in a fairly uniform manner to the slit of the rima glottidis; but as we have already said the larynx above the rima is less regular in shape.

If the pharynx and œsophagus be laid open from behind (Fig. 182) or exposed in vertical section (Fig. 183) the "superior aperture of the larynx" will be disclosed and will be seen to be oval or roundly triangular in outline slanting downwards and backwards. In front and high up the margin of the aperture is formed by the projecting epiglottis (*e*). On each side the margin is continued by a fold of mucous membrane (*ar. e. f.*) which passes obliquely downwards and backwards from the base of the epiglottis, and ends in the cartilage of Santorini and tip of the arytenoid of that side. Each of these *ary-epiglottic folds* as they are called, just before it reaches the cartilage of Santorini, is marked with a rounded projection (*W*) caused by the presence of a small nodule of cartilage, the cartilage of Wrisberg. The cartilages of Santorini also cause rounded projections (*S*), between which the margin of the upper aperture of the larynx is completed by a fold of mucous membrane passing from the tip of one arytenoid to that of the other; this when the cartilages are dragged apart is stretched straight but when they are drawn together is folded into a notch (*i*).

The cavity of the larynx into which this aperture thus sloping rapidly backwards from the level of the epiglottis to that of the tips of the arytenoids opens, does not narrow uniformly to the glottis. A little above the true vocal cords (Figs. 183, 184 *c.v.*) the mucous membrane is thrown on each side into a somewhat thick transverse fold which, stretching from the base of the epiglottis in front to the arytenoid behind, projects horizontally inwards towards the middle line but does not reach so far as do the vocal cords. These are called the *ventricular bands or false*

312   THE VOICE.   [BOOK III.

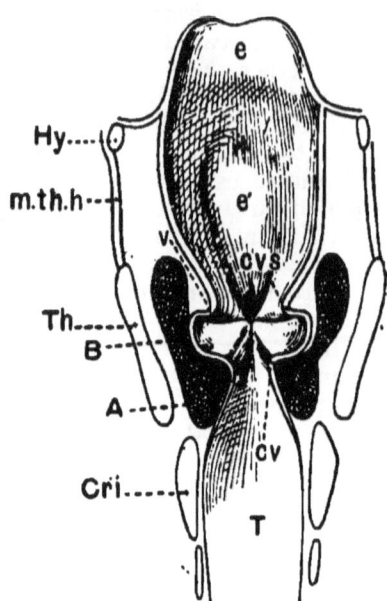

FIG. 184. DIAGRAM OF THE LARYNX IN VERTICAL TRANSVERSE SECTION.

Hy. Hyoid bone. *Th.* Thyroid cartilage; *Cri.* Cricoid cartilage; *m. th. h.* thyrohyoid membrane, all cut across.

e. epiglottis, *e'* its cushion. *c.v.s.* ventricular bands, *c.v.* vocal cords, with *v.* the ventricles of the larynx between them. *T* the trachea.

A. The internal thyro-arytenoid muscle, cut across; it is seen to form the bulk of the wedge-shaped projection of which the vocal cord is the extreme edge. *B* the external thyro-arytenoid, cut across

vocal cords (Figs. 183, 184 *c.v.s.*), and their existence gives rise to two pouches or bags, *the ventricles of the larynx,* (*v.*) one on each side, between the ventricular band above and the vocal cord below.

§ 903. The mucous membrane lining the larynx is a continuation of that lining the trachea (§ 319), and as a whole presents the same features namely a ciliated epithelium of several cells deep, among which are many goblet cells, resting on a dermis largely composed of retiform or even adenoid (§ 259) tissue and containing many elastic fibres running for the most part in a longitudinal direction; but special modifications occur in particular places.

Over the vocal cords themselves the cilia are absent; traced upwards from the trachea they disappear just below the glottis but appear again just above it; the epithelium occupying this narrow interval and thus covering the vocal cords is a thin stratified epithelium, the upper cells of which are flattened and devoid of cilia, an epithelium in fact like that of the pharynx. In each vocal cord the elastic fibres of the dermis undergo a large development, and are arranged as a compact band running along the length of and forming the chief part of the cord, the individual fibres and bundles of fibres running on the whole horizontally, though not regularly so, but interlacing in various planes; each vocal cord is in fact a cord of elastic tissue mixed up with some retiform tissue, wedge-shaped in transverse section and covered with a layer of non-ciliated epithelium.

In the ventricles of the larynx, over the ventricular bands and on the posterior surface of the epiglottis the mucous membrane is rich in adenoid tissue which is often aggregated into distinct follicles; these parts are apt to become swollen or "œdematous" (§ 303) by the accumulation of fluid in the lymph spaces. Numerous small glands, chiefly mucous but in part albuminous, are present here and indeed over the larynx generally; the vocal cords are said to be destitute of them, but this is disputed.

The thyroid and cricoid cartilages are formed wholly of hyaline cartilage, and in old persons or even in middle life may be found partially ossified. The arytenoid cartilage is also chiefly hyaline, but parts of the surface and especially the processus vocalis are of the yellow elastic variety. The cartilages of Santorini and Wrisberg, as well as a small nodule of cartilage (cartilage of Luschka) which is sometimes found imbedded in the front part of the vocal cord, are all formed entirely of yellow elastic cartilage.

§ 904. If a small mirror, warmed in order to avoid the condensation of moisture upon it, be placed in an appropriate slanting position, namely, at about an angle of 45° with the horizon, in the back of the pharynx with its upper margin resting against the base of the uvula and be adequately illuminated, a view of the interior of the larynx may be obtained. Such a mirror with its various appurtenances is called a laryngoscope. The details of the view thus gained will of course vary with the exact position and inclination of the mirror, but the following may be taken as the average appearance (Fig. 185).

In front (reversed of course in the mirror image) will be seen the edge of the back of the tongue (L), and immediately in front of this the top of the epiglottis (e.) These parts will of necessity appear much fore-shortened, and peering out from underneath the top edge of the epiglottis may be seen the swelling at its base known as the "tubercle" or "cushion of the epiglottis" (e'). The curved sides of the epiglottis will be seen sweeping away to the right and to the left, and emerging from near the end of each will be visible the ary-epiglottic fold ($ar.\,ep.\,f.$) on which are obvious first the round swelling due to the cartilage of Wrisberg (w) and next that due to the cartilage of Santorini (s). If at the time when the view is being taken, the voice is being uttered and

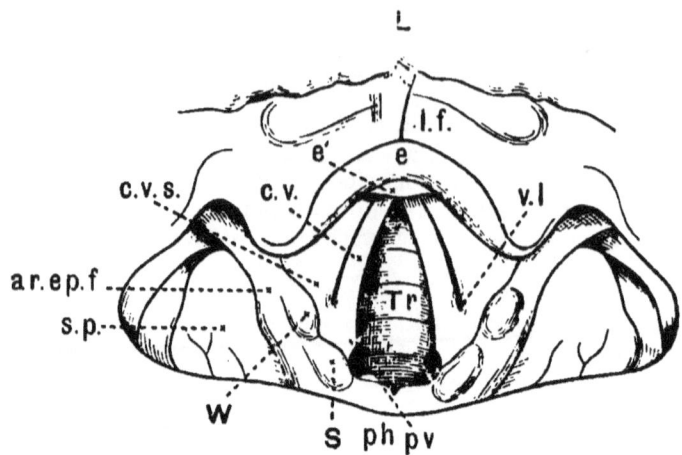

FIG 185 DIAGRAM OF A LARYNGOSCOPIC VIEW OF THE LARYNX (magnified twice).

L. the base of the tongue. *e.* the epiglottis, seen foreshortened with *e′* its cushion. *ar.ep.f.* the ary-epiglottic folds. *W.* the Capitulum Wrisbergi, *S.* Capitulum Santorini; the mucous membrane between the arytenoids is stretched straight, the notch being merely indicated. *c.v* vocal cords, *c.v.s.* ventricular bands, *v.l.* the opening into the ventricle of the larynx seen between them. The former, bounding the widely open glottis of more or less triangular form, through which a view of the trachea (*Tr.*) is obtained are seen to end in the processus vocales (*p.v.*).

On each side of the larynx is seen *s.p.* the pyriform recess. *ph.* the hind wall of the pharynx. *l.f* the median glosso-epiglottic fold.

especially if a high note is being given (Fig. 186 *A*) the two cartilages of Santorini are in close apposition, and the mucous membrane between is folded up. If no voice is being uttered and especially if a deep inspiration be taken (Fig. 186 *B* and *C*), the cartilages of Santorini are far apart and the mucous membrane between them appears as a ridge completing at the hind part the rim of the aperture to the larynx; there may also be seen on each side lying immediately to the median side of the prominence of the cartilage of Santorini a shallower prominence due to the top of the arytenoid itself, shewn at *a* in Fig. 186 *B*. Between the two phases of complete apposition and of the widest separation of the tubercles of Santorini, intermediate phases may from time to time be seen, such as those shewn in Fig. 185, Fig. 186 *B*.

These several structures define the superior aperture of the larynx which in the laryngoscopic view, owing to the fore-shortening, is no longer seen as a slanting orifice with a long fore and aft diameter but appears as a rhomboidal space with the transverse diameter generally the longer one. If no voice is being uttered, and the breathing be gentle and quiet, the glottis may be seen within this aperture as a slit, more or less in the form of an elongated

CHAP. VII.] SPECIAL MUSCULAR MECHANISMS. 315

isosceles triangle with the apex dipping beneath the cushion of the epiglottis, the sides formed by the vocal cords, and the base by the arytenoids with the membrane between them. In a favourable view (Fig. 185) the vocal cords (*v. c.*) may be seen to be attached to the processus vocales and the distinction between the membranous and cartilaginous glottis observed. On the outside of each vocal cord, separated from it by the mouth of the corresponding ventricle of the larynx and reaching to the side of the laryngeal aperture, may be seen the ventricular band (*c. v. s*). By their white colour the vocal cords present a strong contrast to the rest of the larynx.

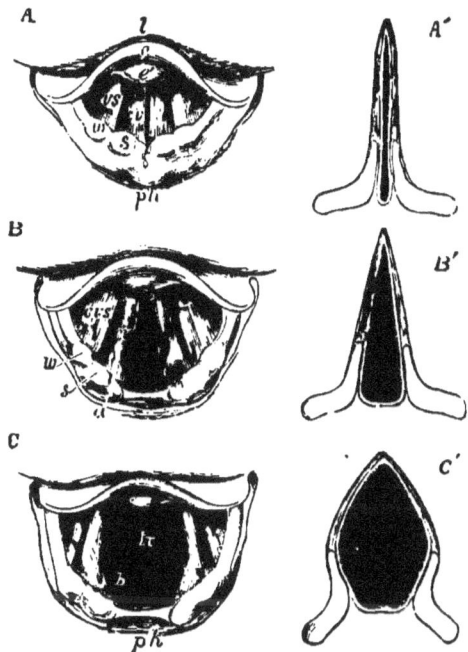

FIG. 186. THE LARYNX AS SEEN BY MEANS OF THE LARYNGOSCOPE IN DIFFERENT CONDITIONS OF THE GLOTTIS. (From Quain's Anatomy, after Czermak.)

A while singing a high note; *B* in quiet breathing; *C* during a deep inspiration. The corresponding diagrammatic figures *A'*, *B'*, *C'*, illustrate the changes in position of the arytenoid cartilages, and the form of the rima vocalis and rima respiratoria in the above three conditions.

l the base of the tongue; *e* the upper free part of the epiglottis; *e'* the tubercle or cushion of the epiglottis; *ph.* part of the anterior wall of the pharynx behind the larynx; *w* swelling in the aryteno-epiglottidean fold caused by the cartilage of Wrisberg; *s* swelling caused by the cartilage of Santorini; *a* the summit of the arytenoid cartilage; *cv* the vocal cords; *cvs* the ventricular bands; *tr* the trachea with its rings; *b* the two bronchi at their commencement.

If the voice, and especially if a high note, be uttered the view changes (Fig. 186 *A*), besides an alteration in the form of the laryngeal aperture, the vocal cords are seen to be brought close together and nearly parallel so that the glottis becomes a mere slit. If no voice is being uttered and a deep inspiration be taken changes of another kind may be observed (Fig. 186 *C*); the glottis becomes a wide aperture with the form of a truncated rhomboid, the obtuse angle on each side marking the attachment of the vocal cord to the processus vocalis; through this wide opening the tracheal rings are clearly visible, and indeed with an appropriate position of the mirror the bifurcation of the trachea into the bronchi may under favourable circumstances be observed. When changes in the voice or in the breathing are being made, the white glistening vocal cords may be seen to come together or to go apart like the blades of a pair of scissors.

§ **905.** Laryngoscopic observation then teaches that the larynx is used not only for the utterance of voice, for phonation, but also for breathing; and indeed in speaking of respiration (§ 336) we called attention to this; but the former is its more important use and we may chiefly dwell on this, referring incidentally to the respiratory functions.

In order that the membranous edges of an aperture may be readlly thrown into sonorous vibrations by a blast of air, the edges should be brought near together and the aperture reduced to a mere slit. Hence the fundamental condition for the formation of the voice, and indeed speaking generally of voices of all kinds, is the approximation and consequent more or less parallelism of the vocal cords.

In the voice, as in other sounds (cf. § 841), we distinguish three fundamental features: (1) Loudness. This depends on the strength of the expiratory blast. (2) Pitch. This depends on the rapidity of the vibrations, and this we may in a broad way consider as determined on the one hand by the length and on the other hand by the tension of the vocal cords. What we may call the natural length of the vocal cords is constant, or varies only with age; and the influence of this factor bears on the general range of the voice, not on the particular note given out at any one time. The tension of the vocal cords on the contrary is very variable, and the pitch of any particular note uttered depends in the main on this; hence great importance attaches to the mechanisms by which changes in the tension of the vocal cords are brought about. But, as we shall see, the problems connected with the compass of a voice and with changes of pitch are very complex; in considering these things we have to do with much more than mere variations in the tension of the vocal cords along the whole of what we have called their natural length. These matters however we shall deal with later on, and may for the present consider tension as the main factor of changes in pitch.

(3) Quality. This, as we have seen (§ 841), depends on the number and character of the partial tones accompanying any fundamental note sounded, and is determined by a variety of circumstances, chief among which are, on the one hand the form, thickness and other physical qualities of the cords, and on the other hand, the disposition of the resonance chamber, or parts of the respiratory passage other than the glottis itself.

We may confine ourselves in the first instance to the conditions which determine the mere utterance of the voice and to the mechanisms which affect the tension of the vocal cords, and hence the pitch of the voice. The problems therefore which we have to attack are, first, By what means are the cords brought near to each other or drawn asunder as occasion demands? and secondly, By what means is the tension of the cords made to vary? We may speak of these two actions as narrowing or widening of the glottis, adduction or abduction of the edges of the glottis, and tightening or relaxation of the vocal cords. We may first dwell on the muscular aspects of the mechanisms by which these results are brought about, taking the nervous factors into consideration afterwards. The change of form of the glottis is best understood when what we have already (§ 901) said is borne in mind, namely, that each arytenoid cartilage is, when seen in horizontal section (Fig. 186), somewhat of the form of a triangle, with a median, an external, and a posterior side, the processus vocalis being placed in the anterior angle at the junction of the median and external sides. When the cartilages are so placed that the processus vocales are approximated to each other and the internal surfaces of the cartilages nearly parallel, the glottis is narrowed (Fig. 186 A'). When on the contrary the cartilages are wheeled round on the pivots of their articulations, so that the processus vocales diverge, and the internal surfaces of the cartilages form an angle with each other, the glottis is widened (Fig. 186 B', C'). Moreover the two cartilages may to a certain extent be bodily drawn together, or dragged apart, the two hind angles, between the median and posterior sides, being now close together, now apart.

§ 906. The muscles of the larynx though small, are numerous and complicated, and are so disposed in respect to their origins and insertions and to the sweep of their fibres, that the effect of the contraction of one muscle will depend upon whether or no and how far other muscles are thrown into contraction at the same time; moreover in the case of some of the muscles at least the effect is different according as the whole muscle or a part only contracts.

The first muscle to which we may call attention is the *transverse arytenoid* (M. arytenoideus posticus s. transversus) (Fig. 187). This is a relatively thick muscle covering the hind surfaces of both arytenoid cartilages; the fibres starting from the outer edge of one cartilage run transversely across to the outer edge of the other cartilage, and the belly of the muscle occupies the concave hind surfaces of the two cartilages together with the intervening space.

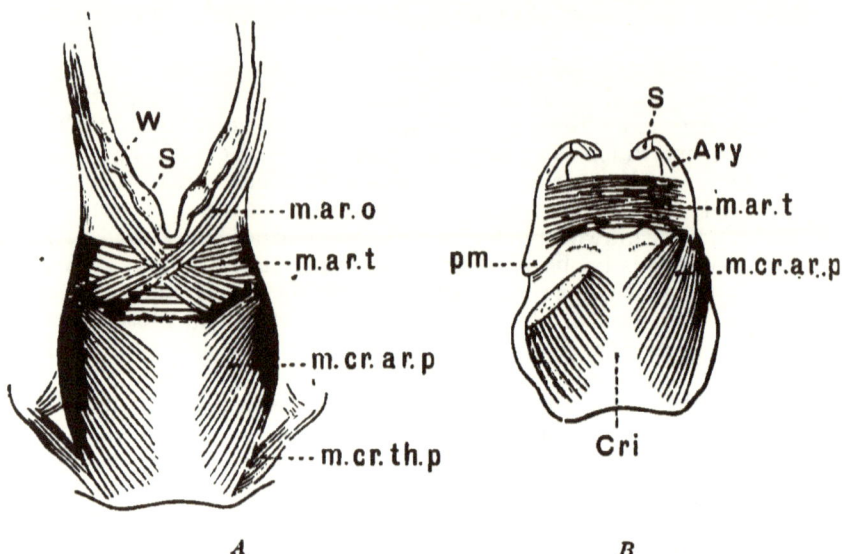

Fig. 187. Diagram of the Transverse and Oblique Arytenoid and of the Posterior Crico-arytenoid Muscles.

A. shews the three muscles in position in reference to the aperture of the larynx; *B.* shews the attachments of the transverse arytenoid and posterior crico-arytenoid.

m. ar. t. transverse arytenoid muscle. *m. ar. o* oblique arytenoid muscle. *Ary.* Arytenoid cartilage. *p m.* processus muscularis of arytenoid cartilage. *W.* prominence of cartilage of Wrisberg (in *B*, it marks the cartilage itself). *m. cri. ar p.* posterior crico-arytenoid muscle. *Cri.* Cricoid cartilage. *Ary.* Arytenoid cartilage. *S.* prominence of cartilage of Santorini *m cr. th. p.* is the small posterior crico-thyroid muscle.

The effect of the contraction of this muscle is to bring the two cartilages closer together and so to narrow the glottis; indeed if in an animal it be divided, the glottis remains widely open behind. It is an important closer of the glottis, adductor of the vocal cords. When it is not contracting the cartilages come apart through the elastic reaction of their connections.

Most important is a mass of muscular fibres, which starting from the lower part of the re-entering angle of the thyroid pass horizontally but inclined somewhat upwards to the arytenoids at about the level of the vocal cords. The whole mass is described as forming two muscles. The outer or lateral part ending in the outer edge of the arytenoid and upper part of its processus muscularis is called the *external thyro-arytenoid* (M. thyro-arytenoideus

CHAP. VII.] SPECIAL MUSCULAR MECHANISMS. 319

externus) (Figs. 188, *m.th.ar.e.* 184 *B*). The direction of the muscle as a whole is horizontally backwards, though inclined outwards and upwards, but the constituent individual bundles run in various

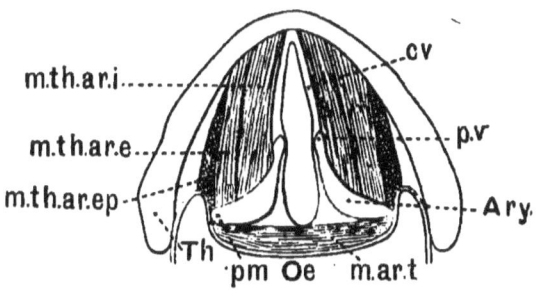

FIG. 188. DIAGRAM TO ILLUSTRATE THE THYRO-ARYTENOID MUSCLES.

The figure represents a transverse section of the Larynx through the bases of the arytenoid cartilages.

Ary. Arytenoid cartilage. *p. m.* processus muscularis. *p. v.* processus vocalis. *Th.* Thyroid cartilage. *c.v.* vocal cords. *Oe.* is placed in the œsophagus.

m. th. ar. i. internal thyro-arytenoid muscle. *m. th. ar. e.* external thyro-arytenoid muscle. *m. th. ar. ep.* part of the thyro-ary-epiglottic muscle cut more or less transversely.

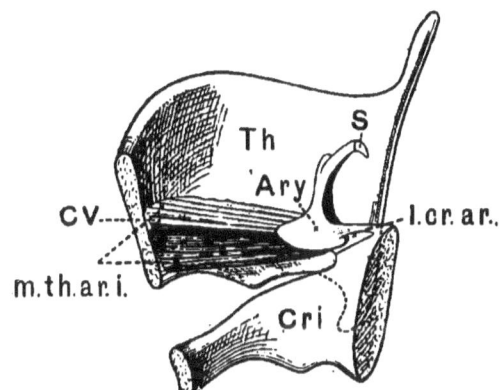

FIG. 189. THE INTERNAL THYRO-ARYTENOID MUSCLE.

The left halves of the thyroid and cricoid have been removed so as to shew the right arytenoid in position.

Th. Thyroid. *Cri.* Cricoid. *Ary.* Arytenoid. *S.* Cartilage of Santorini. *l. cr. ar.* the crico-arytenoid ligament. *m. th. ar. i.* the internal thyro-arytenoid muscle, with *c.v.* the vocal cord.

ways and some even pass vertically into the ventricular bands. To the inner or median side of this external muscle, between it and the corresponding vocal cord, lies the inner muscle which running from the re-entering angle of the thyroid to the processus vocalis and outer surface of the arytenoid forms a wedge-shaped mass, the thin edge of which is covered by the actual vocal cord. It is called the *internal thyro-arytenoid* (M. thyro-arytenoideus internus s. M. vocalis) (Figs. 188, 189, *m.th.ar.i.* 184 *A*) and has by some authors been subdivided into a median and lateral division. The general direction of the muscle is horizontally backwards, but, as in the external muscle, the constituent bundles run in various directions and some are said to end or begin in the vocal cord itself. One most important action of these two muscles is undoubtedly to bring the arytenoids nearer to the thyroid and so to slacken the vocal cords; but they produce other effects, and their contractions, especially those of the external muscle, help under circumstances to bring the vocal cords together and so to narrow the glottis. They also, as we shall see, produce changes in the form and thickness of the cords.

Of less importance than any of the above is a small muscle which starting from the processus muscularis of one arytenoid passes (Fig. 187 *A*, *m.ar.o.*) obliquely upwards towards the summit of the other arytenoid, crossing its fellow obliquely at the back of the transverse arytenoid muscle, which it thus partially covers; some of the fibres seem to end in the cartilage of Santorini but most of them are continued to the thyroid, the ary-epiglottic fold, and the base of the epiglottis. It is called the *oblique arytenoid* (M. arytenoideus obliquus) or it may be regarded as part of a flat, irregular muscle, the thyro-ary-epiglottic muscle (Fig. 188 *m. th. ar. ep.*). Its action is to approximate the two arytenoids and so to help in closing the glottis. It, with the transverse arytenoid and the external thyro-arytenoid muscles, may be looked upon as forming together a sort of sphincter of the larynx; their combined contractions certainly tend to close the glottis.

A relatively large and very important muscle is the *posterior crico-arytenoid* (M. crico-arytenoideus posticus) (Fig. 187 *m. cri. ar. p.*). This, starting from the lower part of the hind surface of the cricoid near to the median line, passes obliquely upwards to be inserted into the outer edge of the arytenoid just below the insertion of the transverse arytenoid muscle, at the upper part of the processus muscularis. Its chief action is by wheeling the outer corners of the arytenoids backwards and towards the middle line, to throw the processus vocalis outwards and so to widen the glottis; it is in fact the special, we may perhaps say the only, dilator of the glottis, or abductor of the cords.

The above muscle meets its antagonist in the lateral *crico-arytenoid* (M. crico-arytenoideus lateralis s. anterior) (Fig. 190 *m. cr. ar. l.*), which taking origin from a large portion of the upper border of the cricoid cartilage in its lateral parts in front of the thyro-cricoid articulation, passes upwards and backwards to be inserted into the processus muscularis and outer side of the arytenoid in front of and below the insertion of the posterior crico-arytenoid. Its main action is to wheel the outer corner of the arytenoid forwards and inwards and thus, by converging the processus vocales, to adduct the cords and to narrow the glottis; but it may, under circumstances, have other effects.

The last muscle to which we need call attention, and which in some respects stands apart from the rest, is the *crico-thyroid* (M. crico-thyroideus anticus). This (Fig. 191 *cr. th.*) starts

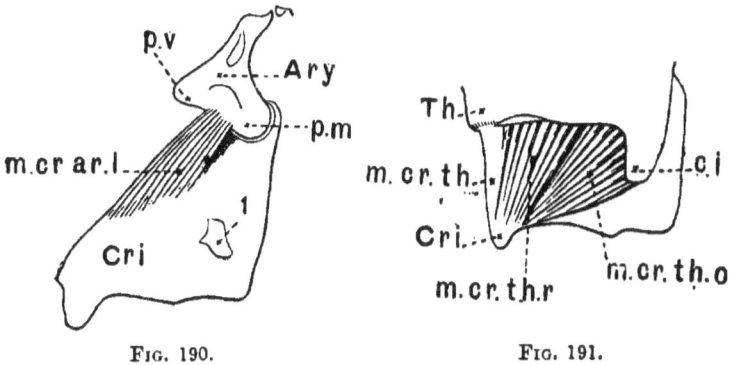

Fig. 190.   Fig. 191.

Fig. 190. THE LATERAL CRICO-ARYTENOID MUSCLE.

Ary. Arytenoid; *p.v.* processus vocalis. *p.m.* processus muscularis. *Cri.* Cricoid.
1. surface for articulation of lower cornu of thyroid. *m. cr. ar. l.* the lateral crico-arytenoid muscle.

Fig. 191. THE CRICOID-THYROID MUSCLE.

Th. Thyroid; *c. i.* its inferior cornu. *Cri.* Cricoid; *m. cr. th. r.* the straight part,
m. cr. th. o. the oblique part of the crico-thyroid muscle. *m. cr. th.* crico-thyroid membrane.

from the front lateral surface of the cricoid, near its lower border, and passing obliquely backwards and upwards is inserted into the lower edge and inner lateral surface of the thyroid. It is sometimes subdivided into a front part (*cr. th. r.*) the fibres of which run more directly upwards (M. cr. thy. rectus) and a lateral part (*cr. th. o.*) the fibres of which run in a more oblique direction (M. cr. thy. obliquus). The action of the muscle is a somewhat complicated one, but the effect of its contractions as a whole is, if the thyroid be regarded as the more moveable of the two cartilages, to pull the thyroid downwards and forwards over the front part of the cricoid, or, if the thyroid be supposed to be the more fixed, to rotate the cricoid on its transverse axis, pulling upwards the front part and tilting downwards the hind part on which the arytenoids sit; the latter is probably its real action. Upon either view, its contractions increase the distance between the reentering angle of the thyroid and the processus vocalis and so stretch the vocal cord; it is in fact the main tightener of the vocal cords.

There are other small muscles in the larynx as well as muscles connecting the larynx with surrounding parts; but it is not necessary for us to dwell on them here. Meanwhile it is obvious from what we have said that narrowing or widening the glottis, and slackening or tightening the vocal cords, are brought about by the above muscles acting somewhat as follows.

§ 907. *Narrowing of the glottis; adduction of the vocal cords.* The glottis is narrowed by the combined contraction of the three muscles which we spoke of above as forming a sort of sphincter for the larynx, namely, the transverse arytenoid, the oblique arytenoid and the (external) thyro-arytenoid. These produce their effect chiefly by bringing the two cartilages near to each other in the middle line, and in this action the transverse arytenoid muscle is the most potent. Hence this muscle may be regarded as the most effective of the constrictors of the glottis.

The glottis is also narrowed by the lateral crico-arytenoid, but this produces its effect by rotation of the arytenoid cartilages; it pulls the processus muscularis forwards and so throws the processus vocalis inwards.

Widening of the glottis; abduction of the vocal cords. The chief if not the only agent for the widening of the glottis is the posterior crico-arytenoid. This, pulling the outer edge of the arytenoid backwards, throws the processus vocales outwards, and so abducts the vocal cords. It has been argued that the transverse arytenoid acting alone or in concert with the above, or at least in the absence of any contraction of the other members of the sphincter group, would also wheel the outer edge of the arytenoid in the same way and so also abduct the vocal cords; but the evidence seems to be against this view.

Tightening of the vocal cords. This is especially effected by one muscle on each side, namely by the crico-thyroid which, by bringing the thyroid and the front part of the cricoid nearer to each other, increases the distance between the thyroid and the arytenoids when the latter are fixed. Supposing the transverse arytenoid and posterior crico-arytenoid to fix the arytenoids, the direct effect of the contraction of the crico-thyroid is to tighten the vocal cords. There is besides a special action of the internal thyro-arytenoid by which this muscle becomes, in contrast to the external thyro-arytenoid, a tightener of the cord; of this action we shall speak later on.

Slackening of the vocal cords. This is effected by the whole sphincter group just mentioned, but more especially by the external thyro-arytenoid and to a certain extent by the internal thyro-arytenoid; these acting alone, produce an effect the reverse of that of the crico-thyroid, bringing the arytenoid cartilages nearer to the thyroid cartilage, and so shortening the distance between the processus vocales and that body.

These several acts, however, the widening or narrowing of the glottis, the tightening or slackening of the vocal cords, are only the gross acts, so to speak, of the movements of the larynx. When a voice of any kind has to be uttered the cords must be approximated and to a certain extent tightened; and for the carrying out of even these gross acts not one muscle only but more than one, and often several are brought into play; the movements which give rise to any kind of voice are combined and coordinated movements. But, as we shall see presently, when this or that particular kind of voice is being uttered or when changes in the voice are being effected, the above words, widening, tightening and the like, very imperfectly describe what is taking place in the larynx; changes of a very complex nature are brought about, and for these the greatest nicety of combination is necessary.

§ 908. We may now turn to the nervous mechanisms of the larynx. Fibres of the superior laryngeal branch of the vagus nerve are distributed to the mucous membrane of the larynx, and serve as the afferent channels by which impulses from the exquisitely sensitive surface pass upwards to the central nervous system.

The same superior laryngeal nerve also contains motor fibres for the crico-thyroid muscle; and in this respect this muscle, the chief tensor of the vocal cords, stands apart from all the rest of the muscles of the larynx, for these are all supplied by the recurrent laryngeal branch. These motor fibres, both of the superior and of the recurrent laryngeal nerves, though running in the trunk of the vagus are generally believed to belong not to the vagus proper but to the spinal accessory nerve, and to the division of that nerve which we (§ 617) called the bulbar accessory nerve; but on this point opinions are not agreed.

There are some reasons for thinking that the superior laryngeal contains afferent fibres not only for the crico-thyroid but also for at least some of the muscles whose motor fibres come from the recurrent laryngeal; and it has been suggested that these afferent fibres of the superior laryngeal convey the afferent impulses of the 'muscular sense;' but this needs further investigation.

In dealing with the nervous mechanism we must now distinguish between the larynx as a part of the mechanism of breathing and as an organ of voice. During breathing the glottis is open, and at least during at all deep or laboured breathing undergoes as we have previously said (§ 336) an increased widening during inspiration followed by narrowing during the succeeding expiration. In many animals this rhythmic respiratory movement is very marked; but careful laryngoscopic observations shew that in man during quite quiet breathing there is no appreciable change in the width of the glottis.

Much difference of opinion has been expressed as to whether the width of the glottis thus permanently maintained during quiet breathing is identical with that assumed after death. But careful laryngoscopic measurements shew that during life the glottis is distinctly wider than after death; the average width during quiet breathing, is in man about 14 mm., in woman about 12 mm.; after death in man 5 mm., in woman 4 mm. This points to a continued 'tonic' contraction of some or other of the dilators of the glottis; and the muscle especially concerned in this action appears to be the posterior crico-arytenoid Whether this tonic dilator action, whose centre lies in the bulb, close to or forming part of the general respiratory centre, is automatic in nature or maintained in a reflex manner by afferent impulses, we need not stay now to discuss; nor need we dwell on the question whether the widened glottis is the result merely of the action of the dilator muscle, or whether it is the balance of a struggle between antagonistic muscles; though analogy would perhaps lead us to expect the latter to be the case, the evidence appears to be in favour of the former view.

The rhythmic alternation of widening and of narrowing observed in laboured breathing is, through the activity of the bulbar nervous mechanism, brought about by the various muscles spoken of above, the sphincter group being especially used for narrowing. When occasion requires, a powerful action of this group leads, as in the first step of a 'cough' (§ 391), to complete closure of the glottis; and further security in the act is obtained by the narrowing of the vestibule or space above the vocal cords, the ventricular bands being brought together by the thyro-aryepiglottic assisted by other muscles.

Both the continued patency and the rhythmic changes are carried out by means of the recurrent laryngeal nerves. When in a living animal both these nerves are divided, the glottis becomes narrowed, assuming what may be considered its natural dimensions, namely, those proper to it after death, when all muscular contractions have ceased. Owing to the narrowing the entrance and exit of air into and out of the lungs is less easy than before, and a certain amount of dyspnœa, especially obvious if the breathing be hurried, may result; but the extent to which this occurs differs much in different kinds of animals and indeed in different individuals. It need hardly be said that when both the recurrent nerves are divided the rhythmic widening and narrowing wholly cease, the glottis remaining immobile; the voice also is lost. When the nerve is divided on one side only, the glottis becomes deformed; when an attempt to utter voice is made, the vocal cord on that side remains farther away from the middle line than its fellow, owing to the failure of the adductor muscles on that side, and no voice is produced, since the approximation and parallelism of the vocal cords can no longer be effected. On the other hand during a deep inspiration the glottis is deformed by the vocal cord on that side being nearer the middle line than its fellow, owing to the failure of the posterior crico-arytenoid on that side. When the peripheral portion of one recurrent nerve is stimulated, the vocal cord of the same side is approximated to the middle line; when both nerves are stimulated, the vocal cords are brought together and the glottis is narrowed; though the nerve is distributed to both dilating and constricting, to abductor and adductor muscles, the latter overcome the former when the nerve is artificially stimulated. But this is true only when the stimulus is adequately strong; if the stimulus be weak, the abductors alone are thrown into contraction and the glottis is widened. We may, in this connection, add the remark that ether paralyses the adductors before the abductors, and has this effect, even after division of both recurrent nerves; the more general respiratory function of the larynx, the maintenance of a wide passage by means of the abductors, is preserved, while the more special function of phonation, the narrowing of the glottis by the adductors, is lost. A like differentiation of the two functions is shewn, in a reverse way, by the clinical experience that while functional nervous disorders, such as hysteria, are marked by failure of the adductors alone, the characteristic loss of voice being due to this, the first effect of structural changes in the bulb or other parts of the nervous mechanism is to bring about failure of the abductors; indeed the condition of the larynx as shewn by the laryngoscope may be used as an aid to the diagnosis of commencing organic disease.

§ 909. When the larynx is used for voice the recurrent laryngeals are brought into play in order to produce the essential condition of voice, the approximation of the vocal cords. The vocal cords having been adequately approximated, low notes may be uttered without any further change in the larynx; in their natural position of rest the vocal cords are sufficiently tense to permit their being thrown into vibrations when brought near enough together and subjected to a sufficient blast. In order however to utter notes at all high, the tension of the cords must be increased; and this as we have said is brought about chiefly by means of the superior laryngeal nerves and the crico-thyroid muscles. Hence when these nerves are divided or fail through disease, high notes can no longer be uttered; and the division or failure of the nerve even on one side only will bring about this result.

When in using the voice a change has to be made from a higher to a lower note, while the action of the crico-thyroid ceases or is lowered, that of the antagonistic thyro-arytenoid comes into play, and the recurrent laryngeal nerve is again used.

§ 910. Utterance of the voice is a conspicuously voluntary act and in the vast majority of cases an eminently skilled act. Hence we find, as we have already (§ 655 — 659) seen, an area in the motor region of the cerebral cortex devoted to phonation. This in the monkey (Fig. 126) lies at the lowest part of the ascending frontal convolution wedged in between the sylvian fissure and the lower end of the precentral fissure; in man as we have seen (Fig. 131) the more highly developed area for 'speech' is situate at the posterior end of the third frontal convolution, having as we have also seen an importance on the left side of the brain which it does not possess on the right.

Stimulation of the area in question in the monkey or of the corresponding area in the dog leads to adduction of the cords and closure of the glottis, the resulting movement being bilateral. As in the case of other areas, the effect is more pure, the laryngeal movement is less mixed with other movements, when the stimulation is strictly limited to a certain part of the whole area, in this case to the lower part. So far adduction only has been the experimental result of stimulation of the cortex in the monkey and the dog, an area for abduction having been found in the cat alone; and as we have said adduction is the salient movement in phonation. But stimulation of the cortex near the pure centre for phonation leads to an acceleration in the rhythm of and exaggeration of the laryngeal respiratory movements, as indeed of the respiratory movements as a whole; though the respiratory laryngeal movements are in the main worked by a bulbar mechanism, they can be influenced by cortical changes.

As in the case of the other cortical motor areas, the path from the cortical area for phonation to the muscles whose actions it governs runs in the pyramidal tract through the internal capsule. Moreover in the bulb there appears to be a subordinate nervous mechanism, with which the impulses or influences descending the pyramidal tract make connection before they issue as coordinate motor impulses along the laryngeal nerves; and indeed by local electrical stimulation of the bulb, in the floor of the fourth ventricle, adduction or abduction of the cords may be brought about. The bulbar mechanism for abduction is placed higher up than that for adduction, and stimulation of either side produces in both cases bilateral movements.

§ 911. So far we have mainly spoken of the voice as the result of two gross acts, the narrowing of the glottis and the tightening of the cords. We must now say a few words on some other changes in the larynx, especially in reference to the various qualities and kinds of voice. Many of the features of the voice are conferred upon it by means of the parts of the respiratory passage above or below the vocal cords, by what we have spoken of as the resonance chamber or tube; these we shall deal with in treating of 'speech,' and may here confine ourselves, in the main, to changes in the larynx itself. It should be noted however that whenever voice is uttered the larynx is more or less firmly fixed by the extrinsic laryngeal muscles, such as the thyro-hyoid, sterno-thyroid, pharyngeal muscles and others. The exact position in which it is fixed will depend on the pitch of the notes which are uttered; it is raised for high notes and lowered for low ones, and may be fixed either above or below or at the natural position of rest.

We are accustomed to classify voices according to the range of pitch within which the voice can sing truly and with ease, and we thus distinguish, in ascending scale, such voices as bass, barytone and tenor in the male, alto, mezzo-soprano and soprano in the female. Could we consider the vocal cord as a membranous edge, possessing a form and nature which was constant or varied only with age, so that the rapidity of its vibrations, and hence the pitch of the voice, depended solely upon its length, fixed by the growth of the individual, and upon its varying tension, determined by muscular contraction, the result being influenced by the varying width of the glottis, the structural basis of the distinction between the several kinds of voice would be simple enough; the bass and the contralto voices would have long vocal cords, and the other voices in each sex would be in ascending scale successively shorter. The vocal cord, however, is not of such a permanent nature; it undergoes under the influence of muscular contraction changes other than those of tension affecting its whole length; its form may be altered and the positions or attitudes which it may assume cannot be described as simply those of greater or less distance from its fellow along its whole length. It is in section as we have said wedge-shaped; and the projecting angle of the wedge may be an open broad one, or a narrow acute one; the vibrating cord may be thick or thin; and its vibrations will vary accordingly.

The change from thick to thin is apparently brought about by muscular contraction; it has been suggested that a partial contraction of some of the fibres of the thyro-arytenoid muscle, external and also internal, more particularly of the bundles which take a more or less vertical direction, produce the result; but the exact mechanism is by no means clear, though special examination of the larynx shews that such a change of thickness may take place.

Again, there are reasons for thinking that contraction of the internal thyro-arytenoid muscle as a whole affects the form and the physical condition of the vocal cord, of which it furnishes so to speak the body; the strand of elastic fibres which forms the surface of the cord lies upon the muscle somewhat after the fashion of a fascia, and when the muscle, which in a state of rest is somewhat curved with the concavity towards the glottis, thickens and shortens in its contraction, carrying with it the overlying layer of elastic fibres, it brings the whole cord into a different form and different physical condition; and this must affect the character of the vibrations. Again, it is maintained that some of the fibres of the internal thyro-arytenoid running forward from the processus vocalis and outer surface of the arytenoid are inserted into the layer of elastic fibres at varying distances from the thyroid. If some of these bundles were to contract by themselves they might render the front part of the cord tense and the hind part relatively lax, or might modify in particular parts that general tension of the whole cord which was being effected by the crico-thyroid.

Further, the closure of the glottis, the adduction of the cords may take place in different ways, according as this or that muscle or part of a muscle is being especially used. While the vocal cords are being sufficiently approximated to allow the expiratory blast to throw them into vibrations the cartilaginous glottis may remain fairly open, or may be nearly or be quite closed; and each of these conditions must affect the voice in a different way. Again, we have seen that the two vocal cords are close together at their insertion into the thyroid, and diverge from the middle line on each side so that the membranous glottis, when the larynx is at rest, is an isosceles triangle. We might infer from this that when the cords are adducted, the glottis must always remain an isosceles triangle with the angle at the apex, next to the thyroid, becoming more and more acute as adduction proceeds, and that the parts of the cords in front, nearer the thyroid, must come into actual contact before the parts behind, nearer the processus vocales, do. But the laryngoscope shews that the form of the membranous glottis is very varied; it may be open behind and closed in front, or closed both behind and in front and open, even widely so, in the middle, or may be along almost its whole length a slit with parallel sides, and in that case either very narrow, a mere linear cleft, or of appreciable width. And though the exact mechanisms are obscure, we cannot doubt but that these several phases result from special muscular contractions.

§ 912. We might dwell on other changes which may by help of the laryngoscope be observed in the larynx during the production of the voice, all shewing that muscular contractions may produce complex and varied changes in the larynx besides simple adduction or abduction and general tension or slackening of the vocal cords; but we have said enough for our present purpose, which is to insist that in the production of voice the mere dimensions of the larynx, and we might add other natural inborn features, serve but as the playground for muscular skill; it is the latter much more than the former which determines the characters and the powers of the voice. A laryngoscopist, even the most experienced, would probably hesitate from a mere inspection of the larynx to predicate the nature of the singing voice. He could not even predicate the possession of a singing voice of any kind. Of two larynges, provided they were both of normal structure, he would be unable to say which belonged to the man who could and which to the man who could not sing; for the power to sing is determined not by the build of the larynx but by the possession of an adequate nervous mechanism through which finely appreciated auditory impulses are enabled so to guide the impulses of the will that these find their way with sureness and precision to the appropriate muscular bundles. And what is true of the difference between singing and not singing at all is in a similar way true of the difference between singing low and singing high, as well as of the difference between singing superbly and singing indifferently well. The physiological difference between a bass voice and a tenor voice, between a contralto and a soprano probably lies not so much in the mere natural length of the vocal cords as in the constitution of the nervous and muscular mechanism; experience shews that cords of the same natural length may in one individual be the instrument of a bass, in another of a tenor voice, or in one individual of a contralto, in another of a soprano voice. Again, though the "magnificent organ" of a distinguished artist may have certain inborn qualities which lighten the labours of the nervous mechanism, it is the latter which is the real basis of the artist's fame; the former may be so slight or so abstruse as to escape observation, and a larynx, the notes of which have charmed the world, may yield through the laryngeal mirror a picture of the most commonplace kind.

That the build of the larynx is thus wholly subordinate to the nervous mechanism is further illustrated by the fact that the same larynx may and indeed does produce different kinds of voice. The difference in the kind of voice which may be brought about by the nervous system working the same larynx in different ways is strikingly shewn by comparing what is called the chest voice and the head voice. In the former, which deals with relatively low notes, the sounds are full and strong, and the lower resonance chamber which is supplied by the trachea, bronchi and indeed by the whole chest, is thrown into powerful and palpable vibrations; hence the name 'chest voice.' The latter, which deals with relatively high notes, is thin and poor, being deficient in partial tones, is not accompanied by the same conspicuous vibrations of the chest but is accompanied by vibrations of the head; hence the name 'head voice.'

It is obvious that the dispositions of the larynx must be very different in the two voices; but what the differences exactly are has been and still is a matter of controversy, and indeed extended laryngoscopic observation leads to the conclusion that the change from the one voice to the other is not effected in precisely the same way by all larynges. The evidence however seems to shew that in the chest voice the vocal cords are relatively broad and thick, and that the membranous glottis is open along its whole length. The cords will of course vary as to their tension through the range of the voice, being more tense with the higher notes, and the width of the glottis is not always the same; but it is probable that throughout the voice the cords, in producing the fundamental tone of any note sung, vibrate along their whole length, and also through their whole breadth, the partial tones being due of course, as in other musical instruments, to vibrations of segments. In order to throw into vibration along their whole length such relatively broad and thick cords a powerful blast of air is needed, and hence the transmission of the vibrations downwards to the chest.

When the same larynx shifts to the head voice the vocal cords appear to become narrow, thin and always distinctly tense. In some cases the membranous glottis is closed before and behind, so that the cords are free to vibrate in their middle portion only, and here the slit is sometimes a relatively wide elliptical space; in other cases the glottis seems to be open along its whole extent, though reduced to a mere linear slit; but it is probable that in all cases the cords vibrate along a part only and not along the whole of their length. In order to throw into adequate vibrations the thin edges now presented, a less powerful blast is required, and the vibrations are no longer felt in the chest, though they are transmitted through the pharyngeal passages to the head.

As subsidiary conditions we may mention that in the chest voice the superior aperture of the larynx is widely open, the transverse diameter being perhaps especially long, while in the head voice the aperture is constricted, at times remarkably so. In the chest voice the epiglottis is usually depressed so as to hide from sight, in the laryngoscopic view, the front part of the cords, while in the head voice it is usually raised, but many variations in the attitude of the epiglottis may be observed. In the head voice the cartilaginous glottis seems always to be completely closed, whereas in the chest voice it is found in some cases to be closed, in other cases to be more or less open.

Making all allowance for discordance of opinion as to what are the exact conditions of each kind of voice, and admitting the imperfection of our knowledge as to both the purpose and the mode of production of many of the differences observed, we may at least draw the conclusion that in the case of each kind of voice a certain general disposition of the mechanism is made, that a certain 'setting' of the machine takes place, by which the quality of the voice is determined, and that the machine thus set is played upon so as to produce a series of notes differing in pitch but all retaining the same particular quality. The setting of the machine in the chest voice is such that the notes produced by it are lower notes reaching up to a certain pitch only, the setting not being adapted for higher notes, and conversely the setting of the head voice allows of the production of high notes only, being incompatible with the production of low notes.

It may be urged that the setting for the chest voice is really the natural disposition of the larynx and that of the head voice a strange and artificial condition into which the larynx is forced (and indeed the latter is in certain cases called a "falsetto" voice, which term however has a technical meaning not always coincident with head voice); but a closer examination of voices shews that there is in reality no one natural condition of the larynx, and that there are other dispositions or settings of the larynx besides those of the chest voice and the head voice, these being so to speak extreme cases.

When a singer sings a series of notes in an ascending scale it will be observed that beginning with the lowest notes the voice during a certain range remains through all the notes of the same quality, differing in pitch only, but that at or about a certain note, the voice in passing from one note to the next above is not merely raised in pitch but absolutely altered in quality, and further maintains the new features in the succeeding higher notes. This sudden change is spoken of as the 'break' in the voice, and a range of notes of differing pitch but of the same quality which is thus separated by a 'break' from another range of notes of a different quality is called a 'register.' Laryngoscopic observations, especially recent ones in which photographic aid has been called in, shew that during a register there is a particular 'setting' of the larynx, which is maintained throughout the whole register, the chief change observable being an increasing tension of the cords as the notes rise in pitch, and that at the break there is a sudden shifting of the setting, the new setting being maintained during the ensuing register with changes which as before are chiefly directed to a tension of the cords increasing with the rising notes.

In most voices the ear may recognize two such breaks, separating three registers, lower, middle, and upper, the lower and upper being usually the chest and head voice described above. But some voices are marked by three breaks separating four registers, the differences being distinctly recognisable by the ear, and there are some reasons for thinking that a break, that is a change in the setting of the larynx, may take place without being evident to the ear, though visible by help of the laryngoscope. We may add one part of the training of a singing voice consists in rendering the break. the transition from one register to another, as little obvious to the ear as possible.

It would be beyond the scope of this work to enter upon the details of the several registers of the different kinds of voices, beyond the little we have said touching the chest voice and head voice; these are matters of great difficulty subject to much controversy, and indeed, as we have already said, observations tend to shew that exactly the same disposition does not obtain for the same register in all larynges: it seems probable that two larynges may gain the same end by two different manœuvres, may produce the same kind of voice by different dispositions of the larynx. In any case the subject is one of extreme complexity, and we have ventured to dwell on it, even so long as we have, because it affords a striking illustration of what we have more than once insisted upon, the complicated character so often belonging to the muscular contractions by which the animal body gains its ends, and the delicately adjusted coordination of efferent nervous impulses needed to secure for the effort a complete success. We have so repeatedly, in previous parts of this work, insisted on the importance of afferent impulses to the coordination of complex movements that it is hardly necessary here to do more than to point out that the connection of the use of the voice with auditory sensations affords striking instances of the truth of what we have insisted upon. Auditory sensations are at least as important for the proper management of the voice as are visual sensations for the movements of the eyes, and more important than are visual sensations for the movements of the body and limbs. Indeed they are in a way essential to the very utterance of the voice; the dumbness which is so conspicuous a concomitant of congenital deafness is in most cases due not to deficiency in the muscular apparatus or even in the nervous mechanism on what we may call its motor side, but to the lack of afferent impulses from the auditory nerve. And in popular language we recognise this dependence of the management of the laryngeal muscles on auditory sensations when we talk of such or such one, who is deficient in this respect, as "having no ear."

§ 913. The ventricles of the larynx appear to be useful in allowing the vocal cords sufficient room for their vibrations; they also supply a secretion by which the vocal cords are kept adequately moist. The purpose of the ventricular bands is not exactly known; it has been suggested that they may at times exert a damping action by being brought down to touch the vocal cords; but this is very doubtful. The epiglottis, the position of which as we have seen varies in different kinds of voice, has also an influence on the character of the voice; and further influences which we shall consider under 'speech' are exerted by the form of the pharynx and the mouth.

§ 914. At the age of puberty a rapid development of the larynx takes place, leading to a change in the range of the voice. The peculiar harshness of the voice when it is thus 'breaking' seems to be due to a temporary congested and swollen condition of the mucous membrane of the vocal cords accompanying the active growth of the whole larynx. The change in the mucous membrane may come on quite suddenly, the voice 'breaking' for instance in the course of a night.

SEC. 2. SPEECH.

§ 915. All sounds as we have seen (§ 840) may be divided into musical sounds, in which the vibrations are regular, and noises in which the vibrations are irregular; but we have also seen that the distinction between the two is not a sharp one. The vibrations into which the air in the larynx is thrown by the vibrations of the vocal cords in ordinary voice are on the whole regular; the sound so produced is a musical sound. The vibrations of the glottis may however vary as to the degree of their regularity; and under certain circumstances they may be so irregular that the sound becomes an undeniable noise; as for instance in the sound which we call a 'cry' or a 'shriek.'

The sounds produced in the larynx like other musical sounds consist of partial tones added to a fundamental tone, and are in many cases very rich in partial tones. By modifying the shape of the passage leading through the pharyngeal, the buccal, and to a certain extent the nasal cavities, to the opening of the mouth, which we have spoken of as a resonance tube or chamber, and which, for reasons which we shall see, we may now call the *vowel chamber*, we are able to render loud and prominent one or other of the partial tones of a sound which is produced by the larynx and thus to affect its quality as it leaves the mouth.

We are also able, quite independently of the larynx (and indeed independently of breathing), to create sounds by means of parts of the mouth or other portions of the vowel chamber. These are for the most part noises but, as for instance in whistling, may be musical sounds.

In speech we make use on the one hand of laryngeal sounds, more or less modified in quality by the vowel chamber, and on the other hand of sounds generated in various parts of that chamber; our speech in fact consists of a basis of musical sounds with an addition of noises.

§ 916. One great feature of speech is that it is "articulate;" it consists of syllables jointed together, the parts of speech which we call words being formed of two or more syllables, or at times of one only. In the great majority of syllables we recognize two kinds of sounds which we call *vowels* and *consonants*. Though it is easy to say which is a vowel and which is a consonant, it is difficult to frame a definition which shall be free from all objections. It has been said that vowels are formed by the voice, that is by the vibrations of the vocal cords (hence the name vowel, *vocalis*), and consonants by the mouth, lips or other parts of the chamber above the larynx; but as we shall see, on the one hand the vowels, as indeed the name which we have adopted for the chamber indicates, are formed by help of that chamber, and on the other hand many consonants are formed by help of the voice. The word 'consonant' expresses the view that what we call consonants are always sounded with some vowel or other and cannot be sounded alone by themselves; but several consonants can be so sounded; hence the name is inappropriate. We may make the distinction that whereas in a vowel the form assumed by the resonance tube merely modifies the sound produced by the larynx, in a consonant a change of form in the same tube creates a noise which may exist by itself or may mingle with the sound produced by the larynx; but this is not exact, since as we shall see such a consonant as M may be used (as for instance in 'bottom,' in which though we write we do not sound the second o) in such a way that the form of the mouth only modifies a laryngeal sound, and the utterance may be continued indefinitely, like that of a vowel. Indeed we employ M and certain other consonants in two ways; we use M as a true consonant in company with a vowel as in 'my' or, as in the above instance, we may use it by itself, it alone forming a syllable. In this latter function M may conveniently be called a *sonant*, the sounds of speech being divided into 'sonants' and true 'consonants.' Again, it may be said that in the formation of a vowel, the whole of the vowel chamber is employed, in the formation of a consonant a particular part only. Lastly in M and similar consonants the very assumption of the form of the vowel chamber, which modifies the laryngeal sound, enters into the formation of the whole sound; this is not the case in a vowel; hence we may say that a vowel results from the mere relative position of, but a consonant from some action or movement of, the parts of the vowel chamber. We may however leave these definitions and turn at once to the mode of formation of the several vowels and consonants, or rather to the more common of these, since each language has its own vowels and sounds, and while some are common to all, some are special to a few, or even to one. We may merely remark that in speech the vowels bear the brunt of the work, they carry the 'accent,' and the consonants are, so to speak, built upon them as on a foundation. Some consonants (sonants) however may be used like vowels to carry accent.

§ 917. *Vowels.* With the utterance, either in singing or

speaking, of each vowel the vowel chamber is moulded into a particular shape. Taking the most common vowels U, O, A, E, I pronounced in the way in which most nations pronounce them and so corresponding to our *oo*, *o*, broad *a* (*ah*), *e* as in bet, and *ee*, we find the following.

In U the vowel chamber is large with a narrow opening at the mouth. The larynx is depressed, or at least not raised above the position of rest, the tongue is flattened, especially in front, and the lips are protruded so as to reduce the mouth to a narrow round opening. The form of the vowel chamber, with a wide pharyngeal and a narrow buccal orifice, may be compared to that of a round flask without a neck. In A, the mouth is opened wide, the larynx is somewhat raised, the tongue flattened and at the same time driven somewhat backwards towards the hind wall of the pharynx, so that the entrance from the pharynx to the larynx is narrowed. The vowel chamber thus assumes a form which may be compared to that of a funnel, the wide end being at the mouth and the narrow end at the larynx. In O the shape is intermediate between that of U and that of A, the exact shape depending on the kind of O which is being uttered.

In I the shape is very different. The larynx is raised and the tongue is carried forwards and upwards in such a way that it touches the teeth and the hard palate at the sides and nearly so in the middle leaving only a narrow canal in the middle line. At the same time the lips instead of being protruded are drawn back and the soft palate is raised high up. In this way there is developed above the larynx a relatively large pharyngeal space which communicates with the exterior by a narrow canal; the form of the vowel chamber may now be compared to that of a round flask with a long narrow neck. In E, and the other vowels between A and I, the shape of the resonance is correspondingly intermediate; in passing from A to I, the tongue is brought forwards and upwards, the buccal orifice narrowed and the larynx raised.

In each of the above cases what we have called the vowel chamber acts as a resonance chamber; that is to say in each case, owing to the shape of the cavity (in relation to the nature of its walls), the air in the chamber is more readily thrown into vibrations by certain tones than by others, and when a sound containing those particular tones is sounded into the chamber, those particular tones are reinforced and rendered loud and prominent. The shape of the vowel chamber in uttering U is such that the cavity acts as a resonator towards a particular tone, namely, the bass f, or more probably the bass b, and while the laryngeal sound with its fundamental and partial tones is passing through it, reinforces and renders loud the tone b, occurring as a constituent of the whole sound. And similarly with the other vowels. In fact vowel sounds are musical sounds in which a particular constituent tone is reinforced and rendered loud out of proportion to the other tones; in the case of some vowels two tones are so reinforced. The tone thus reinforced is generally a partial tone, but may be the fundamental tone. When the vowel is sung or spoken in notes of different pitch the particular partial tone which is reinforced will occupy different positions in the series of partial tones; it may be the first, second, third or other partial tone according to the pitch of the fundamental tone.

That the vowel chamber does act in this way as a resonator for a particular tone is shewn by moulding the cavity into the proper form for uttering a particular vowel, and bringing before the mouth a series of sounding tuning-forks of different pitch; it will be found that it is the sound of one tuning-fork and one in particular which is reinforced and made louder, namely, the one whose pitch corresponds to the fundamental tone of the particular vowel cavity; in the case of the vowel U for instance it will be the tuning-fork having the pitch b. On the other hand that what we recognize as vowel sounds do result from the reinforcement in a musical sound of a particular tone or of particular tones may be shewn by setting into vibration a series of tuning-forks of different pitch, in imitation of a musical sound with its constituent tones, and then in turn reinforcing the sound of particular tuning-forks by the help of artificial resonators. When this is done the reinforcement of the appropriate tone gives rise to a vowel sound, the reinforcement of b giving rise to U and so on. The curves moreover described by the vowel sounds in the phonograph, in which the vibrations of the air transmitted to a thin plate or membrane are made to write on a recording surface, are in form such as would be described by sounds in which particular constituent tones were reinforced in the manner described. Again, as we said in dealing with hearing (§ 850), when a note is sung into the open piano, the particular strings of the piano corresponding to the constituent tones of the note sung are thrown into sympathetic vibration; in the sound thus returned by the piano the nature of the vowel on which the note was sung may be recognized; the string corresponding to the characteristic tone of the vowel is thrown into appropriately strong vibrations.

The nature of the vowel sounds is especially well illustrated by the kind of speech which we call *whispering*. In this, in contrast to audible speech, no musical sounds are generated by the vocal cords. A laryngeal sound is generated but it is a noise, not a musical sound, and is caused by the friction of the air as it passes through the glottis, which assumes a peculiar form, the processus vocales projecting inwards towards each other, leaving the cartilaginous glottis as well as the greater part of the membranous glottis more or less open. This noise, like the musical sound of audible speech, is modified by the parts of the mouth and pharynx, and in it we may recognize vowels and consonants. The noise of the whisper, though weak, contains multifarious vibrations, contains among other and irregular vibrations, the regular vibrations corresponding to the several vowels. When we whisper a vowel, we 'set' the vowel chamber so that it may reinforce the set of vibrations of the particular pitch characteristic of the vowel; and a well trained ear may recognize in the whispered vowel the dominant tone.

A vowel then is essentially a musical sound of a special quality, due to the dominance of a particular tone reinforced by the conformation of the vowel chamber. There are many subsidiary questions connected with the formation of vowels but into these we cannot enter here. We will merely add that in uttering a true diphthong, the conformation of the vowel chamber proper to the initial vowel is changed rapidly but gradually and without any obvious break into that proper to the final vowel.

§ 918. *Consonants.* These as we have already said in some cases so far resemble vowels in that they are modifications of the voice, that is to say of laryngeal sounds, caused by the disposition of the parts of the vowel chamber, the disposition however being in all such cases different from that giving rise to a vowel, and moreover the very assumption of the disposition taking part in the generation of the whole sound. Such consonants however are relatively few, the great majority of consonants are caused by changes which set up vibrations in some part or other of the vowel chamber or in the larynx itself; they are noises which may or may not be accompanied by voice, the vibrations producing a different consonant according as they are or are not accompanied by voice. Such vibrations may be set up in several ways. In the case of many consonants vibrations are set up by the passage being closed and then suddenly opened at a particular part; such consonants are spoken of as *explosives*. In the case of these consonants the noise which is their essential part cannot be prolonged, it is momentary in duration, though when it is accompanied by voice the latter may continue for some time. In the case of other consonants the noise is caused by the rush of air through a narrow space and the consonants may be described as 'frictional', or the noise may be produced by the vibratory movements of particular parts. Both these kinds of consonants can naturally be prolonged for an indefinite time, and have been called *continuous* consonants.

On the other hand the characters of a consonant are also dependent on the particular part of the passage at which they are generated; and this also may be used as a means of classification. Some consonants are produced at the lips by the movement or position of the lips in reference to each other or to the teeth; these are called *labial* and *labio-dental* consonants. Others again are produced by the movement or position of the tongue in reference to the teeth or to the teeth and hard palate; these are called *dental*. Yet others are produced by the movement of the back of the tongue in relation to the soft palate and pharynx or fauces; these are called *guttural* consonants, the name being also applied to certain consonants which are essentially noises generated in the larynx itself. These names are useful for a general broad classification; the term dental however is used to include consonants which are formed by the tongue in relation to, not the teeth, but the front part of the hard palate; hence it is to that extent open to objection. There are also other classifications into which we cannot enter here.

§ 919. When the various languages of the world are examined the number of consonants, that is to say of sounds used in speech and having the characters on which we are dwelling, is found to be very large; and concerning the nature and mode of formation of many of them much discussion has taken place. We must content ourselves here with very briefly indicating the chief facts concerning the mode of formation of the most important and common.

The group of consonants represented by M, N, NG, are very closely allied to vowels. In each of these as in a vowel the larynx is thrown into vibrations; but instead of the vibrations passing out by the mouth through a passage which has assumed a form belonging to this or that vowel, the passage to the mouth is closed, and the vibrations find their way out through the nasal cavity which acts as a resonance chamber. When a vowel is sounded the soft palate either completely shuts off the nasal cavity from the vowel chamber, or at least offers such resistance that an insignificant proportion of the expiratory blast passes into the nasal cavities. A vowel may be sung powerfully, and yet a flame exposed to the nostrils only will shew no movements; in the case at least of some vowels however a piece of cold polished steel will become dim, shewing that some air is passing through the nostrils. When the communications between the nasal and pharyngeal cavities are sufficiently free, and the other conditions are favourable for the nasal cavities to act as a resonance chamber, the vowel sounds are apt to take on a nasal character; and this occurs more readily when the vowel is said than when it is sung. In the group of consonants in question the nasal cavities become all important, the passage through the mouth being blocked. In M the passage is closed by shutting the lips, in N by the application of the tongue to the front of the hard palate and upper teeth, in NG by the application of the tongue to the soft palate.

While in the above group no new vibrations are added to the laryngeal vibrations, in the ordinary L which like them is based on laryngeal sounds, new vibrations, constituting a noise, are added. The passage is not completely, only partially closed; the front of the tongue is pressed against the hard palate in such a way that the passage is blocked in the middle but the air escapes through narrow channels on each side. It is the noise caused by the rush of air through these narrow spaces which added to the voice produces the sound we distinguish as L. In certain forms of L, for instance in the Welsh ll, the noise is not accompanied by laryngeal vibrations..

R is also allied to the above in so far as it too needs voice, and is based on laryngeal vibrations. But these vibrations are modified in a special way; they are rendered intermittent by vibratory movements started in some part of the passage, there being different kinds of R according as the interruption takes place at the tongue, or at the fauces. The common R is produced by the vibrations of the point of the tongue raised against the front of the hard palate, and the guttural R by the vibrations of the uvula against the root of the tongue. In the feeble English R there appear to be no vibrations; the vowel chamber is simply narrowed in front by the tip of the tongue.

It will be observed that L and the common R resemble each other to a considerable extent in the position of the tongue; the chief difference is that in L the tongue is not itself the subject of muscular movement, and the vibrations are produced by the friction of the expiratory blast through the narrow channel, whereas in R the vibratory interruptions are produced by the movements of the tongue. If in pronouncing L the tongue is suddenly set in movement, or in pronouncing R the tongue is suddenly arrested in its movements while in approximation to the palate, the one consonant is changed into the other; and, as is well known, certain persons, for instance the Chinese, are apt to use the one instead of the other.

The *explosives* differ according to the part where the interruption takes place; and in each kind of interruption the sound is different and receives a different name according as voice is used or no. P is uttered when the lips, being first closed and an expiratory blast driven against them, are suddenly opened. During this act no voice is uttered. If voice is uttered the P becomes B. These are labial explosives. In T, the interruption which is suddenly removed is caused by the application of the tongue to the front part of the hard palate in the case of the English, to the teeth in the case of most other languages; it is called a dental explosive, dental being used in the wide meaning stated above. With T, there is no voice; if voice be added the sound becomes D. Since P differs from B, and T from D, only in the absence or presence of voice, the removal or addition of voice will at once convert in each case the one consonant into the other, and by certain nations P and B are used the one for the other, as are also T and D.

It will be observed that B and D, both with voice, have certain relations to M and N respectively. In B and M the action is labial, in D and N dental, voice being present in all; the difference is that in M and N, the action consists in the establishment of an obstruction in the buccal passage, in B and D in the sudden removal of an obstruction. If there be, as in nasal catarrh, an adequate obstruction to the exit through the nasal passages of the expiratory blast which creates the sound, it becomes difficult if not impossible to establish the obstruction in the mouth, since in that case there is no exit at all for the expiratory blast. Hence in nasal catarrh there is a tendency for the effort to pronounce M to result in B, and that to pronounce N in D; 'name' becomes 'dabe.'

If the tongue be brought to the back instead of to the front of the hard palate the consonant K (hard C) is uttered; if voice be added the sound is G (hard). These are guttural explosives. Allied to them is the brief sound which in certain cases inaugurates a vowel and which, due to the sudden opening of the closed glottis, is immediately followed by the vibrations of the cords and so by the true vowel sound. This is the *spiritus lenis* as distinguished from H or the *spiritus asper*, of which, formed as it is in a different manner, we shall speak directly.

Certain other consonants are continuous, and like L are formed by the rush of air through a constriction formed somewhere in the passage; they are frictional in origin. They differ however from the ordinary L in that they are not always accompanied by voice; like the explosives they may be uttered without any vibrations of the vocal cords, and when these do accompany the frictional sound the consonant is altered in its characters and receives another name. As in the case of the explosives, in forming the different members of the group, the vibrations giving rise to the sound are started in different parts of the passage, at the lips, at the teeth or hard palate, or at the fauces.

When the constriction is caused by the lip being brought into contact with the teeth (and generally the lower lip and upper teeth are used), so as to reduce the outlet to a narrow space, the vibrations started at the constriction give rise to F, when no voice is uttered at the same time. If voice be also uttered the F becomes V. If the teeth take no part in the constriction, and this be made exclusively by the two lips, the vowel chamber at the same time assuming the shape proper to the vowel U, the sound if voice be uttered is *W*; the English *W* and the allied French *ou* (in *oui*) may be regarded as the vowel *U* (in two different forms) turned into consonants. The sound which is formed by the two lips alone, in the absence of voice is the English (North Country) *Wh*.

When the constriction is formed between the tongue and the teeth in such a way that the tip of the tongue protrudes between the partially open rows of teeth the sound is called *Th*: a 'hard' Th as in 'thin' if without voice, a 'soft' *Th* as in 'this' if with voice. The effect of this manœuvre does not differ greatly from that of forming F, and certain persons in attempting to pronounce a hard Th give utterance to *F*, as in the cockney 'nuffin.'

When the constriction takes place between the tongue and the teeth in such a way that a narrow channel is formed between the upper incisors and the tip of the tongue curved into a groove the sound is called S (soft C) if without voice, and Z if with voice. If the constriction be formed a little farther back behind the front teeth by the approximation of the tongue to the front of the hard palate, the sound uttered without voice is Sh; if voice be added the sound becomes the French j, which we represent by z as in 'azure' or by g as in 'badger.'

If instead of being formed by the teeth the constriction be carried farther back from the region of the teeth and hard palate to that of the soft palate and fauces, a guttural aspirate is formed. Without voice this is the hard Ch as in the Scotch 'loch,' with voice the soft Ch.

Y appears to be the vowel I (ee) used as a consonant, much in the same way that, as stated above, W is U used as a consonant.

Lastly, a consonantal sound may be formed by the glottis itself supplying a constriction but in such a way that the vocal cords are not thrown into musical vibrations. When in uttering a vowel we begin with the glottis not closed as in the *spiritus lenis* but open, and send through the glottis an expiratory blast which creates irregular vibrations by friction before the cords are brought into a proper position for their regular vibrations, the result is the aspirate H, the *spiritus asper*. The particularly powerful H of Arabic is produced by bringing the processus vocales into contact with or near to each other but so as to leave the cartilaginous glottis widely open and the membranous glottis more or less open; at the same time the ventricular bands are approximated, and the superior aperture of the larynx is forcibly constricted. The expiratory blast driven through the series of irregular passages gives rise to the irregular vibrations which constitute the sound, the vocal cords being motionless or at least not giving rise to the regular vibrations of voice.

We have seen that in whispering no true voice is uttered, no regular vibrations are generated in the vocal cords, though the passage of the air through the glottis produces vibrations which serve as the basis of the whisper, being modified by the vowel cavity so as to form vowels. To these vibrations may be added the vibrations of the consonants, so that a whisper becomes complete though feeble speech. Since the irregular glottic vibrations of a whisper are very weak compared with the relatively powerful true vocal vibrations, the distinction between consonants with voice and without voice is in a whisper largely obscured; it is difficult for instance in a whisper to distinguish between P and B.

SEC. 3. ON SOME LOCOMOTOR MECHANISMS.

§ 920. The skeletal muscles are for the most part arranged to act on the bones and cartilages as on levers, examples of the first kind of lever being rare, and those of the third kind, where the power is applied nearer to the fulcrum than is the weight, being more common than the second. This arises from the fact that the movements of the body are chiefly directed to moving comparatively light weights through a great distance, or through a certain distance with great precision, rather than to moving heavy weights through a short distance The fulcrum is generally supplied by a (perfect or imperfect) joint, and one end of the acting muscle is made fast by being attached either to a fixed point, or to some point rendered fixed for the time being by the contraction of other muscles.

There are indeed few movements of the body in which one muscle only is concerned; in the majority of cases several muscles act together in concert; the movements of the larynx which we have just studied afford a striking illustration of this. The relations of the muscles which thus act together are many and varied. When one muscle is contracting, the contractions of another muscle or of other muscles may, as just stated, serve to secure a fixed point, or may enforce the effect of the first muscle, or, and this is perhaps the most common case, may give a special direction to the action, the movement effected being the resultant of the forces employed in combination. Many muscles are, either partially or wholly, antagonistic in action to each other, such for instance as the flexors and extensors, and such muscles as those of the face which act bilaterally in opposite directions on parts placed in the middle line; and the relations of these antagonistic muscles seem to be specially complex. When a muscle contracts it is, as we saw in treating of nerve and muscle, of advantage that the muscle should at the moment of contraction be already "on the stretch;" this is provided by the anatomical disposition of the parts assisted probably, as we saw (§ 597), by skeletal tone, but is also further secured by the action of its antagonists, which moreover, after the muscle has contracted, assist its return to its proper position of rest. When a point has to be fixed the two sets of muscles which act as antagonists on the point are both thrown into contraction, in proportion to their relative effect. If the action of one muscle (or set of muscles) is to be dominant its antagonist may take no part in the action, being neither contracted nor relaxed; but there are reasons for thinking that, in many cases at all events, the action of one muscle though remaining dominant is tempered and guarded so to speak by a concomitant feebler action of its antagonist; there is no satisfactory evidence of the occurrence of a relaxation in the antagonist. These several phases are governed by the nervous system, and the behaviour of antagonistic muscles and groups of muscles affords many instances of what we have so often insisted upon, namely, that nearly all the various movements of our body are coordinate movements, and that in many cases the coordination is extremely complex.

§ 921. The *erect posture*, in which the weight of the body is borne by the plantar arches, is the result of a series of contractions of the muscles of the trunk and legs, having for their object the keeping the body in such a position that the line of gravity falls within the area of the feet. That this does require muscular exertion is shewn by the facts that a person when standing perfectly at rest in a completely balanced position falls when he becomes unconscious, and that a dead body cannot be set on its feet. The line of gravity of the head falls in front of the occipital articulation, as is shewn by the nodding of the head in sleep. The centre of gravity of the combined head and trunk lies at about the level of the ensiform cartilage, in front of the tenth thoracic vertebra, and the line of gravity drawn from it passes behind a line joining the centres of the two hip-joints, so that the erect body would fall backward were it not for the action of the muscles passing from the thighs to the pelvis assisted by the anterior ligaments of the hip-joints. The line of gravity of the combined head, trunk and thighs falls moreover a little behind the knee-joints, so that some, though little, muscular exertion is required to prevent the knees from being bent. Lastly, the line of gravity of the whole body passes in front of the line drawn between the two ankle-joints, the centre of gravity of the whole body being placed at the end of the sacrum, hence some exertion of the muscles of the calves is required to prevent the body falling forwards.

§ 922. In *walking* advantage is taken of this forward position of the centre of gravity, and the tendency to fall forwards is utilised to swing each leg in turn forwards after the fashion of a pendulum. In each step there is a moment at which the body is resting vertically on one leg, say the right, while the other is inclined obliquely behind. The two legs and the plane of the ground form a right-angled triangle, of which the left leg is the hypothenuse, the right angle being between the right leg and the ground. At a certain moment the foot of the right leg will be flat on the ground and the line of gravity will pass through its heel. But the centre of gravity is moving forwards; even if there had been no previous steps, and so no momentum, the body and with it the centre of gravity, unless prevented by muscular effort, would have fallen forward; we may therefore speak of the line of gravity as travelling forwards; it passes from the heel to the toe (of the right foot). If the body were simply falling forwards the centre of the hip-joint would move downwards as well as forwards, describing a circle with the leg as a radius. But at the moment of which we are speaking the (right) leg is somewhat flexed, both at the ankle and still more at the knee. And, as the line of gravity is travelling forward from the heel to the toe, the active part of the performance intervenes. The foot is raised from the ground from the heel forwards, until it is only the ball of the great toe which is resting on the ground, and the whole leg is, by muscular effort, straightened. In this act the right leg acts as a lever, the ball of the great toe serving as a fulcrum; and the effect of the act is to prevent the centre of gravity, or the hip-joint, from moving downwards, and to carry it forwards only in more nearly a straight line. In thus carrying the hips (and body) forward the leg has changed its position; from being vertical and flexed with the whole sole resting on the ground, it has become inclined forwards obliquely, extended straight, with the toes only resting on the ground. It has assumed the same posture as that of the left leg at the moment at which we started.

Even at that moment the left leg was behind the line of gravity, and unless it moved would become more and more so as the changes in the right leg went on; hence if left to itself it would swing forward much as a pendulum which had been raised up would swing forward when let go. And during the changes in the right leg which we have just described the left does swing forward, its movement being chiefly determined like that of a pendulum by gravity, though it may be assisted by direct muscular effort, and is certainly so guided, being for instance slightly flexed during the transit. It swings forward in front of the line of gravity and is thus brought to the ground, the toes in proper walking making contact first and the heel later, though many people who wear shoes bring the heel down at least as soon as the toes. It swings we say in front of the line of gravity; but that line of gravity is travelling forwards, so that in a very short time the body is resting vertically on the left leg, with the line of gravity falling at the left heel. That is to say, the left leg has now assumed the position which the right leg had when we began; meanwhile, as we have seen, the right leg has assumed the former position of the left leg; the step is completed, and the movements of the next step merely repeat those of the one which we have described.

It is obvious from the above that in walking there are in each step periods when both feet are touching the ground, and periods when one or the other foot is raised from the ground, but there is no period when both feet are off the ground. This is shewn in the diagram, Fig. 192, which represents two steps.

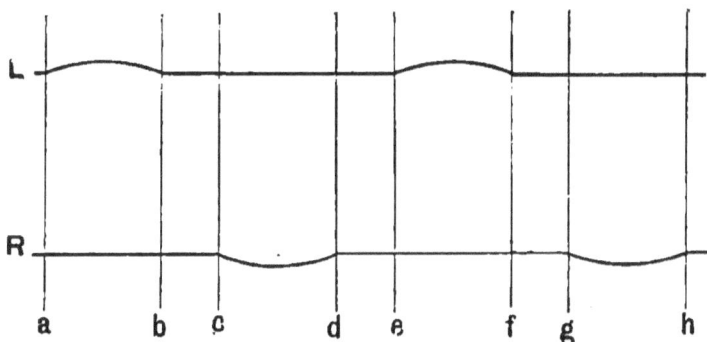

FIG. 192. DIAGRAM TO ILLUSTRATE THE CONTACT OF THE FEET WITH THE GROUND IN WALKING.

R, the right foot. L, the left foot. In each case the curved line represents the time when the foot is not in contact with the ground, and the straight line when it is in contact.

During $a - b$, the left leg (L) leaves the ground as indicated by the curving of the line. During $b - c$ both feet are on the ground. During $c - d$ the right leg (R) is above but the left (L) is still on the ground. During $d - e$ both are on the ground and the double step is completed, the next step beginning again at e with the left leg leaving the ground.

We have said that the centre of gravity is in walking prevented from moving downwards as well as forwards, as it would do in the act of falling forwards. It does not however describe a straight line forwards, it, and with it the top of the head, rises and falls at each step of each leg, and hence describes a series of consecutive curves not unlike the line of flight of many birds.

Since in standing on both feet the line of gravity falls between the two feet, a lateral displacement of the centre of gravity is necessary in order to balance the body on one foot. Hence in walking the centre of gravity describes not only a series of vertical, but also a series of horizontal curves, inasmuch as at each step the line of gravity is made to fall alternately on each standing foot. While the left leg is swinging, the line of gravity falls within the area of the right foot, and the centre of gravity is on the right side of the pelvis. As the left foot becomes the standing foot, the centre of gravity is shifted to the left side of the pelvis. The actual curve described by the centre of gravity is therefore a somewhat complicated one, being composed of vertical and horizontal factors.

The natural step is the one which is determined by the length of the swinging leg, since this acts as a pendulum; and hence the step of a long-legged person is naturally longer than that of a person with short legs. The length of the step however may be diminished or increased by a direct muscular effort, as when a line of soldiers keep step in spite of their having legs of different lengths. Such a mode of marching must obviously be fatiguing, inasmuch as it involves an unnecessary expenditure of energy.

In slow walking, which Fig. 192 may be taken to illustrate, there is an appreciable time during which, while one foot is already in position to serve as a fulcrum, the other, swinging, foot has not yet left the ground. In fast walking this period is so much reduced, that one foot leaves the ground the moment the other touches it; hence there is practically no period during which both feet are on the ground together; this might be shewn by omitting $b-c$ and $d-e$ in Fig. 192.

When the body is swung forward on the one foot acting as a fulcrum with such energy that this foot leaves the ground before the other, swinging, foot has reached the ground, as shewn in Fig. 193, there being an interval, $b-c, d-e$ in the figure, during

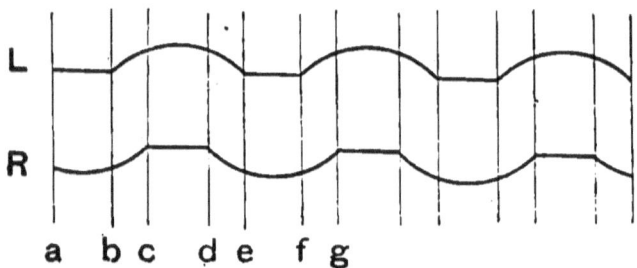

FIG 193 DIAGRAM TO ILLUSTRATE RUNNING.

L, the line of contact with the ground of the left, R of the right foot; in each case the curved portion of the line represent the time during which the foot leaves the ground.

which neither foot is on the ground, the person is said to be *running*, not walking.

In jumping this propulsion of the body takes place on both feet at the same time; in hopping it is effected on one foot only.

BOOK IV.

THE TISSUES AND MECHANISMS OF
REPRODUCTION.

REPRODUCTION.

§ **923.** MANY of the individual constituent parts of an animal body are capable of reproduction, *i.e.* they can give rise to parts like themselves; or they are capable of regeneration, *i.e.* their places can be taken by new parts more or less closely resembling themselves. The elementary tissues undergo during life a very large amount of regeneration. Thus the old epithelium scales which fall away from the surface of the body are succeeded by new scales from the underlying layers of the epidermis; old blood-corpuscles give place to new ones; worn-out muscles, or those which have failed from disease, are renewed by the accession of fresh fibres; divided nerves grow again; broken bones are united; connective tissue seems to disappear and appear almost without limit; new secreting cells take the place of the old ones which are cast off; in fact, with the exception of some cases, such as cartilage, and these doubtful exceptions, all those fundamental tissues of the body which do not form part of highly differentiated organs are, within limits fixed more by bulk than by anything else, capable of regeneration. To that regeneration by substitution of molecules, which is the basis of all life, is added a regeneration by substitution of mass.

In the higher animals regeneration of whole organs and members, even of those whose continued functional activity is not essential to the well-being of the body, is never witnessed, though it may be seen in the lower animals; the digits of a newt may be restored by growth, but not those of a man. And the repair which follows even partial destruction of highly differentiated organs, such as the retina, is in the higher animals very imperfect.

In the higher animals the reproduction of the whole individual can be effected in no other way than by the process of sexual generation, through which the female representative element or ovum is, under the influence of the male representative or spermatozoon, developed into an adult individual.

We do not purpose to enter here into any of the morphological problems connected with the series of changes through which the ovum becomes the adult being; or into the obscure biological inquiry as to how the simple, all but structureless ovum contains within itself, in potentiality, all its future developments, and as to what is the essential nature of the male action. These problems and questions, which are fully discussed in other works, do not properly enter into a work on physiology, except under the view that all biological problems are, when pushed far enough, physiological problems. We shall limit ourselves to a brief survey of the more important physiological phenomena attendant on the impregnation of the ovum, and on the nutrition and birth of the embryo, incidentally calling attention to some of the leading structural features of the parts concerned.

CHAPTER I.

IMPREGNATION.

SEC. 1. ON SOME STRUCTURAL FEATURES OF THE FEMALE ORGANS.

The uterus and its appendages.

§ 924. THE uterus is a somewhat flask-shaped, hollow, muscular organ, of which the wider upper part, or *body*, is continued above on each side as a narrow tube, the *Fallopian tube*, the roof between being called the *fundus*, while the narrow lower end, *neck* or *cervix*, projects into the cavity of the vagina, and ends in a transverse slit-like orifice, the *os uteri*, distinguished as the *os externum* from the *os internum* or constriction marking the junction of body and cervix.

The Fallopian tube, thin and slender where it joins the uterus, into the cavity of which its canal opens by a very narrow orifice, runs in the free margin of the broad ligament in a horizontal course which is near the uterus straight but afterwards wavy, and ends in a trumpet-shaped mouth opening free into the peritoneal cavity. Like that of the alimentary canal, the wall of the tube consists of an outer muscular coat, covered over the larger part of its circumference by peritoneum, and of an inner mucous coat. The muscular coat is composed of plain muscular fibres, for the most part disposed circularly, with more scanty longitudinal or oblique bundles on the outside. The mucous coat consists of a vascular dermis lined with a single layer of columnar ciliated cells. No glands are present, or if present are inconspicuous, but the mucous membrane is thrown into a large number of irregular folds, giving the surface a glandular appearance.

The margin of the trumpet-shape end or mouth is not even but cut up radially into four or five large and many small processes, or *fimbriæ*; one of the larger processes, more conspicuous than the rest and directly connected with the ovary by a ligament is called the "ovarian fimbria." The inner surface of each of the larger fimbriæ is rendered irregular by folds of the mucous membrane, continuations of the folds of the canal of the tube, which are so arranged, especially in the ovarian fimbria, that the fimbriæ form shallow grooves converging to the opening of the canal of the tube in the centre of the trumpet; muscular fibres, disposed longitudinally, also pass into these larger fimbriæ. The epithelium on the fimbriæ is like that of the canal of the tube, columnar and ciliated; at the edges of the fimbriæ it passes into the flat epithelioid plates of the peritoneum. When examined in the living body the end of the tube is seen to curl round the adjoining ovary so that the fimbriæ embrace that organ from below.

§ 925. The uterus, the body of which is covered by peritoneum except at the attachment of the ligaments or peritoneal folds which keep it in place, also consists of a muscular coat of plain muscular fibres and a mucous coat, but both coats are much more complex than in the tube. The muscular coat is more or less distinctly divided into an outer thin and an inner thick and much more conspicuous layer, between which there is a large development of blood vessels, especially of veins, which here form large venous sinuses. In the thin outer layer the fibres are disposed longitudinally on the outside and circularly or obliquely on the inner side. The thick inner layer consists of fibres which are disposed in the main circularly, but still to a large extent obliquely and indeed in various directions.

The mucous coat is very thick, the thickness being due to the presence of numerous generally simple glands, each having a wide lumen and tortuous course. The epithelium lining the glands and the surface between their mouths is a single layer of tall columnar ciliated cells with intervening goblet cells, the cilia being very tender, easily destroyed and overlooked. The dermis of the mucous membrane is very vascular, and is characterised by the presence of a large number of spindle-shaped cells, some of which at least appear to be epithelioid plates lining the numerous lymph spaces present in the connective tissue. Another special feature of the dermis is that the line of demarcation between it and the adjoining inner layer of the muscular coat is not a sharp one; bundles of muscular fibres radiate from the one into the other, so that the two are irregularly interlaced together. The thick inner layer of the muscular coat, distinguished from the outer thin one, has thus some of the characters of a greatly hypertrophied muscularis mucosæ.

The above description applies to the body of the uterus. In the cervix or neck, the muscular fibres have a more regular, less interlaced disposition, and a marked difference is observed in the mucous membrane. The glands suddenly become shorter, branched, and changed in character, the cilia of their epithelium being often absent, while the surface is thrown up into a number of ridges, which radiate in an arborescent manner from two main longitudinal folds, one on the anterior, the other on the posterior surface.

The uterine glands secrete a considerable quantity of viscid mucus, alkaline or neutral in reaction. Since the cavity of the uterus is so narrow and slit-like as to be almost potential rather than actual, this mucus is found chiefly at the cervix and in the vagina.

§ 926. At the os uteri there is a somewhat sudden transition from the columnar or finally cubical epithelium, arranged in a single layer, to the epithelium lining the vagina, which like that of the mouth and pharynx is a stratified epithelium consisting of several layers of nucleated cells, the uppermost of which are flattened. The lining membrane of the vagina, in fact, though it is spoken of as a mucous membrane, is in many points allied to the skin ; and its dermis, like that of the skin, is rich in elastic elements, and bears numerous simple or compound papillæ, containing vascular loops and nervous endings. A special feature is the presence of numerous permanent ridges, *rugæ*, which contain venous plexus, supported by interlacing bundles of muscular fibres. Though the matter is disputed the evidence goes to shew that the vaginal walls do not possess glands and do not themselves secrete mucus; such fluid as is present has an acid reaction.

Outside this mucous coat is a muscular coat of plain muscular fibres, with an outer longitudinal and inner circular layer, and outside this again a fibrous coat. Between the fibrous and muscular coat, as well as between the muscular coat and mucous lining, in the situation of the submucosa, are numerous venous plexuses ; the walls of the vagina, in fact, are very vascular, and largely of the nature of erectile tissue.

§ 927. The uterus is mainly supplied with blood by the uterine arteries, one on each side, coming from the anterior division of the internal iliac artery, but it also receives blood from a branch of the ovarian artery, coming direct from the aorta. The blood finds its way back from the uterus by the uterine and ovarian veins, that from the venous plexuses of the vagina falling into the internal iliac.

The lymphatics of both uterus and vagina are abundant, numerous plexuses, lymph capillaries and lymph spaces being present in both the muscular and mucous coats.

Concerning the nerves of the uterus, we will speak later on.

The Ovary.

§ 928. In the embryo, at a particular spot on each side of the cavity which is called the body cavity, and which becomes in the adult the peritoneal cavity, the epithelium lining the cavity undergoes a remarkable differentiation. Elsewhere a single layer of flat cells, it here thickens into several layers, the cells being cubical or rounded, and some of them being distinguished from the rest by their size and other features. This thickened area is called "germinal epithelium," the special cells which it contains are "primordial ova," and the patch of differentiated epithelium with the underlying connective tissue becomes the ovary.

In the course of development the line of demarcation between the epithelial and the mesoblastic or connective tissue elements is broken up by the ingrowth of the one into the other; and this takes place in such a way that nests of undifferentiated epithelial cells surrounding a primordial ovum, or it may be several ova, are marked out by investments of connective tissue. These nests become what are called *Graaffian follicles*. A typical Graaffian follicle consists at an early stage of a central cell, which by its relatively large size, ample cell-body and conspicuous nucleus is marked out as an ovum, surrounded by a layer of smaller cubical or columnar or rounded cells, which in turn are enveloped by mesoblastic elements on their way to be developed into vascular connective tissue. The patch of germinal epithelium thus becomes developed into a *germinal layer*, consisting of a number of Graaffian follicles embedded in connective tissue and covered over towards the peritoneal cavity by a single layer of epithelium cells, which by the cubical form and more deeply staining cell-substance of its cells is still distinguished as germinal epithelium from the rest of the epithelium lining the peritoneal cavity, the cells of which take on the form of flat epithelioid plates (§ 289).

§ 929. In the adult the ovary is an oval swelling, which bulges out from the back of the broad ligament on each side of the uterus, between that organ and the mouth of the Fallopian tube. It is placed with its long axis nearly horizontal, and along the middle of its front surface is attached to the broad ligament by what is called the *hilus*, the attachment passing at the median end into the ligament of the ovary connecting the ovary with the uterus, and at the opposite outer end becoming connected with the ovarian fimbria of the mouth of the Fallopian tube. Elsewhere the surface projects free into the peritoneal cavity, and is covered with an epithelium which differs from the ordinary epithelium lining the peritoneal cavity by the constituent cells being cubical and granular; this epithelium is in fact the remains of the germinal epithelium spoken of above. At the attachment of the ovary these cubical cells pass suddenly into the epithelioid plates of the ordinary peritoneal epithelium.

The body of the ovary consists on the one hand of Graaffian follicles of different sizes and in different phases, and on the other hand of a stroma, or connective tissue basis, in which the former are embedded. The smaller, immature follicles are aggregated in a zone, lying beneath the superficial epithelium of the free surface of the ovary, but separated from it by a layer of stroma, the *tunica albuginea ;* this 'cortical layer,' or 'germinal zone,' corresponds to the germinal layer spoken of above. The larger follicles are distributed throughout the thickness of the ovary, but are absent from the hilus and from a wedge-shaped strand of stroma, which at the hilus passes from the broad ligament into the ovary.

In this strand the connective tissue is for the most part ordinary connective tissue, made up of the usual fibrillated bundles, and carries the blood vessels passing to the ovary from the broad ligament. Some plain muscular fibres are present among the connective tissue bundles; and here and there in this situation are seen in sections of the ovary tubes cut in various planes and lined with short cubical or flat cells; these are parts of the "parovarium" derived from the Wolffian body. Elsewhere, between and among the follicles and in the tunica albuginea, the connective tissue has peculiar characters; it consists entirely of spindle-shaped nucleated cells interwoven together; these have a superficial resemblance to plain muscular fibres, but are true connective tissue elements; they are also present to a certain extent in the strand just mentioned.

§ 930. In a Graaffian follicle which is large but not yet fully developed and mature, one may recognize the following parts. On the outside the follicle is defined by an envelope, *theca,* of stroma, with its spindle-shaped cells and blood vessels. This is limited internally by a delicate structureless membrane, the *membrana propria,* and within this lies an epithelium consisting of two or more layers of cells constituting the *membrana granulosa ;* the cells next to the membrana propria are columnar, the rest are cubical or spheroidal. At one part the membrana granulosa expands into a rounded mass of cells, bulging inwards towards the centre of the follicle; and in the midst of this mass of cells, called the *cumulus* or *discus proligerus,* lies the highly developed cell which is the *ovum.* The outline of the ovum is defined by an investing membrane, which if the follicle be old enough has a distinct double contour and is called the *zona pellucida* or, because it is marked by radiating lines or striæ, *zona radiata ;* the cells of the cumulus touching this are columnar in shape, and arranged in a radiate manner. Within this membrane lies the cell body of the ovum, which in the larger follicles has become granular by the formation of yolk or *vitellus.* Placed, more or less excentrically, in the cell substance is a large transparent spherical nucleus, the *germinal vesicle,* in which may be distinguished a nuclear membrane, a nuclear network with more fluid nuclear contents, and either one or more than one conspicuous nucleolus or *germinal spot.*

If the follicle be one well advanced in development, a considerable portion of its interior will be occupied by fluid, into which the cumulus proligerus projects. If the follicle be a smaller one the fluid may be very scanty, or absent altogether, there being no distinction between the cumulus and the rest of the membrana granulosa, of which the former is merely a part. If the follicle be very small, as in the case of the majority of those forming the germinal zone, the membrana granulosa will be reduced to a single layer of cells surrounding the ovum, the zona pellucida will be absent, and the investment of stroma will be very delicate, but the ovum will still be recognized by its large cell-substance and its already conspicuous nucleus or germinal vesicle.

§ 931. In a ripe or nearly ripe follicle of a human ovary, the ovum measures about ·2 mm. in diameter The cell body may be considered as consisting on the one hand of undifferentiated protoplasmic cell-substance which, as in so many other cases, presents a reticulate appearance under the influence of reagents, and on the other hand of a multitude of refractive spherules or granules of varying size imbedded in the former and constituting the yolk proper. These yolk granules, which are more scanty in the immediate neighborhood of the germinal vesicle and at the periphery of the cell-substance than elsewhere, are, if we may judge by analogy from the results of the examination of eggs such as that of the hen in which the yolk is massive, largely composed of three chemical substances, namely, of *vitellin*, which is a proteid having the general characters of the globulin group but also special features of its own, of *lecithin* (§ 71) and of the ill-understood substance or substances known as *nuclein* (§ 29). The chemical history of the yolk is however at present very imperfectly known; but we may conclude that the yolk granules if they do nothing else, at least serve as food material of a highly nourishing character for the real living substance of the ovum. The zona radiata may probably be regarded as a cuticular product of the cells of the cumulus, the surfaces of which next to it are irregular, presenting numerous delicate processes. In some animals at least the striæ are due to minute canals occupied by minute threads of a protoplasmic nature which probably serve to establish continuity between the cell-substance of the follicular cells and that of the ovum; but it appears doubtful whether, at least in the ripe follicle, these canals are present in the human ovum, and it is maintained that in the ripe ovum not only is the inner surface of the zona smooth but that a space containing fluid always exists between it and the surface of the ovum. There can be no doubt that the cells of the cumulus do in some way feed the ovum; but if no canals are present in the zona radiata, we may infer that the material is conveyed through the zona in a dissolved state. and that the yolk granules are formed by a subsequent activity of the cell substance of the ovum comparable to that which leads to the appearance of granules in a secreting cell; the ovum transforms and deposits in itself in the form of yolk granules for future use the surplus material which it receives from the follicle. One purpose of the zona radiata appears to be to afford mechanical protection to the ovum, but it also probably exercises some influence on this transmission of material; and in this respect the likeness which it presents to the striated border of the villi of the intestine is suggestive. Outside the zona radiata between it and the cells of the cumulus may be more or less distinctly recognized a layer of granular material which, secreted apparently by the cells of the cumulus, seems to be the analogue of the albumen or 'white' of the hen's egg. We know little definite concerning the nature of the fluid filling the ripe follicle, but it is probably of the nature of lymph; in ovarian cysts, which are abnormal developments of follicles, special forms of proteids have been found.

§ 932. The growth of a follicle from its earliest phase onwards consists, then, partly in an increase in the ovum itself; the cell-substance of this becomes larger and increasingly loaded with yolk, the nucleus or germinal vesicle and its nucleoli or germinal spots become increasingly conspicuous, and during part of the time the zona pellucida acquires increasing thickness; but the growth also consists in an enlargement of the walls of the follicle and an increase of the cells of the membrana granulosa, as well as in a distension of the enlarging follicle by fluid which appears to be secreted by those cells.

In an ordinary epithelium, such as that of the skin or mucous membrane, each constituent cell is nourished by the lymph which reaches it from the blood vessels of the underlying dermis, and each cell influences or is influenced by its fellows only so far as to secure that all the cells should work together in due order. In the germinal epithelium certain cells, namely the ova, acquire special properties and are set apart for a special end. By the introduction of what special material, or by the bringing to bear of what special influences they are thus set apart, is a problem too wide for us to consider here; they are thus set apart. And it would appear that having been thus set apart, they pursue a career different from that of other cells. They do not multiply in the same simple way that other cells multiply; as far as can be ascertained, the number of ova in an ovary is early established, and not subsequently increased. And their nutrition is of such great value that the lives of neighbouring cells are devoted to securing it; the ordinary cells of the germinal epithelium surrounding an ovum are told off to nourish it, and so form the membrana granulosa of a follicle. The nutritive assistance given by these subsidiary cells is twofold: they secrete the lymph-like follicular fluid, by which the ovum is indirectly nourished; and those cells which form the cumulus appear to convey material in a more direct way to the ovum, loading its cell-substance with substances which form the yolk.

§ 933. The cumulus, with the ovum within it, is placed, as we have said, excentrically in the cavity of the follicle, and indeed is so placed that the ovum lies in the region of the follicle nearest to the surface of the ovary. As the follicle in course of growth becomes larger and more distended, it not only occupies a considerable portion of the whole bulk of the ovary, but encroaches on the germinal zone owing to the wall of the follicle near which the ovum lies becoming more and more superficial in position. This part of the wall also becomes thinner and thinner, and ultimately bursts, the ovum with the cumulus being discharged from the interior of the follicle, and received, as we shall see, into the mouth of the adjoining Fallopian tube.

Before this ejectment occurs, changes other than those of mere growth or enlargement take place in the ovum itself. Into those changes we cannot enter here, and must merely content ourselves with stating that the nucleus and other parts of the ovum-cell behave very much as if the ovum were about to divide; there are movements and transformations of the nuclear elements resembling in many ways those seen in a cell about to multiply by division; but these do not issue in an actual division of the ovum, they lead merely to the casting out of a portion of the old nucleus, the germinal vesicle, in the form of bodies known as 'polar globules,' and to the conversion of the remainder into a new nucleus, the *pro-nucleus*. To carry on these changes to a division of the ovum and cellular multiplication, a new influence must be brought to bear, that of the male element in impregnation.

§ 934. The bursting of the follicle in the discharge of the ovum is accompanied by rupture of the small blood vessels of the follicular walls and, at times but not always, blood escapes into the cavity of the collapsed follicle. A new growth now takes place; the cells of the membrana granulosa multiply rapidly, the connective tissue elements of the walls also multiply, sending vascular ingrowths into the interior, until the follicle becomes a solid mass of cells traversed by radiating septa of stroma, the centre being occupied, when hæmorrhage into the follicle has taken place, by the effused blood, the hæmoglobin of which is soon converted into hæmatoidin. After a while the growth of cells and stroma is arrested, and fatty degeneration of the cells takes place. The fat is coloured with a yellow pigment, *lutein*, about which there has been much dispute but which appears to be distinct from bilirubin; and this gives a yellow colour to the whole mass, which is now called a *corpus luteum*. The degeneration is followed by absorption, and ultimately a mere scar is left.

Whenever an ovum is discharged changes of this kind follow, but if the ovum be impregnated they are generally exaggerated. The membrana granulosa is much folded during its excessive growth, and the whole corpus luteum becomes a large and striking object; in sections of an ovary taken at a time when a corpus luteum succeeding pregnancy is at its maximum of growth, a large part of the whole bulk of the ovary will sometimes be seen to be taken up by it. In all cases the corpus luteum of pregnancy lasts a much longer time, passing more slowly through its several phases, especially those of degeneration, than does that of an unimpregnated ovum; this difference is much more constant than that of mere size.

SEC. 2. MENSTRUATION.

§ 935. From puberty, which may be said to occur at from 13 to 17 years of age, to the climacteric, which may be said to arrive at from 45 to 50 years of age, the exact time in each case varying considerably and being apparently determined by various conditions, the human female is subject monthly to a discharge from the vagina known as the 'menses,' 'catamenia,' and by many other terms. The discharge is the result of changes in the lining membrane of the uterus, and these are accompanied by changes in the ovary leading to the escape of an ovum from its Graaffian follicle into the Fallopian tube, as well as by certain general changes in the body at large. A similar change in the uterus, associated with a like escape of ova from the ovary, occurs in the lower animals, being repeated at intervals differing in length in different animals, and is usually accompanied by sexual excitement and changes in the external genital organs; the phenomena are then spoken of by such names as 'heat,' 'rut,' &c.

Leaving aside for the present the causation of the phenomena, we may regard as at least a conspicuous event in menstruation the escape of an ovum from its Graaffian follicle. The whole ovary at this time becomes congested, the blood vessels becoming so dilated and filled with blood that we may almost speak of the condition as one of erection; and the ripe follicle, whose ovum is about to escape, bulges from its surface. The most projecting portion of the wall of the follicle, which has previously become excessively thin, is now ruptured, apparently by the mere distension of the cavity, and the ovum, now lying close under the projecting surface of the follicle, escapes, invested by some of the cells of the discus proligerus, into the Fallopian tube. Much discussion has taken place as to how the entrance of the ovum into the Fallopian tube is secured. We have seen (§ 924) that probably under ordinary circumstances the ovary is embraced by the trumpet-shaped fimbriated mouth of the Fallopian tube, and the contact is probably rendered more complete by the turgid and congested condition of both organs; it is possible that the plain muscular fibres present in the mouth of the tube may assist, and indeed gliding movements of the mouth of the tube over the ovary have been observed in animals. It has, however, been asserted that the turgescence of the tube does not occur until after the ovum has become safely lodged in the tube, and it is argued that the ovum is carried in the proper direction by currents set up by the action of the ciliated epithelium lining the tube, currents whose direction and strength seem, as shewn by experiment, to be adequate to carry into the uterus particles present in the peritoneal fluid ; and the groove in the ciliated surface of the ovarian fimbria especially connected with the ovary, suggests itself as the natural path for the ovum.

Arrived in the tube, the ovum travels downwards very slowly, by the action probably of the cilia lining the tube, though possibly its progress may occasionally be assisted by the peristaltic contractions of the muscular walls. The stay of the ovum in the Fallopian tube may extend to several days ; the channel, as we have seen, is a narrow one, especially at the entrance into the uterus. The escape of the ovum is followed by changes in the follicle and rest of the ovary, which we have already described.

§ 936. The discharge of the ovum is accompanied, or rather, in part at least preceded, not only by a congestion or erection of the ovary and Fallopian tube, but also by marked changes in the uterus ; and it is to these that the obvious phenomena of menstruation are due. While the whole organ becomes congested and enlarged, the mucous membrane undergoes special changes. Not only are the blood vessels and lymphatic vessels and spaces distended with their respective fluids, and the glands turgid so that the whole membrane becomes red, thick and swollen, but also, according to some observers, a proliferation of the epithelial cells, and to a certain extent of the connective tissue elements, takes place. The change does not extend to the cervix, which thus presents a contrast to the body of the uterus. From the distended blood vessels blood escapes not only into the dermic connective tissue between the glands but also into the lumina of the glands and especially on to the free surface of the mucous membrane, the extravasation coming apparently chiefly from the capillaries and being due to a kind of diapedesis (§ 184) rather than to rupture. In this way there takes place from all parts of the swollen surface a hæmorrhagic discharge, often considerable in extent, which together with a mucous secretion furnished by the glands and containing a number of cells resembling leucocytes, constitutes the menstrual or catamenial flow. The blood as it passes through the vagina becomes somewhat altered, probably by the influence of the other constituents of the discharge, and when scanty coagulates but slightly ; when the flow however is considerable, distinct clots may make their appearance. The swollen and changed mucous membrane then undergoes a rapid degeneration, and is shed, passing away generally as mere detritus but sometimes in distinct masses, forming the latter part of the menstrual flow, which with the diminution of the hæmorrhage becomes less and less coloured. The amount of mucous membrane which is thus shed appears to vary extremely in different cases; according to some authors the loss may be so complete, that the bases only of the uterine glands are left, and from the epithelial cells lining these the regeneration of the new membrane is said to take place. The escaped ovum, if no spermatozoa come in contact with it, dies, the uterine membrane returns to its normal condition, and no trace of the discharge of an ovum is left, except the corpus luteum in the ovary.

§ 937. It is difficult to resist the conclusion that there is a causal connection between the changes in the ovary leading to the escape of an ovum, and the changes in the mucous membrane of the uterus leading to the menstrual flow; looking at the matter from a teleological point of view, we may say that the changes in the uterus appear to be preparatory for the reception of the ovum. But the exact connection between the two sets of changes is far from clear. For it is by no means certain that menstruation, in the human subject at all events, is always accompanied by a discharge of an ovum; indeed it is stated that menstruation has, in certain cases, continued after what appeared to be complete removal of both ovaries. On the other hand the fact that impregnation may follow upon a coitus effected in the interval between two menstrual periods, at which time presumably no ovum is present in the uterus for the spermatozoa to act upon, has been used as an argument that the act of coitus itself may lead to the escape of an ovum independent of menstrual changes. Since however the time during which both the ovum and the spermatozoon may remain in the female passages alive and functionally capable is considerable, probably extending to some days, coitus effected either some time after or some time before the menstrual escape of an ovum might lead to impregnation and subsequent development of an embryo; hence no great stress can be laid upon this argument. And a somewhat similar argument drawn from the experience that pregnancy may occur while menstruation is suspended by lactation or otherwise may be met by the reflection that the uterine mucous membrane is by the circumstances unfitted to respond to the uterine changes.

In any case it would seem that it is not so much the mere escape of the ovum, as the changes in the ovary of which the escape of the ovum is the culminating point, which we must regard as the cause of the uterine changes. If we allow ourselves to admit such a causal connection, we may conclude that in these phenomena of menstruation we have to deal with complicated reflex actions affecting not only the vascular supply, but apparently in a direct manner, the nutritive changes of the organs concerned. Our studies on the nervous action of secretion render it easy for us to conceive in a general way how the several events are brought about. It is no more difficult to suppose that the stimulus of the enlargement of a Graaffian follicle causes nutritive as well as vascular changes in the uterine mucous membrane, than it is to suppose that the stimulus of food in the alimentary canal causes those nutritive changes in the salivary glands or pancreas which constitute secretion. In the latter case we can to some extent trace out the chain of events; in the former case we hardly know more than that the maintenance of the lumbar cord is sufficient, as far as the central nervous system is concerned, for the carrying on of the work. In the case of a dog in which the spinal cord had been completely divided in the thoracic region while the animal was as yet a mere puppy, 'heat' took place as usual. And, though 'heat' is something different from human menstruation the two are probably, in this respect, analogous.

SEC. 3. THE MALE ORGANS.

The Testis.

§ 938. We have seen that the ovum is an epithelium cell which, like other epithelium cells, is eventually shed; though it is prepared and nurtured in a way very different from that in which other epithelium cells are cared for, we may in a broad way speak of the discharge of ova as a 'secretion.' The history of the male element, the spermatozoon, much more strikingly suggests the idea of secretion. The semen containing the spermatozoa is formed in epithelium lined tubules, the *seminiferous* or *seminal tubules, tubuli seminiferi*, which have all the appearance of secreting tubules, and is carried off thence by conducting tubules uniting into a duct, the two sets of tubules making up together the secreting gland which is called the *testis*. The distinction which we have more than once pointed out between the secreting and the conducting portion of a gland is remarkably salient in the testis. Not only do the seminiferous tubules, which alone form the secreting portion, differ markedly from the rest of the testis in structure, but they also have a wholly different origin. The male embryo developes on each side of the body cavity a patch of germinal epithelium, which is at first apparently identical with that of the female embryo, and contains primordial ova. As in the female, the epithelium cells are separated into clusters by mesoblastic growths, but these clusters become not Graaffian follicles with enduring ova, but tubules from which the primordial ova disappear, and which eventually become the seminiferous tubules, the secreting portion of the testis. The rest of the testis, the conducting portion, is derived from a wholly different source, namely, the Wolffian body; the two portions become connected to form the whole organ. In the female the Wolffian body also joins the germinal epithelium to form part of the whole ovary; but the connection is not a functional one, and in the adult ovary remnants only of the Wolffian body are found at the hilus (§ 929).

§ 939. In the adult the testis proper is an oval body surmounted by a cap; the latter, which is the coiled up main duct of the organ, receives the name of *epididymis*.

The testis proper is surrounded by a lamellated capsule of connective tissue, the *tunica albuginea*. On the outside, this capsule is further covered by the *tunica vaginalis*, the visceral layer of the lining of the serous cavity in which the testis lies; this, like other serous layers, such as that of the peritoneum, consists of a connective tissue basis (closely adherent to the capsule beneath), covered with epithelioid plates. From the capsule numerous septa, like the capsule lamellated in texture, radiate through the body of the testis converging to the hind and upper part of the oval, where they join into an irregular network, the *corpus Highmori*. The more or less conical converging chambers defined by these septa are occupied by groups of relatively large tubules, the seminiferous tubules, which may be compared to the uriniferous tubules of the kidney, except that they are much larger, and have a very wide lumen. Beginning at the periphery of the testis, and frequently anastomosing at their commencement, the tubules, supported by a scanty reticulum of connective tissue, pursue a somewhat tortuous course converging to the corpus Highmori. Each seminiferous or seminal tubule consists of a basement membrane lined by a special epithelium, whose characters we will discuss presently.

As they approach the corpus Highmori, the tubules somewhat suddenly change in character, they become narrower, their epithelium is reduced to a single layer of low cubical cells, and their course becomes straight; they are now called *vasa recta*. In the corpus Highmori itself these vasa recta anastomose freely with each other and form a network of irregular but narrow passages, lined with a single layer of flat squamous epithelium, the basement membrane being lost or fused with the connective tissue of the walls. From this network, which is called the *rete vasculosum* or *rete testis*, there issue on the farther side a number (twelve to twenty) of ducts, with relatively thick walls, the *vasa efferentia*. The terminal portion of each vas efferens is coiled up into a *conus vasculosus*, and the several coni vasculosi join at intervals to form a single duct, the *epididymis* which, though of great length, is coiled up into the compact mass spoken of above as forming a cap to the testis proper. When it issues from the coil the duct receives the name of *vas deferens*, and leaving the scrotum in which the testis hangs pursues its course to the penis.

The vasa efferentia and the epididymis which they by joining form, in contrast to the channels of the rete testis have well-defined walls of connective tissue, strengthened by muscular fibres, disposed for the most part circularly, and are lined by columnar epithelium cells, each of which bears a tuft of cilia. If the tubes be examined along their course from the rete to the end of the epididymis, it will be seen that the walls become stouter as the tube increases in size, and that the epithelium cells become conspicuously tall, with cilia of remarkable length. The walls of the vas deferens are still stouter, an inner circular and an outer longitudinal, with often a third innermost longitudinal layer of muscular fibres making their appearance; the epithelium here consists of several layers of cells, the uppermost of which are columnar, but the cilia disappear.

§ 940. The seminal tubules form the true secreting portion of the testis; the vasa recta mark the junction of the germinal with the Wolffian elements, and from thence onward all the rest of the organ is conducting in nature; it is in the seminal tubules that the spermatozoa make their appearance, and we may now turn our attention to these.

If a section of ripe testis be examined under a moderately high power of the microscope, no great difficulty will be found in observing the following features. Each tubule is defined by a basement membrane, generally consisting in man of more than one lamina, imbedded in which, as in the case of so many other basement membranes, the constituent nuclei may be seen. Upon the basement membrane rest several layers of epithelium cells; the cells are obviously not all alike, and some are obviously undergoing changes. The centre of the lumen of the wide tube (200μ in diameter) is occupied by bundles of spermatozoa; and it may be seen that each spermatozoon consists of an enlarged head, and a narrow tapering tail, and further that the bundles of spermatozoa are so arranged that the heads seem plunged among the epithelium cells while the tails converge towards the centre of the lumen. There can be no doubt that the spermatozoa are formed in some way out of the epithelium cells lining the tubule; but the exact way in which they are formed has been, and still is, the subject of much discussion. We must be content with briefly describing what appears to be the best supported view.

A spermatozoon consists in the first place of a homogeneous (or for the most part homogeneous) highly refractive *head*, which in man is a flattened, somewhat curved oval (about 5μ long), but which differs in shape and size in different animals. From the hind pole of this oval head there proceeds a delicate filament (50μ in length in man) tapering to a point; this is the *tail* which, obviously of a different nature from the head, possesses the power of spontaneous movement, and in many respects resembles a cilium. The portion immediately succeeding the head is thicker than, and in other ways differs from the rest of the tail; it is often distinguished as the *middle piece, body*, or *neck*. In the newt the tail for the greater part of its length consists of a delicate membrane with a thickened edge wound spirally, like a miniature spiral fin, round a central thread; and something similar is seen in some mammalia but not in man. We may at once say that there is every reason to regard the head as a specialized nucleus, or part of a nucleus, and the tail as a remnant of a cell-body.

§ 941. In a seminal tubule, a cell of the outer layer next or near to the basement membrane undergoes the changes in the nucleus known as karyomitosis or karyokinesis, and gives rise to two cells, one of which remains to fill the place of the mother-cell in the outer layer, while the other advances inwards towards the centre of the lumen. The latter undergoes repeated karyomitosis, and so gives rise to a number of cells characterized by the cell-substance being small in bulk relatively to the nucleus. Whether each division of the nucleus is from the first accompanied by a complete division of the cell-body, or whether a number of nuclei are formed in a coherent mass of cell-substance, which is subsequently partitioned out among the nuclei, is a question which we may leave on one side; the important thing is that a mother-cell by mitosis gives rise to a brood of small daughter-cells. Each daughter-cell soon assumes an oval or club-shaped appearance, with the nucleus at one end. Changes then take place in the nucleus; into these we cannot enter, but may say that they appear to consist in the ejection or separation of a part of the nucleus, and a transformation of the rest, so that what was an ordinary nucleus becomes the differentiated head of a spermatozoon. At the same time the cell-body is transformed into the middle piece from which the tapering tail subsequently grows out.

In a transverse section of a tubule it is easy to recognize that while some of the nuclei in the outer layer of cells are undergoing this mitosis, others are completely at rest; the latter are regarded by many not as simply nuclei undergoing a temporary rest before they once more undergo mitosis, but as nuclei belonging to other kinds of cells which do not themselves give rise by division to spermatozoa but which possessing a branched cell-substance, are arranged at intervals along the circumference of the tubule so as to radiate inwards, like the spokes of a wheel, towards the centre of the lumen, thus furnishing a framework in the spaces of which the active, dividing, spermatozoa-forming cells are lodged. And it is maintained that these cells perform an important subsidiary function, by assisting in some way or other in the development of the spermatozoa; it is urged that the cells about to become spermatozoa are for a while lodged in the cell substance of these branching cells, and there undergo their full development, and that the grouping of the spermatozoa into radiating bundles is due to their aggregation in connection with these radiating cells, and not as might otherwise be supposed to each bundle having a common origin from a single mother cell. In any case the spermatozoa from one of the mother cells of the outer layer seem to be supported and held together by a fine reticulum, either belonging to the cells just spoken of or supplied by the cell bodies of cells which like the spermatozoa arise from the division of mother cells, but which do not become developed into spermatozoa. For there seems reason to believe that besides the cell-division which gives rise to the spermatozoa, other cell-divisions are taking place; and the products of these may fulfil the above purpose, or being eventually disintegrated may be lost among the more fluid parts of the semen which serve as a nutritive and mechanical vehicle for the spermatozoa, or may pass bodily into the secretion as undifferentiated nucleated cells.

It may well be imagined that the transformations needed for the development of the potent spermatozoa are of a special kind, and that the changes within the seminal tubule are in more ways than one unlike those taking place in an ordinary epithelium; but concerning the details of the changes there is at present great diversity of opinion. The important fact for our present purposes is that in the seminal tubules spermatozoa are developed out of some or other of the lining epithelium cells, and further are developed in such a way that a specialized nucleus becomes the head, while the body (middle piece) and tail appear to be of the nature of cell-substance.

§ **942**. The tail of a spermatozoon may be regarded, as we have said, as a single cilium, the movements of which are of an undulatory character, the waves travelling from the middle piece to the end of the tail.; and the statements previously made (§ 94) concerning ciliary action may be applied generally to the movement of a spermatozoon. The motion is apparently not a very rapid one, for it has been calculated that a half vibration takes at least a quarter of a second. It has also been calculated that a spermatozoon progresses at the rate of about 2 or 3 mm. a minute.

When discharged semen is left to itself the movements continue for some (24 or 48) hours, but they appear to last much longer in the female passages. Spermatozoa have been observed in movement when removed from the neck of the living human uterus 5 or even 7 days after coitus; and in some of the lower animals the duration of vitality may be enormously long. Making all allowance for any possible direct nutrition of the living substance of the spermatozoon by means of the fluid of the semen, we must conclude that the energy of the movement is derived from the expenditure of what we may venture to call the contractile material stored up in the middle piece and tail of the organism at its formation; the material of the head we may suppose to be devoted entirely to the work of impregnation. So small a store must be soon exhausted; hence it is difficult to suppose that vigorous movements can be continued for very long periods; and probably the activity of the spermatozoa is largely dependent on the circumstances by which it is surrounded; it may remain motionless in one medium, and become active when the medium is changed. The spermatozoon is probably quiescent so long as it remains in the seminal tubes, but we have no exact information as to whether or no movements begin in the epididymis and vas deferens without exposure to air; and it is possible that after coitus the beginning and maintenance of its vigorous movements may largely depend on the condition of the secretions in the vagina and uterus. In this connection it may be noted that the movements of a spermatozoon-like ciliary movements are favoured by fluids having a weak alkaline reaction, whereas almost any degree of acidity (unless used to neutralize excessive alkalinity) arrests them; and the mucous secretion of the uterus while it is alkaline at the neck of the uterus becomes acid as it passes down the vagina. Hence it might be inferred that those spermatozoa only which rapidly find their way into the os uteri manifest vigorous movements; but it would be dangerous to lay too great stress on this.

§ 943. The semen contains a relatively large quantity of solid matter, and this in turn is to a great extent furnished by the spermatozoa; indeed the spermatozoa form so large a portion of the semen that the chemical substances present in the former are dominant in the latter. The head of a spermatozoon appears to be largely composed of the body or group of bodies known as nuclein or nucleo-albumin, a result which supplies chemical evidence of the nuclear nature of the spermatozoan head; and nuclein forms a considerable portion of the solid matter of the whole semen. Lecithin is also present in the semen in considerable quantity; otherwise the chemical features of the secretion, which are as yet imperfectly known, present no special interest. The crystals found in dried semen are not as was once thought of a proteid nature but are compound phosphates containing an organic base. As discharged in coitus the semen proper from the testicle is mixed with the prostatic and other secretions.

From the testicle itself various forms of proteid of the globulin class have been extracted; and glycogen is not unfrequently present.

§ 944. The testis is peculiarly rich in lymphatics; they occur between the tunica vaginalis and tunica albuginea, are abundant in the latter, in the reticulum supporting and in the septa separating the seminal tubules and their continuations, as well as in the connective tissue binding together the coni vasculosi and the coils of the epididymis. It should be added that in many animals masses of polyhedral granular nucleated cells, sometimes assuming the form of occluded tubules, are found intercalated among the proper seminal tubules. They are especially abundant in the testicle of the boar; they appear to take origin from the Wolffian body, but the purpose of their presence is wholly obscure.

Accessory Organs.

§ 945. The vas deferens passing from the scrotum through the inguinal ring into the abdomen, finds its way to the under surface of the neck of the bladder, and here, while maintaining the general features of its structure though somewhat dilated, is joined by the two *vesiculæ seminales*. Each vesicula is virtually a long tubular diverticulum of the vas deferens, so coiled up as to present a sacculated appearance, and its walls, though thinner than those of the vas deferens itself, have the same general structure as they have; the internal surface, much folded, is like that of the vas deferens lined by columnar non-ciliated epithelium. The cavity serves as a temporary receptacle for the semen, though some secretion, and in some animals a decided quantity, takes place from its interior. In certain animals the secretion clots, and then appears to contain a substance identical with or allied to fibrinogen; in these animals the clot which is thus formed by the mixture of the male secretion with the bloody secretion of the rutting female helps to secure the retention of the former within the female passages.

The vas deferens and the duct or end of the vesicula seminalis on each side join to form the common *ductus ejaculatorius*, which passing through part of the *prostate gland*, opens into the prostatic portion of the urethra. The prostate gland is composed of twisted tubular alveoli lined with columnar epithelium, and of branched ducts lined with shorter cubical cells; its striking feature is the presence of plain muscular fibres which are especially abundant in the smaller septa, wrapping round the individual alveoli. The prostatic portion of the urethra is lined with stratified epithelium, the upper cells of which are flattened, and into this the cubical epithelium of the prostatic ducts and the columnar epithelium of the ejaculatory duct pass. The secretion of the prostate presents no special features, except that it is apt to contain peculiar concentric corpuscles; but the fact that the prostate remains undeveloped in castrated animals suggests that the secretion plays some part in coitus.

§ 946. *Erectile Tissue.* The walls of the urethra consist of a mucous membrane lined from the prostate onwards to near the mouth with a stratified columnar epithelium, and strengthened by plain muscular fibres disposed circularly and longitudinally. It contains numerous small glands, and at its commencement two large mucous glands, the glands of Cowper, affording a thick mucous secretion. The tube thus constituted receives for some little space in front of the prostate no special support, and is here spoken of as the membranous urethra, but further on is supported by a median column of erectile tissue, the *corpus spongiosum*, in the axis of which it runs, as well as by two lateral columns of erectile tissue, the *corpora cavernosa*, which meet in the middle line above the medially placed corpus spongiosum and urethra. The *glans penis* surrounding the end of the urethra is also essentially a mass of erectile tissue. The structure of all these bodies is in the main the same, the corpora cavernosa being distinguished by possessing very stout capsules of fibrous tissue, and in some minor features.

The erectile tissue in each of these structures consists of an irregular labyrinth formed by trabeculæ composed of connective tissue, with abundant elastic elements mixed up with a large but variable amount of plain muscular tissue. The spaces of the trabeculæ are lined by spindle-shaped epithelioid plates, resting in some cases on a layer of plain muscular fibres, and are venous sinuses, into which blood finds its way chiefly through the terminal capillaries of the numerous arteries lying in the trabeculæ but also in some cases by minute arteries opening directly into the spaces; from the sinus the blood finds its way out into smaller regular veins. In the corpora cavernosa, and to a less extent in the corpus spongiosum, the small arteries in the trabeculæ are extremely twisted up and looped, bulging into the venous sinuses as arterial coils, the so-called 'helicine arteries.'

When the arteries supplying these masses of erectile tissue, namely, the branches of the pudic arteries and dorsal artery of the penis are constricted, and when the plain muscular fibres of the trabeculæ are in a state of contraction, whereby the venous spaces are largely closed, the greater part of the blood flowing through the arteries finds its way by ordinary capillaries into the efferent veins, little blood passes into the venous sinuses, and the whole tissue is relatively small in bulk. When on the other hand the arteries are dilated and in addition the muscular bundles of the trabeculæ are relaxed, a large quantity of blood passes into the venous sinuses, these become greatly distended with blood; the whole mass of erectile tissue becomes turgid, and in proportion to the resisting nature of the outer envelope, as is especially seen in the corpora cavernosa, hard and rigid.

§ 947. In the dog and cat, fibres from the anterior roots of the second and first, or sometimes from the third, sacral nerves form the *nervi erigentes*, which passing to the pelvic plexus are distributed to the penis and to other organs; in the monkey the fibres are supplied by the seventh lumbar and first sacral, sometimes also by the second sacral nerves. They receive this name because stimulation of them leads to erection of the penis; and this results from a vaso-dilator action on the arteries supplying the erectile tissue. Erection of the penis is hence to a large extent a vaso-dilator effect. But not wholly so; the entrance of the blood from the dilated arteries into the venous sinuses is facilitated by the relaxation of the muscular bundles in the trabeculæ, whose contraction would offer an obstacle to the spaces becoming filled. Further the filling of the venous sinuses tends of itself to compress the large longitudinal veins running in the centre of the corpora cavernosa and thus to increase the distension already begun; moreover contractions of the striated muscles, the *transversus perinaei*, and the *bulbo-cavernosus*, between the bundles of which the veins pass, also tend to check the outflow and so to increase the erection. In the dog even powerful stimulation of the nervi erigentes will not produce complete erection; the factors just mentioned are absent, and the blood, though it more or less fills the venous sinuses, flows freely away by the veins.

The dilating action of the nervi erigentes and the nervous impulses leading to the subsidiary acts in erection may be set going as part of a reflex action, by stimulation of the glans penis. Of such a reflex act the centre lies in the lumbar spinal cord and erection, with emission of semen, has been witnessed in a dog after division of the spinal cord in the thoracic region. But erection also takes place as the result of emotions, in which case we may suppose that impulses descending from the brain affect the lumbar centre in a direct manner; and indeed erection has been experimentally brought about by stimulation of certain parts of the brain.

The antagonistic act, namely, constriction of the blood vessels and retraction of the penis may, in the cat, be brought about by stimulation of fibres coming from the upper lumbar (and possibly the lower thoracic) region, and reaching their destination by way of the sympathetic.

§ **948.** The emission of semen, for which act erection is preparatory, is carried out by a succession of agencies. The epididymis with its coni vasculosi may be regarded as a reservoir filled by the secretory activity of the seminal tubes; hence its relatively enormous length. It is possible that the act may begin with an increase of secretory activity on the part of the seminal tubes, bearing perhaps especially on the fluid parts of the semen, by which the epididymis becomes overfilled; we have no positive evidence of this. Nor have we evidence of any pressure, either intrinsic by means of the plain muscular fibres which are said to occur scantily in the septa of the testis, or extrinsic through the cremaster or other muscles, being brought to bear on the contents of the seminal tubes. Hence we may conclude provisionally that the act begins with a propulsion of the contents of the distended epididymis by means of peristaltic contractions of the muscular walls of that tube. In any case the flow of fluid having reached the vas deferens, is carried along that tube by the peristaltic contractions of its much stouter and much more muscular walls. In the monkey stimulation of the anterior roots of the second and third lumbar nerves leads to a powerful contraction of the vas deferens, sweeping down it in a single wave.

One effect, possibly a chief effect, of the flow along the vas deferens is to fill and distend the vesiculæ seminales; or we may suppose that preparatory feeble contractions of the epididymis fill and distend both the vas deferens and the vesiculæ seminales, and that the act really begins with a more powerful contraction of both these distended organs by which their contents are rapidly ejected into the prostatic urethra; at the same time contractions of the muscular fibres of the prostate discharge the secretion of that gland into the urethral canal. So far plain muscular fibres only are brought into play; but the act is completed by the aid of striated muscles, namely, by forcible contractions of the levator ani, of the constrictor urethræ including the external sphincter of Henle, of the ischio-cavernous muscle, which starting from the ischium on each side embraces the root of the penis, and of the bulbo-cavernosus muscle (or ejaculator urinæ) which starting from the perinæum embraces the beginning of the urethra and corpus spongiosum. A contraction begins in the external sphincter ani, extends to the levator ani and then passes to the other muscles, progressing in a wave-like manner from behind forwards, and is repeated in a more or less distinctly rhythmic manner until all the semen is ejected from the urethra.

These expulsive contractions, especially the last named, appear like erection to be carried out by the help of a centre in the lumbar region of the cord, and for them afferent impulses generated in the sensitive surface of the glans penis are more essential than for simple erection. In the dog stimulation of the internal pudic nerve throws the whole group of striated muscles just named into successive contractions as described, but each muscle may be made to contract separately by stimulation of its own individual branch.

The semen being received into the vagina, the walls of which, and especially the external appendages of which, are at the time in a state of turgescence resembling the erection of the penis, but less marked, lies, probably, at the far end of the vagina in a pool into which the os uteri dips; and it is possible that contractions of the round ligaments (which contain striated muscular fibres) by tilting the cervix backwards assist in bringing the os uteri into the semen. In this manner the spermatozoa find their way into the uterus and so into the Fallopian tube, where (probably in its upper part) they come in contact with the ovum. In the rabbit spermatozoa may reach the ovary within two hours after coitus. In the case of some animals impregnation may take place at the ovary itself. The passage of the spermatozoa is most probably effected mainly by their own vibratile activity; but in some animals a retrograde peristaltic movement travelling from the uterus along the Fallopian tubes has been observed; this might assist in bringing the semen to the ovum, but inasmuch as these movements are probably parts of the act of coitus and impregnation may be deferred till some time after that event, no great stress can be laid upon them.

CHAPTER II.

PREGNANCY AND BIRTH.

SEC. 1. THE PLACENTA.

§ 949. THE spermatozoa travelling up the female passages come in contact with the ovum. Making their way through the cells of the discus, which by this time are undergoing degenerative changes, and piercing the zona pellucida, they enter the vitellus; it is stated that as a rule one spermatozoon only actually reaches the vitellus. Here the tail, which by its vibratile activity has thus brought the spermatozoon to its destination, ceases to move and soon disappears; but the head (and we have seen that the head is a prepared and, so to speak, purified nucleus, a male pronucleus) unites with the pronucleus of the ovum to form the nucleus of the now impregnated ovum.

As the result of this action of the spermatozoon on the ovum, the latter, instead of dying as when impregnation fails, awakes to new nutritive activity. It undergoes segmentation, the one cell becomes by cell-division a mass of cells, which, passing through a series of remarkable morphological changes, into the details of which we cannot enter here, developes into an embryo.

§ 950. No sooner, however, have these changes begun in the ovum than correlative changes, brought about probably by reflex action, but at present most obscure in their causation, take place in the uterus. The mucous membrane of this organ, whether the coitus, which was the cause of the impregnation, took place at a menstrual period or at some time in the interval, undergoes changes which though more intense are at first not unlike those of menstruation; it becomes congested, and a rapid growth takes place, characterized by a proliferation of the epithelial and other tissues. Unlike what takes place in menstruation, however, this new growth does not give way to hæmorrhage and immediate decay; it remains, and may be distinguished as a new temporary lining to the uterus, the so-called *decidua*. Into this decidua the ovum, on its descent from the Fallopian tube, in which it has already undergone some developmental changes, is received; and in this it becomes embedded, the new growth closing in over it. Meanwhile the rest of the uterine structures, especially the muscular tissue, become also much enlarged; as pregnancy advances a large number of new muscular fibres are formed.

As the ovum, now developing into the embryo and its appendages, continues to increase in size, it bulges into the cavity of the uterus, carrying with it the portion of the decidua which has closed over it. Henceforward, accordingly, a distinction is made in the now rapidly developing decidua between the *decidua reflexa*, or that part of the membrane which covers the projecting ovum, and the *decidua vera*, or the rest of the membrane lining the cavity of the uterus, the two being continuous round the base of the projecting ovum. That part of the decidua which intervenes between the ovum and the nearest uterine wall is spoken of as the *decidua serotina*. As the embryo with its appendages continues to enlarge, carrying with it the decidua reflexa, the latter becomes pushed against the decidua vera, gradually obliterating the cavity of the uterus, except at the cervix: about the end of the third month, in the human subject, the two come into complete contact all over, and ultimately the distinction between them is lost.

The changes through which the mucous membrane becomes the decidua are seen in their simplest form in the decidua vera at some little distance from the ovum, on the sides for instance of the uterus. When this portion of the decidua is examined, it is found that the glands have become very much enlarged and tortuous with irregular lateral bulgings. The change is greatest in the middle third of the length of the glands; this region is seen in sections to form a zone in which the channels of the glands (the epithelium lining which is now cubical rather than columnar and is devoid of cilia) appear as a number of irregular spaces giving the zone a spongy texture. The mouths and necks of the glands though enlarged and altered are not so markedly changed, and the basal portions have undergone still less transformation. The dermic connective tissue between the glands has also become hypertrophied, though not in the same proportion as the glands; it is very vascular, and a number of new cells make their appearance in it.

Similar changes take place in the decidua reflexa, account being taken of the fact that here the glands are by the intercalation of the growing ovum cut off as it were from their bases. The changes, however, undergone by the decidua reflexa, and by the decidua vera in general, are of subordinate interest compared with those which take place in that part of the decidua vera which is called the decidua serotina. In the decidua vera generally, and in the decidua reflexa, the hypertrophy having reached a certain stage ceases, or even gives place to a retrograde process. This change, which may be said to occur at about the fifth month, is most marked and begins earliest in the decidua reflexa, which is soon reduced to a mere membrane, the glands gradually disappearing; later on the decidua vera, though it continues to grow with the expanding uterus, becomes much altered in its more superficial portions. In the decidua serotina, on the other hand, the changes in the mucous membrane which are at first hardly more than those of mere enlargement or hypertrophy, assume new characters and lead to a special union between maternal tissues and tissues belonging to the growing embryo, a union which gives rise to the structure known as the *placenta* or 'after-birth.'

§ 951. During the development of the ovum while some of the cells, arising by cell-division from the primordial cell, become the embryo proper, others form the appendages of the embryo; to the latter belongs the double bag which encloses the embryo, and which consists of an inner bag, the *true amnion*, and an outer bag, the *false amnion*. The latter over the whole of its surface is in contact with the decidua, and develops a number of branched villi, consisting, like the rest of the membrane, of an epithelium (epiblast) resting on a dermic (mesoblastic) basis; these villi are imbedded in or applied to the decidual surface. The false amnion, bearing villi, often called the *chorion*, is at first devoid of blood vessels; but a diverticulum of the hinder part of the developing alimentary canal of the embryo, called the *allantois*, grows out rapidly into the space (containing fluid) between the false and the true amnion, and soon applies itself to the former. As it grows, two arteries, continuations of the primitive aorta, the allantoic arteries, subsequently called umbilical arteries, make their appearance. These carry the blood of the embryo to the villi of the chorion; from thence it is returned at first to two veins, but ultimately to a single vein running in company with the umbilical arteries, and called the umbilical vein.

At first all the villi over the whole surface of the chorion except at two opposite poles are thus supplied with blood, but ultimately the supply is restricted to that part of the chorion which is applied to the decidua serotina. Here the villi become developed into large and conspicuous vascular tufts, whereas over the rest of the chorion they soon atrophy.

The decidua serotina at first resembles the rest of the decidua in consisting of enlarged and tortuous glands with hypertrophied dermic tissue between them; and as in the rest of the decidua, while the changes in the basal zone are relatively speaking not great, the middle portions of the glands become a zone of spongy texture. But this primary condition, in which the decidua serotina may still be recognized as a mucous membrane, with its glands and intervening connective tissue altered but yet extant, gives way before complex changes, the details of which have been and still are the subject of much discussion, changes by which the whole region, stretching from the basal portion of the uterine glands, or even from the uterine muscular coat, to the connective tissue which carries the capillary loops in which the umbilical arteries end, is so altered that it becomes difficult to say which are maternal, which are embryonic elements, which structures are of glandular and true epithelial origin, which of connective tissue or epithelioid origin.

§ 952. When fully developed, the placenta is a cake-like mass, more or less distinctly divided into lobes, or cotyledons, which occupies normally the roof of the uterus between the mouths of the Fallopian tubes. The umbilical vein with the twisted umbilical arteries, supported by the jelly-like immature connective tissue, 'Wharton's jelly,' of the umbilical cord, stretching upwards from the underlying fœtus (such being the name given to the embryo when its development is advanced), reach the placenta at about its middle, and radiate thence over its surface.

In vertical sections taken through the uterine wall and placenta the following structures may be observed. Next to the now greatly hypertrophied and very vascular uterine muscular coat, separated from it by a thin layer of connective tissue, comes a layer of a cellular nature which possesses few blood vessels of its own, but which is traversed by conspicuous arteries whose course is very tortuous, "curling arteries," passing from the muscular coat to the parts beyond the layer and by corresponding veins whose course is straighter. This layer, of which the uterine surface is smooth, but the other surface very irregular, marked with rounded projections, is often spoken of as the "decidual layer" of the placenta, and may be regarded as the transformed uterine mucous membrane. But the transformation is very great. Some authors recognize in it the basal remnants of the uterine glands; these however are greatly obscured by the presence of cells having the appearance of epithelium cells, which may be spoken of under the general name of *decidual cells*, but which appear to be not all of the same kind, some of them being multinucleated giant cells, and about the origin of which there is much diversity of opinion ; indeed some authors maintain that the whole of the layer is a new formation, not originating even in part from the uterine glands, the whole of these having been absorbed to make way for it.

Next to this decidual layer comes what may be called the placenta proper. This, except on the under surface, where covered by the amnion and supported by connective tissue are found the larger branches of the umbilical vessels, is in the main made up of two elements, namely the branching, in fact extremely arborescent cauliflower-like *fœtal villi*, and of the irregular spaces, *intervillous spaces* left between the villi. Though the matter is one which has been and still is much disputed the evidence seems to be in favour of the view that these intervillous spaces are, in the living body, filled with maternal blood, that they are in reality blood sinuses belonging to the maternal circulation. Since the villi branch out in all directions they present in a thin vertical section the appearance of an irregular labyrinth with many outlying islets, and if before the section is made the intervillous spaces be injected as they may be from the maternal blood vessels, the injection material appears likewise as an irregular labyrinth filling up the interstices of, and surrounding the islets of, the fœtal labyrinth. The placenta is in fact a labyrinth of fœtal villous tissue fitting into a corresponding labyrinth of maternal intervillous spaces, each branch of a fœtal villus projecting into and being bathed by the blood of a maternal sinus.

Into this labyrinth of blood sinuses blood is poured by the curling uterine arteries, each of which as it passes through the decidual layer on its way from the muscular coat loses most of the elements of its coats until these are reduced to a single layer of epithelioid plates, and suddenly opens by a more or less round mouth into a blood sinus without the intervention of any capillaries. From the sinuses the blood escapes by more irregular orifices into veins, which pursuing a straighter though oblique course through the decidual layer, pass to the uterine muscular coat; much of the returning blood flows by a vein or rather a plexus of veins, which takes a circular course around the edge of the placenta.

§ 953. The sinuses of the placental labyrinth appear to be lined by epithelioid plates continuous with the lining of the maternal arteries and veins; and, though on this point there is difference of opinion, we may probably look upon the sinuses as being greatly transformed maternal capillaries. The body of each villus consists of arteries branching into capillaries, and so ending in returning veins, all supported by an immature though abundant connective tissue; but the nature of the wall of the villus, that which forms the partition separating the blood of the fœtal capillary from the blood in the intervillous space or maternal sinus has been the subject of much controversy. The view which has perhaps the greater support is that the basement membrane or surface of the fœtal connective tissue bears two (some say three) layers of epithelium, of which the inner one is fœtal, a derivative of the epiblast of the chorion, and the outer one is maternal, having probably the same origin, whatever that may be, as the epithelioid plates lining the rest of the sinus. In any case there can be no doubt that an epithelium of some kind or other does separate the fœtal from the maternal blood, and it is worthy of notice that the cell-substance of the cells of this epithelium has the appearance of being 'active' cell-substance engaged in metabolic labours.

The placenta then, taken as a whole, presents in the first place a mechanical arrangement by which the fœtal blood carried to the villus by the umbilical artery is brought in an ample manner into close proximity to the maternal blood carried to the intervillous space in a very direct way by the curling uterine arteries. But this is not all. The partition between the fœtal and the maternal blood is not an inert membrane serving a mechanical purpose only; the epithelium of which it is in part composed exerts, we must believe, an important influence on the interchange between the fœtus and the mother. Moreover the decidual layer consists of other also apparently active cellular structures, which we may conclude exert an influence on the maternal blood as it passes through their midst on its way to and from the intervillous spaces, as also on the blood during its stay in the intervillous spaces which adjoin the decidual layer. In the early stages of embryonic life, before the metabolism of the embryonic tissues has become specialized, this part of the placenta is prominent, it forms a large portion of the whole attachment of the embryo to the mother and is obviously the seat of important changes, one object of which appears to be the nourishment in a special manner of the embryonic tissues. In the later stages, as the fœtus becomes more and more capable of transforming for its own uses food of a more common kind, this part of the attachment loses its predominance, and is reduced to the decidual layer having the structure we have described. This however remains to the end of intra-uterine life and assists the epithelium of the villi in the metabolic labours whereby the embryonic blood is adequately nourished at the expense of that of the mother.

§ 954. It appears then that in the transformation of the decidua into a placenta all obvious traces of the glands, unless it be what we have called the basal remnants, have disappeared; the uterine mucous membrane has been replaced by the decidual layer, and by the system of blood sinuses into which the fœtal villi project. We may in the fate of the uterine glands in pregnancy trace a certain analogy with what takes place in menstruation. In menstruation there is an enlargement of the uterine glands, which appears to have for its object an increased secretory activity; this is followed by a shedding of portions of the membrane, sometimes of the whole of it with the exception of the basal portion, from which new growth takes place: and we may look upon this shedding as a violent act of secretion. In pregnancy a similar but more marked enlargement amounting to considerable hypertrophy at first takes place, and this too we may perhaps regard as having for its object increased secretory activity, destined in this case for the nutrition of the embryo. In this activity the whole decidua at first shares, the influence of the decidua reflexa being direct, that of the vera more indirect; but very early the nutrition of the embryo is concentrated towards the region of the decidua serotina. There is evidence that in the formation of the placenta the hypertrophied glandular mucous membrane, having done its work in nourishing by secretory activity the embryo at an early stage, is, at least in its more superficial portions, absorbed, eaten as it were, by the advancing chorionic vascular tufts. This is introductory to the special vascular arrangements of the placenta, the uterine glands making way for the system of blood sinuses; but even in the full-grown placenta we may recognize, as we have said, that the interchange between mother and foetus is effected not in a wholly mechanical manner by the mere bringing into close juxtaposition the maternal and foetal blood, but also by an activity which we may venture to call secretory on the one hand of the epithelium covering the villi, and on the other hand of the decidual cells, whatever may be the exact origin and nature of each of these kinds of cell.

As the nutrition of the embryo becomes more and more concentrated in the altered decidua serotina or placenta, the decidua vera and reflexa, having played their part, are done away with. They are not, however, shed abruptly as in menstruation; they are returned piecemeal by absorption into the maternal system; they atrophy until the whole reflexa and the superficial part of the vera is reduced to a mere membrane adherent to the expanded chorion, while the basal portion of the vera remains to grow up after the birth of the foetus into a normal mucous membrane.

The serotina having become the maternal portion of the placenta continues its functions during the whole of the intra-uterine life of the embryo. When the term of the maternal nutrition of the embryo is ended and birth takes place, there is a sudden disruption of tissue along the line of the decidual layer, either where this joins the muscular coat, the whole mucous coat being subsequently renewed, or at some little distance from it, the 'basal remnants' of the glands being left to grow up into the new mucous lining; and the transformed serotina, like the changed mucous membrane of menstruation but even more suddenly and abruptly, is shed as the "after-birth." With the placenta there are also shed the so-called 'membranes,' that is to say the amniotic membranes together with the membranous remnants of the vera and reflexa, which have become adherent to and fused with these. Hence ultimately the whole decidua, the whole transformed mucous membrane of the pregnant uterus, like the changed mucous membrane of the menstruating uterus is, though in a different manner, cast off.

We may add that the form and structure of the placenta and the mode of connection between the mother and the embryo differ in different placental animals; in all cases, however, the blood of the chorionic villi of the embryo are bathed in sinus-like blood-spaces of the mother. In all cases too there is a development around the villi of epithelial structures of a secretory character; in ruminant animals collections of such cells form what is called 'uterine milk.' It is in these cells belonging to the border line between mother and infant, whether they are of maternal or of embryonic origin, that the glycogen, which is so often present in the placenta, is placed, and the presence of this substance may be taken as a token of the metabolic activity of these cells.

At times, in the human subject, in what is called "extra-uterine gestation," the embryo undergoes considerable development, not in the cavity of the uterus, but outside it, generally in the Fallopian tube. In such cases the nutrition of the embryo is effected by a vascular connection between the chorionic villi and the mucous membrane of the tube, and even apparently with the adjoining peritoneum. This shews that the uterus is not essential to at least a certain development of the embryo. We may add that in such cases though the muscular walls of the uterus hypertrophy and some changes take place in the uterine mucous membrane, there is no expansion of the cavity, and no true decidua is formed; the actual presence of the ovum in the cavity is necessary for the full sequence of events, a fact which is interesting in reference to the causation of the changes (§ 950).

SEC. 2. THE NUTRITION OF THE EMBRYO.

§ 955. In a hen's egg a very small part only of the whole egg, namely, a minute collection of cells called the blastoderm, is actually developed into the chick and its appendages; by far the greater part of the mass included within the egg-shell, namely the 'yolk' and the 'white,' is mere nutritive material. Through the porous egg-shell the oxygen of the air has adequate access to the contents within, and through the same egg-shell carbonic acid can escape. The yolk and the white supply all the food needed by the developing chick until it is hatched, and either directly or indirectly by means of the allantoic vessels the tissues of the embryo and its appendages breathe through the shell.

In the mammal the supply of yolk is insignificant; almost from the first the developing ovum receives nutritive material from the mother. We have seen that within the ovary the ovum is fed by the cells of the Graaffian follicle; and a similar mode of feeding is continued for some little time in the uterus. The repeated cell division of the ovum produces a compact mass of cells, the 'mulberry mass,' and this in turn is converted into the 'blastodermic vesicle,' which consists of a cellular membrane investing fluid contents; during this conversion a considerable increase in the total bulk of the ovum takes place, water and nutritive material passing into the ovum from the mother, probably from the cells lining the Fallopian tube. Received within the uterus and covered up by the decidua, the developing embryo is supplied with food and oxygen by the cells of the uterine mucous membrane with which it lies in contact, very much in the same way that the growing ovum was supplied by the cells of the Graaffian follicle; and the same uterine cells carry away the scanty waste matters of the embryo's nutritive activity.

The amount of food which the embryo needs and receives is at first small but continually and rapidly increases; the amount of oxygen which the embryo needs is at first insignificant, but the need of oxygen also increases continually and rapidly, though especially during the early stages it is limited by the fact that the processes going on in the embryonic tissues are largely synthetic, directed to the building up of the tissues, and such processes consume very little oxygen compared with the processes leading to expenditure of energy in movement and heat. Hence the simple method of nutrition and respiration by means of the direct contact of the cells of the uterine mucous membrane is exchanged for the special vascular mechanism of the placenta, by which the embryo lives upon and breathes through the uterine blood of the mother. From an early period up to birth the placental circulation is the chief, we may almost, say the only means by which the embryo breathes and is fed; but the details of the placental events are changing during the whole of this time. The embryo, all the while increasing in bulk, passes through phase after phase; the structural features of one day give way to those of the next, its morphological history being as it were a series of dissolving views; and each new structural phase entails new functional events both in the embryo itself and in the placenta. This is perhaps especially seen in the earlier stages at a time when the placental circulation has been established in its main outlines, but in the embryo most of the future organs are still in a shadowy inchoate condition. At this epoch, of the total bulk of blood coursing from the embryo towards the tissues of the mother and back again, the greater part is at any one moment to be found in the placenta and only a small part in the tissues of the embryo itself; later on the blood is equally divided between the placenta and the embryo; and still later the embryo has the larger share, and it is the smaller part which is at any one moment flowing through the chorionic villi of the placenta. There can be no doubt that in the earlier phase the influences which the placental structures exert on the fœtal blood are in many ways different from those which are exerted later on. We find that during the earlier phases the cellular placental elements are correspondingly prominent, indicating that much labour of the kind for which cells are necessary is being then carried on, whereas in the later stages the placental mechanism approaches though it never quite reaches the more mechanical condition of a simple membrane separating the fœtal and maternal blood. We cannot enter at all fully here into the successive phases; we must confine ourselves chiefly to the main features of what is going on during the latter months of gestation when the placental circulation is in full swing.

§ **956.** At this time the somewhat rapid strokes of the fœtal heart drive the fœtal blood through the umbilical arteries to the capillaries of the chorionic villi, from whence it is returned by the umbilical vein. From experiments on lambs and other animals it would appear that the blood pressure in the umbilical artery is moderately high (40 to 80 mm. Hg.) and that in the umbilical vein very considerable (15 to 30 mm. Hg.), higher than the venous pressure in the mother in a vein of corresponding size: the difference between the arterial and venous pressure is therefore relatively less than in the mother. Corresponding to this the velocity of the blood flow is relatively low. The number of red corpuscles in a given bulk of fœtal blood, which was of course at first very scanty, has by this time much increased, but as a rule remains up to the end less than that of the mother, though this has become diminished by the pregnancy. In many cases no marked distinction of colour can be observed between the blood in the umbilical arteries and that in the umbilical vein, but such difference as can be noted is in the direction of the blood in the vein being brighter than that in the arteries, and at times this is conspicuously the case. If, for instance, the fœtus at the time of observation happens to make prolonged movements, the contrast between the dark blood of the umbilical arteries and the bright blood of the umbilical vein may become striking. An examination of the gases of the blood shews that the blood in the vein contains more oxygen and less carbonic acid than that of the arteries; the former for instance has been found to contain from 7 to 20 p.c. of oxygen and 40 p.c. of carbonic acid, the latter 2 to 6 p.c. of oxygen and 40 p.c. of carbonic acid. Hence the blood in the umbilical vein is essentially arterial blood, and that in the umbilical arteries essentially venous blood. It may be observed that while as regards the amount of carbonic acid the blood of the fœtus runs parallel to that of the mother, the arterial blood of the fœtus (in the umbilical vein) contains less oxygen than that of the mother. This is not due alone to the relatively smaller amount of hæmoglobin, for as shewn by experiment the hæmoglobin of the fœtal arterial blood is far from being saturated with oxygen, whereas as we have seen (§ 355) that of the adult is, or very nearly so. We may add that the fœtal blood left to itself uses up its free oxygen rapidly, very much more rapidly than does adult blood.

The maternal blood is conveyed, as we have seen, to the placental sinuses by arteries which open directly into the sinuses. Hence, though independently of any influence exerted by the fœtal blood the blood returned from the sinuses by the uterine veins is venous blood, rendered venous by the maternal tissues themselves, yet the blood in the sinus to which the capillaries of the villi are exposed may be regarded as rather arterial than venous, and in any case contains more oxygen and less carbonic acid than does the fœtal blood arriving by the umbilical arteries. Seeing that the relatively narrow uterine arteries open out suddenly in the wide placental sinuses the flow in the latter must be slow; the flow in the fœtal vessels is also as we have seen not rapid; hence ample time is given for the interchange of gases. The change which is thus effected is probably carried out by diffusion, the amount of change being determined by the relative percentages of the gases in the maternal and fœtal blood. At least we have no more evidence in the case of this placental respiration than we had in the case of the pulmonary respiration that the interchange is in any way assisted by cellular activity of a secretory kind. The placental respiration of the mammal seems in fact exactly to repeat the branchial respiration of the fish; in the former the fœtus breathes by means of the maternal blood in the same way that in the latter the fish breathes by means of the water in which it lives.

It follows from the above that the fœtus may be asphyxiated in two ways: on the one hand by interference with the access of fœtal blood to the placenta, as when the cord is tied, and on the other hand by the maternal circulation being arrested, or by the maternal blood being wanting in oxygen. When the mother is asphyxiated the fœtus is asphyxiated too, the oxygen passing from the fœtal blood to that of the mother. In such a case, owing to the more imperious demands of the maternal blood, the store of oxygen in the fœtal blood is sooner exhausted and asphyxia is more rapidly developed than in the case when the cause lies in the fœtus, not in the mother, and the oxygen simply disappears from the fœtal blood as it is slowly used up by the fœtal tissues; for the rate of fœtal oxidation though it increases continually during the intra-uterine life, especially in the later stages, is slow compared to what it becomes some time after birth.

§ 957. The fœtus not only breathes but also feeds and probably excretes by means of the placenta; the blood returning by the umbilical vein is not only richer in oxygen and poorer in carbonic acid but also richer in nutritive material and poorer in waste products than the blood of the umbilical arteries. In dealing however with the nutrition of the embryo we must bear in mind a special condition under which the embryo lives. As we have said the embryo proper becomes at an early date invested with the double membranous bag of the amnion, consisting of the inner amnion and outer (false) amnion. Between the two there is at first a space, into which as we have seen the allantois grows in order to become the placenta; but, as the fluid, which from the first is present within the inner bag, increases in amount, without any corresponding increase in the fluid between the inner and outer bag, the (true) amnion in its expansion after the formation of the placenta reaches and unites with the false amnion which by this time is known as the chorion. The whole interior of the uterus is lined, next to the decidua, by a membrane apparently simple but composed of united amnion and chorion, and within this, surrounding and supporting the embryo, lies the amniotic fluid, which at first scanty rapidly increases in amount until in the later stages of pregnancy it may amount to 800 c.c. or even much more.

In the roof of the uterus, in the region of the placenta, the amniotic fluid is in close proximity not only to the branching umbilical arteries and veins of the fœtus, but also to many of the maternal blood vessels being separated from the maternal blood by nothing more than the thin wall of the blood vessel and the membrane just spoken of. The fluid is also over the rest of the internal surface of the uterus, in close proximity to the blood vessels of the maternal decidua, and indeed in the later stages, when the decidua apart from the placenta has largely retrograded, to the blood vessels of the uterine mucous membrane. The conditions therefore are favourable for the transudation of material from the blood of the mother into the amniotic cavity; and we have experimental evidence that not only water but various substances may pass in this way from the one to the other. If indigo-carmine (§ 416) be injected into the veins of the mother, none passes by the umbilical vein into the tissues of the fœtus; these remain wholly uncoloured. Yet the amniotic fluid becomes deeply tinged with the pigment, which obviously must have passed directly from the mother into the amniotic cavity. Hence we may conclude that though the amniotic fluid is at first derived exclusively from the fœtus, and during the whole time is partly derived from the same source, it is also, and especially in the later stages, largely derived by direct transudation from the mother.

Into this amniotic space the passages of the fœtus, the mouth, anus, &c. open, and it serves as we shall see as a repository for the excretions of the fœtus. Into it is discharged such urine as the fœtus secretes, into it are shed the fœtal epidermic scales, and appendages such as hairs, and into it may be discharged the contents of the alimentary canal, known as the *meconium*. Now, hairs, epidermic scales, in the case of hoofed mammals portions of shed hoofs, and at times meconium have been found in the fœtal stomach; they arrived there by the fœtus swallowing the amniotic fluid; we have other evidence that the fœtus in the uterus may execute swallowing movements, and if these are executed they must lead to swallowing of the amniotic fluid, since this will pass into the mouth and pharynx whenever the mouth is opened. If these swallowing movements occur frequently, and there is some evidence that they do, nutritive material contained in the fluid and derived directly from the mother, might thus be conveyed to the fœtus; the latter might be nourished by means of the amniotic fluid. But, even making all allowance for any possible nourishment in this way, we may probably regard it as insignificant compared with that which is carried on by the placental and umbilical vessels; we may assume that the food of the fœtus reaches it mainly by passing from the maternal sinuses into the capillaries of the chorionic villi.

§ 958. Judging from analogy we may conclude that the food of the fœtus consists, like that of the adult, of proteids, fats, carbohydrates and salts conveyed in water. In attempting to understand how these materials pass from the blood of the maternal sinus to the blood of the fœtal villus, we have to face problems of the same kind as those which we met with in considering absorption from the alimentary canal (§ 312).

Here as there diffusion and filtration play their parts; but here also as there the passage of material does not follow the laws of diffusion and filtration which regulate the passage of material through non-living membranes. We have evidence that diffusible substances pass readily from mother to fœtus and from fœtus to mother. When sugar is injected in considerable quantity into the vessels of the mother, it is found in excess in the tissues of the fœtus. When such a drug or poison as atropin is injected into the mother it passes to the fœtus, and manifests its presence there by dilation of the pupils. Not only may the fœtus be killed by injection of strychnine into the mother, but the mother may be killed by the injection of strychnine carefully restricted to the fœtus. Again, if curare, which is inert towards the fœtus at least up to a certain dose, be injected into the fœtus, the mother is affected by the drug, the fact that the drug does not poison the fœtus assisting in its transmission to the mother; this result is especially worthy of notice since curare has a very low diffusible power. The influence of diffusion seems to be further illustrated by the fact that if large quantities of sugar or other diffusible substance be injected into the blood vessels of the mother, while the thickened plasma of the maternal blood is diluted by the entrance of water, as shewn by the diminished proportion of red corpuscles, that of the fœtus as shewn by the same method undergoes concentration; water passes from the fœtal blood to meet the needs of the maternal blood.

Nevertheless that in the passage of nutritive material from the mother to the fœtus, and of waste products from the fœtus to the mother, we have to deal with something more than ordinary diffusion, is shewn by the fact that the specific gravity of the fœtal blood differs from, being definitely above, that of the maternal blood; if diffusion had its full power the specific gravities of the two bloods would soon become equalized. Although exact information concerning the matter is at present very limited or almost wholly wanting, it is probable that the epithelium cells of the placenta, either those of the villi or the 'decidual' cells or both, play a part not unlike that played by the epithelium of the alimentary canal or even play a more important part. Whether the proteids of the maternal blood undergo a change analogous to peptonification in passing to the fœtus, whether the mother furnishes fat to the fœtal blood, and if so how, — to these and other questions which suggest themselves no very satisfactory answer can at present be given. With regard to fat, leaning on the analogy of the conclusion at which (§ 542) we arrived, that in the adult the fat of the food is probably not taken up by the tissues as fat during the nutrition of the tissues by the blood, we may perhaps suppose that the mother does not supply the fœtus with fat as such. We have already referred to the significant presence of glycogen in the placenta; and it would almost seem as if the placenta exerted at one and the same time on the material passing from the mother to the fœtus influences comparable not only with those exerted by the walls of the alimentary canal but also with those subsequently exerted by the hepatic cells on the material which passes by way of the portal vein from the intestines to the right side of the heart. Again the very phrase "uterine milk" suggests that the placenta epithelial cells exercise a secretory and metabolic influence comparable to that of the mammary gland. But how far these analogies are false or true future research must determine; and putting aside for a while the special problems thus suggested we may, in a broad way, say that the fœtus lives on the blood of its mother, very much in the same way that all the tissues of any animal live on the blood of the body of which they are the parts.

§ 959. For a long time all the embryonic tissues are 'protoplasmic' in character; that is to say, the gradually differentiating elements of the several tissues remain still embedded in undifferentiated material; and during this period there must be a general similarity in the metabolism going on in various parts of the body. As differentiation becomes more and more marked, it obviously would be an economical advantage for partially elaborated material to be stored up in various fœtal tissues, so as to be ready for immediate use when a demand arose for it, rather than for a special call to be made at each occasion upon the mother for comparatively raw material needing subsequent preparatory changes. Accordingly, we find the tissues of the fœtus at a very early period loaded with glycogen. The muscles are especially rich in this substance, but it occurs in other tissues as well. The abundance of it in the former may be explained partly by the fact that they form a very large proportion of the total mass of the fœtal body, and partly by the fact that, while during the presence of the glycogen they contain much undifferentiated substance, they are exactly the organs which will ultimately undergo a large amount of differentiation, and therefore need a large amount of material for the metabolism which the differentiation entails. It is not until the later stages of intra-uterine life, at about the fifth month, when it is largely disappearing from the muscles, that the glycogen begins to be deposited in the liver. By this time histological differentiation has advanced largely, and the use of the glycogen to the economy has become that to which it is put in the ordinary life of the animal; hence we find it deposited in the usual place. We do not know how much carbohydrate material finds its way into the umbilical vein; and we cannot therefore state what is the source of the fœtal glycogen; but it is at least possible, not to say probable, that it arises, in part at all events, from a splitting up of proteid material in the fœtal body.

§ 960. Concerning the rise and development of the functional activities of the embryo, our knowledge is almost a blank. We know scarcely anything about the various steps by which the primary fundamental qualities of the living matter of the ovum are differentiated into the complex phenomena which we have attempted in this book to expound. We can hardly state more than that while muscular contractility becomes early developed, and the heart probably, as in the chick, beats even before the blood-corpuscles are formed, movements of the fœtus are in the human subject first felt about the sixteenth week; they probably occur before but are not easily recognised, while from that time onward they increase and subsequently become very marked. They are often spoken of as reflex in character, and some of them are undoubtedly of this nature. When the uterus of a pregnant animal is prematurely opened, various reflex movements of the fœtus may be excited by appropriate stimulation, different kinds of animals differing in this respect as they do with regard to the powers of the new-born animals. Such reflex movements may be witnessed before the placental circulation has been interrupted, but they are increased if the fœtus be made to breathe. We have already referred to swallowing movements; and may add that an immature fœtal animal may be made to bite by introducing the finger into its mouth. Some of these normal intra-uterine movements appear however to be not reflex but automatic if not voluntary in nature. Movements of the limbs, apparently automatic, have been observed in fœtuses in which the brain has not been developed. We may add that in the human subject the occurrence of intra-uterine convulsions is fully acknowledged.

§ 961. The digestive functions are naturally, in the absence of all food from the alimentary canal, in abeyance. Though pepsin may be found in the gastric membrane at about the fourth month, it is doubtful whether a truly peptic gastric juice is secreted during intra-uterine life; trypsin appears in the pancreas somewhat later, but an amylolytic ferment cannot be obtained from that organ till after birth. The date however at which these several ferments make their appearance in the embryo appears to differ in different animals. The excretory functions of the liver are developed early, and about the third month bile-pigment and bile-salts find their way into the intestine. The quantity of bile secreted during intra-uterine life accumulates in the intestine and especially in the rectum, forming, together with material secreted by the walls of the alimentary canal and some desquamated epithelium, the so-called meconium. Human meconium is found to contain about 20 p.c. of solids. These consist of a considerable quantity of cholesterin (·7 p.c.), some fatty acids, bile salts with bile pigments, both largely unaltered, and calcium and sodium salts; the ash is rather more than 1 p.c. Though bile contributes normally to form the meconium, it is not essential, for a considerable quantity has been found in the fœtus in cases where the liver has been absent.

The distinct formation of bile is an indication that the products of fœtal metabolism are no longer wholly carried off by the maternal circulation; and to the excretory function of the liver there are now added those of the skin and kidney. Since in man, and in many other animals, such substances as are secreted by the kidney find their way at an early date into the cavity of the amnion, the determination of the history of the renal secretion is a matter of difficulty, for as we have seen the amniotic fluid is derived in part at least directly from the mother, and substances present in it may or may not have been discharged into it by the fœtus. The amniotic fluid varies not only in quantity but also in specific gravity (1·002 to 1·086) and in composition, and there does not seem to be any definite relation between its specific gravity and the quantity in which it occurs, or between its specific gravity and the size or age of the fœtus. It may be said to contain on the average about 1·6 p.c. of solid matter, of which about ·2 are proteids, ·8 extractives and ·6 salts. The proteids are serum albumin and probably paraglobulin, mucin or a mucin-like body being also present. Sugar appears to be sometimes present, sometimes absent. The most important constituent is perhaps urea, which seems to be always present. Since this is found at quite an early stage, before any secretion from the fœtal kidney could take place, it may be thus considered as derived from the mother and comparable in origin to the urea found in serous fluids; but since urine containing urea is found in the fœtal bladder at least as early as the seventh month, we may conclude that during the later stages of pregnancy, and possibly much earlier, part of the urea of the amniotic fluid comes from the fœtal kidney. In some animals, *ex. gr.* ruminants, the cavity of the allantois remains for a long time permanent and filled with fluid, instead of as in man becoming at an early date obliterated in its distal portion. In these animals the kidneys discharge their secretion into this allantoic sac, and in the contents of the sac is found the body allied to urea, allantoin, so called from its having been first discovered in this situation. Traces of allantoin have also been found in human amniotic fluid, which result suggests that this substance is at an early stage formed by the kidney but subsequently gives place to the permanent urea.

There is no evidence that any sweat is secreted by the fœtus in the uterus; and indeed if any such secretion does take place this can only be for the discharge of solid matter, and not as in the adult for the discharge of water; but the epidermic scales are undoubtedly shed, and may be detected in the amniotic fluid.

§ 962. About the middle of intra-uterine life, when the fœtal circulation is in full development, the blood flowing along the umbilical vein (see Fig. 194) is chiefly carried by the ductus venosus into the inferior vena cava and so into the right auricle. Thence

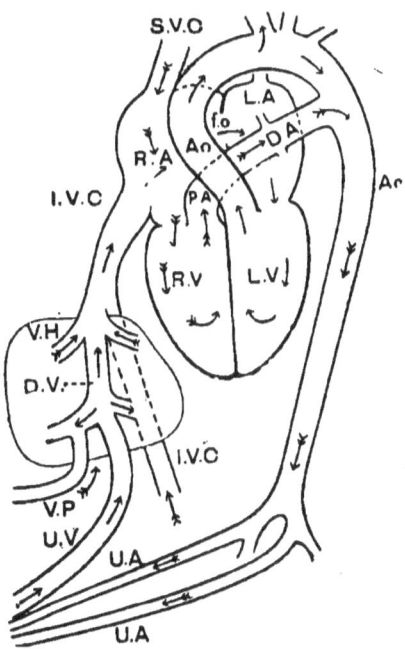

Fig. 194. Diagram to illustrate the Fœtal Circulation.

It will be understood that the figure is purely diagrammatic and constructed simply to shew in a convenient manner the general course taken by the blood.

The winged arrow indicates venous, the plain arrow arterial, or, in parts, mixed blood.

U.V. The umbilical vein, passing in part to the liver (indicated in outline), joined by blood from the alimentary canal along the mesenteric, becoming the portal vein V.P., but chiefly flowing on by the ductus venosus D.V. (into which fall the hepatic veins V.H.) into the inferior vena cava, I.V.C.

This chiefly arterial but still mixed blood passes through the right auricle R.A., the foramen ovale f.o. to the left auricle L.A., thence to the left ventricle L.V. and so by the arch of the aorta Ao. to the arteries of the head and upper limbs.

The venous blood of the head and upper limbs passes from the superior vena cava S.V.C. through the right auricle to the right ventricle R.V. and thence by the pulmonary artery P.A. and ductus arteriosus D.A. to the descending aorta, and so to the umbilical arteries U.A.

it appears to be directed by the valve of Eustachius through the foramen ovale into the left auricle, passing from which into the left ventricle it is driven into the aorta. Part of the umbilical blood, however, instead of passing directly to the inferior cava, enters with the blood carried by the portal vein into the hepatic circulation, from which it returns to the inferior cava by the hepatic veins. The inferior cava also contains blood coming from the lower limbs and lower trunk. Hence the blood which passing from the right auricle into the left auricle through the foramen ovale is distributed by the left ventricle through the aortic arch, though chiefly blood coming direct from the placenta, is also blood which on its way from the placenta has passed through the liver and blood derived from the tissues of the lower part of the body of the fœtus. The blood descending as fœtal venous blood from the head and limbs by the superior vena cava appears not to mingle largely with that of the inferior vena cava, but to fall into the right ventricle, from which it is discharged through the ductus arteriosus (Botalli) into the aorta below the arch, whence it flows partly to the lower trunk and limbs, but chiefly by the umbilical arteries to the placenta. A small quantity only of the contents of the right ventricle finds its way into the lungs. Now the blood which comes from the placenta by the umbilical vein direct into the right auricle is, as far as the respiration of the fœtus is concerned, arterial blood; and the portion of umbilical blood which traverses the liver probably loses at this epoch very little oxygen during its transit through that gland, the liver being at this period much more a simple excretory than an actively metabolic organ. Hence the blood of the inferior vena cava, though mixed, is on the whole arterial blood; and it is this blood which appears to be sent by the left ventricle through the arch of the aorta into the carotid and subclavian arteries. Thus the head of the fœtus is provided with blood comparatively rich in oxygen. The blood descending from the head and upper limbs by the superior vena cava is distinctly venous; and this passing from the right ventricle by the ductus arteriosus is driven along the descending aorta, and together with some of the blood passing from the left ventricle round the aortic arch falls into the umbilical arteries and so reaches the placenta. The fœtal circulation then appears to be so arranged that, while the most distinctly venous blood is driven by the right ventricle back to the placenta to be arterialized, the most distinctly arterial (but still mixed) blood is driven by the left ventricle to the cerebral structures, which, we may conclude, have more need of oxygen than have the other tissues. Contrary to what takes place afterwards, the work of the right ventricle is in the fœtus greater than that of the left; and, accordingly, that greater thickness of the left ventricular walls, so characteristic of the adult, does not become marked until close upon birth.

§ 963. In the later stages of pregnancy the mixture of the various kinds of blood in the right auricle increases preparatory to the changes taking place at birth. But during the whole time of intra-uterine life the amount of oxygen in the blood passing from the aortic arch to the brain is sufficient to prevent any inspiratory impulses being originated in the bulbar respiratory centre. This, during the whole period elapsing between the date of its structural establishment, or rather the consequent full development of its irritability and the epoch of birth, remains dormant; the oxygen-supply to its substance is never brought so low as to set going the respiratory molecular explosions. As soon however as the intercourse between the maternal and umbilical blood is interrupted by separation of the placenta or by ligature of the umbilical cord, or when, as by the death of the mother, the umbilical blood ceases to be replenished with oxygen by the maternal blood, or when in any other way blood of sufficiently arterial quality ceases to find its way by the left ventricle to the bulb, the supply of oxygen in the respiratory centre sinks, and when the fall has reached a certain point an impulse of inspiration is generated and the fœtus for the first time breathes. This action of the respiratory centre may be assisted by adjuvant impulses reaching the centre along various afferent nerves, such as those started by exposure of the body to the air, or to cold; but these are subordinate, not essential. A retarded first breath may be hurried on by dashing water on the face of the new-born infant; but so long as the placental circulation is intact, stimulation, even varied and strong, of the fœtal skin, though it may give rise to reflex movements of the limbs and other parts will not call forth a breath; whereas, on the other hand, upon the cessation of the placental circulation, the fœtus may make its first respiratory movements while it is still invested with the intact membranes and thus sheltered from the air and indeed from all external stimuli.

§ **964.** When the first breath is taken, as under normal circumstances it is, with free access to the atmosphere, and the lungs become inflated with air (we dwelt in dealing with respiration, § 326, on some features of this first breathing), the scanty supply of blood which at the moment was passing from the right ventricle along the pulmonary artery returns to the left auricle brighter and richer in oxygen than ever was the fœtal blood before. With the diminution of resistance in the pulmonary circulation caused by the expansion of the thorax, a larger supply of blood passes into the pulmonary artery instead of into the ductus arteriosus, and this derivation of the contents of the right ventricle increasing with the continued respiratory movements, the current through the latter canal at last ceases altogether, and its channel shortly after birth becomes obliterated. The obliteration is ultimately secured by proliferation of the internal coat, in which even before birth the sub-epithelial layer is unusually developed, a thrombus (§ 33) at times helping, but before this takes place the closure seems to be assisted by the mechanical arrangements of the parts. Corresponding to the greater flow into the pulmonary artery, a larger and larger quantity of blood returns from the pulmonary veins into the left auricle. At the same time the current through the ductus venosus from the umbilical vein having ceased, the flow from the inferior cava has diminished; and the blood of the right auricle finding little resistance in the direction of the ventricle, which now readily discharges its contents into the pulmonary artery, but finding in the left auricle, which is continually being filled from the lungs, an obstacle to its passage through the foramen ovale, ceases to take that course. Any return of blood from the now vigorous and active left auricle into the right auricle is prevented by the valve which, during the latter stages of intrauterine life, has been growing up in the left auricle over the foramen ovale. At birth the edge of this valve is to a certain extent free so that, in case of an emergency, as when the pulmonary circulation is obstructed, a direct escape of blood into the left auricle from the overburdened right auricle can take place. Eventually, in the course of the first year, adhesion takes place, and the separation of the two auricles becomes complete. With its larger supply of blood and greater work the left ventricle acquires the greater thickness characteristic of it during life. Thus the fœtal circulation, in consequence of the respiratory movements to which its interruption gives rise, changes its course into that characteristic of the adult.

SEC. 3. PARTURITION.

§ 965. Owing to the growth of the fœtus, and also to the accumulation of the amniotic fluid, the uterus towards the end of pregnancy has become much distended and has risen into the cavity of the abdomen, displacing the abdominal viscera. The expansion of the uterus during pregnancy is a complex process in which the mechanical effects of the increasing internal pressure are mingled with those of growth. Though the uterine walls are, as we have said, much thickened by the addition of new muscular fibres as well as by the increase in length, breadth, and thickness of the individual fibres, and also enlarged by the vascular development, they become somewhat thinned again towards the end of pregnancy by reason of the great distension of the cavity. The Fallopian tubes and the round ligaments share in the uterine enlargement, in so far that their muscular tissue is increased; but the mucous membrane of the former does not alter, and the only changes taking place in the ovary are those concerned with the corpus luteum left by the shed ovum. The walls of the vagina are congested, soft and hypertrophied. Previous to labour the fœtus occupies in the womb a position which it assumes at a quite early date, namely, one in which the head is directed downwards towards the pelvis; this is at least the normal position, though deviations from it not infrequently occur.

From an early date waves of contraction, at times rhythmical, sweep over the enlarged uterus and towards the end of pregnancy become more marked. As a rule these are "insensible" contractions, that is to say the mother is not conscious of them, though at times they may be distinctly felt; and in all cases they are temporary, producing no permanent effect on the uterus or its contents. Though, as shewn by the cases of premature labour and abortion, whether occurring from natural causes or induced artificially, the uterine muscles are capable at even an early date of carrying out the systematic contractions which lead to the expulsion of the fœtus, they do not in normal parturition enter upon this phase of activity until a certain time has been run. In the human subject the period of gestation generally lasts from 275 to 280 days, $i.\,e.$ about 40 weeks, the general custom being to expect parturition at about 280 days from the last day of the last menstruation. Seeing however that, in many cases, it is uncertain whether the ovum which developes into the embryo left the ovary in connection with the last menstruation or with the first one missed or during the intervening weeks, an exact determination of the duration of pregnancy is difficult if not impossible.

In some animals the period of gestation is longer, in others shorter than in man, being in the mare about 350 days, in the cow about 280 days, sheep about 150 days, dog about 60 days, rabbit about 30 days.

Immediately preceding labour a secretion of mucus, coming from the os uteri and at times mixed with blood, is often a sign or 'show' that the efficient uterine contractions are about to begin.

§ **966.** The onset of labour is marked by rhythmically repeated contractions of the uterus which most distinctly affect consciousness and are recognized as "labour pains." The first effect of these is the opening up or widening of the os uteri constituting the "first stage of labour." The contractions may perhaps be spoken of as "peristaltic" in character, but the arrangement of the bundles of muscular fibres in the body of the uterus is a complex one, and the gross effect of the contractions is to exert pressure, probably of a fairly uniform kind, on the uterine contents, that is on the amniotic fluid or "waters" enclosed in the "membranes" and surrounding the fœtus. These membranes are the amnion, the chorion and the decidua, the first being easily separated from the other two along the loose connective tissue joining it to the chorion, and thus forming an inner and outer sheet or membrane. Over the os uteri the decidua consists of decidua reflexa only; and here the membranes with the contained fluid act as a hydraulic plug directing the force of the uterine contractions towards expanding the mouth. As labour goes on a special character of the uterine contractions becomes prominent. In the contractions of which we spoke above as occurring during pregnancy before labour really commences, the relaxation of each muscular fibre following upon a contraction is a complete one. But in labour the muscular fibres while with each pain they contract and relax, are all the while becoming permanently and progressively thicker and shorter. This change by which the uterine wall becomes progressively thicker and more compact has been spoken of under the not very desirable term "retraction," as distinguished from "contraction," but appears to be a sort of tonic contraction or perhaps rather a residue of contraction like that seen in skeletal muscles under certain conditions; at each recurring "pain" the shortening of the muscular fibres starts so to speak from this more permanent shortening instead of from complete relaxation, and the return is to it not to complete relaxation.

This more permanent tonic contraction or "retraction" does

not however affect the whole uterus; it is, broadly speaking, confined to the body and absent from the cervix. Indeed in the latter region all contractions are wanting, the muscular fibres appear to be inhibited, and the walls yielding to the pressure exerted upon them become thinner instead of thicker; as the pressure increases the fibres possibly become lamed or paralysed. In this way a distinction is established between an "upper segment" of the uterus corresponding to the body, the walls of which become thicker and shorter through the continued and progressive "retraction," and a "lower segment," corresponding to the cervix but possibly including the lower part of the body, the walls of which become stretched and thinner, the line of demarcation between the two segments being often called " the retraction ring." As the pressure in the body of the uterus continues and waxes greater, the mouth becomes wider and wider, until the head of the fœtus begins to pass through it into the vagina, the walls of which like those of the "lower segment" have meanwhile become stretched and thinner; and as the fœtus is thus leaving the uterus the progressive tonic contraction adapts the uterine walls to the lessening cavity. Sometimes the membranes are ruptured, with escape of the "waters," before the head has left the uterus, at other times they form a bulging loose cushion preceding and making way for the fœtus.

When the os uteri has become fully expanded and is ready to allow the head of the fœtus to pass through it into the vagina, the intrinsic contractions of the uterus begin to be assisted by an extrinsic force, by contractions of the abdominal walls which thus exert on the uterus and its contents a pressure very similar to that exerted on the rectum in defæcation (§ 275). These contractions, which mark the onset of the "second stage" of labour, are rhythmical in nature like those of the uterus itself, and synchronous with them. The expulsive power of the uterus is thus greatly increased, and the head of the fœtus followed by the rest of its body is driven through the vagina and then through the vulva, these playing apparently a wholly passive part in the matter, and the child is thus literally "thrust upon the world."

At the very beginning of labour there takes place at the internal os a cleavage of the decidua vera between a deeper less altered and a superficial more altered layer, so that the latter, attached to the chorion and thus forming part of the "membranes," separates from the uterine surface. This separation, the beginning of which is the cause of the "show" spoken of above, and which is considered to be a mechanical effect of the uterine contractions but which must be prepared for by histological changes, during the early stages of labour extends upwards for two or three inches only; but, at the last, it is carried right through the "decidual layer" of the placenta. Hence, after the expulsion of the fœtus, the uterus contains within its cavity, separated from and now foreign to itself, the placenta and membranes, the latter consisting of amnion, chorion, the whole of the remains of the decidua reflexa and a variable part of the decidua vera; and, under normal conditions, these are by the last expulsive efforts ejected with or immediately after or soon after the child. As a rule the membranes are ruptured and the amniotic fluid escapes before the head extrudes, but at times the child is born still surrounded by the intact membranes with their contained fluid, it comes into the world in its "caul."

When the placenta and membranes have left the uterus (they not unfrequently are lodged for a while in the vagina), the tonic contraction or "retraction" spoken of above, which during the whole of labour has been following up the advance of the fœtus, and progressively lessening the uterine cavity, continues its work and serves an important purpose. When the last pain of labour, by which the emptied uterus is gathered up into a small hard ball, passes away, the walls under normal conditions do not wholly relax, a permanent tonic contraction keeps the walls thick and in contact, thus closing the uterine cavity; and over this compact and closed uterus waves of rhythmical contraction, the "after-pains," still for a while pass without altering its permanent condition. By this continued contraction or retraction, not only the open, torn ends of the vessels of the decidua but all the vessels throughout the thickness of the uterine walls are so compressed that all extensive bleeding is prevented. Should this continued contraction give away to relaxation, hæmorrhage or "flooding" follows. This retraction or tonic contraction, whatever be its exact nature, which is so conspicuous in the uterus but which perhaps may be recognized in a lesser degree as mere ordinary tonic contraction in other rhythmically contracting organs, in the bladder, in the intestine, and even in the heart, appears to serve more than one purpose in the work of the uterus; by continually lessening the uterine cavity it renders more efficient during labour the rhythmic uterine "pains," by compressing the blood vessels during labour it gradually shuts off the extravagant blood supply now no longer needed, and by continuing that compression after labour and by closing the uterine cavity it prevents hæmorrhage and wards off the evil effects which the free entrance into the uterine cavity of foreign organisms might bring about. And probably it is on account of its great usefulness that this peculiar form of muscular activity is so prominent in the uterus.

Even before labour proliferation of the epithelioid cells may be observed in the lining membrane of the uterine vessels; these are rapidly increased after labour is completed, and form part of the healing processes which follow. The tonic contraction of which we have been speaking is maintained until the blood vessels are permanently closed by these nutritive healing processes. After birth the muscular elements of the uterus dwindle, many of the fibres undergoing fatty degeneration, and thus the mucous and muscular walls are gradually brought back to their natural condition. During the early days of this process of involution a discharge, the lochia, takes place from the internal surface of the uterus.

§ 967. The whole process of parturition may be broadly considered as a reflex act, the nervous centre of which is placed in the lumbar cord. In a dog, whose thoracic cord had been completely severed, parturition took place as usual; and the fact that, in the human subject, labour will progress in a natural manner while the patient is unconscious from the administration of chloroform, though it is often retarded and sometimes arrested, shews that in woman also the contractions both of the uterus and of the abdominal muscles are involuntary, however much the latter may be assisted by direct volitional efforts.

Observations on animals shew that even in a virgin uterus and in one which is not enlarged by pregnancy movements can be excited in a reflex manner through the central nervous system and may occur rhythmically in an apparently spontaneous manner; but the latter are often absent or are so slight as readily to escape observation, and the former are often feeble and excited with difficulty. In a pregnant animal on the other hand, especially if pregnancy be advanced, powerful rhythmic expulsive movements repeatedly occur in the apparent absence of all extrinsic stimuli and are very readily provoked by the stimulation of various afferent nerves. They may also be induced by direct stimulation of the spinal cord at any part of its whole length as well as of various regions of the brain; the analogy with the movements of the urinary bladder leads us to suppose that the impulses thus started in the brain and upper part of the spinal cord do not pass directly to the uterus but throw into activity the reflex centre in the spinal cord. Movements of the uterus are readily excited when the blood ceases to be duly arterialized, extrusion of the fœtus being a common result when a pregnant animal is asphyxiated; and though the venous blood may act in part as a direct stimulus to the uterine muscles the contractions are mainly due to the blood exciting the nervous centre. Drugs such as ergot which increase uterine contractions probably in like manner produce their effect chiefly at least through their action on the nervous centre. The ready way in which the uterus enlarged by pregnancy responds by reflex contraction to the stimulation of various afferent nerves is illustrated in the human subject by the means usually adopted to secure after the birth of the child that continued contraction by which hæmorrhage is avoided. Should for any reason such a contraction fail to take place, it may be secured by applying cold or pressure to the abdominal walls or by introducing a hand or some foreign body into the vagina, or, what perhaps best illustrates the reflex nature of the matter, by applying the child to the nipple; in the latter case the relatively feeble afferent impulses generated in the mammary nerves by the sucking of the child are especially potent in producing by reflex action contraction of the uterine muscles.

§ 968. The nerves of the uterus reach that organ chiefly along the broad ligament in company with the blood-vessels, are partly medullated, partly non-medullated, and are derived from the pelvic plexus lying between the rectum and the vagina. The pelvic plexus, on which as also on the nerves passing to the uterus, numerous small ganglia are scattered, is a continuation on each side of the body of the medially placed hypogastric plexus, but it is joined by branches coming directly from the sacral nerves. In the lower animals (dog) the roots which supply fibres to the uterus are on the one hand the upper lumbar, which traverse the sympathetic strand known as the hypogastric nerve, and on the other hand probably the first and second sacral. In the human subject the corresponding roots are probably the upper lumbar and third, fourth and second sacral.

Stimulation, in the dog, either of the hypogastric nerve or of the sacral nerves produces contractions in the pregnant uterus; it is stated that the mode of contraction is different in the two cases, in the latter the longitudinally disposed fibres, in the former the circularly disposed fibres being especially thrown into action, it will be remembered that a like difference has been stated to obtain in the case of the rectum (§ 276). Moreover, while the fibres passing by the hypogastric nerve are vaso-constrictor towards the uterine arteries, it is said that those passing by the sacral nerves are vaso-dilator. It would be hazardous at present however to insist on any sharp distinction between the two sets of fibres as to the kind of muscular contraction which they bring about; and we may conclude that when the lumbar centre, excited in a reflex action, sends out efferent impulses, these, whatever be their exact nature, pass along both sets of fibres to the uterine muscles.

§ 969. Though we may speak of even the distinctly uterine portion of the act of parturition as reflex in nature, we are hardly justified in considering the rhythmical contractions of the uterus during parturition as simple reflex acts exactly comparable to the contractions of the skeletal muscles in an ordinary reflex movement of the limbs. The peculiar rhythmic character of the contractions, each 'pain' beginning feebly, rising to a maximum, then declining, and finally dying away altogether, to be succeeded after a pause by a similar pain just like itself, pain following pain like the tardy long-drawn beats of a slowly beating heart, suggests that the cause of the rhythmic contraction is seated, like that of the rhythmic beat of the heart, in the organ itself. And this view is supported by the fact that contractions of the uterus, similar to those of parturition, have been observed in animals even after complete destruction of the spinal cord; in such cases the movements may be induced or increased by asphyxia, the venous blood acting, under these conditions, directly on the muscular fibres; and a like result has been obtained with certain drugs. The same view moreover is not only indirectly supported by the occurrence of the rhythmic but futile contractions, which as we have said take place during a large part of the period of pregnancy, but is strongly confirmed by the observation in animals that rhythmic movements may take place in a uterus wholly removed from the body, and that even in a uterus which is not pregnant. We may therefore probably apply to the uterus arguments similar to those which we used (§ 429) in connection with the movements of the bladder in micturition; and indeed many analogies may be drawn between the two acts. We may regard the efferent impulses which issue from the lumbar centre, not so much of the nature of directly excitor impulses as of impulses of an augmentor kind, increasing and developing the intrinsic beats of the uterus itself.

§ **970.** Though under normal circumstances efficient uterine contractions do not set in until the full period of gestation is completed, yet by reason of changes in the uterus or its contents, occurring from natural causes or induced artificially, the full swing of movements may, at almost any time, though at some times more readily than at others, be brought about. On the other hand it may be delayed for a considerable time beyond the proper term. We may be said to be in the dark as to why the uterus, after remaining for months subject only to futile contractions, is suddenly thrown into powerful and efficient action, and within it may be a few hours or even less gets rid of the burden which it has borne with such tolerance for so long a time. None of the various hypotheses which have been put forward can be considered as satisfactory. There is no evidence for the view, based on the occurrence of contractions in consequence of an asphyxiated condition of the blood, that the onset of labour is caused by a gradual diminution of oxygen or accumulation of carbonic acid in the blood, reaching at last to a climax. Nor are there sufficient facts to connect parturition with any condition of the ovary resembling that accompanying menstruation. Nor can much stress be laid on the supposition that the real exciting cause is the separation of the decidua from the permanent uterine wall, the separation being the outcome of the preceding processes of growth, since the actual separation itself seems to be caused by the initial contractions of labour, and the histological changes which precede it are only one set of changes among many others all having their goal in labour. We can only say that labour is the culminating point of a series of events, and must come sooner or later, though its immediate advent may at times be decided by accident; but it would not be profitable to discuss this question here.

The action of the abdominal muscles in parturition, at least so much as takes place independently of the will, is, in contrast to that of the uterine muscles, obviously a reflex act of a more ordinary kind carried out by means of the spinal cord; and we may suppose that, though the mere contractions of the uterus may serve as a possible source, the necessary stimulus is supplied by the pressure of the fœtus in the vagina; in support of this it may be noted that the action becomes much intensified towards the end of labour as the stress and strain caused by the advancing head tell more and more on the external skin.

§ 971. Hence as we have said the whole act of parturition may with reason be considered as a reflex one. Whether it be wholly a reflex or in a certain sense an automatic one, the act can readily be inhibited by other contemporary actions of the central nervous system. Thus emotions very frequently become a hindrance to the progress of parturition; as is well known, the entrance into the bedroom of a stranger often causes for a time the sudden and absolute cessation of 'labour' pains, which previously may have been even violent. Judging from the analogy of micturition, we may suppose that this inhibition of uterine contractions is brought about by an inhibition of the centre in the lumbar cord leading to a sudden cessation of the augmentor action of which we spoke above as far as the uterus itself is concerned, and in a more direct way to a cessation of the contractions of the abdominal muscles. Some observations tend to shew that a region of the bulb exerts such an inhibitory influence; but the matter needs fuller investigation.

CHAPTER III.

THE PHASES OF LIFE.

§ **972.** THE child has at birth, on an average, rather less than one-third the maximum length, and about one-twentieth the maximum weight, to which in future years it will attain. The composition of the body of the new-born babe, as compared with that of the adult, will be seen from the following table, in which the details are more full than those given in § 519; the figures in brackets are more recent observations.

	Weight of organ in percentage of Body-weight		Weight of organ in adult, as compared with that of new-born babe taken as 1.
	New-born babe.	Adult.	
Eye	·28	·023	1·7
Brain	14·34 (12·28)	2·37 (2·25)	3·7 (3·34)
Kidneys	·88	·48	12
Skin	11·3	6·3	12
Liver	4·39 (5·03)	2·77 (3·05)	13·6 (11·05)
Heart	·89 (·73)	·52 (·49)	15 (12·1)
Stomach and Intestine	2·53	2·34	20
Lungs	2·16	2·01	20
Skeleton	16·7	15·35	26
Muscles, &c.	23·4	43·1	48
Testicle	·037	·08	60

It will be observed that the brain and eyes are, relatively to the whole body-weight, very much larger in the babe than in the adult. This disproportion is a very marked embryonic feature, and has a morphological or phylogenic, as well as a physiological or teleological, significance. Inasmuch as the smaller body has relatively the larger surface, the skin is naturally proportionately greater in the babe; but the same disproportion is observed in the kidneys, these like the skin increasing in weight twelve times only between birth and full growth, whereas the whole body increases twenty times. The heart and the liver according to the newer observations behave very similarly, and even according to the older observations lag considerably behind the whole frame, whereas the lungs and the alimentary canal almost exactly keep pace with it, and the skeletal framework, in spite of its being specifically lighter in its earlier cartilaginous condition, maintains throughout life very nearly the same relative weight. The muscles on the contrary grow more than twice as fast as the whole body ; the great increase in these covers the relative decrease of the other parts, and it is largely by the laying on of flesh and fat that the babe gains the bulk of the man.

§ 973. We usually measure growth by taking account of two sets of changes, changes of stature and changes of weight; and we may study both these changes in more than one way.

If we measure the height at intervals we may plot out the curve of growth of stature, and when we do this we find that the curve rises rapidly at first but afterwards more slowly, shewing that the increment is decreasing, and at about the twenty-fifth year ceases to rise at all. From thence to about fifty years of age the height remains stationary, after which there may be a decrease, especially in extreme old age. The curve moreover is not regular, but indicates by its changes that the increment of height in a given time is now smaller, now greater.

The curve of weight is, on the whole, at first very similar to that of height, rising in a somewhat similar way and shewing similar irregularities; but instead of ceasing to rise at about the twenty-fifth year it continues to rise, though marked with many irregularities, and may continue to do so until about the fortieth year. After the sixtieth year a decline of variable extent is generally witnessed. It should be noted that in the first few days of life, so far from there being an increase, there is an actual decrease of weight, so that, even on the seventh day the weight still continues to be less than at birth ; and a similar post-natal loss of weight is observed in animals. If we take the curve of growth from the impregnation of the ovum onwards this post-natal loss of weight will appear as an abrupt change in the curve due to the so to speak violent act of birth. It should be added that the curves both of height and weight exhibit differences dependent on sex, circumstances, race, climate and the like.

We may also study the progress of growth by measuring the increment of growth in a given time, in a year for instance, and plotting out the curve of the yearly increment. When we do this we obtain very instructive results. We find that the yearly increment decreases very rapidly during the first two or three years, then remains nearly stationary or even rises, and at about the seventh or eighth year undergoes a marked fall.

This fall, however, is temporary only; the curve soon rises again and with some irregularities attains a maximum between the twelfth and fifteenth year, from which point onwards it falls rapidly with some minor irregularities. These marked variations in the increment of growth which are obviously connected with and preparatory to the important change which we call puberty, are seen in the curves both of stature and of weight, the changes in weight occurring however rather later than those of stature, and both being somewhat different in boys from what they are in girls. Both are also influenced by the conditions of life; but a study of the curves of growth of young people living under various surroundings, while it teaches the great importance of properly administering to the wants of youth, at the same time illustrates the recuperative elasticity of the bodily frame; it may often be observed that the ill effects of adverse circumstances, provided they be not too great, are soon recovered from under the influence of a happy change; food and comfort will turn the abnormal fall in the curve of growth of a starved waif into a sharp rise.

Lastly, we may study growth by observing the actual rate of growth, by measuring the magnitude of the fraction of the total weight which is added to the weight in a given time; we take weight because this is the most significant element of growth. When this method is adopted, an examination of such statistics as are available with regard to man, confirmed by the results of careful observations on young animals, tends to shew that the rate diminishes continually from birth onwards, the diminution being rapid at first but slower afterwards, and being broken by various irregularities. In other words, the power of growth diminishes continually though somewhat irregularly throughout life, and a like diminution apparently obtains in intra-uterine existence. It seems as if the impetus of growth given at impregnation gradually dies out.

§ **974.** The saliva of the babe, very scanty at first and not abundant until teething begins, is active on starch though less so than in the adult, and its gastric juice, unlike that of many new-born animals, has good peptic powers, and its pancreas good tryptic powers, though apparently the pancreatic action on starch is feeble. From this we may infer that its digestive processes are in general identical with that of the adult though ill suited for any large amount of starch in the food; and they are feeble, since the fæces of the infant contain a considerable quantity of undigested food (fat, casein &c.), as well as unaltered bile-pigment, and undecomposed bile-salts.

The heart of the babe, as shewn in the preceding Table, is, relatively to its body-weight, larger than the adult, and the frequency of the heart-beat much, greater, viz. about 130 or 140 per minute, falling to about 110 in the second year, and about 90 in the tenth year. Corresponding to the smaller bulk of the body, the whole circuit of the blood system is traversed in a shorter time than in the adult (12 seconds as against 22); and consequently the renewal of the blood in the tissues is exceedingly rapid. Relatively to the body-weight there is also considerably more blood in the babe than in the adult. The respiration of the babe is quicker than that of the adult, being at first about 35 per minute, falling to 28 in the second year, to 26 in the fifth year, and so onwards. The respiratory work, while it increases absolutely as the body grows, is, relatively to the body-weight, greatest in the earlier years. It is worthy of notice, that the absorption of oxygen is said to be during these earlier years relatively more active than the production of carbonic acid; that is to say, there is a continued accumulation of capital in the form of a store of oxygen-holding explosive compounds (cf. § 358). This, indeed, is the striking feature of infant metabolism. It is a metabolism directed largely to constructive ends. The food taken represents, undoubtedly, so much potential energy; but before that energy can assume a vital mode, the food must be converted into tissue; and, in such a conversion, morphological and molecular, a large amount of energy must be expended. The metabolic activities of the infant are more pronounced than those of the adult, for the sake, not so much of energies which are spent on the world without, as of energies which are for a while buried in the rapidly increasing mass of flesh. Thus the infant requires over and above the wants of the man, not only an income of energy corresponding to the energy of the flesh actually laid on, but also an income corresponding to the energy used up in making that living sculptured flesh out of the dead amorphous proteids, fats, carbohydrates and salts, which serve as food. Over and above this, the infant needs a more rapid metabolism to keep up the normal bodily temperature This, which is no less, indeed slightly (·3°) higher, than that of the adult, requires a greater expenditure, inasmuch as the infant with its relatively far larger surface, and its extremely vascular skin, loses heat to a proportionately much greater degree than does the grown-up man. It is a matter of common experience that children are more affected by cold than are adults. The bodily temperature is moreover less stable in the infant than in the adult, and departures from the normal temperature have not the grave significance they have in the adult.

This rapid metabolism is however not manifest immediately upon birth. During the first few days, corresponding to the loss of weight mentioned above, the respiratory activities of the tissues are feeble; the embryonic habits seem as yet not to have been completely thrown off, and, as was stated in § 376, new-born animals bear with impunity a deprivation of oxygen, which would be fatal to them later on in life.

Associated probably with these constructive labours of the

growing frame is the prominence of the lymphatic system. Not only are the lymphatic glands largely developed and more active (as is probably shewn by their tendency to disease in youth), but the quantity of lymph circulation is greater than in later years. Characteristic of youth is the size of the thymus body, which increases up to the second year, and may then remain for a while stationary, but generally before puberty has suffered a retrogressive metamorphosis, so that very often hardly a vestige of it remains behind. The thyroid body is also relatively greater in the babe than in the adult; the spleen, on the other hand, relatively to the body-weight does not change greatly, though rather smaller in the adult. As we have already said the recuperative power of infancy and early youth is very marked.

The quantity of urine passed, though scanty in the first two days, rises rapidly at the end of the first week, and in youth the quantity of urine passed is, relatively to the body-weight, larger than in adult life. This may be, at least in quite early life, partly due to the more liquid nature of the food, but is also in part the result of the more active metabolism. For not only is the quantity of urine passed, but also the amount of urea and of some other urinary constituents excreted, relatively to the body-weight, greater in the child than in the adult. The presence of uric acid, of oxalic acid, and according to some, of hippuric acid in unusual quantities is a frequent characteristic of the urine of children. It is stated that calcic phosphates, and indeed the phosphates generally, are deficient, being retained in the body for the building up of the osseous skeleton.

§ 975. It would be beyond the scope of this work to enter into the psychical condition of the babe or the child, and our knowledge of the details of the working of the nervous system in infancy is too meagre to permit of any profitable discussion. It is hardly of use to say that in the young the whole nervous system is more irritable or more excitable than it is in later years; by which we probably to a great extent mean that it is less rigid, less marked out into what, in preceding portions of this work, we have spoken of as nervous mechanisms. In new-born puppies and some other animals stimulation of the various cerebral areas does not give rise to the usual localized movements; these do not appear until some time after birth; but in this respect differences are observed in different kinds of animals corresponding to the well known differences between different kinds of animals in the powers possessed at birth; the human babe as regards the latter is intermediate between the puppy and the young guinea-pig. As we have seen, the fibres of the various tracts in the central nervous system acquire their medulla at different epochs; there is experimental evidence in support of the view, otherwise probable, that the assumption of functional activity follows in the same order; and the pyramidal tract is as we have seen the one in which the fibres are very late in acquiring their medulla. It has been asserted that in a new-born animal stimulation of the vagus produces no cardiac inhibition and that this does not appear for several days; other observers however have obtained positive results and that even in the uterus; probably in this respect also animals differ. In the human infant the sense of touch, both as regards pressure and temperature, appears well developed, as does also the sense of taste, and possibly, though this is disputed, that of smell. The pupil (larger in the infant than in the man) acts fully, and normal binocular movements of the eyes have been observed in an infant less than an hour old. The eye is from the outset fully sensitive to light, though of course visual perceptions are imperfect. Auditory sensations on the other hand, seem to be dull, though not wholly absent, during the first few days of life; this may be partly at least due to absence of air from the tympanum and to a tumid condition of the tympanic mucous membrane. As the child grows up his senses sharpen with constant exercise, and in his early years he possesses a general acuteness of sight, hearing, and touch, which frequently becomes blunted as his psychical life becomes fuller. Children however are said to be less apt at distinguishing colours than in sighting objects; but it does not appear whether this arises from a want of perceptive discrimination or from their being actually less sensitive to variations in hue. A characteristic of the nervous system in childhood, the result probably of the more active metabolism of the body, is the necessity for long or frequent and deep slumber.

§ 976. Dentition marks the first epoch of the new life. At about seven months the two central incisors of the lower jaw make their way through the gum, followed immediately by the corresponding teeth in the upper jaw. The lateral incisors, first of the lower and then of the upper jaw, appear at about the ninth month, the first molars at about the twelfth month, the canines at about a year and a half, and the temporary dentition is completed by the appearance of the second molars usually before the end of the second year.

About the sixth year the permanent dentition commences by the appearance of the first permanent molar beyond the second temporary molar; in the seventh year the central permanent incisors replace their temporary representatives, followed in the next year by the lateral incisors. In the ninth year the temporary first molars are replaced by the first bicuspids, and in the tenth year the second temporary molars are similarly replaced by the second bicuspids. The canines are exchanged about the eleventh or twelfth year, and the second permanent molars are cut about the twelfth or thirteenth year. There is then a long pause, the third or wisdom tooth not making its appearance till the seventeenth, or even twenty-fifth year, or in some cases not appearing at all.

§ **977.** Shortly after the conclusion of the permanent dentition (the wisdom teeth excepted) the occurrence of puberty marks the beginning of a new phase of life; and the difference between the sexes, hitherto merely potential, now becomes functional. In both sexes the maturation of the generative organs is accompanied by the well-known changes in the body at large; but the events are much more obvious in the typical female than in the aberrant male. Though in the boy, the breaking of the voice and the rapid growth of the beard which accompany the appearance of active spermatozoa, are striking features, yet they are after all superficial; and though, as we have seen (§ 973), the curves of his increasing weight and height undergo before and at this period, characteristic variations, the general events of his economy pursue for a while longer an unchanged course; the boy does not become a man till some years after puberty; and the decline of his functional manhood is so gradual that frequently it ceases only when disease puts an end to a ripe old age. With the occurrence of menstruation, on the other hand, at from thirteen to seventeen years of age, subsequent to the acceleration of growth noted above § 973, which indeed appears preparatory to it, the girl almost at once becomes a woman, and her functional womanhood ceases suddenly at the climacteric in the fifth decennium. During the whole of the child-bearing period her organism is in a comparatively stationary condition. The variations in the yearly increment of the girl before puberty though not so marked are more complex than those of the boy, and she reaches the maximum of yearly increment sooner than does he; during this whole period indeed she precedes him in growth and she has nearly reached her maximum, while he is still continuing to grow. Her curve of weight from the nineteenth year onward to the climacteric, remains stationary, being followed subsequently by a late increase, so that while the man reaches his maximum of weight at about forty, the woman is at her greatest weight about fifty.

Of the statical differences of sex, some, such as the formation of the pelvis, and the costal mechanism of respiration, are directly connected with the act of child-bearing, while others have only an indirect relation to that duty; and indications at least of nearly all the characteristic differences are seen at birth. The baby boy is heavier and taller than the baby girl, and the maiden of five breathes with her ribs in the same way as does the matron of forty. The woman is lighter and shorter than the man, the limits in the case of the former being from 1·444 to 1·740 metres of height and from 39·8 to 93·8 kilos of weight, in the latter from 1·467 to 1·890 of height, and from 49·1 to 98·5 kilos of weight. The muscular system and skeleton are both absolutely and relatively less in woman, and her brain is lighter and smaller than that of man, being about 1272 grammes to 1424. Her metabolism, as measured by the respiratory and urinary excreta, is also not only absolutely but relatively to the body-weight less, and her blood is not only less in quantity but also of lighter specific gravity and contains a smaller proportion of red corpuscles. Her strength is to that of man as about 5 to 9, and the relative length of her step as 1000 to 1157.

§ 978. From birth onward (and indeed from early intra-uterine life) the increment of growth as we have seen, though undergoing certain variations, continues to diminish. At last a point is reached at which the curve cuts the abscissa line, and the increment becomes a decrement. After the culmination of manhood at forty and of womanhood at the climacteric, the prime of life declines into old age. The metabolic activity of the body, which at first was sufficient not only to cover the daily waste but to add new material, later on is able only to meet the daily wants, and at last is too imperfect even to sustain in its entirety the existing frame. Neither as regards vigour and functional capacity, nor as regards weight and bulk, do the turning-points of the several tissues and organs coincide either with each other or with that of the body at large. We have already seen that the life of such an organ as the thymus is far shorter than that of its possessor. The eye is in its dioptric prime in childhood, when its media are clearest and its muscular mechanisms most mobile, and then it for the most part serves as a toy; in later years, when it could be of the greatest service to a still active brain, it has already fallen into a clouded and rigid old age. The skeleton reaches its limit very nearly at the same time as the whole frame reaches its maximum of height, the coalescence of the various epiphyses being pretty well completed by about the twenty-fifth year. Similarly the muscular system in its increase tallies with the weight of the whole body. The brain, in spite of the increasing complexity of structure and function to which it continues to attain even in middle life, early reaches its limit of bulk and weight. At about seven years of age it attains what may be considered as its first limit, for though it may increase somewhat up to twenty, thirty, or even later years, its progress is much more slow after than before seven. The vascular and digestive organs as a whole may continue to increase even to a very late period. From these facts it is obvious that though the phenomena of old age are, at bottom, the result of the individual decline of the several tissues, they owe many of their features to the disarrangement of the whole organism produced by the premature decay or disappearance of one or other of the constituent bodily factors. Thus, for instance, it is clear that were there no natural intrinsic limit to the life of the muscular and nervous systems, they would nevertheless come to an end in consequence of the nutritive disturbances caused by the loss of the teeth. And what is true of the teeth is probably true of many other organs, with the addition that these cannot, like the teeth, be replaced by mechanical contrivances. Thus the term of life which is allotted to a muscle by virtue of its molecular constitution, and which it could not exceed were it always placed under the most favourable nutritive conditions, is, in the organism, further shortened by the similar life-terms of other tissues; the future decline of the brain is probably involved in the early decay of the thymus.

Two changes characteristic of old age are the so-called calcareous and fatty degenerations. These are seen in a completely typical form in cartilage, as, for instance, in the ribs; here the cell-substance of the cartilage corpuscle becomes hardly more than an envelope of fat globules, and the supple matrix is rendered rigid with amorphous deposits of calcic phosphates and carbonates, which are at the same time the signs of past and the cause of future nutritive decline. And what is obvious in the case of cartilage is more or less evident in other tissues. Everywhere we see a disposition on a part of the living substance of the tissue to fall back upon the easier task of forming fat rather than to carry on the more arduous duty of manufacturing new material like itself; everywhere almost we see a tendency to the replacement of a structured matrix by a deposit of amorphous material. In no part of the system is this more evident than in the arteries; one common feature of old age is the conversion by such a change of the supple elastic tubes into rigid channels, whereby the supply to the various tissues of nutritive material is rendered increasingly more difficult, and their intrinsic decay proportionately hurried.

Of the various tissues of the body the muscular and nervous are however those in which functional decline, if not structural decay, becomes soonest apparent. The dynamic coefficient of the skeletal muscles diminishes rapidly after thirty or forty years of life, and a similar want of power comes over the plain muscular fibres also; the heart, though it may not diminish, or even may still increase in weight, possesses less and less force, and the movements of the intestine, bladder, and other organs, diminish in vigour. In the nervous system, the lines of resistance, which, as we have seen, help to map out the central organs into mechanisms, and so to produce its multifarious actions, become at last hindrances to the passage of nervous impulses in any direction, while at the same time the molecular energy of the impulses themselves becomes less. The eye becomes feeble, not only from cloudiness of the media and presbyopic muscular inability, but also from the very bluntness of the retina; the sensory and motor impulses pass with increasing slowness to and from the central nervous system, and the brain becomes a more and more rigid mass of nervous substance, the molecular lines of which rather mark the history of past actions than serve as indications of present potency. The epithelial glandular elements seem to be those whose powers are the longest preserved; and hence the man who in the prime of his manhood was a 'martyr to dyspepsia' by reason of the sensitiveness of gastric nerves and the reflex inhibitory and other results of their irritation, in his later years, when his nerves are blunted, and when therefore his peptic cells are able to pursue their chemical work undisturbed by extrinsic nervous worries, eats and drinks with the courage and success of a boy.

§ 979. Within the range of a lifetime are comprised many periods of a more or less frequent recurrence. In spite of the aids of a complex civilisation, all tending to render the conditions of his life more and more equable, man still shews in his economy the effects of the seasons. The birth-rate for instance shews an increase in winter, and most people gain weight in winter and lose weight in summer. Careful observations of school children shew that these increase in length rapidly in the spring but hardly at all in the autumn, and very slowly in the winter, while their increase in weight is most marked in autumn, being very slight or even negative in the spring, and not great in winter. Some of these apparent effects of the season are the direct results of varying temperature, but some probably are habits acquired by descent, and in some again the connection is a very indirect or possibly not a real one. Within the year, an approximately monthly period is manifested in the female by menstruation, though there is no exact evidence of even a latent similar cycle in the male. The phenomena of recurrent diseases, and the marked critical days of many other maladies, may be regarded as pointing to cycles of smaller duration than that of the moon's revolution, save in the cases in which the recurrence is to be attributed rather to periodical phases in the disease-producing germ itself, than to variations in the medium of the disease.

§ 980. Prominent among all other cyclical events is the rhythmic rise and fall in the activities of the central nervous system; most animals possessing a well-developed nervous system, must, night after night, or day after day, or at least time after time, lay them down to sleep. The salient feature of sleep is the cessation or extreme lowering of the psychical activity of the brain and of the nervous processes which serve as the basis of that activity. When sleep is at its height, the afferent nervous impulses which external agents set going in the afferent somatic nerves such as those of the special senses, are no longer the starting points of complex cerebral processes; not only do they fail to excite consciousness and to leave their mark on memory, but they may be unable to call forth even a simple reflex movement. And yet they are not wholly without effect; for though a set of feeble afferent impulses may produce no visible reaction and leave no impression on the mind of the sleeper, yet impulses of the same kind, if made stronger in proportion to the depth of the sleep, may be followed by their wonted cerebral consequences, and may thus awake, as we say, the sleeper. It would seem as if the afferent impulses met in their course with an unwonted resistance to their progress, as if the wheels of the cerebral machinery worked stiffly so that the lesser shocks of molecular change which otherwise would have moved them, were broken and wasted upon them. Corresponding to this block or lessened inroad of afferent impulses, the outflow of efferent impulses is stopped or largely diminished; the body gives no sign of the working of a conscious will, the eyelids drop and the head nods, and the various actions by which the erect posture is maintained are let go for lack of the governing motor impulses. And psychological self-inquiry tells us that in complete sleep this absence of outward signs of cerebral activity has its fellow in the absence of inward marks; the interval between falling asleep and awakening is a blank and gap in the history of the mind.

We say 'complete sleep' since there are many degrees of sleep, the state which we call that of dreaming being one of them; and between the most perfect wide-awakefulness and that deepest slumber which refuses for a long time to give way before even the strongest stimuli, no clear line of demarcation can be drawn. When we fall asleep the tie between 'ourselves' and the external world is not suddenly snapped, we do not by one step pass from consciousness to unconsciousness; and the same when conversely we awake; as the world vanishes from us or comes back to us, the afferent impulses of sight, of sound and of other kinds, for a period which may be brief but always exists, produce, before they cease or begin appreciably to affect us at all, effects in ascending or descending scale which we call unreal. And the outward signs of sleep may vary from one in which volition is present and even dominant, to one in which even the simplest reflex movements of the skeletal muscles are with difficulty evoked, and the maintenance of some skeletal tone (§ 597) and of breathing afford, so far as the skeletal muscles are concerned, almost the only token that the central nervous system is alive. But we cannot enter here into the psychology of sleep and dreams.

Though the phenomena of sleep are largely confined to the central nervous system and especially to the cerebral hemispheres, the whole body shares in the condition. The pulse and breathing are slower, the intestine, the bladder, and other internal muscular mechanisms are more or less at rest, and the secreting organs are less active, some apparently being wholly quiescent; the secretion of mucus attending a nasal catarrh is largely diminished during slumber, and the sleeper on waking rubs his eyes to bring back to his conjunctiva its needed moisture. The output of carbonic acid, and the intake of oxygen, especially the former, is lessened; the urine is less abundant and the urea falls. Indeed the whole metabolism and the dependent temperature of the body are lowered; but we cannot say at present how far these are the indirect results of the condition of the nervous system, or how far they indicate a partial slumbering of the several tissues.

Thoracic respiration is said to become more prominent than diaphragmatic respiration during sleep, and a rise and fall of the respiratory movements, resembling if not identical with the Cheyne-Stokes rhythm of respiration (§ 375), is frequently observed. During sleep the pupil is constricted, during deep sleep exceedingly so; and dilation, often unaccompanied by any visible movements of the limbs or body, takes place when any sensitive surface is stimulated; on awaking also the pupils dilate. The eyeballs have been generally described as being during sleep directed upwards and converging, or according to some authors, diverging; but others maintain that in true sleep the visual axes are parallel and directed to the far distance. The eyes of children have been described as continually executing during sleep movements, often irregular and unsymmetrical and unaccompanied by changes in the pupils. The contraction of the pupils is worthy of notice, since it shews that the condition of sleep is not merely the simple and direct result of the falling away of afferent impulses; when the eyes are closed in slumber the pupils ought, since the retina is then quiescent, to dilate; that they are constricted, the more so the deeper the sleep, shews that important actions in the brain, probably in the middle portions of the brain, are taking place.

We are not at present in a position to trace out the events which culminate in this inactivity of the cerebral structures. The analogies between ordinary sleep and winter sleep or hibernation (§ 540) are probably real; the chief difference appears to be that in the latter the diminished activity is due to an extrinsic cause, cold, and in the former to intrinsic causes, to changes in the organism itself; but we saw in treating of hibernation, that intrinsic changes prepared the way for the action of external cold. It has been urged that during sleep the brain is anæmic, and though observations have yielded conflicting results, the evidence seems to be in favour of this view; but even if this anæmia is a constant accompaniment of sleep, it must, like the vascular condition of a gland or any other active organ, be regarded as an effect, or at least as a subsidiary event, rather than as a primary cause. Nor can the view which regards sleep as the result of a change in the mechanical-arrangements of the cranial circulation, such as either a retardation or acceleration of the venous outflow, be considered as satisfactory. The essence of the condition is rather to be sought in purely molecular changes; and the analogy between the systole and diastole of the heart, and the waking and sleeping of the brain, may be profitably pushed to a very considerable extent. The sleeping brain in many respects closely resembles a quiescent but still living ventricle. Both are so far as outward manifestations are concerned at rest, but both may be awakened to activity by an adequately powerful stimulus.

Both, though quiescent, are irritable, in both the quiescence will ultimately give place to activity, and in both an appropriate stimulus applied at the right time will determine the change from rest to action. Just as a single prick will under certain circumstances awake a ventricle, which for some seconds has been motionless, into a rhythmic activity of many beats, so a loud noise will start a man from sleep into a long day's wakefulness. And just as in the heart the cardiac irritability is lowest at the beginning of the diastole and increases onwards till a beat bursts out, so is sleep deepest at its commencement after the day's labour; thence onward slighter and slighter stimuli are needed to wake the sleeper. For judging of the depth of ordinary nocturnal sleep by the intensity of the noise required to wake the sleeper, it may be concluded that, increasing very rapidly at first, it reaches its maximum within the first hour; from thence it diminishes, at first rapidly, but afterwards more slowly.

We cannot, however, at present make any definite statements concerning the nature of the molecular changes which determine this rhythmic rise and fall of cerebral irritability. The fact that the products of metabolic activity when they accumulate within a tissue appear to become in the end an obstruction to that activity, has suggested the idea that the presence in the cerebral tissue of an excess of the products of nervous metabolism is the cause of sleep; and a parallel has been drawn between the sleep of cerebral tissue and the diminished irritability of muscular tissue attending muscular fatigue, in which the products of muscular metabolism have been supposed (§ 86) to play an important part. Indeed lactic acid has been especially pointed to in this connection; but there is no solid reason for attributing any such importance to this particular substance; and if during the rest of sleep this or any other metabolic product is washed out of the nervous tissue by the blood stream we should expect a greater, not a less supply of blood to the brain during sleep. Besides, if the mere accumulation of metabolic products of any kind were the cause of sleep, it is not clear why we should ever have any hope of waking. More perhaps may be said in favour of the conception that during the waking hours the expenditure of oxygen exceeds the income and that the quiescence, which we call sleep, comes from the exhaustion of the body's store of oxygen, more especially of that 'intramolecular' oxygen of which we spoke (§ 358), in dealing with the respiration of the tissues. But to this view must be added some hypothesis, such as the byplay of some inhibitory mechanism, whereby the respiratory centre is not roused to increased activity by this lack of oxygen (for as we have seen the breathing shares in the slumber of the body) though continuing to play with an amount of energy, which permits a gradual restoration of the lost store of oxygen and so finally brings on the awakening which ends the sleep. And the necessity for such a complication indicates that the explanation is, at present at least, inadequate.

The phenomena of sleep shew very clearly to how large an extent an apparent automatism is the ultimate outcome of the effects of antecedent stimulation. When we wish to go to sleep we withdraw our automatic brain as much as possible from the influence of all extrinsic stimuli. We lie down in order to relieve the skeletal muscles and indeed the heart too, from the labour entailed by the erect posture; we put off the tight garments which continually spur the skin; we empty the bladder to avoid the stimulus of its distension; and we choose for sleep the night and a quiet place, drawing the curtains, in order that our eyes may be withdrawn from light and our ears from sounds. In this connection may be quoted the interesting case which is recorded of a lad whose sensory tie with the external world was, from a complicated anæsthesia, limited to that afforded by a single eye and a single ear; the lad could be sent to sleep at will, by closing the eye and stopping the ear.

§ 981. The cycle of the day is however manifested in many other ways than by the alternation of sleeping and waking, with all the indirect effects of these two conditions. There is a diurnal curve of temperature, apparently independent of all immediate circumstances, the hereditary impress of a long and ancient sequence of days and nights. Even the pulse, so sensitive to all bodily changes, shews, running through all the immediate effects of the changes of the minute and the hour, the working of a diurnal influence which cannot be accounted for by waking and sleeping, by working and resting, by meals and abstinence between meals. And the same may be said concerning the rhythm of respiration, and the products of pulmonary, cutaneous and urinary excretion. There seems to be a daily curve of bodily metabolism, which is not the product of the day's events. Within the day we have the narrower rhythm of the respiratory centre with the accompanying rise and fall of activity in the vaso-motor centres. And lastly, there stands out the fundamental fact of all bodily periodicity, that alternation of the heart's systole and diastole which ceases only at death. Though, as we have seen, the intermittent flow in the arteries is toned down in the capillaries to an apparently continuous flow, still the constantly repeated cycle of the cardiac shuttle must leave its mark throughout the whole web of the body's life. Our means of investigation are, however, still too gross to permit us to track out its influence. Still less are we at present in a position to say how far the fundamental rhythm of the heart itself, that rhythm which is influenced, but not created, by the changes of the body of which it is the centre, is the result of cosmical changes, the reflection as it were in little of the cycles of the universe, or how far it is the outcome of the inherent vibrations of the molecules which make up its substance.

CHAPTER IV.

DEATH.

WHEN the animal kingdom is surveyed from a broad standpoint, it becomes obvious that the ovum, or its correlative the spermatozoon, is the goal of an individual existence; that life is a cycle beginning in an ovum and coming round to an ovum again. The greater part of the actions which, looking from a near point of view at the higher animals alone, we are apt to consider as eminently the purposes for which animals come into existence, when viewed from the distant outlook whence the whole living world is surveyed, fade away into the likeness of the mere byplay of ovum-bearing organisms. The animal body is in reality a vehicle for ova; and after the life of the parent has become potentially renewed in the offspring, the body remains as a cast-off envelope whose future is but to die.

Were the animal frame not the complicated machine we have seen it to be, death might come as a simple and gradual dissolution, the 'sans everything' being the last stage of the successive loss of fundamental powers. As it is, however, death is always more or less violent; the machine comes to an end by reason of the disorder caused by the breaking down of one of its parts. Life ceases not because the molecular powers of the whole body slacken and are lost, but because a weakness in one or other part of the machinery throws its whole working out of gear.

We have seen that the central factor of life is the circulation of the blood, but we have also seen that blood is not only useless, but injurious, unless it be duly oxygenated; and we have further seen that in the higher animals the oxygenation of the blood can only be duly affected by means of the respiratory muscular mechanism, presided over by the respiratory centre in the bulb. Thus the life of a complex animal is, when reduced to a simple form, composed of three factors; the maintenance of the circulation, the access of air to the hæmoglobin of the blood, and the functional activity of the respiratory centre; and death may come from the arrest of either of these. As it has been put, death may begin at the heart or at the lungs or at the brain. In reality, however, when we push the analysis further, the central fact of death is the stoppage of the heart, and the consequent arrest of the circulation; the tissues then all die, because they lose their internal medium. The failure of the heart may arise in itself, on account of some failure in its nervous or muscular elements, or by reason of some mischief affecting its mechanical working. Or its stoppage may be due to some fault in its internal medium, such for instance as a want of oxygenation of the blood, which in turn may be caused by either a change in the blood itself, as in carbonic oxide poisoning, or by a failure in the mechanical conditions of respiration, or by a cessation of the action of the respiratory centre. The failure of this centre, and indeed that of the heart itself, may be caused by nervous influences proceeding from the brain, or brought into operation by means of the central nervous system; it may, on the other hand, be due to an imperfect state of blood, and this in turn may arise from the imperfect or perverse action of various secretory or other tissues. The modes of death are in reality as numerous as are the possible modifications of the various factors of life; but they all end in a stoppage of the circulation, and the withdrawal from the tissues of their internal medium. Hence we come to consider the death of the body as marked by the cessation of the heart's beat whenever that cessation is one from which no recovery is possible; and by this we are enabled to fix an exact time at which we say the body is dead. We can, however, fix no such exact time to the death of the individual tissues. They are not mechanisms, and their death is a gradual loss of power. In the case of the contractile tissues, we have apparently in rigor mortis a fixed term, by which we can mark the exact time of their death. If we admit that after the onset of rigor mortis recovery of irritability is impossible, then a rigid muscle is one permanently dead. In the case of the other tissues, we have no such objective sign, since the rigor mortis of other tissues manifests itself chiefly by obscure chemical signs. And in all cases it is obvious that the possibility of recovery, depending as it does on the skill and knowledge of the experimenter, is a wholly artificial sign of death. Yet we can draw no other sharp line between the seemingly dead tissue whose life has flickered down into a smouldering ember which can still be fanned back again into flame, and the handful of dust, the aggregate of chemical substances into which the decomposing tissue finally crumbles.

Moreover, the failure of the heart itself is at bottom loss of irritability, and the possibility of recovery here also rests, as far as is known at present, on the skill and knowledge of those who attempt to recover. So that after all the signs of the death of the whole body are as artificial as those of the death of the constituent tissues.

INDEX.

Aberration, spherical, iv. 48
,,   chromatic, iv. 50
Abscissa line, mode of measuring curves on, 233 note
Absorption from alimentary canal, 481, 511
Achroodextrin, 359
Acid-albumen, formation of, 19, 97, 368
Acid, benzoic, an antecedent of hippuric acid, 672
,,   ,, renal action on, 673
,, butyric &c. in fat, 772
,, carbonic, clotting retarded by its presence, 21
,,   ,, development of, in rigor mortis, 100
,,   ,, set free during muscular contraction, 103, 151, 806
,,   ,, in expired and inspired air, 550
,,   ,, amount of daily excretion of, 551, 793
,,   ,, percentage of, in expired air, 552, 576
,,   ,, its relations in the blood, 570
,,   ,, causes of its escape from the blood, 577
,,   ,, 'fixed' and 'loose' in blood serum, 571
,,   ,, its exit from the lungs, 575
,,   ,, in the blood, effect of excess of, 600
,,   ,, thrown off by the skin, 696
,,   ,, lessened output of, in sleep, iv. 413
,, cholalic, 747
,, hippuric, chemical composition, 650
,,   ,, how formed in kidney, 672
,, hydrochloric, in gastric juice, 418
,, lactic, in the blood, 51

Acid, lactic, isomeric variations of, 99 note
,,   ,, its effect on the heart, 303
,,   ,, fermentation, 577
,, uric, chemical composition, 649
,,   ,, a result of special metabolism, 757
Acids, fatty, 428, 473
,,   ,, in milk, 781
,,   ,, in sebum, 695
,, organic, in urine, 651
,, various, in spleen-pulp, 743
Action, currents of, 111, 124, 1044
,, peristaltic, in plain muscular tissue, 159
,,   ,, of the stomach, 459
,,   ,, of intestine, 465
,,   ,, increased in asphyxia, 469
,,   ,, of uretur, 664, 680
,,   ,, of bladder, 682
,,   ,, of gall-bladder, 708
,, reflex, 179, 683
,,   ,, purposive nature of, 181
,,   ,, of spinal cord, 902—919
,, automatic, 183, 920
Addison's disease, 767
Adenoid tissue in lymphatic glands, 45
,,   ,, multiplication of leucocytes in, 45, 490
,,   ,, structure of, 443
,,   ,, seat of interchange between blood and lymph, 494
Adipose tissue, structure of, 769
After-birth, formation of, iv. 376
After-images, iv. 75
,,   ,, negative and positive, iv. 128
After-pains, services effected by, iv. 398
Age, vascular changes due to, 343
,, old, phenomena of, iv. 410
Agminated follicles, 489
"Ague-cake," 743
Air, tidal and stationary in lungs, 535
,, complementary, supplemental and residual, 536

INDEX.

Air, expired, changes of, 550, 552
Albumin, acid- and alkali-, 19, 97
Albumose, clotting prevented by, 29
" distinguished from peptone, 369
Alcohol, changes in proteids produced by, 24
" physiological action, 348
" use in diet, 837
Alimentary canal, vaso-constrictor nerves of, 322, 438
" structure, 377
" mucous membrane of, 378
" glands of, 379
" circular and longitudinal coats of, 456, 464
" spontaneous movements of, 464, 465, 468
" nerve supply to coats, 466, 467
" peristaltic action, 464
" mutually destructive juices of, 476
" absorption from, 481, 511
Alkaloids, vegetable, their kinship to urea, 823
Allantoin, iv. 390
Allantois, growth of the, iv. 376
Altitudes, high, diminished oxygen pressure in, 575
Alveolus of lung tissue, 527, 529, 532
" of gland, 380, 386
" " lymphatic, 492
" " mammary, 777
" of pancreas, 391
Alvergniat's gas-pump, 555
Amblyopia, 1078
Ammonia, mode of conversion into urea, 756, 757
Amnion, true and false, iv. 376, 385
Amniotic fluid, iv. 385
" " function, iv. 386
" " composition, iv. 390
Amœbæ, 3—7
Amœboid movements of white corpuscles, 38, 42, 166, 337, 490
Amphibia, ending of nerve fibres in muscles of, 120
" urinary tubule in, 643
" double vascular supply to kidney, 666
Ampullæ of the semicircular canals, 1009
Amylolytic action of saliva, 361
Anabolic changes of living substance, 42
Anacrotic pulse, usually pathological, 265, 271
Anæmia, lessened number of red corpuscles in, 34
" respiration impeded by, 629
Anelectrotonus, 129
" relation to irritability, 133
"Animal starch," 730

Animals, cold-blooded, temperature of, 809
" warm-blooded, " "
810
Annulus of Vieussens, 299
Anode, 60
Aorta, proportion of sectional area of capillaries to the, 199
" comparative blood-pressure in, 239
" closure of valves of, 246
" absence of nerves from, 272
Aphasia, connection of with cortical lesion, 1053, 1056
" phenomena of, 1057
Apnœa, how produced, 603, 604
Aqueduct of Sylvius, 930
Aqueous humour, iv. 168
Arachnoid membrane, 1125
Arantii, corpus, 197
Area, cortical motor, in dog, 1035
" " " in monkey, 1038
" " " in anthropoid ape, 1043
" " " in man, 1052
" " " mode of action of, 1056
" " " for movements of the eyes, 1079
" " " for speech, 1053, iv. 326
" " " for movements of larynx, iv. 325
" " " for respiratory movements, iv. 326
" " for vision, 1081
" " for smell, 1087
" " for cutaneous sensations, 1091
" diabetic, 730
" visual, iv. 81
" tactile, iv. 277
Areolar tissue, 188
Aristotle, experiment of, iv. 305
Arterial pressure, 203, see also Blood, pressure of.
" " tracings of, 206
" " heart-beat in inverse ratio to, 305
" " as affected by tonic contraction, 307
" " " by quantity of blood, 341
" " " by exercise, 350
" " vaso-motor action on, 319, 327
" scheme, model of, 215
" tone, 308
" " intrinsic nature of, 329
Arterialization, effect of deficient, 622
Arteries, effect of ligature on, 28, 202
" structure of, 193, 194
" elasticity and contractility of, 198, 214, 306

INDEX. 421

Arteries, pulse in, 201, 209, 227
,, changes of calibre in, 306, 622
,, supply of vaso-motor nerves to, 307, 319
,, effect on blood-pressure of their contractility, 320
,, intrinsic tone of muscular wall of, 329
,, as affected by age, 343, iv. 411
,, basilar, 627
,, of the brain, 1131
,, bronchial, 534, 617
,, of the eye, iv. 165
,, helicine, in erectile tissue, iv. 371
,, hepatic, 707
,, radial, 257—260
,, renal, 644, 657
,, ,, in amphibia, 666
,, splenic, 737
,, umbilical, iv. 383
,, ,, venous blood in the, iv. 384
Arterioles, 193
Artificial pulse, tracings of, 216, 258
Arytenoid cartilages, iv. 308
Ash of muscle, salts in, 102
,, ,, nerves, salts in, 123
Asphyxia, phenomena of, 599, 606, 611, 625
,, increased peristaltic action in, 469
,, its effect on the vascular system, 624
,, how produced, 599
Astigmatism, iv. 48
Ataxy, locomotor, 1103
Atrophy, acute, of liver, decrease of urea in, 755
Atropin, cardiac inhibition counteracted by, 300
,, salivary secretion arrested by, 400, 417
,, its action on sweating, 700, 701
,, its effect on the pupil, iv. 44
,, ,, on accommodation, iv. 45
Auditory epithelium, structure, iv. 234
,, hairs, iv. 233
Auerbach, plexus of, 441
Aura, forms of, preceding epileptiform attacks, 1087, 1093
Auricle, histology of, 275
Automatism of heart and brain, analogy of, 1115
Automatic action, 183, 920
Axis-cylinder of nerve fibre, 115, 116
,, ,, of nerve fibre, impulses transmitted by, 118
,, ,, of nerve fibre, disintegration of, after severance, 144
,, ,, process, in nerve cells of spinal cord, 177, 178

Babe, the, composition of as compared with adult, iv. 405
,, digestive processes of, iv. 405

Babe, nervous system of the, iv. 407
Bacteria, ingestion of by white corpuscles, 46
,, results of their presence in alimentary canal, 428, 477
Bacterium photometricum, its reaction towards light, iv. 115
Bands, bright and dim, in muscle tissue, 29
,, ,, ,, in cardiac tissue, 275
Banting, his method of reducing fatness, 844
Basis, molecular, of chyle, 500
Bat, the, gastric glands of, 412
Beats, in musical sounds, iv. 230
Beehive, temperature of, 810
Bertini, columns of, 636
Bidder's ganglia in heart of frog, 278, 279
Bile, characters and composition, 421
,, secretion of, 434, 435
,, action on food, 424
,, ,, on fats, 475
,, antagonistic to peptic action, 424, 476
,, storage of in gall-bladder, 435
,, resorption of under pressure, 439
,, osmotic passage of fat facilitated by, 517
,, manufactured by hepatic cells, 713
,, relation of to formation of urea, 749
,, fœtal secretion of, iv. 389
Bile-acids, their formation, 747
Bile capillaries, 710
,, ducts, 708
,, pigments, 38, 423, 744
,, salts, 423
Bilin, its composition, 424
Bilirubin, its relations with hæmatin, 35, 744
,, composition of, 423
,, formation of, 746, 748
Biliverdin, its composition, 423
Birth, changes of the lungs at, 537
,, ,, of circulation at, iv. 393
Bladder, muscles of, 680
Blastodermic vesicle, iv. 382
Blind spot of retina, iv. 110
Blindness, total and partial, 1077
Blood, the, 13—53
,, an internal medium of interchanges, 13, 186, 191
,, clotting, 15—30
,, ,, circumstances affecting, 20
,, ,, causes of, 26
,, its relation to vascular walls, 27, 339
,, corpuscles, see Corpuscles, blood-
,, 'laky,' how formed, 32, 560
,, ,, results of injection of, 671
,, chemical composition, 49—51
,, specific gravity, 49

INDEX.

Blood, quantity and distribution, 52
,, ,, in a part, mode of measuring, 314
,, ,, results of changes in, 341
,, rate of flow in vessels, 220
,, ,, its dependence on vasomotor action, 319
,, amount driven by each heart-beat, 201, 253
,, time occupied by one circulation, 254
,, quality of, its effect on heart-beat, 303
,, ,, ,, on peripheral resistance, 340
,, ,, as affected by exercise, 349
,, ,, ,, by deficient aeration, 600
,, ,, in infant, iv. 406
,, œdema due to changes in, 509
,, respiratory changes in, 553—571
,, gases of, 557
,, ,, how measured, 555
,, entrance of oxygen into by diffusion, 526
,, exit of carbonic acid from, 575
,, relations of oxygen in, 557
,, ,, of carbonic acid in, 570
,, ,, of nitrogen in, 571
,, venous, its influence on muscular irritability, 148
,, ,, colour of, 565
,, ,, spectrum of, 565
,, ,, an excitant of respiratory centre, 598
,, ,, its slowing effect on heart-beat, 621
,, ,, arterial in colour during hibernation, 820
,, ,, in umbilical arteries, iv. 384
,, arterial, colour of, 565
,, constancy of percentage of sugar in, 727
,, proteids of the, 823
,, course of, in fœtus, iv. 390
,, circulation of, see Circulation.
,, platelets, 47
,, ,, in relation to inflammation, 337
,, pressure, arterial and venous compared, 202, 209, 215
,, ,, how measured, 202, 207
,, ,, mode of registering, 204
,, ,, in arteries, 207
,, ,, in veins, 203, 207
,, ,, in capillaries, 208
,, ,, nature and causes of, 211
,, ,, its relation to peripheral resistance, 215
,, ,, ,, ,, to heart-beat, 304, 351

Blood pressure, as affected by cardiac inhibition, 296
,, ,, ,, by stimulation of depressor, 324
,, ,, ,, by stimulation of sciatic, 325
,, ,, ,, by action of drugs, 340
,, ,, ,, by changes in amount of blood, 342
,, ,, ,, by respiration, 614
,, ,, in the kidneys, 657, 660, 665
,, ,, in the brain, 1134
,, ,, in umbilical vein and artery, iv. 383
,, serum, constituents of, 19
,, supply, its influence on muscular irritability, 146
,, vessels, their influence on fluidity of blood, 28
Blushing, its cause, 332, 347
Body, the, characteristics of in life and death, 1, 2
,, average composition of, 788
,, metabolic processes of, 703
,, changes during starvation, 789
,, potential and actual energy of, 801
,, expenditure of energy by, 803
,, sense of position of, 1012
Bois-Reymond, du, his 'key,' 61
,, ,, ,, on muscle-currents, 109, 110
Bone, presence of lymph in, 486
Bowman, glands of, iv. 247
Brachial plexus, constrictor and dilator fibres in, 315
Brain, the, reaction of, 123, 1138
,, its action on spinal reflexes, 916
,, embryonic, 929
,, ,, primary vesicles of, 929
,, ,, transformations of, 930
,, ,, correspondence of with adult brain, 935
,, cerebral hemispheres, 930
,, ,, ,, relation of to crossed side of body, 1093
,, ,, ,, their action not identical, 1118
,, ,, ,, results of removal of in frog, 999
,, ,, ,, results of removal of in birds, 1003
,, ,, ,, results of removal of in mammals, 1005, 1008
,, ,, ,, lateral ventricles, 931
,, ,, ,, corpus striatum, 932
,, ,, ,, ,, lenticularis, 932
,, ,, ,, ,, callosum, 932

INDEX. 423

Brain, cerebral hemispheres, fornix, 932
,, ,, ,, choroid plexus, 934
,, cerebellum, development of, 930, 935
,, ,, peduncles of, 936, 993
,, ,, corpus dentatum of, 984
,, ,, relations of to the spinal cord, 1107
,, ,, histology, 1022
,, crura cerebri, 930
,, grey matter of, chief collections of, 952
,, ,, ,, nature and relations of, 967
,, ,, ,, central, 953, 1021
,, ,, ,, nuclei of cranial nerves, 95—365
,, ,, ,, superficial of cerebrum, 970, 1026
,, ,, ,, ,, of cerebellum, 970, 1022
,, ,, ,, intermediate, of crural system, 970
,, ,, ,, corpus striatum, 971
,, ,, ,, optic thalamus, 971
,, ,, ,, crus, pes and tegmentum, 972, 984
,, ,, ,, internal capsule, 974
,, ,, ,, nucleus lenticularis, 974, 987
,, ,, ,, nucleus caudatus, 974
,, ,, ,, nucleus amygdalæ, 978
,, ,, ,, globus pallidus, 974
,, ,, ,, putamen, 974
,, ,, ,, pulvinar, 981
,, ,, ,, substantia nigra, and red nucleus, 981
,, ,, ,, pons, 982
,, ,, ,, upper olive, 982
,, ,, ,, other collections of, 983
,, ,, ,, corpora quadrigemina, 983
,, ,, ,, corpora quadrigemina anterior, their connection with vision, 1076
,, ,, ,, corpora geniculata, 983
,, fibres of, arrangement, 984
,, ,, pedal and tegmental systems, 984
,, ,, pedal system, longitudinal fibres of, 987
,, ,, pyramidal tract, 987
,, ,, anterior cortical, 989
,, ,, posterior cortical, 990
,, ,, from the nucleus caudatus to the crus, 990
Brain, fibres of tegmental system, longitudinal fibres, 991
,, ,, ,, cortical, 991
,, ,, ,, optic radiation, 991
,, ,, ,, superior peduncles of cerebellum, 993, 1111
,, ,, ,, fillet, 993
,, ,, ,, longitudinal posterior bundle, 993
,, ,, ,, tracts from the corpora quadrigemina, 994
,, ,, transverse or commissural, 994
,, ,, ,, corpus callosum, 994
,, ,, ,, anterior and posterior white commissures, 995
,, ,, ,, fornix, 995
,, ,, ,, middle peduncles of cerebellum, 995
,, summary of relations of parts, 996
,, histological features, 1021
,, histology of central and other grey matter, 1021
,, ,, of superficial grey matter of cerebellum, 1022
,, ,, of the nuclear layer of cerebellum, 1022
,, ,, of the cells of Purkinjé, 1022
,, ,, of cerebral cortex, 1026
,, ,, of cortex, pyramidal cells of, 1027
,, ,, of cortex, layers of, 1029
,, ,, of parietal, occipital and frontal regions, 1030
,, cortex, experimental interference with, 1035
,, ,, localization of function in, 1036
,, ,, results of removal, 1049
,, ,, sources of energy of, 1115
,, ,, psychical processes in the, 1117, 1118
,, optic tracts, 1074
,, ,, ,, endings of, 1075, 1076
,, splanchnic functions, 1114
,, membranes of, 1125
,, cerebro-spinal fluid of, 1126
,, arteries of, 1131
,, venous sinuses of, 1133
,, circulation in, 1134
,, supply of blood to, 1136
,, its condition during sleep, iv. 414
Breaking of the voice at puberty, iv. 332
Breath, the first, cause of, iv. 393
,, ,, effect of, iv. 393
Breathing, normal rate of, 542
,, male and female, differences between, 543

Breathing, an involuntary act, 584
," laboured, nervous mechanism of, iv. 324
Bright's disease, œdema of, 509
Broca's convolution, 1053
Bronchia of mammalian lung, 529
Bronchioles ,, ,, 529, 532
Brownian movements in molecular basis of chyle, 500
Bruch, membrane of, iv. 23
Brunner, glands of, 450
Buffy coat of clotted blood, 16
Bulb, transformation of cord into, 937
," decussation of pyramids in, 937
," olivary and restiform bodies, 937
," calamus scriptorius, 937, 943
," fasciculus gracilis and f. cuneatus, 939, 947
," sensory decussation in, 942
," reticular formation in, 944
," posterior fissure of, 943
," arcuate fibres, 944
," grey matter of, 944, 946
," olive, inferior, 945
," olivary nuclei, 945
," antero-lateral nucleus, 946
," fibres of, 948
," tubercle of Rolando, 948
," inter-relations of parts, 949
," restiform body, 950
Bulbus arteriosus, absence of nerves in, 277
Burdach, column of, 870
Burdon-Sanderson, his stethometer, 540

Calamus scriptorius, 937, 943
Calorimeters, 804
Calcic phosphate, insolubility of curd dependent on, 375
,, ,, in milk, 782
Calories, combustion of food expressed in, 802
Canalis auricularis in lower vertebrates, 277
,, cochlearis, iv. 199
,, reuniens, iv. 200
Canal, cerebro-spinal, structure of, 1125
Canals, semicircular, result of injury to, 1009
,, structure, iv. 198
Capacity, vital, 538
Capillaries described, 13
,, their permeability, 13, 191, 199
,, structure, 190
,, blood-interchanges effected in, 13, 186, 191
,, calibre, 192, 334
,, plasmatic layer in, 335
,, proportion of sectional area of, to aorta, 199, 210
,, measurements of blood-pressure in, 208

Capillaries, disappearance of pulse in, 209
,, peripheral resistance in, 211
,, lymph, their structure, 483
,, of the kidney, 644
,, bile, 710
Capillary circulation, normal phenomena of, 210, 334
,, as affected by inflammation, 336
Capsule, Malpighian, 634, 639
,, of kidney, 646
,, Glisson's, 705
,, of spleen, 735
,, Tenon's, iv. 167
Capsules, supra-renal, structure, 765
,, ,, histology of, 766
Carbo-hydrates in white corpuscles, 41
,, ,, in food stuffs, 356
,, ,, various forms of, 359
,, ,, in food, presence of glycogen dependent on, 717, 720
,, ,, formation of fat from, 773
,, ,, as food, potential energy of, 802
Carbon, inspired, retained in bronchial glands, 534
Carbon monoxide, asphyxia from, 609
Carbonic acid, see Acid, carbonic.
Cardio-graphic tracings, 244, 247
Cardio-inhibitory centre, 295
Casein, 367
,, precipitation of, 374
,, a constituent of milk, 781
,, its formation in mammary gland, 781
Cartilage, passage of lymph in, 486
,, thyroid, iv. 308, 313
,, cricoid and arytenoid, iv. 308, 313
,, of Santorini, iv. 309, 313
,, of Wrisberg, iv. 311, 313
,, of Luschka, iv. 313
Cavities, serous, 487
Cells, of adenoid tissue, 443
,, albuminous in salivary glands, 387, 389
,, ,, changes in, 407
,, auditory epithelium, iv. 208
,, cardiac muscular tissue, in frog, 275, 292
,, ,, ,, ,, in mammal, 276
,, central, of gastric glands, 383, 419
,, of cerebellum, 1022
,, of cerebral cortex, 1026
,, ciliary, 162
,, ,, action of chloroform on, 166
,, ,, in trachea, 531
,, of Claudius, iv. 215
,, columnar of the villi, 445

INDEX.

Cells, columnar, fat absorbed by, 516
,, connective tissue, 187
,, of Corti, iv. 218
,, of Deiters, iv. 218
,, cylinder, of auditory epithelium, iv. 208
,, ,, of olfactory mucous membrane, iv. 246
,, demilune, 388
,, differentiation of, during development of ovum, 6
,, endocardial, 197
,, epithelioid of capillaries, 191
,, ,, of arteries, 193
,, ,, of lymph vessels, 483
,, epithelium, 163
,, fat, 770
,, ganglionic, 173
,, ,, of cardiac ganglia, 279
,, goblet, of villus, 446
,, ,, in glands of Lieberkühn, 449
,, ,, of tracheal mucous membrane, 531
,, of grey matter, 177
,, hair, inner and outer, of auditory epithelium, iv. 208, 217
,, of Hensen, iv. 214
,, hepatic, continuous activity of, 437
,, ,, structure, 708
,, ,, work of, 713
,, ,, changes in, 719
,, ,, glycogen in, 720
,, ,, action on hæmoglobin, 746
,, of lymph capillaries, 483
,, of mammary gland, 778
,, mucous, of cardiac glands of the stomach, 382
,, ,, of salivary glands, 385, 409
,, ,, "loaded" and "unloaded," 387
,, nerve, of spinal ganglia, 173
,, ,, of splanchnic ganglia, 176
,, ,, spiral, 176, 279
,, ,, of ganglia of heart, 278
,, ,, of central nervous system, 177
,, ovoid, or parietal, of gastric glands, 383, 412
,, of pancreas, 391, 405
,, of parotid, 407
,, 'prickle,' of epidermis, 688
,, of œsophagus, 392
,, pulmonary, of newt, 527
,, of Purkinjé, 277, 1023
,, pyramidal, large and small, 1027
,, of sebaceous glands, 692
,, secreting, series of events in, 413, 419, 785
,, of trachea, 531
,, unipolar and multipolar, 279

Cells, of villi, 445
,, ,, double function of, 523
Cell-action in absorption, 523
Cell-substance, milk partly formed from, 779, 784
,, ,, sebum formed from, 693
Cellulose, digestion of, in large intestine, 478
,, a food stuff for the herbivora, 836
Cement substance in capillaries, 190, 192
Centres, nervous, cardio-inhibitory, 205
,, for deglutition, 453, 458
,, ,, lactation, 787
,, ,, micturition, 683
,, ,, ,, inhibition of, 915
,, ,, movements of eyeball, iv. 151
,, ,, parturition, iv. 399
,, ,, phonation, iv. 326
,, ,, pupil-constrictor, iv. 37
,, for respiration, 585, 586
,, ,, automatic action of, 587, 601
,, ,, regulation of, 596
,, ,, as affected by blood supply, 598, 627
,, ,, activity of, increased by exercise, 602
,, for sweating, 701
,, possible, for thermal changes, 816
,, trophic, for nutrition of nerves, 853
,, vaso motor, 323
,, ,, limits of, 326
,, visual, higher and lower, 1083
,, for vomiting, 461
Cerebellar tract, 874, 890, 950, iv. 264
Cerebellum, see Brain, cerebellum.
Cerebral operations, time taken by, 1120
Cerebrin in nerve tissue, 121
Changes, anabolic and katabolic in living substance, 41, 822
,, ,, nervous regulation of, 829
,, diurnal, in functions, iv. 416
Chauveau and Lortet, their hæmatochomometer, 223
,, and Marey, their mode of measuring endocardiac pressure, 240
Chest, expansion and contraction of, during respiration, 536
Chest-voice, how produced, iv. 329
Cheyne-Stokes respiration, 605
Chiasma, optic, decussation of fibres in, 1073
Chloral, its effect on stimulation of depressor, 325
Chlorides, their presence in serum, 50
Chloroform, its effect on ciliary action, 166
Cholesterin, composition of, 422

426 INDEX.

Cholesterin, a constituent of bile, 422
,, its presence in blood, 50
,, ,, in red corpuscles, 51
,, ,, in nerve substance, 121
,, ,, in gall stones, 121
,, ,, in milk, 781
,, ,, in the lens, iv. 29
Chondrin, action of gastric juice on, 374
Chordæ vocales, iv. 306
Chorion, the, iv. 376
Choroid, development of, iv. 4, 5
,, structure of, iv. 21
,, pigment of, iv. 23
,, blood supply of, iv. 165
Chromogens, their presence in urine, 652
Chyle, characters of, 499
,, molecular basis of, 500
,, passage of fat into, 512, 517
,, presence of sugar in, 513
,, absence of peptone in, 514
,, elaboration of, in the villus, 517
Chyme, how formed, 471
Cilia, 102
,, of tracheal mucous membrane, 531
,, of bronchial passages, 533
,, of urinary tubule of frog, 641
Ciliary movements, 55, 162
,, ,, circumstances affecting, 163
,, processes, iv. 23
,, plexuses, aqueous humour furnished by, iv. 168
,, zone, iv. 26
,, muscle, iv. 28
Circulation of the blood, main facts of 201
,, capillary, 210, 334
,, hydraulic principles of, 211, 212
,, aids to, 219
,, methods of measuring its velocity, 220
,, time occupied by circuit, 225, 254
,, constant and variable factors of, 343
,, changes in, 344
,, causes of irregularity in, 345
,, ,, of cessation of, 346
,, effect of asphyxia on, 626
,, slowing of during hibernation, 820
,, renal, 644, 657
,, placental, iv. 383
,, early fœtal, iv. 383
,, late fœtal, iv. 390
,, changes of, taking place at birth, iv. 393
Circus movements, 1017
Cisterna magna lymphatica of frog, 487
Clarke's column of spinal cord, 869

Claustrum, the, 1029
Clotting of blood, 15—30
,, retarded by cold, 16
,, ,, by addition of saline solutions, 16, 22
,, ,, by presence of oil, 21
,, ,, by carbonic acid in blood, 21
,, ,, by injection of albumiose, 29
,, causes of, 26
,, in the living body, 28
,, favoured by presence of foreign bodies, 21, 28, 48
,, of fluids other than blood, 23
,, of muscle plasma in rigor mortis, 40, 98, 99
,, small amount of fibrin required for, 50
,, of lymph, 497
Coagulation of proteids by heat, 18
Cochlea of ear, iv. 179
,, structure, iv. 202
,, its several parts, iv. 211
,, basilar membrane of, iv. 212
,, measurements of parts of, iv. 222
Cohnheim's areas, 91
Coitus, behaviour of spermatozoa after, iv. 311
Cold, its influence on clotting, 1621
,, ,, on irritability of muscle and nerve, 146
,, ,, on vaso-constrictor action, 348
,, ,, on skin action, 674
,, great, lowering of metabolism by, 819
,, terminal organs for sensation of, iv. 281
,, sensations of, due to changes of skin-temperature, iv. 278
Colostrum, composition of, 783
Colour sensations, many kinds of, iv. 84
,, ,, mixing of, iv. 85, 87, 91
,, ,, characters of, iv. 88
,, ,, Young-Helmholtz's theory of, iv. 91
,, ,, Hering's theory of, iv. 93
,, ,, due to metabolic changes, iv. 94
,, ,, in relation to intensity of stimulus, iv. 107
,, ,, unequal change of, under diminishing light, iv. 108
,, vision, variations in, iv. 99
,, blindness, iv. 99
,, ,, different kinds of, iv. 100
,, ,, dichromic in nature, iv. 100
,, ,, Young-Helmholtz's theory of, iv. 101

Colour blindness, Hering's theory of, iv. 102
" " absolute, iv. 105
Colours, complementary, iv. 89
" primary, iv. 91
" of arterial and venous blood, 553—565
Columns of spinal cord, 858
Combustion of various foods, rates of, compared, 802
Connective tissue, structure of, 187
" " corpuscles, 189
" " "loose," 188
" " action of gastric juice on, 374
" " retiform or reticular, 443
" " in relation to lymphatic vessels, 481
" " " to lymph spaces, 485, 486
Constant current, its action, 126
" " as compared with induction shock, 134
Commissure, inferior optic, 1074
Commissures of cord, anterior and posterior, 858, 863
Cones of retina, iv. 62 et sup.
" " approximate dimensions of, iv. 81
Coni vasculosi of testis, iv. 365
Conjunctiva, structure of, iv. 173
Consciousness, its connection with nervous action, 911
Consonants and vowels, iv. 334
" manner of formation, iv. 337
" classification of, iv. 338
Constriction of arteries, 307
" of pupil a reflex act, iv. 37
Constrictor fibres, 312
Contour, double, of nerve fibre, 114
Contractile tissues, the, 54—167
" material of muscle tissue, 153
Contractility, 56, 166
Contraction of muscle, movements of body due to, 54
" simple and tetanic, 58
" graphic method of recording, 58
" simple, phenomena of, 68
" tetanic, 78, 82
" of skeletal muscles tetanic in character, 81, 1066
" wave of, 86
" visible changes of muscle during, 91
" nature of act of, 94, 151
" chemical changes due to, 102
" heat given out during, 104
" making and breaking, 126
" influence of nature of stimulus on, 136
" prolonged, of red muscle, 140

Contraction, as influenced by load, 141
" idio-muscular, 143
" exhausting effects of the products of, 150
" result of chemical changes in, 151
" of unstriated muscle, 159
" peristaltic, 159
" spontaneous, 160
" tonic, 162
" relation of amœboid movements to, 166
" of arterial muscular fibre, 313
" of heart, auricular and ventricular synchronous, 252
" features of, 286
" of villus, 519
Contractions, peristaltic, in plain, muscular tissue, 159
" " of ureter, 664, 680
" " of bladder, 682
" " of the stomach, 459
" " of the intestine, 465
" " of gall-bladder and duct, 708
" rhythmical, of spleen tissue, 741
" " of the uterus during pregnancy, iv. 395
" " " during 'labour,' iv. 396
" " " after parturition, iv. 398
" " " intrinsic nature of, iv. 400
" of the abdominal walls, 464, iv. 397
Contrast visual, simultaneous, iv. 126
" " successive, iv. 128
Conus medullaris, 878
Convulsions, anæmic, how produced, 607
Co-ordination of movements, machinery of, in birds, 1009
" " machinery of, in mammals, 1011
" " machinery of, in man, 1012
" " parts of middle brain concerned in, 1019
" of ocular movements, iv. 148
" " " nervous mechanism governing, iv. 151
Cord, spinal, 169, 849—928
" " general features, 856
" " white matter, structure, 859
" " " tracts of, delimitation of, 869, 872
" " neuroglia, 859, 860
" " " structure, 862

Cord, spinal, grey matter, 859, 861, 880, 896
,, ,, ,, ,, nature of, 895
,, ,, central canal, and substantia gelatinosa, 863
,, ,, grouping of nerve cells, 865
,, ,, ascending and descending degeneration, 870, 872
,, ,, connections of nerve roots, 876
,, ,, special features of the several regions, 878—895
,, ,, variation in sectional area of white matter, 879
,, ,, variation in sectional area of grey matter, 880
,, ,, course of pyramidal tracts in the, 888
,, ,, cerebellar tract, 890
,, ,, median posterior tract, 891, 894, 900
,, ,, antero-lateral ascending tract, 895
,, ,, commissural tracts and transverse connections, 900
,, ,, reflex actions of, 180, 902—919
,, ,, ,, complexity of, 908
,, ,, ,, relations to consciousness, 911
,, ,, ,, in man, 912
,, ,, ,, inhibition of, 915
,, ,, ,, time required for, 818
,, ,, automatic actions of, 920—928
,, ,, action in disease, 927
,, ,, hypothetical segmentation of, 896, 1067
,, ,, motor mechanisms of, 1102
,, ,, lymphatic arrangements of, 1125
Corium, 686
Cornea, nutrition of, 486
,, structure of, iv. 25
,, blood supply to, iv. 165
,, nerve fibres of, iv. 272
Cornu, anterior, of cord, 858
,, ,, nerve cells of, 177, 865
,, ,, connection of afferent with efferent fibres in, 182
,, posterior, 858
,, ,, nerve cells of, 867
Corpora geniculata, 983, 1075, 1076
,, quadrigemina, grey-matter of, 983
,, ,, connections of, 994
,, ,, connection of with vision, 1076
,, ,, considerations touching the, 1112
,, cavernosa, iv. 370
Corpus Arantii, 197

Corpus callosum, 932, 933, 994
,, luteum, iv. 358
,, spongiosum, iv. 370
,, striatum, 932, 971, 974, 992
,, subthalamicum, 982
Corpuscles of blood, not an essential part of clot, 16
,, ,, relations with the plasma, 27
,, connective tissue, 189
,, ,, their relations with epithelioid lymphatic plates, 485
,, cartilage, presence of glycogen in, 729
,, colostrum, 783
,, Malpighian, of spleen, 738
,, red and white, relative proportions, 38
,, ,, ,, composition, 39—50
,, ,, ,, capillary walls permeable by, 191
,, red, microscopic appearance, 31
,, ,, structure, 32
,, ,, chemical composition, 33
,, ,, method of counting, 34
,, ,, their life and death, 35
,, ,, their destruction in the liver, 36
,, ,, as oxygen bearers, 33, 35, 566
,, ,, formed in red marrow of bones, 37, 38
,, ,, their passage through the capillaries, 335
,, ,, diapedesis of, 338
,, ,, proportion of in fœtal blood, iv. 384
,, white (see also Leucocytes), their connection with clotting, 29
,, ,, appearance and structure of, 38, 39
,, ,, composition of, 39, 41, 51
,, ,, type of all living tissue, 41, 44
,, ,, amœboid movements of, 38, 42, 166, 335, 337
,, ,, origin, 44
,, ,, work, 46
,, ,, granulation in, 47
,, ,, behaviour in inflammation, 336—338
,, ,, their migration, 337
,, Pacinian, iv. 268
,, Grandry's, iv. 270
,, touch, iv. 270

Cortex, *see* Brain, Cortex.
Corti, organ of, iv. 205, 213, 214, 234
,, ,, features of, iv. 222
,, rods of, iv. 214
,, ,, inner and outer, iv. 216
,, cells of, iv. 218
Coughing, 631
Cowper, glands of, iv. 370
Cramp, abolished by anelectrotonus, 133
Crassamentum, or clot, 15
Cream, 782
Cretinism, 764
Crico-arytenoid muscle, the posterior, iv. 320
Crico-thyroid muscle, iv. 321
Cricoid cartilage, iv. 308
Cristæ acusticæ, iv. 202
,, ,, structure, iv. 206
Croaking of frog, connection of, with corpora quadrigemina, 1113
Crura cerebri, *see* Brain.
Crying, 631
Crypts or glands of Lieberkühn, 448
Curd, 375
Curdling of milk, phenomena of, 374, 781
Currents of action, in a muscle, 111
,, ,, in a nerve, 124
,, ,, as started by excitation of cortex, 1044
,, of rest in a muscle, 107
,, ,, ,, in a nerve, 123
,, ,, ,, in electrotonus, 131
,, electrical, constant and induced, 59, 60
,, ,, interrupted or faradaic, 65
,, electrotonic, 130
Curves, mode of measuring, 233, note
Cutaneous sensations, *see* Sensations, cutaneous.
Cutis vera, 686
Cyanogen compounds, their relation to urea, 649, 759
Cycle, cardiac, described, 229, 246
,, ,, duration of phases, 249, 250

Death, a gradual process, 1
,, slow clotting of blood after, 27
,, of blood corpuscles, 38, 46
,, from failure of heart's action, 346
,, from high temperature, phenomena of, 818
,, phenomena of, iv. 417
Decidua, formation of, iv. 375
,, reflexa, iv. 375
,, ,, absorption of, iv. 380
,, vera, iv. 375
,, ,, absorption of, iv. 380
,, serotina, iv. 375
,, ,, its transformation into the placenta, iv. 376
,, expulsion of after parturition, iv. 398

Decussation of the pyramids, 889, 939—942, 1046
,, ,, superior or sensory, 943
,, ,, of optic fibres, 1073
,, ,, in relation to movements of pupil, iv. 39
Defæcation, how effected, 463
Degeneration of severed nerve, 144
,, of muscle after severance of nerve, 145
,, of constrictor prior to dilator fibres in severed nerve, 315
,, of nerve fibres in mixed nerve, 853
,, ascending and descending tracts in spinal cord, 870—875
,, calcareous and fatty, iv. 410
Deglutition, how effected, 452
,, different stages of, 454
,, a reflex act, 455
,, movements of œsophagus in, 457
Deiters cell, 860, iv. 218
Dentition, temporary, iv. 408
,, permanent, iv. 408
Depressor nerve, 323, 351
Dermis, structure of, 686
Descemet or Desmours, membrane of, iv. 26
Despretz signal, 73, 74
Dextrins, characters of, 359
Dextrose, reactions of, 359
,, appearance of, in liver after death, 716
,, as food of muscle, 824
Diabetes, natural and artificial, 730
Diapedesis of red corpuscles, 338
Diaphragm, lymph stomata on tendon of, 488
,, method of recording movements of, 541
,, its movements during respiration, 543
,, tetanus of, produced by stimulation of vagus, 590
Diastole of heart's action, 229, 232
Dicrotism in pulse tracings, 264
,, causes of, 267
,, less marked in rigid arteries, 269
,, natural, 733
Diet, average, 802
,, normal, composition of, 833
,, need of the three classes of food stuffs in, 835
,, value of alcohol in, 837
,, vegetable, physiological value of, 839, 841
,, modifications of, with regard to size of body, 842
,, ,, ,, to climate, 843

Diet, modifications of, to labour, 845
,, ,, ,, to mental work, 846
"Differential capacity, extreme," 538
Diffusible substances, absorption of, 520
Diffusion, laws of, 521
,, passage of gases in the tissues by, 526
,, in air of lungs, 535
Digestion, tissues and mechanisms of, 355—525
,, of living tissues, 420
,, muscular mechanisms of, 452
,, effects on, of presence of bacteria, 428, 477
,, main products of, 511
,, course taken by products of, 512
,, gastric, 365
,, ,, circumstances affecting, 371
,, ,, gross effect of, 471
,, ,, time needed for, 472
,, pancreatic, 426
,, salivary, 358
,, infantine, iv. 404
Digitalis, physiological action of, 676
Dioptric mechanisms, iv. 1
,, apparatus, simple form of, iv. 6
,, ,, imperfections in, iv.47
"Discrimination period," 1122
Discus proligerus, iv. 355
Distance, judgment of, iv. 160
Distension of lungs after birth, cause of, 537
Diuretics, their mode of action, 664, 676
Diurnal curves of functions, iv. 416
Division of labour, physiological, 6
Dog, pancreas of, 401
,, submaxillary gland of, 401
,, ,, nerve supply to, 397
,, succus entericus of, 430
,, nerves of alimentary canal, 466
,, saliva of, 470
,, composition of hæmoglobin in blood of, 560
,, cortical motor area in, 1035
Dropsy, character of lymph in, 498
Drowning, 608
Duct, thoracic, structure of, 482
Ducts, salivary, 387
,, hepatic and cystic, 708
Ductus endolymphaticus, iv. 200
,, ejaculatorius, iv. 370
,, venosus, fœtal blood carried to the heart by, iv. 390
,, arteriosus, fœtal circulation through the, iv.392
,, ,, obliteration of after birth, iv. 393
Dudgeon's sphygmograph, 256
Dura mater, 1125
Dyspnœa, at high altitudes, 575
,, nature of, 598, 600

Dyspnœa, cardiac, 627
,, its effect on the kidney, 661
,, sweating caused by, 701

Ear, structure, iv. 176—179
,, embryonic history of, iv. 176
,, otic vesicle, iv. 177
,, general relation of parts, iv. 177
,, cochlea, iv. 179
,, general use of parts, iv. 180
,, tympanum, conducting apparatus of, iv. 181, 190
,, ,, muscles of, iv. 194
,, auditory ossicles, iv. 181
,, ,, ,, attachments of, iv.187
,, Eustachian tube, iv. 196
,, vestibule, parts of, iv. 198
,, ,, perilymph cavity of, iv. 200
,, of rabbit, vaso-motor control of circulation in, 308
Egg-albumin, coagulation of, 367
,, ,, its conversion into acid albumin, 368
Elastic fibres in connective tissue, 189
,, membrane of arteries, 193
Elasticity, diminished, in exhausted muscles, 149
,, of arteries, as affecting circulation, 212
,, ,, as affecting dicrotism, 269
,, of lungs, amount of pressure exerted by, 537
Elastin, in yellow elastic fibres, 190
Electric changes during muscle contraction, 106
,, ,, in a nerve impulse, 123
,, stimuli described, 59
,, spark, vision by illumination of, iv. 74
,, currents, their development by retinal processes, iv. 122
,, organs of certain animals, 120, 155
Electrotonus, features of, 128
Electrotonic currents, 130
Embryo of mammal, undifferentiated protoplasm in, 36
,, development of red corpuscles in, 37
,, glycogen in muscles of, 101, 729
,, development of adipose tissue in, 773
,, ,, of nerve fibre, 870
,, ,, of brain, 929
,, growth of, iv. 376
,, respiration of, iv. 382
,, nutrition of, iv. 382—394
,, supply of oxygen to the, iv. 382
Emetics, various, action of, 462
Emission of semen, iv. 372
,, ,, the striated muscles concerned in, iv. 373

INDEX. 431

Emission of semen, the nervous centre for, iv. 373
Emotions, as affecting respiration, 588, 597
,, respiratory mechanism a means of expressing, 630
,, their effect on secretion of urine, 677
,, ,, ,, of saliva, 398
,, ,, on splanchnic functions, 1114
Emulsion of fats, action of bile and pancreatic juice on, 425, 474, 516
End-bulbs, structure of, iv. 268
End-plates of nerves, probable action of urari on, 58
,, the two parts of, 119
,, ,, their analogy with electric organs of animals, 120, 155
Endocardium, its structure, 197
Endolymph of semicircular canals, in relation to coordinate movements, 1011
,, secretion of, in otic vesicle, iv. 177
,, nature of, iv. 210
Energy, potential, of bodies living and dead, 1
,, of living body expended in work, 2, 805
,, of dead body shewn as heat, 3
,, renewed and set free by different tissues, 6
,, set free by breaking down of living substance, 1
,, of muscle and nerve, 151—164
,, of the body, income of, 801
,, ,, ,, expenditure of, 803
,, potential, of various diets, 801, 833
Entoptic phenomena, iv. 51
Ependyma, the, 953, 1021
Epidermis, structure of, 687
,, prickle cells of, 688
Epididymis, iv. 365
,, action of in emission, iv. 372
Epiglottis, the, iv. 308
,, cushion of the, iv. 313
Epithelioid or endothelial cells of capillaries, 190, 191
Epithelium, ciliated, 162
,, cells, their action in absorption, 523
,, renal, secretion by the, 665
,, auditory, 176
,, germinal, iv. 354
Equilibrium, nitrogeneous, 795
,, sense of, 1012
Erect posture, how preserved, iv. 347
Erectile tissue, structure and action of, iv. 370
Erection of penis, *see* Penis.
Eructation, composition of gases of, 473
Erythrodextrin, 359

Eserin, pupil-contraction caused by, iv. 43
Evaporation, temperature of body regulated by, 699
Eupnœa, 598
Eustachian valve in adult life, 229
,, in fœtal life, iv. 391
Eustachian tube, iv. 183, 196
Excretin a fæcal constituent, 480
Excretion, tissues of, 8
Exercise, effect of, on the muscles, 148
,, ,, on vascular mechanism, 349, 350
,, ,, on respiration, 627
,, ,, on the secretion of urea, 805
,, ,, on the production of heat, 813
,, production of carbonic acid increased by, 806
,, visual discrimination increased by, iv. 83
,, tactile perceptions increased by, iv. 277
Exhaustion of muscle and nerve tissue, 143
,, of muscles, 149
,, auditory effects of, iv. 228
Expiration, how effected, 547
Extractives, various, of spleen pulp, 743
,, of the supra-renal body, 766
,, of the thymus, 768
,, their value in diet, 800, 837
Eye, the, nature of movements caused by stimulation of cortex, 1079
,, general structure, iv. 2
,, development of, iv. 5
,, structure of the investments of, iv. 21—30
,, changes in during accommodation, iv. 18
,, sclerotic coat, iv. 21
,, choroid coat, iv. 21
,, suprachoroidal membrane, iv. 22
,, ciliary processes, iv. 23
,, iris, iv. 24
,, cornea, iv. 25
,, ciliary zone, iv. 26
,, ,, muscle, iv. 26, 28
,, lens, iv. 28
,, ,, changes of curvature of, iv. 31
,, suspensory ligament, iv. 29
,, humour of, vitreous, iv. 4, 29, 170
,, ,, aqueous, iv. 6, 168
,, diagrammatic, iv. 10
,, accommodation of, iv. 13—20
,, ,, for far and near objects, iv. 13
,, ,, changes during, iv. 17
,, ,, mechanism of, iv. 31
,, ,, ,, nervous, iv. 45
,, ,, associated movements in, iv. 46
,, ,, imperfections of iv. 47.

432   INDEX.

Eye, constrictor influences on, iv. 34—40
,, dilator influences on, iv. 41—43
,, emmetropic, iv. 16, 47
,, myopic, iv. 16, 47
,, hypermetropic, iv. 17, 47
,, presbyopic, iv. 17
,, pupil, movements of, iv. 34—46
,, ,, constriction and dilation of, iv. 34
,, ,, nerves supplying, iv. 35
,, ,, constriction of, a reflex act, iv. 37
,, ,, changes in through action of cervical sympathetic, iv. 40
,, ,, nature of dilating mechanism, iv. 41
,, ,, action of drugs and other agencies, iv. 43
,, retina, development of, iv. 2
,, ,, a part of the brain, iv. 4
,, ,, structure of, iv. 55—70
,, ,, junction of optic nerve with, iv. 55
,, ,, layers of, iv. 56
,, ,, neuroglial elements, iv. 57
,, ,, nervous elements, iv. 59
,, ,, rods and cones, iv. 59—62
,, ,, ,, ,, function of, iv. 119
,, ,, ,, ,, possible differences between, iv.151
,, ,, rods, presence of visual purple in the, iv. 118
,, ,, inner nuclear layer, iv. 62
,, ,, layer of ganglionic cells,iv.64
,, ,, ,, of optic fibres, iv. 64
,, ,, macula lutea and fovea centralis, iv. 66
,, ,, blood vessels of, iv. 67
,, ,, pigment epithelium, iv. 68, 118
,, ,, stimulation of, by other agencies than light, iv. 76
,, ,, visual areas of, iv. 81
,, ,, intrinsic light of, iv. 97
,, ,, colour-blindness of periphery, iv. 106
,, ,, blind spot of, iv. 110, 156
,, ,, photochemistry of, iv. 115
,, ,, effect of compression of the eyeball on the, iv. 108
,, ,, corresponding or identical points, iv. 137, 153
,, ,, lines of separation, iv. 139
,, retinal structures, fatigue of, iv. 113
,, retinal processes, electric currents developed by, iv. 122
,, muscles, ocular, iv. 143, 145
,, nutrition of, iv. 164—171
,, arrangement of blood vessels, iv. 164
,, vaso-motor changes in, iv. 165
,, lymphatics and lymph-spaces,iv.164
,, protective mechanisms, iv.172—175

Eye, protective mechanisms, eyelids and their muscles, iv. 172
,, ,, ,, conjunctiva and its glands, iv. 173
,, ,, ,, Meibomian and lachrymal glands, iv. 174
,, in old age, iv. 411
Eyeball, rotation of, iv. 134, 140
,, movements of, iv. 139
,, primary position of, iv. 139, 141
,, muscles of, iv. 143
,, simultaneous movements, iv.148

Fæces, composition of, 479
Fainting, a result of cardiac inhibition, 297, 345
Fallopian tube, iv. 351
,, ,, reception of ovum by the, iv. 358, 360
Falsetto voice, iv. 331
Falx cerebri, iv. 87
Fasciculus gracilis and f. cuneatus, 939
Fat, its presence in chyle, 499
,, amount absorbed during digestion 512
,, mode of absorption, 517
,, formation of, 725, 772, 797
,, history of, 769—776
,, increase of, in cell substance, 770
,, disappearance of, from cells, 771
,, nature of, in adipose tissue, 772
,, limits to construction of, 775
,, its potential energy as food, 802
,, storage of, in the body before hibernation, 821
Fats in white corpuscles, 41, 42
,, in the blood, 50
,, in chyle, 449
,, in nerve tissue, 115, 121
,, in food stuffs, 356
,, action of gastric juice on, 365, 471
,, ,, of bile, 425
,, ,, of pancreatic juice, 429
,, emulsification of, during digestion, 425, 474, 516
,, course taken by, in digestion, 512, 516
,, change of, in the lacteal radicle, 517
,, various, melting points of, 775
,, ,, of milk, 781
,, as food, metabolism lessened by, 785, 797
Fat-cells, structure of, 770
Fattening, aids to, 844
Fatigue, its effect on muscular irritability, 139, 148
,, sense of, its nature, 149
,, retinal, negative images produced by, iv. 128
,, auditory, iv. 228
Feet, sweating in dogs and cats only present in the, 700, note
Fehling's fluid, as test for dextrose, 359

INDEX. 433

Fenestra ovalis and fenestra rotunda of ear, iv. 180
Fenestrated membrane of arteries, 194
Ferment, fibrin, efficient cause of co-agulation, 24, 26
„ the amylolytic, in saliva, 362
„ „ destroyed by gastric juice, 365
Ferments, organized and unorganized, 362, note
„ their presence in urine, 652
Ferrein, pyramids of, 638
Fever, metabolism heightened by, 818
Fibres, muscular, *see* Muscle.
„ nerve, *see* Nerves.
„ elastic, in connective tissue, 190
„ „ in dermis, 687
„ of brain, *see* Brain, fibres of.
Fibrillæ of muscle substance, 91
„ gelatiniferous, 188
„ of dermis, 686
Fibrin, of the blood, 15
„ its development during clotting, 16
„ its proteid nature, 17
„ structure, 18
„ causes of its appearance, 20
„ action of gastric juice on, 367, 370
„ in clotting lymph, 497
Fibrin-ferment, 24, 26
Fibrinogen, its precipitation from plasma, 23
„ its conversion into fibrin, 25, 26, 30
Fick, his spring manometer, 255
„ his pneumatograph, 540
Fimbria, ovarian, iv. 352
Fingers, clubbed, in phthisis, 831
Flatulence, 473
Flavours, sense of smell appealed to by, iv. 259
„ localization of seat of perception of, iv. 263
Flourens, 'nœud vital' of, 585
Fluid, serous, 23
„ „ its identity with lymph, 487, 499
„ in diet, 838
„ cerebro-spinal, its sources, 1126
„ „ characters of, 1128
„ „ renewal of, 1129
„ „ purposes served by, 1130
„ amniotic, iv. 385
„ „ its functions, iv. 386
„ „ composition, iv. 390
Fluidity of living blood, 26
„ of blood in the vessels after death, 27
Follicles, solitary and agminated, 489
„ Graafian, iv. 354
„ „ structure, iv. 355
„ „ ovum nourished by, iv. 357

Fœtus, nourishment and respiration of, iv. 382
„ swallowing movements executed by, iv. 386, 389
„ transmission of food-material to the, iv. 386
„ growing differentiation of tissue in, iv. 388
„ movements of, iv. 389
„ digestive functions, iv. 389
„ circulation in, iv. 390
„ expulsion of, iv. 397
Fontana, spaces of, iv. 27
Food, amœboid absorption of, 3
„ carried to the tissues by the blood, 8
„ its gradual change into living substance, 42
„ ingestion of, by white corpuscles, 42, 46
„ its effect on vascular mechanism, 351
„ effect of its presence on gastric secretion, 402
„ „ „ on pancreatic secretion, 433
„ „ „ on bile secretion, 435
„ „ „ on stomach movements, 465
„ „ „ on intestinal movements, 468
„ as acted on by saliva, 358
„ „ „ „ by gastric juice, 365
„ „ „ „ by bile, 424
„ „ „ „ by pancreatic juice, 429
„ „ „ „ by succus entericus, 430
„ peptogenous, 419
„ changes of, in the alimentary canal, 470
„ as income compared with output of material, 791
„ potential energy supplied by, 801
Food-stuffs, classification of, 355
„ changes of, in the body, 632
„ relative digestibility of, 717
„ fatty and carbohydrate, 797
„ peptones and salts, 799
„ various, in normal diet, 833—5
Foramen of Monro, 931, 933
„ of Majendie, 1127
„ ovale, course of fœtal circulation through the, iv. 391
„ „ gradual occlusion of the, iv. 394
Fornix, development of the, 932, 995
Fovea centralis, the, iv. 66
„ „ region of distinct vision in, iv. 80
Freezing, its effect on muscle, 98
Friction, peripheral, as affecting circulation, 211, 214
Frog, rheoscopic, 112
„ capillary circulation in, 209

434 INDEX.

Frog, brainless, phenomena shewn by, 56, 180, 999
,, cisterna magna lymphatica in, 487
,, lymph-hearts of, 502, 509
,, winter, possible source of glycogen store in, 827
Fuscin, its presence in the retina, iv. 69, 116

Galen, veins of, 1133
Gall-bladder, changes in bile effected by the, 421
,, ,, storage of bile in the, 435
,, stones, cholesterin, present in, 121
,, ,, composition of, 431
Galvanic battery described, 60
Ganglia, spinal, 171, 173
,, ,, of the posterior root, 853
,, of splanchnic system, 176, 179, 183, 184
,, cardiac, of lower vertebrates, 277
,, ,, of frog, 278, 284, 295
,, ,, of tortoise, 278, 288
,, ,, of mammal, 278
,, ,, relations of the, 284
Ganglion stellatum, 297
,, spirale of cochlea, iv. 205, 219
Ganglion-cells, their structure, 173, 175
Gases, absorption of, by liquids, 557
,, their presence in blood, 557, 620
,, ,, ,, in urine, 653
,, various, their effects on respiration, 609
Gaskell, his method of recording heart-beat, 280
Gastric juice, normal composition of, 364
,, ,, the amylolytic ferment destroyed by, 365
,, ,, its action on fats, 365
,, ,, artificial, how prepared, 366
,, ,, its action on proteids, 367-376
,, ,, nature of its action, 373
,, ,, secretion of, 402
,, ,, ,, influenced by absorption of food, 404
,, ,, formation of free acid in, 418
,, ,, changes in its character as digestion proceeds, 471
Gelatin, composition and properties, 188
,, in food-stuffs, 355
,, action of gastric-juice on, 374
,, its effect as food, 798
Germinal vesicle, iv. 355
,, epithelium, iv. 354
,, ,, the origin of the secreting part of the testes, iv. 364
Gestation, extra-uterine, iv. 381
,, human, period of, iv. 395
Giddiness, a result of disarrangement of co-ordinating machinery, 1015
Gland, prostate structure of, iv. 370
Glands, albuminous, 389, 407
,, of alimentary canal, 379

Glands, alveoli of, 380, 386
,, of Brunner, 450
,, buccal, 390
,, cardiac, of stomach, 381
,, ceruminous, 693
,, common features of, 380
,, of Cowper, iv. 370
,, gastric, 379—385
,, ,, secretion from, intermittent, 402
,, ,, of newt, 411
,, ,, of bat, 412
,, ,, changes in central cells of, 411
,, hibernating, 821
,, lachrymal, iv. 174
,, lenticular, 489, 491
,, of Lieberkühn, 442, 448, 451
,, lymphatic, multiplication of leucocytes in, 45
,, ,, structure, 489, 492
,, mammary, structure of, 777, 786
,, ,, at birth, 784
,, Meibomian, 693, iv. 174
,, of Moll, iv. 174
,, mucous, 387, 480
,, storage of granular matter in, 413
,, of the œsophagus, 393
,, oxyntic, of frog, 419
,, Pacchionian, 1126
,, parotid, human, 389
,, ,, double nerve supply to, 401
,, ,, changes of during secretion, 407
,, pyloric, of stomach, 381, 384
,, salivary, venous pulse in, 271
,, ,, general structure of, 385
,, sebaceous, 691
,, solitary, of intestine, 487
,, sublingual, mammalian, 389
,, submaxillary, of dog, 387
,, ,, ,, ducts of, 388
,, ,, ,, double nerve supply of, 310, 396, 401
,, ,, ,, effect of stimulation of chorda, 399
,, ,, ,, ,, of servical sympathetic, 401
,, ,, ,, cell changes in, 409
,, ,, of man, principally mucous, 389
,, sweat, 690
,, thermal changes in, 808
Glisson's capsule, 705
Globin, the proteid constituent of hæmoglobin, 568
Globulins, a group of proteids, 19
,, their changes to acid- and alkali-albumin, 97
Globus pallidus, 974
Glomeruli, the, of kidney, structure, 639
,, , secretion by, 665

INDEX. 435

Glomeruli, special substances excreted by, 666
,, effect of blood-pressure on the, 669
,, complexity of their action, 671
Glossopharyngeal nerve, 398
Glottis, the, iv. 310
,, changes in, during utterance of voice, iv. 317, 327
,, narrowing and widening of the, iv. 322
Glycerin, its effect on the hepatic cell, 734
Glycin, a product of metabolism, 673, 747
Glycocholic acid, 423
Glycogen, its presence in white corpuscles, 41
,, ,, in muscle substance, 101, 728
,, ,, ,, embryonic, 729
,, ,, in unstriated muscle, 159
,, ,, in hepatic cells, 719
,, ,, in the placenta, 729, iv. 381, 387
,, ,, in the testis, iv. 369
,, ,, in the fœtus, iv. 388
,, characters of, 714
,, its conversion by the liver into sugar, 715
,, storage of, in the liver, 717, 723
,, manufacture of, 722, 723
,, hepatic, of hibernating animals, 821
Goitre, its nature, 762
,, its connection with cretinism, 764
Golgi, organ of, iv. 300
Goll, column of, 869
Goltz and Gaule, their maximum manometer, 239
Gout, accumulation of uric acid in the blood in, 757
Gowers, tract of, 874
Graafian follicle, iv. 354
,, ,, structure, iv. 355
,, ,, nutrition of the ovum effected by, iv. 357
Grandry's corpuscles, iv. 207
Granules in white corpuscles, 39, 42, 47
,, of resting glandular cells, 407
,, ,, albuminous cells, 407
,, ,, mucous cells, 409
,, in leucocytes, 518
,, in hepathic cells, 719
Growth, human, curve of, iv. 404
Grützner's method of preparing fibrin, 372
Guanin, presence of in urine, 650
Gudden's commissure, 1074
Gustatory sensations, *see* Sensations, gustatory.

Gymnema sylvestre, gustatory sensations affected by, iv. 263

Hæmacytometer described, 34
Hæmadromometer of Volkmann, 220
Hæmatachometer of Vierordt, 222
,, of Chauveau and Lortet, 223
Hæmatin, 34, 568
,, its relations with bilirubin, 35, 744
,, oxygen-holding power of, 568
,, iron-free, 568, 744
Hæmatoblasts described, 37
,, development of, 46
Hæmatoidin, 745
Hæmatoporphyrin, 745
Hæmin, crystals of, 570
Hæmoglobin, 33, 559
,, an oxygen-bearer, 35, 38, 566
,, its proportion in red corpuscles, 50
,, in red muscle, 98
,, crystals of, 560
,, spectroscopic features of, 560
,, ,, ,, of reduced, 561
,, reduced, change of colour in, 563
,, absorption of oxygen by, 564
,, its combination with gases other than oxygen, 566
,, products of decomposition of, 567
,, its respiratory functions, 568, 629
,, its relation to biliburin, 35, 744
,, of fœtal arterial blood, iv. 384
Hæmorrhage, its effect on blood-pressure, 341
Haidinger's brushes, iv. 54
Hair-cells, iv. 209
,, inner, iv. 214, 217
,, outer, iv. 218
Hairs, structure of, 691
,, auditory, iv. 208
Hallucinations, ocular, iv. 133
,, auditory, iv. 241
Hamulus of cochlea, iv. 204
Head voice, how produced, iv. 329
Hearing, sensations of, 1088
,, mechanisms of, iv. 176
,, binaural, iv. 243
Heart, visible movements, 228
,, changes in, during cardiac cycle, 229
,, auriculo-ventricular valves, action of, 229, 230
,, auricular systole, 229
,, ventricular systole, 230
,, auriculo-ventricular valves, action of, 230
,, semilunar valves, action of, 231
,, change of form, 232
,, cardiac impulse, 234
,, sounds of, 235
,, pressure exerted by (endocardiac pressure), 238
,, ,, ,, graphic record of, 240—246
,, negative pressure in, 239, 248

INDEX.

Heart, pressure in ventricle, the phases, 246
,, duration of cardiac phases, 249, 251
,, summary of events in, 251
,, work done by, 253
,, structure of cardiac muscular tissue in frog, 275
,, ,, ,, in mammal, 276
,, nerves and ganglia of, 277
,, ,, ,, relation of to rhythmic power, 284
,, sequence of events in beat of, 282
,, power of independent rhythm in the several parts, 282—284
,, characters of the contraction of muscular fibres of, 285—288
,, rhythmic power resident in muscular tissue, 288
,, causation of normal sequence in, 289
,, inhibition of beat of in frog, 290
,, ,, ,, in mammal, 296
,, augmentation of beat of in frog, 293
,, ,, ,, in mammal, 298
,, inhibitory and augmentor fibres in frog, 292, 294
,, ,, ,, in mammal, 297—300
,, inhibition, reflex, of, 295
,, centre of inhibition of (cardio-inhibitory centre), 295
,, inhibition and augmentation of beat, nature of, 300
,, inhibition of, suspended by atropin, 301
Heart-beat, regulation of, 273
,, intrinsic regulation of, 286, 344, 921
,, development of normal, 280
,, influences other than nervous affecting, 308
,, relations of with vaso-motor system, 351
,, normal rate of, 346
,, slowing effects of venous blood on, 621
,, during asphyxia, 626
,, of the babe, iv. 405
,, death marked by cessation of, iv. 418
Heat, given out by contracting muscle, 104
,, loss of energy in the form of, 803
,, bodily, measurement of, 804
,, sources and distribution of, 807
,, modes of loss of, 809
,, regulation of, by variations in loss, 810
,, production of, 812
,, ,, increased by labour, 813
,, regulation of, by the nervous system, 814
,, increased production of, by injury to parts of the brain, 816
,, great, effects of, 818

Heat and cold, sensations of, iv. 278—280
,, ,, separate terminal organs for, iv. 291
,, ,, epidermal seat of sensations of, iv. 293
Helicotrema of cochlea, iv. 204
Helmholtz's magnetic interruptor, 67
Hemianopsin, 1078
Hemiplegia, crossed phenomena of, 1093
Henle's sheath of nerve fibre, 119, 175
,, loop of in urinary tubule, 637
,, ,, its structure, 642
,, sphincter of, 681
,, layer of, in hair follicle, 691
Hensen's body, iv. 218
Hepatic artery, 707
,, cells, 708
,, ,, their work, 713
,, ,, changes of, 719
,, ,, glycogen in, 720
,, plexus, 731
,, vein, temperature of blood in, 808
Hering, his theory of colour vision, iv. 93—95
,, ,, colour-blindness explained by, iv. 102
,, ,, as to simultaneous and successive contrasts, iv. 129
Hermann on muscle currents, 110
Hibernation, phenomena of, 820
Hiccough, 630
Highmori corpus of testis, iv. 365
Hippocampus, structure of, 1031
Hippuric acid, its presence in urine, 650
,, how formed in the kidney, 672
Histohæmatin in red muscle, 98
Hopping, how effected, iv. 346
Horny layer of epidermis, 689
Horopter, the, iv. 153
Horse, sweat of, 695
Humour, aqueous, iv. 168
,, ,, how furnished, iv. 169
,, vitreous, iv. 170
Hunger and thirst, sensations of, iv. 285
Huxley's layer in hair follicle, 691, 692
Hydrochloric acid, free, in gastric juice, 365
Hydrogen, evolution of, in small intestine, 478
Hyperpnœa, 598
Hypoxanthin, presence of, in urine, 650

Illusions, visual, iv. 157
,, tactile, iv. 305
Images, retinal, formation of, iv. 6—12
,, ,, in relation to sensations excited by, iv. 12
,, ,, entoptical, iv. 52
Impulses, nervous, 57, 123
,, ,, electrical changes accompanying, 123

INDEX. 437

Impulses, nervous, nature of, 154
,, ,, dependence on strength of stimulus, 137
,, cardiac, how caused, 234
,, ,, modes of recording, 244
,, afferent and efferent, 850
,, ,, their paths along the cord, 1094, 1095, 1106
,, ,, relays in course of, 1096, 1106
,, ,, crossing of, 1099
,, sensory, different paths for different, 1101
,, ,, transmitted by grey matter and intermediate tracts, 1104
,, ampullar, 1012
,, painful, probable course of, 1097
,, motor, effect of afferent impulses on the coordination of, 1013, iv. 150
,, visual, 1070, iv. 1, 102
,, auditory, how excited, iv. 176, 180
,, ,, development of, iv. 232
,, volitional, time required for transmission, 1063
,, ,, course of, in man along the pyramidal tract, 1065
Impurities in expired air, 552
Income and output of material in nutrition, 791
Incus, the, iv. 181
Indentations of nerve medulla, 117
Indol, a product of bacterial action, 428, 477
Induction coil, construction of, 62
Infancy, characteristics of, iv. 407
Inflammation, phenomena of, 336
,, œdema due to, 508
Infundibulum of mammalian lung, 529, 530
Infusoria, ciliary motions in, 165
Inhibition of heart-beat in frog, 290
,, ,, ,, in mammal, 296
,, ,, ,, fainting a result of, 297, 345
,, ,, ,, effect of atropin on, 300
,, of secretion of saliva, 398
,, of respiration, 591
,, of reflex actions, 915
,, of parturition, iv. 402
Inhibitory nerves, 184
,, fibres in vagus of frog, 293
,, ,, ,, of mammal, 296
,, ,, cardiac,' continuous action of, 297
,, ,, their analogy with vasodilator fibres, 313
Inogen, or 'contractile material of muscle,' 153

Insect muscles, fibrillæ of, 90
Inspiration, mechanism of, 543
,, movements of, 544
,, laboured, phenomena of, 546
Intercostal muscles, their work in respiration, 545
Interfibrillar substance of muscle, 90, 91
Intermediate line, in muscle fibre, 90, 92
Intermittence, cardiac, 304
Interruptor, magnetic, the, 66
Interlobular and intralobular veins and veinlets of liver, 706, 712
Intestine, general plan of structure, 441
,, absorption of fats in the, 516
,, ,, of diffusable substances and water, 520
,, small, mucous membrane of, 441
,, ,, retiform and adenoid tissue, 442
,, ,, glands of Lieberkühn, 442, 448
,, ,, the villi, 445
,, ,, the columnar epithelium, 445
,, ,, goblet cells of the villi, 446
,, ,, structure of the body of a villus, 447
,, ,, glands of Brunner, 450
,, ,, movements of, 462
,, ,, changes of food in the, 473, 474
,, ,, fermeutative changes of food in the, 478
,, ,, fluidity of food maintained in the, 522
,, large, structure of, 450
,, ,, structure of rectum, 451
,, ,, movements of, 462
,, ,, changes of food in, 478
,, ,, digestion of cellulose in the, 479
" Intrinsic light " of retina, iv. 97
Iodine, coloration of starch by, 360
Iris, development of the, iv. 4
,, structure, iv. 24
,, muscular and vascular changes in the, iv. 34
Iron, its presence in hæmatin, 568
,, ,, in hæmoglobin, 560
,, ,, in bile, 422, 745
,, ,, in the spleen, 742
Irradiation, iv. 126
Irritability, muscular and nervous, 56—83
,, ,, independent, 57, 145
,, diminution and disappearance of, after death, 83
,, as affected by electrotonus, 128
,, circumstances determining, 143
,, centrifugal loss of, in severed nerve, 144

INDEX.

Irritability, influence of temperature on, 145
" " " blood-supply, 146, 148
" " " functional activity, 148
" presence of oxygen a condition of, 150, 154, 597
" prolonged, of heart, 282
Irritants, inflammatory action of on tissues, 336
Islets, extra-vascular, 185

Jaundice, how caused, 440, 748
Judgments of distance and size, how formed, iv. 161
" of solidity, iv. 162
Juice, gastric, *see* Gastric juice.
" pancreatic, *see* Pancreatic juice.
" intestinal, *see* Succus Entericus.
Jumping, how effected, iv. 346

Karyomitosis of leucocytes, 490
" of cells of Malpighian layer, 688
Katabolic changes in living tissue, 41, 43
" " heat liberated by, 807
Katelectrotonus defined, 129
Kathode or negative electrode, 60
Keratin, its nature, 689
Key, galvanic, various forms of, 61
Kidney, the, structure of, 634
" tubuli uriniferi of, 636, 639, 641
" mammalian and amphibian compared, 643
" duplexity of its mechanism, 656
" vasomotor mechanisms of, 657
" relations, various, of flow of blood through, 660
" vaso-constrictor nerves of, 662
" vaso-dilator nerves of, 663
" effect of chemical changes in the blood, 664
" secretion by the renal epithelium, 665
" double vascular supply to in amphibia, 666
" work of the epithelium of the tubules, 671
" and skin, mutual relations of secretory activity of, 674
" its relations to water absorbed by the alimentary canal, 675
" influence of central nervous system on, 677
" structure of pelvis of, 678
" fœtal, urea secreted by, iv. 390
Kilogram-meters, daily work of heart estimated in, 254
" energy of food and body and day's work estimated in, 803
Knee-jerk, 913, 926
Krause's membrane, 90
Kreatin, its presence in the blood, 51
" chemical composition, 102

Kreatin, in unstriated muscle, 159
" in the thyroid, 762
" the product of muscle metabolism, 751
Kreatinin, its presence in urine, 650
" the urinary form of kreatin, 752
" difficulties presented by its presence in urine, 752
Kymograph, Ludwig's, for recording blood-pressure, 208

Labour, physiological division of, 6
" circumstances governing capacity for, 628
" increased production of heat from, 813
" diet with reference to, 845
"Labour," the events of, iv. 396
" first stage of, iv. 396
" second stage of, iv. 397
" causes determining its onset, iv. 407
Labyrinth of ear, bony and membranous, iv. 179
" " perilymph cavity of, iv. 200
" " connections of auditory nerve with, iv. 201
" " general structure of cochlear, iv. 202
" " minute structure of, iv. 211
" " probable functions of,iv.233
" " vestibular, parts of, iv. 198
" " " minute structure, iv. 206
" " " probable functions, iv. 238
" " the otoconia and otoliths, iv. 210
" " transmission of impulses through the, iv. 232
Lachrymal gland, structure of, iv. 174
Lactalbumin, 781
Lactation, nervous centre for, 787
Lacteal radicle of intestinal villus, 447, 484
" " passage of fat into, 512, 518
Lacteals, the, absorption by, 481
" " chyle contained by, in fasting animals, 497, 499
" " passage of products of digestion into, 512
Lactic acid, its presence in the blood, 51
" " isomeric variations of, 99, note
" " its effect on the heart, 303
" " fermentation, 477
" " a product of muscular metabolism, 826
Lactoprotein, 781
Lactose, ready fermentation of, 782
" its formation in the mammary gland, 786
"Laky" blood, how formed, 32, 560
Lamina cribrosa, iv. 55

Laryngeal nerves, iv. 323
Laryngoscope, larynx as seen by the, iv. 313
Larynx, the its conditon in respiration, 548
,, cartilages of, iv. 308
,, superior aperture, iv. 309
,, ventricles, iv. 312
,, ,, uses of, iv. 332
,, minute structure of parts of, iv. 312
,, muscles of, iv. 317, 322
,, nervous mechanisms of, iv. 323
,, respiratory movements of, iv. 323
,, cortical area for movements of, iv. 326
Laughter, mechanism of, 631
Lecithin, in stroma of red corpuscles, 33
,, in white corpuscles, 41
,, in the blood, 50
,, in muscle substance, 101
,, in nervous tissue, 121
,, in milk, 781
,, in yolk of egg, iv. 356
,, its composition, 121, 148
Lens, the, development of, iv. 4
,, origin and structure, iv. 28
,, mechanisms for changing curvature of, iv. 31
,, action of the suspensory ligament on, iv. 32
Lenticular glands, 489, 491
Leucin, composition of, 428
,, in intestinal contents, 476
,, a product of nitrogenous metabolism, 755
,, its conversion into urea, 756
Leucocytes, in the lymphatic system, 44
,, their origin, 45
,, in connective tissue, 189
,, in reticular tissue of intestine, 444, 448
,, abundant in solitary follicles, 489
,, their multiplication by division, 490
,, pigment occasionally found in, 495
,, their presence in the villi, 518
,, ,, in the Malpighian layer, 688
,, ,, among epidermic cells, 698
,, of spleen, peculiarities of, 739
Leucocythæmia, increase of white corpuscles in, 46
Levatores costarum, their work in respiration, 546
Lieberkühn, glands or crypts of, 442
,, ,, in small intestine, 448
,, ,, succus entericus probably furnished by, 449

Lieberkühn, glands in large intestine, 451
Life, processes of, compared with those of death, 1
,, its existence possible without organs, 3
,, periodic events of, iv. 412
,, factors of, iv. 417
Ligamentum denticulatum, 1125
,, nuchæ, elastic fibres of, 190
Ligament, suspensory, iv. 29
Light, as stimulus to visual apparatus, iv. 35, 71
,, "intrinsic," of retina, iv. 97
,, changes in retina produced by, iv. 71
,, sensitiveness of living matter to, iv. 115
,, decomposition of, iv. 84
Lips, tympanic and vestibular, of cochlea, iv. 211
Lissauer's zone, 875
Listing, diagrammatic eye of, iv. 0
,, his law, iv. 141
Liver, the, destruction of red corpuscles in, 36
,, nerve supply to, 435
,, blood supply to, 436
,, ,, ,, quality of as affecting bile secretion, 437
,, structure of, 705
,, lobules, 705
,, Glisson's capsule, 707
,, of frog, 711
,, ,, lymphatics of, 712
,, ,, storing of glycogen in, 717, 722
,, mammalian, 712—720
,, nerves of, 731
,, nervous control of glycogenic function, 731
,, "acute yellow atrophy" of, 748, 755
,, presence of urea in, 755
,, conversion of leucin into urea in, 756
,, heat set free in, 808
,, its action on lactic acid, 826
,, fœtal, deposition of glycogen in, iv. 388
Living substance, food and waste of, 3
Locomotor mechanisms, iv. 342
Locus cæruleus of brain, 963, 982
Loose connective tissue, 188
Ludwig, his stromuhr, 220
,, his mercurial gas pump, 554
Lungs, the, their function chiefly mechanical, 526
,, structure of, in newt, 527
,, ,, in frog, 528
,, ,, in mammal, 529
,, ,, infundibula of, 530
,, ,, the trachea and bronchi, 530

Lungs, structure of the bronchia and bronchioles, 532
,, ,, the alveoli, 532
,, lymphatics of, 533
,, nerves of, 534
,, entrance into and exit of air from, 535
,, air, tidal and stationary in, 535
,, ,, complementary, supplementary and residual, 536
,, results of opening into pleural chamber, 536
,, condition of, before birth, 537
,, elasticity of, pressure exerted by, 537
,, respiratory changes in, 572—577
,, effects of inflation and suction, 592
,, first inflation of, iv. 393
Lutein, a constituent of the corpus luteum, iv. 352
" Luxus consumption " of food, 477, 796
Lymph, nature and movements of, 496—500
,, the, a medium of exchange between blood and tissues, 13, 14, 191, 500, 509
,, salts present in, 41
,, migration of white corpuscles into, 45, 337
,, coagulable, in inflammation, 337
,, its universal presence in the body, 486
,, circulation of, 487
,, microscopical characters of, 497
,, clotting of, 497
,, chemical composition of, varying, 498
,, total diurnal flow, 500
,, movements of, 500
,, its flow increased by muscular movements, 501
,, transudation of, nature of the process, 503
,, its passage from the pleural cavity to the vessels, 533
,, its functions in the eye, iv. 167
Lymph capillaries, compared with blood capillaries, 481
,, ,, their structure, 483
Lymph corpuscles, 497
Lymph hearts in amphibia and reptilia, 509
Lymph knots, leucocytes in, 495
Lymph sinus, 490
Lymph spaces, structure, 484
,, ,, passage of the white corpuscles into, 45
,, ,, in the dermis, 687
Lymphatic glands, structure of, 489—495
,, ,, their influence on lymph, 498
,, system, 481—495
,, ,, prominence of, in infancy, iv. 407

Lymphatic arrangements of brain and cord, 1125—1130
,, spaces of the brain, 1126
Lymphatics of the eye, iv. 166—171
,, ,, lung, 533
,, ,, liver, 712
,, perivascular, 484

Macula lutea, iv. 66
Maculæ acusticæ, iv. 201
,, ,, structure of, iv. 210
Magnetic interruptor, 66
Majendie, foramen of, 1127
Making and breaking currents, 60
,, ,, shocks, 65
,, ,, contractions with the constant current, 126
Male breathing, diaphragmatic character of, 543
,, organs of reproduction, iv. 364
Malleus, the, iv. 182
Malpighi, pyramids of, 635
Malpighian capsule of kidney, 634, 643
,, ,, blood supply of, 644
,, layer of dermis, structure of, 687
,, corpuscles of spleen, 735, 738
Maltose, 359
Mammary gland, structure of, 777
,, ,, changes in, during secretion, 778
,, ,, dormant, characters of, 780
,, ,, at birth, 784
,, ,, relations of, to the nervous system, 787
Manometer, for measuring blood-pressure, 204
,, minimum and maximum, 238
,, endocardiac pressure shewn by, 239, 247
,, Fick's, 255
,, ,, vaso-motor actions observed by, 320
Marey's pneumograph, 540
,, tambour, 241
Marrow, red, formation of red corpuscles in, 37, 38
,, yellow, of bones, 771
Mastication, how effected, 452
Massage, metabolism excited by, 844
Meatus, auditory, external, iv. 180
,, ,, internal, iv. 201
Meconium, iv. 386
,, sources of, iv. 389
,, chemical composition, iv. 389
Medulla of nerve-fibre, structure, 116, 117
,, loss of in augmentor fibres, 300
,, ,, in vaso-constrictor fibres 317
,, retention of in vaso-dilator fibres, 318

Medulla oblongata, cardiac effect of stimulation of, 295
,, ,, centre for nerves of taste in, 321
,, ,, ,, for vaso-motor impulses in, 321–323
,, ,, ,, for secretion of saliva, 398
,, ,, ,, for deglutition, 455
,, ,, ,, for vomiting, 461
,, ,, ,, for respiration, 585
,, ,, effect on blood-pressure of successive sections of, 326
,, ,, antero-lateral nucleus in, 327
,, ,, diabetic area of, 730
see also Bulb.
Meibomian glands, 693, 1312
Meissner, plexus of, 441
Melting point of various fats, 772
Membrana pupillaris, absorption of before birth, iv. 5
,, tympani, iv. 179, 191
,, tectoria, iv. 221
Membrane, elastic, of arteries, 193
,, fenestrated, 194
,, hyaloid, iv. 29
Membranes of the brain and cord, 1125
Ménière's disease, 1015
Menstruation, iv. 360–363
,, causation of, iv. 362
Mercurial gas pump, Ludwig's, 554
,, ,, Pflüger's, 555
,, ,, Alvergniat's, 555
Metabolic processes of body, 703
Metabolism, defined, 41
,, water of, 44
,, increased by exercise, 349
,, ,, by proteid food, 785, 795, 828
,, of muscle, products of, 752
,, of nervous tissue, 753
,, of glands, 754
,, proteid, its complexity, 759
,, nitrogenous, 794
,, ,, products of, 827
,, of muscle the chief source of heat, 808
,, conducted in the tissues, 822
,, its relation to structural elements, 825
,, course of products of, 826
,, nervous control of, 829
,, rapidity of, in infancy, iv. 406
Metals, retention of, in the liver, 422
Metameres, hypothetical, of spinal cord, 169, 896, 1067
Methæmoglobin, spectrum of, 569, 570
Meynert's commissure, 1074
Micro-organisms, their actions in digestion, 477
,, in expired air, 552
Micro-unit of heat defined, 105, note
Micturition, mechanism of, 680

Micturition, nervous mechanism of, 682
,, centre for, 683
,, voluntary and involuntary, 684
Migration of the white corpuscles, 45
,, ,, ,, in inflammation, 337
,, ,, ,, aided by changes in vascular walls, 339
Milk, action of gastric juice on, 371, 374
,, ,, rennet, 374
,, double mode of secretion of, 779, 784
,, nature of, 780
,, constituents of, 781, 782
,, uterine, 1526, iv. 381, 388
Millon's reagent for detection of protein, 17
Mitral valves, their structure, 197
,, ,, their action, 230
Molecular basis of chyle, 500
,, ,, where elaborated, 517
Moll, glands of, iv. 174
Monro, foramen of, 931, 1127
Morse key, 62
Mouth, lining membrane of, iv. 254
Movements of alimentary canal, 452—469
,, amœboid, 38, 166, 337
,, ,, of lymph corpuscles, 497
,, bilateral, 1047
,, of body, how accomplished, 54
,, Brownian, in chyle globules, 500
,, cardiac, visible, 228
,, ciliary, 55, 162
,, co-ordinating machinery of, 1009
,, of cortical origin, how effected, 1044
,, ,, ,, of dog, 1035, 1056
,, ,, ,, of monkey, 1038, 1051
,, ,, ,, of anthropoid ape, 1043
,, 'forced,' 1111, 1017
,, ,, from injury to optic lobes in frog, 1113
,, fœtal, iv. 389
,, gastric, 464, 465
,, intestinal, 464, 465
,, in living bodies, 2
,, of locomotion, iv. 342
,, muscular, flow of lymph increased by, 501
,, ,, heat given out during, 104, 813
,, ocular, iv. 145
,, peristaltic, 455
,, respiratory, 538
,, sense of, iv. 295
,, skilled, correlation of with pyramidal tract 1048, 1052
,, voluntary, 1034—1069
,, ,, spinal mechanisms for, 914
,, ,, action of motor area in effecting, 1056—1058

Movements, voluntary, as influenced by sensory impulses, 1061
Mucigen, a forerunner of mucin, 411
Mucin, a constituent of saliva, 357
" in perilymph of vestibule, iv. 206
Müller, radial fibres of, iv. 58
Mulberry gall-stones, 431
"Mulberry mass" the, iv. 382
Multipolar cells of splanchnic ganglia, 179
"Muscæ volitantes," iv. 52
Muscarin, its action on cardiac tissue, 301
Muscle, irritability, 56
 " contractious, phenomena of, 68, 77
 " " tetanic, 78
 " wave of contraction, 86
 " gross structure of, 84
 " minute structure of, 88
 " striated, 89, 91
 " " under polarized light, 93
 " mobility of, 95
 " chemistry of, 95—104
 " living and dead, contrast of, 95
 " dead, chemical substances in, 96, 98
 " frozen, 98
 " rigid, acid reaction of, 99
 " living, reaction of, 100
 " constituents of, 101
 " chemical changes due to contraction, 104
 " thermal changes, 103, 104, 106, 151, 807
 " electrical, 106
 " action of the constant current on, 134
 " work done by, as influenced by fatigue, 139
 " " " by load, 141
 " " " by size and form of muscle, 142
 " " " by temperature, 145
 " influence of functional activity on, 148
 " exhaustion of, 149
 " oxygen consumed during contraction, 152
 " contractile material of, 153
 " contraction a chemical process, 154
 " plain, structure of, 156
 " " arrangement of nerves in, 158
 " " chemistry of, 159
 " " characters of contraction of, 159
 " " spontaneous contractions of, 161
 " " tonic contraction of, 162
 " respiration of, 579
 " nutrition of, 148, 824
 " cardiac, 274
 " " unlike skeletal muscle, 285

Muscle, cardiac, histology of, 288
 " " nerve-endings in, 118
 " " vaso-dilator fibres in nerve supply to, 316
 " embryonic, glycogen in, 729
 " relations of metabolism to structural features, 825
 " governance of nutrition of, 923
Muscles, skeletal, result of metabolism in, 751
 " " their proportion in body, 789
 " " tone of, 922
 " " rigidity of, 927
 " " their mode of action, iv. 342
Muscle-currents, 106
 " " negative variation of, 111
 " " velocity of, 106
Muscle-curves, 68
 " " analysis of, 74
 " " variations of, 77
 " " tetanic, 78
Muscle-nerve preparation, 58—83
 " " " muscle current shewn in, 111
 " " " as a machine, 136
Muscle-plasma, 98
Muscle-serum and clot, 98
Muscle-sound, 140
Muscularis mucosæ of stomach, 384
 " " of œsophagus, 392
 " " of intestine, 442
Musical sounds, characters of, iv. 223
Myoglobulin, 98
Myograph described, 68, 71
Myosin in white corpuscles, 40
 " in dead muscle, 96
Myosinogen in white corpuscles, 40
 " in living muscle, 99
Myxœdema, its connection with disease of thyroid, 764

Nasal passages, inspired air warmed in the, 548
Nausea, sensation of, iv. 287
Nerves, irritability of, 56
 " " " tested by constant current, 135
 " end-plates of, probable action of urari on, 58
 " " connection of with muscular fibre, 85
 " " the two parts of, 119
 " " their analogy with electrical organs, 120, 155
 " structure of, 114, 115
 " arrangement of in striated muscle, 118
 " " " in unstriated muscle, 158
 " chemistry of, 121
 " severed, degenerative changes in, 144
 " mixed, 171

INDEX. 443

Nerves, inhibitory, 184
,, vaso-motor, 306—333
,, specific energy of, iv. 287
,, special sensations not caused by stimulation of trunk, iv. 288
,, abdominal splanchnic, 172
,, ,, ,, vaso-constrictor fibres in, 310, 316, 320, 430
,, ,, ,, inhibitory fibres in, 467
,, abducens, course of, 962
,, of alimentary canal, 464—466
,, ,, ,, vaso-motor of, 438
,, auditory, 957, iv. 201
,, brachial, 314
,, cardiac, 277, 278, 279
,, cervical sympathetic, cardiac augmentor fibres of, in frog, 293, 294
,, ,, fibres of, to salivary glands, 401
,, ,, vaso-motor fibres in, 308, 309, 313
,, ,, pupil-dilating action of, iv. 40
,, chorda tympani, vaso-motor fibres in, 312—314
,, ,, ,, secretory fibres to submaxillary gland, 396, 399, 416
,, ,, ,, its connection with sense of taste, iv.265
,, cochlear, endings of, iv. 219
,, cranial, nuclei of, 952—970
,, ,, evolution of, 967
,, depressor, vaso-motor functions of, 323, 351
,, of eyeball, iv. 35
,, facial, 960
,, glosso-pharyngeal, 398
,, ,, ,, its roots, 955
,, ,, ,, its connection with sense of taste, iv. 265
,, hypoglossal, 954, 968
,, of larynx, iv. 323
,, lingual, 396, 397
,, ,, its connection with sense of taste, iv. 265
,, of liver, 435, 731
,, oculo-motor, 965
,, olfactory, iv. 248
,, optic, decussation of, in optic chiasma, 1073
,, ,, development of, iv. 2
,, phrenic, functions in respiration, 584
,, pneumogastric, see Vagus.
,, portio intermedia Wrisbergi, 959
,, pulmonary, 534
,, renal, 646, 663
,, sacral, regulation of bladder action by, 682

Nerves, sciatic, constrictor and dilator fibres in, 314, 315
,, spinal accessory, cardio-inhibitory fibres in, 297, 299, 955
,, spinal, 170
,, ,, anterior and posterior roots of, 171, 850
,, ,, efferent and afferent fibres of, their separate paths, 852
,, splenic, 739
,, of stomach, 402
,, submaxillary, 396
,, trigeminal, 962
,, trochlear, 965
,, vagus, inhibitory action of, 184
,, ,, government of heart-beat by, in frog, 278, 290
,, ,, a mixed nerve, 292
,, ,, of mammal, inhibitory fibres in, 279
,, ,, supply to œsophagus, 394
,, ,, ,, to the stomach, 402, 465
,, ,, ,, to the intestines, 465
,, ,, cardiac augmentor and inhibitory fibres in, 292, 315, 467, 591
,, ,, influence on respiration of, 588
,, ,, ,, on the circulation, 593
,, ,, origin of, 955
,, vestibular, iv. 201
Nerve-cells, of spinal ganglia, 173
,, of splanchnic ganglia, 176
,, of grey matter of cord, 177
,, of cardiac ganglia, 279
,, spiral, 176, 279
,, of cerebellum, 1022
,, of cortex, 1026
,, of central and other grey matter of brain, 1021
,, of cord, grouping of, 865–869
Nerve-endings, in striated muscular fibres, 113
,, ,, ,, of amphibia, 120
,, in epidermis, iv. 272
,, in plain muscular tissue, 153
,, of the skin, iv. 267—273
,, specific terminal organs of, for tastes, iv. 264
,, ,, ,, for pressure, iv. 274
,, ,, ,, cutaneous, iv. 289
,, ,, ,, for heat and cold, iv. 290
Nerve-fibres, their structure, 114
,, medullated, 116
,, non-medullated, 120
,, efferent and afferent, 171,850
,, ,, ,, in spinal cord, 178
,, ,, ,, their connection in anterior cornu, 182

Nerve-fibres, inhibitory and augmentor, 293
,, secretory and trophic, 417
,, trophic, 830
,, vaso-constrictor and vaso-dilator, 312
,, vaso-constrictor, course of, 317
,, vaso-dilator, course of, 318
,, nutrition of, 853, 898
Nervi erigentes, vaso-dilator fibres of, 321
,, ,, action of, on the rectum, 467
,, ,, action on penis, and roots of, iv. 371
Nervous system, central, grey and white matter of, 177
,, ,, regulation of temperature by the, 814, 816
,, ,, metabolism governed by the, 829
,, ,, vicarious action of, after partial injury, 799
Nervous mechanism, co-ordinating, 910, 1009, 1019, iv. 151
Neurilemma, defined, 115, note
,, structure of, 117
Neurin in nervous tissue, 121
Neuroglia, 178
,, of white matter of cord, 859
,, of grey matter of cord, 862
Neurokeratin in nerve medulla, 117, 123, 860
Newt, chief cells of gastric glands in, 411
,, structure of lung of, 527
Nicol prism described, 93
Nitrogen in proteids, 17
,, in expired air, 551
,, relations of, in the blood, 571
,, free, inassimilable by living beings, 792
Nitrogenous waste not increased by muscle contraction, 103,106
,, equilibrium, 795
Node of Ranvier, 115
"Nœud vital" of Flourens, 585
Noises and musical sounds, iv. 223
,, characters of, iv. 227
Nostrils, their work in inspiration, 548
Notch, dicrotic, in pulse tracings, 265
Notes, how produced vocally, iv. 315, 325
Nuclei of cranial nerves, 953—970
Nuclein, in white corpuscles, 40
,, a modified proteid, 43
,, a constituent of milk, 781, 785
,, ,, ,, of semen, iv. 369
,, ,, ,, of yolk, iv. 356
Nucleus of white corpuscles, how shewn, 39
,, of the neurilemma, 117
,, of non-medullary nerve-fibre, 120
,, of a nerve-cell, 174
,, caudatus, 932, 974
,, lenticularis, 932
,, olivary, 945

Nucleus accessory olivary, 945
,, arcuate, 946
,, gracile, 947
,, cuneate, 947
,, amygdalæ, 978
,, red, of the tegmentum, 981
Nucleolus in ganglionic cells, 174
Nutrition, statistics of, 788
,, income and output of material in, 791
,, summary of phenomena of, 822
,, of muscle, 823
,, ,, increased by activity, 148
,, influences determining, 828
,, nervous control of, 829, 915
,, disordered, phenomena of, 831
,, of nerve-fibres, 853, 898
,, of embryo, iv. 382, 395
Odours, perception of, iv. 250
,, discrimination of, iv. 251
Oedema, possible causes of, 496, 507
,, inflammatory, 508
,, of Bright's disease, 509
Oesophagus, structure of, 392
,, muscles of, 393
,, nerve supply to, 394
,, movements of in deglutition, 457
,, force of contraction in the, 458
Oil, clotting of blood prevented by presence of, 21
Old age, phenomena of, iv. 410
,, degenerations characteristic of, iv. 410
Olein, a constituent of animal fat, 772
,, presence of, in blood, 50
Olfactory bulb and tract, structure of, 1085
,, mucous membrane, iv. 246
,, nerve-fibres, terminations of, iv. 248
,, sensations, iv. 250, 251
,, judgments, iv. 252
Olivary body (inferior olive), 937
,, nuclei, 942
Olive, upper, 982
Oncograph, renal, 659
Oncometer, renal, 657, 1135
Ophthalmoscope, principle of the, iv. 120
Optic vesicle, development of, 929, 1074
,, thalamus, 930
,, ,, structure of, 971
,, ,, its nuclei, 980
,, ,, pulvinar, 981
,, ,, results of removal in frog, 1019
,, lobes, results of removal in frog, 1019
,, chiasma, 1073
,, nerve, its decussation in chiasma, 1073

INDEX. 445

Optic nerve, its reflex action in pupil constriction, 1075
,,   ,,   an extension from the brain, iv. 4
,,   ,,   structure, iv. 55
,,   cup, iv. 3
,,   disk, iv. 56
,,   fibres, layer of in retina, iv. 56
,,   ,,   their insensibility to light, iv. 111, 283
Optical systems, simple and complex, iv. 7
Optogram, how obtainable, iv. 117
Ora serrata, iv. 5
Ordeal by rice, its mode of action, 398
Organs, definition of, 8
,,   terminal, for sensations of touch and temperature, iv.287
,,   ,,   for sensations of pressure, iv. 290
,,   ,,   for sensations of heat different from those for sensations of cold, iv.291
,,   ,,   nature of, iv. 293
,,   ,,   of reproduction, female, iv.351—363
,,   ,,   ,,   male, iv. 364–373
Os uteri, expansion of during 'labour,' iv. 396
Ossicles, auditory, iv. 181
,,   attachments of, iv. 187
,,   conduction of vibrations through, iv. 192
Otoconia, iv. 210
Otoliths, iv. 210
,,   their possible functions, iv. 234
Ova, primordial, iv. 354
,,   early establishment of number of, iv. 357
Ovary, early history of, iv. 354
,,   general structure, iv. 354
,,   changes in, after escape of ovum, iv. 358
Ovum, ripe, features of, iv. 356
,,   its nourishment by the Graaffian follicle, iv. 357
,,   escape of the, iv. 358
,,   transference to the uterus, iv. 360
,,   impregnation of the, iv. 374
,,   nutrition of, in the uterus, iv. 382
Oxidation in the tissues, seat of, 580
Oxygen, its absorption by the living body, 2
,,   borne by the blood to the tissues, 13
,,   in proteids, 17
,,   borne by hæmoglobin, 33—35, 572
,,   its presence necessary to muscular irritability, 148—150
,,   consumed during muscular contraction, 152
,,   its entrance to the blood by diffusion, 526

Oxygen, in air expired and inspired, 550
,,   relative proportions of in arterial and venous blood, 553
,,   varying amounts of in venous blood, 557
,,   relations of in the blood, 557
,,   absorption of, by blood not according to 'law of pressures,' 558
,,   its access in the lung to the corpuscle, 573
,,   its relations in laboured breathing and asphyxia, 575
,,   mode of storage in muscle tissue, 579
,,   " intramolecular," 579
,,   ,,   sleep possibly due to exhaustion of store of, iv. 415
,,   effect on respiration of deficiency of, 600
,,   effect of breathing, 609
,,   results of high pressure of, 611
,,   mode of measuring amount consumed, 793
,,   consumption of, as affected by temperature, 815
,,   absorption of, during hibernation, 820
,,   supply of, to the embryo, iv. 382, 388, 392
,,   absorption of, in infancy, iv. 406
Oxyhæmoglobin defined, 564, note
,,   colour of, 566
Oxyntic gland of frog, 419

Pacchionian glands, 1126, 1130
Pacinian corpuscles, structure, iv. 268
,,   ,,   distribution, iv. 269
Pain, sense of, iv. 281—285
,,   ,,   localisation of, iv. 282
,,   ,,   special nerve endings not needed for, iv. 284
" Pains " of labour, iv. 396
,,   their rhythmic character, iv.400
Pallor caused by emotion, 332
Palmitin, a constituent of animal fat, 772
,,   present in blood, 50
Palpitation of heart, its causes, 299, 345
Pancreas, structure of, 390
,,   histological changes during secretion, 404
,,   of dog, 406
,,   its nerve supply, 433
Pancreatic juice, trypsin a constituent of, 414
,,   ,,   its composition, 425
,,   ,,   its action on food-stuffs, 226—230
,,   ,,   ,,   on fats, 474
,,   ,,   ,,   on proteids, 476
,,   ,,   secretion of, 432
,,   ,,   ,,   circumstances affecting, 433

Pancreatic juice, approximate amount daily secreted, 434
Panniculus adiposus, 769
Papillæ of dermis, 686
,, of tongue, iv. 255
,, foliatæ, iv. 255
Paraglobulin a constituent of blood-serum, 19
,, precipitated from plasma, 23
,, in white corpuscles, 40
Parapeptone, 370
Paraplegia, reflex action in, 913
Parotid gland, 389
,, ,, nerve supply to, 401
,, ,, cell-changes in, 407
Parturition, iv. 395—402
,, mechanisms of, iv. 396
,, a reflex act, iv. 399
,, inhibition of, iv. 402
Peduncles of the cerebellum, superior, 993
,, ,, ,, middle, 995
,, ,, ,, inferior, 874, 937, 950
,, ,, ,, general connections of, 1111
Pelvis of kidney, structure of, 678
Pendulum myograph, 70
Penis, erection of, iv. 371
,, ,, nerves concerned in mechanism of, iv. 371
,, ,, striated muscles assisting, iv. 371
,, ,, nervous centre for, iv. 372
Pepsin, the ferment body of gastric juice, 373
,, proteids converted into peptone by, 374
,, secreted by the 'chief' gastric cells, 419
,, in the fœtal gastric membrane, iv. 389
Pepsinogen, an antecedent of pepsin, 415
Peptone formed from proteids by gastric juice, 365, 369
,, ,, ,, by pancreatic juice, 374
,, test for, 369, 370
,, its absence from chyle, 514
,, its course during absorption, 515
,, as food, 799
Perceptions, visual, time required for, 1124
,, ,, and judgments, iv. 155
,, ,, psychical modifications of, iv. 125
,, and judgments, auditory, iv. 231, 245
,, ,, ,, olfactory, iv. 250
,, ,, ,, tactile, iv. 302
Pericardial fluid, its persistent fluidity in pericardial bag, 28
Periodic events of life, iv. 412
Peripheral region, blood-pressure in, 209
,, resistance defined, 211

Peripheral resistance, the part it plays in the circulation, 211—214
,, ,, illustrated by model, 214
,, ,, lowered by action of depressor nerve, 325
,, ,, affected by vaso-motor changes, 319
,, ,, ,, by condition of vascular walls, 339
,, ,, ,, by changes in character of blood, 339
,, zone in capillary contents, 335
,, ,, blood-platelets present in, during inflammation, 337
Peristaltic contractions of unstriated muscle, 159
,, movements of alimentary canal, 455
,, ,, excited by stimulation of vagus, 465
,, ,, influences bearing on, 468
,, ,, of ureter, 680
,, ,, of bladder, 683
Personal equation as affecting reaction-time, 1120
Perspiration, nature and amount of, 694
,, secretion of, 699
,, regulation of temperature by, 811
Pes of crus cerebri, 984
Peyer's patches, absent in large intestine, 451
,, ,, presence of in small intestine, 489
,, ,, structure of, 491
Pflüger's gas pump, 555
Phakoscope, Helmholz's, iv. 21
Phantoms, ocular, iv. 132
,, auditory, iv. 241
,, tactile, iv. 305
Phases of life, iv. 403
Phenol, a bacterial product in digestion, 477
,, compounds of, in urine, 650
Phonation, nervous mechanism of, iv. 325
,, centre for, iv. 326
Phloridzin, temporary diabetes produced by, 732
Phosphates in muscle ash, 102
,, ,, nerve ash, 123
,, ,, urine, 651
Phosphenes, iv. 77
Phosphorus, a constituent of nuclein, 40, 799
,, ,, of serum, 50
,, ,, of nerve tissue, 121
,, ,, of milk, 782
,, its importance in organisms, 799
Photochemistry of the retina, iv. 115
Physiology, divisions of, 3
,, problems of, 9
Physiological unit defined, 6
Physostigmin, its effect on pupil contraction, iv. 43

INDEX.

Physostigmin, its effect on accommodation, iv. 45
Pia mater of cord, 859
,, ,, of brain and cord, 1125, 1126
Pigment, yellow, of serum, 50
,, black, occasionally found in lymphatic glands, 495
,, cells in dark skins, 683
,, epithelium of retina, iv. 68, 118
Pigments, their possible formation from hæmoglobin, 38
,, of bile, 423
,, of urine, 651, 673
Pilocarpin, its action on the sweat glands, 701
Pineal gland, 934
Pitch of sounds, discrimination of, iv. 226
Pituitary body, the, structure of, 764, 935
Placenta, the, glycogen present in, 729
,, formation of, from the decidua serotina, iv. 376
,, structure of, iv. 377
,, vascular events of the, iv. 378
,, shedding of the, iv. 379
,, expulsion of, after parturition, iv. 397
Plasma-corpuscles in connective tissue, 189
Plasmatic layer in capillary contents, 335
Plasmine, properties of, 23
Platelets, blood, 29, 47, 337
Plethysmograph, amount of blood in parts determined by, 314
,, for kidney measurements, 657
,, for measurement of blood-supply to brain, 1135
Pleural cavity, communication of with lympathics, 489
,, ,, result of access of air to the, 536
Plexus, brachial, constrictor and dilator fibres in, 315
,, of Auerbach, 441
,, of Meissner, 441
,, solar, 403, 435
,, renal, 646
,, hepatic, 731
Pneumograph, Marey's, 540
,, tracing of respiratory movements by, 542
Pneumatograph of Fick, 540
Polar globules, ejectment of the, iv. 358
Polarizing current, irritability of nerve affected by, 128
Pons varolii, development of, 230
,, ,, grey matter of the, 982
,, ,, transverse fibres of, 996
Porta hepatica, 705
Portal vein, 707
Posture, erect, how maintained, iv. 343
Potassium salts in cell tissue, 41, 51
,, ,, in muscle tissue, 102
,, ,, in urine, 649
Predicrotic wave, its causes, 270
Pregnancy and birth, iv. 374—402

Pressure, arterial, as compared with venous, 202, 207
,, ,, as affecting pulse-tracings, 270
,, ,, heart-beat in inverse ratio to, 305
,, blood, in the small vessels, peripheral region, 209, 216
,, ,, flow of lymph regulated by, 501—505
,, endocardiac, 238
,, ,, graphic records of, 240—246
,, ,, negative during each cardiac cycle, 239, 240
,, ,, ,, how produced, 248
,, auricular and ventricular compared, 229, 241
,, of salivary secretion, 400
,, of bile secretion, 439
,, pulmonary, 537
,, thoracic, 541
,, ,, negative, 616
,, partial, of gases, 558
,, absorption of oxygen, dependent on, 572
,, ,, results of, 611
,, atmospheric, effect of diminution of, 610
,, ,, increase of, 611
,, of carbonic acid in pulmonary alveoli, 576
,, within the bladder, 683
,, intra-ocular, conditions affecting, iv. 171
,, sensations of, iv. 274
,, ,, modified by temperature, iv. 291
,, sensibility of skin to changes of, iv. 276, 292
Pressures, Henry Dalton law of, 558
Prickle-cells of epidermis, 683
Primitive sheath of nerve-fibre, 115
Primordial utricle, 4
Prism, Nicol, described, 93
Processus vocalis, and muscularis, iv. 309
Prostate gland, structure of, iv. 307
,, ,, secretion of, iv. 307
Proteids, general composition of, 17
,, changes in, produced by alcohol, 24
,, in food-stuffs, 355
,, action of gastric juice on, 365
,, ,, ,, pancreatic juice on, 426
,, ,, classification of, in order of solubility, 367
,, ,, path taken by, during digestion, 514
,, ,, amount of urea increased by, 745
,, ,, a source of fat, 774
,, ,, metabolism of body increased by, 785
,, ,, disruption of, during digestion, 795

448   INDEX.

Proteids, probable molecular composition of, 428
  ,, possible storage of, in the body, 796
  ,, "tissue" or morphotic and "floating" or circulating, 796, 826
Proteid material, potential energy of, expressed in calories, 802
  ,, ,, a constituent of living matter, 43
  ,, ,, the pivot of metabolism, 828
Protoplasm, definition of, 4
  ,, "differentiated," 4
  ,, "undifferentiated" in the embryo, 36
Pseudopodia of the white corpuscles, 38
  ,, amœboid movements by means of, 166
Psychical processes, analysis of, 1120
  ,, ,, duration of, 1122
  ,, ,, visual, complexity of, iv. 157
Ptomaines, their bacterial origin, 417
Ptyalin, a constituent of saliva, 362
Puberty, phenomena of, iv. 409
Pulse, the, 201
  ,, methods of recording, 255
  ,, curves, 256
  ,, ,, from artificial model, 257
  ,, ,, characters of, 260
  ,, wave, changes undergone by, along the arteries, 261
  ,, ,, velocity of, 262
  ,, ,, length of, 263
  ,, dicrotism in the, 264
  ,, venous, 271
  ,, ,, respiratory, 272, 613
Pulvinar, the, 978
  ,, ending of part of the optic tract in, 1075
Punctum lachrymale, iv. 174
Puncture of pleura, result of, 536
Pupil, the, *see* Eye, pupil.
Purkinjé, cells of, their structure, 277
  ,, ,, 1022—1025
  ,, figures of, iv. 111
Purple, visual, iv. 116
  ,, ,, bleaching of by light, iv.117
Pus corpuscles, their formation, 45
Putamen, the, 974
Pyloric glands of stomach, 384
Pylorus, ejection of chyme through the, 472
Pyramid cells of cortex, large and small, 1026, 1029, 1032
Pyramidal tract of cord, 873
  ,, ,, its connection with cerebral cortex, 888
  ,, ,, efferent nature of impulses of, 1044, 1056
  ,, ,, not indispensable for voluntary movements, 1059

Pyramids (in kidney) of Malpighi, 635
  ,, ,, of Ferrein, 638
  ,, of the bulb, decussation of, 937, 1046
Pyrexia, causes of, 817

Radial artery, tracings of the pulse in, 257—260
Radiation, optic, 991
Radical, lacteal, of intestinal villus, 447, 484
  ,, ,, contents of, 518
Ranke's diet table, 802
Ranvier, node of, in nerve-fibre, 115
  ,, ,, division of nerve-fibre takes place at, 119
Reaction-period, subdivision of, 1120
  ,, for vision, iv. 74
Receptaculum chyli, 482
Rectum, structure of, 451
  ,, special movements of, 467
  ,, nervous control of movements of, 467, 468
Recurrent sensations, iv. 132
Reflex actions, general features of, 179
  ,, ,, doubtful if carried out by ganglia, 179
  ,, ,, not always proportioned to stimulus, 180
  ,, ,, often purposive in nature, 181, 905
  ,, ,, vaso-motor, 321
  ,, ,, of the cord, 904
  ,, ,, ,, features of dependent on afferent impulses, 905
  ,, ,, ,, nervous mechanisms of, 908
  ,, ,, ,, "crossed," 908
  ,, ,, ,, their relations to intelligence, 909
  ,, ,, ,, coordination of, 910
  ,, ,, ,, relations of to consciousness, 911
  ,, ,, ,, determined by intrinsic condition of cord, 912
  ,, ,, ,, other than movements, 914
  ,, ,, ,, inhibition of, 915
  ,, ,, ,, inhibitory action of the brain on, 916
  ,, ,, ,, time required for, 918
Refraction of muscular-fibre bands, 24
Refractory period of cardiac contraction, 287
Regeneration of nerve tissue, 145
  ,, of organs in lower animals, iv. 349
"Registers" of the voice, iv. 330
Reissner, membrane of, iv. 205, 211
Relaxation of muscular fibre an essential part of contraction, 151, 167
  ,, ,, ,, a complex vital process, 313

INDEX.

Remak, ganglion of, in heart of frog, 278
Rennet, curdling action of on milk, 374
Rennin, its direct action on casein, 375
,, its formation in gastric cells, 419
Reproduction, tissues and mechanisms of, iv. 349
,, general features of, iv. 389
,, female organs of, iv. 391
,, male organs of, iv. 364
Respiration, 526—631
,, pulmonary, circulation aided by, 219
,, ,, its mechanism, 526, 535–549
,, ,, work of the muscles of the ribs in, 545
,, ,, laboured, muscles of, 546
,, ,, expiration, the expiratory muscles, 547
,, ,, change of temperature of air in, 550
,, ,, change of aqueous vapour in, 550
,, ,, changes in blood caused by, 553
,, ,, chemical aspects of, 582
,, ,, an involuntary act. 584
,, ,, sequence of muscular contractions in, 584
,, ,, centre for, medullary, 585
,, ,, ,, automatic action of, 586
,, ,, ,, influenced by afferent impulses, 588
,, ,, ,, duplexity of its action, 591
,, ,, ,, effects of inflation and suction, 592
,, ,, ,, double action of vagus on, 595
,, ,, ,, nature of action, 596
,, ,, ,, two lateral halves of, 598
,, ,, ,, influenced by character of blood supply, 598
,, ,, ,, ,, by heat, 599
,, ,, ,, ,, by deficiency of oxygen, 600, iv. 393
,, ,, ,, ,, by excess of carbonic acid, 600
,, ,, ,, ,, by other changes in the blood, 602
,, ,, ,, apnœa, phenomena of, 603
,, ,, Cheyne-Stokes, 605
,, ,, affected by changes in atmospheric pressure, 601, 611
Respiration, pulmonary, its effect on arterial pressure, 613
,, ,, artificial, its effect on the circulation, 619
,, ,, impeded, its effect on heart-beat, 621
,, ,, ,, ,, on vaso-motor system, 622
,, ,, phenomena of asphyxia, 624
,, as affected by muscular work, 628
,, regulation of temperature by, 811
,, slowing of during hibernation, 820
,, as affected by sleep, iv. 414
,, facial and laryngeal, 548
,, cutaneous, 696
,, of muscle, 578
,, of other tissues, 580
,, of the embryo, iv. 382
,, placental compared with branchial, iv. 384, 385
Respiratory quotient in herbivora and carnivora compared, 798
Restiform body, its connection with the cord, 950
Rete testis, iv. 365
,, vasculosum of testis, iv. 365
Reticular tissue of intestine, 444
Reticulum, splenic, structure of, 736
Retina, *see* Eye, retina.
Rheometer of Ludwig, 220
Rheoscopic frog, 111
,, ,, current of action shewn in, 124
Rhythm, secondary respiratory, 605
Rhythmic changes of calibre in artery, 307
,, beat of heart, spontaneous nature of, 280
,, contractions of uterus during pregnancy, iv. 395
Ribs, movements of, in respiration, 544, 545
Rigor mortis, its cause, 40
,, ,, characters of, 95
,, ,, development of carbonic acid in, 100
,, ,, conversion of myosinogen into myosin during, 104
,, ,, progressive order of, 147
,, ,, as compared with contraction, 153, 154
,, ,, accession of heat at onset of, 810
Rima glottidis, the, iv. 310
Rings, cartilaginous, of trachea, 531
Ritter Valli law, 144
Rod cells of olfactory mucous membrane, iv. 247
,, ,, of taste-buds, iv. 256
Rods, of retina, iv. 59, 116, 119

450 INDEX.

Rolando, substantia gelatinosa of, 863
,, tubercle of, 998
Rolls formed by red corpuscles, 32
Roots of spinal nerve, 171, 852
,, ,, ,, anterior, 876
,, ,, ,, posterior, 877
,, ,, ,, ,, degeneration from section of, 892
Rosenthal's calorimeter, 804
Round ligament of uterus, contractions of, iv. 373
Roy and Rolleston, their method of recording endocardial pressure, 242
Roy, his perfusion cannula, 281

Saccule of labyrinth, iv. 199
Sacculi of large intestine, 450
,, peristaltic contractions of, 462
"Sago-spleen," cause of, 739
Saline solution, normal, defined, 16, note
Saliva, characters and properties of, 357
,, its properties, 358
,, its amylolytic action, 361, 470
,, characters of parotid, submaxillary, sublingual and mixed, 362
,, amount daily secreted, 395
,, reflex secretion of, 396
,, centre for secretion of, in medulla oblongata, 398
,, of dog, mechanical use of, 470
,, of the babe, iv. 405
Salivary glands, general structure, 385
,, mucous cells of, 387
,, albuminous or serous cells of, 389
,, venous pulse observable in, 271
,, nervous supply of, 390
Salts, neutral, needed for formation of fibrin, 26
,, in food-stuffs, 356
,, absorption of, 513
,, as food, 799
,, importance of, for nutrition of nervous system, 799
,, essential to life of muscle, 824
,, in diet, 836
Santonin, vision as affected by, iv. 105
Santorini, cartilage of, iv. 309, 314
Sarcolemma, structure of, 84
Scalæ of the cochlea, iv. 204
Scaleni muscles, the, their service in respiration, 545
Scheiner's experiment, iv. 9, 49
Schlemm, circular canal of, iv. 27, 165
,, ,, passage of aqueous humour by means of, 170
Schneiderian membrane, iv. 246
Sclerotic coat of eye, development of, iv. 4, 21
Sebaceous glands, change of their cells into sebum, 692
Sebum, secretion of, 693

Secretion of saliva, nervous mechanism of, 395
,, of gastric juice, 402
,, changes in gland constituting act of, 404
,, changes in albuminous cells, 407
,, changes in mucous cells, 408
,, by central cells of stomach, 411
,, special substances elaborated during, 416
,, of pancreatic juice, 432
,, of bile, 434
,, of urine, 656
,, glomerular and tubular in the kidney compared, 666
,, glomerular, its nature, 668
,, of wax of ear, 693
,, of sweat, 695
,, ,, mechanism of, 699
,, of milk, 784
Secretions, carbonic acid in, 580
,, their constituents manufactured by glandular action, 671
Segmentation of the ovum, 6
Self-digestion, 419, 420
Self-induction, 64, 65
Semen, chemical composition of, iv. 369
,, emission of, see Emission of semen.
Semicircular canals, effects of injury to the, 1009
,, ,, their structure, iv. 198
Semilunar valves, their structure, 197
,, ,, their action, 231
,, ,, dicrotic wave as affected by closure of, 268-270
Seminal tubules, see Tubules, seminal.
Seminiferous tubules, ,, ,,
Sensations, special auditory, iv. 223-231
,, ,, ,, limits of, iv. 225
,, ,, ,, fusion of, iv. 229
,, cutaneous, 1090-1108, iv. 267-294
,, ,, importance of contrast in, iv. 292
,, ,, of pressure, iv. 274
,, ,, ,, localization of, iv. 276
,, ,, of heat and cold, iv. 278
,, ,, of pain, iv. 281
,, ,, of touch and temperature, terminal organs necessary for, iv. 287
,, ,, of pressure, terminal organs for, iv. 290
,, ,, of heat, terminal organs for, different from those for cold, iv. 291
,, ,, connection of, with the muscular sense, iv. 802
,, olfactory, iv. 250, 251
,, of taste, 1087, iv. 259-266
,, ,, structure of organs of, iv. 254

INDEX. 451

Sensations of taste, usually accompanied by other sensations, iv. 259
,, ,, caused by electrical or mechanical stimuli iv. 260
,, ,, conditions of, iv. 261
,, ,, localization of, iv. 262
,, ,, distribution of terminal organs for, iv. 263
,, ,, theories as to mode of origin, iv. 264
,, ,, nerves for, iv. 265
,, visual, probable progressive development of, 1083
,, ,, general features of, iv. 71
,, ,, fusion of, iv. 75, 79
,, ,, localization of, iv. 78
,, ,, of colour, iv. 84
,, ,, ,, due to metabolic changes, iv. 94
,, ,, psychological features of, iv. 123
,, ,, their want of agreement with perceptions, iv. 125
,, ,, recurrent, iv. 132
,, afferent, as factors in co-ordination of movement, 1013
,, crossing of, from opposite hemisphere of brain, 1093
,, development of, along the spinal cord, 1101
,, transmission of, within the brain, 1106
,, co-ordination of motor impulses regulated by, iv. 150
Sense, the muscular, iv. 295
,, ,, of movement, of position and of effort, iv. 296
,, ,, afferent impulses forming basis of, iv. 298
Sensibility, general, iv. 283
,, recurrent, 853
Septal nerves of heart of frog, 278
Serous cavities, fluid of, 499
,, fluids, artificial clotting of, 23
,, ,, characters of, 499
Serum left after clotting of fibrin, 15
,, chemical composition of, 18, 49
,, carbonic acid in, 571
Serum-albumin, its characters, 19
,, action of gastric juice and hydrochloric acid on, 368
,, importance of, in nutrition of muscle, 823
Sex, differences of, iv. 409
Sheath, primitive, of nerve-fibre, 115
,, of arteries, 195
Shivering from cold, temperature raised by, 816
Shock, induction, 62
,, in operation, results of, 327
,, nature of, 903
Short-circuiting, 61

Sight, see Vision.
Singing, power of, dependent on nervous mechanism, iv. 329
Sinuses, lymph, of solitary follicles, 490
,, venous, of brain, 1133
,, placental, structure of, iv. 378
,, ,, quality of blood in, iv. 384
Size, judgment of, iv. 158
Skin, structure of, 686
,, as regulator of heat, 811
,, nerve endings, general and special, of the, iv. 267
,, different kinds of sensations experienced through the, iv. 274
,, as 'field of touch,' iv. 276
Sleep, phenomena of, iv. 412
,, afferent impulses as affected by, iv. 413
,, respiration during, iv. 414
,, the brain during, iv. 414
Smell, sensations of, iv. 250—253
,, cerebral structures for, 1085—1087
,, cortical area for, 1087
,, organ of, iv. 246
Snout, pig's, touch-cells of the, iv. 273
Sodium chloride, its action on plasma, 22
,, ,, ,, on mucin, 357
,, glycholate and taurocholate, 423
,, hydrate, its effect on the heart, 303
,, sulphindigotate, excretion of by kidney, 667
,, ,, ,, by liver, 709
Solar plexus, nerve supply to stomach from the, 403
,, ,, ,, to the liver from the, 435
Sole of end-plate of nerve fibre, 119
Solidity, judgment of, iv. 162
Solitary follicles, 489
Somatic nerves, 171
Sound, musical, of contracting muscle, 140
,, waves of, iv. 180
,, complex, analysis of, iv. 234
,, psychical nature of appreciation of, iv. 237
Sounds, musical, characters of, iv. 243
,, appreciation of outwardness of, iv. 242
,, judgment of direction of, iv. 243
,, ,, distance of, iv. 244
Spaces, subdural and subarachnoid, 1125
,, of Fontana, iv. 27
Spectrum, limitations of visibility of, iv. 84
Speech, cortical area for, 1053, iv. 326
,, a skilled movement, 1053
,, movements of, bilateral, 1053
,, causes of various imperfections of, 1057
,, special mechanisms of, iv. 332—341

INDEX.

Speech, sounds made use of in, iv. 333
Spermatozoa, structure of, iv. 366
,, formation of, iv. 367
,, movements of, iv. 368
,, action of on the ovum, iv. 374
Sphincters of stomach, their action during digestion, 459
,, tonic of dependent on cord, 922
Sphincter ani, its nerve-supply, 463
,, vesicæ, 681
,, iridis, iv. 24, 34
Sphygmograph, Dudgeon's, 256
Spinal cord, *see* Cord, spinal.
Spiral cells, 176
Spiral tubule of kidney, 641
Spirometer, 538
Splanchnic nerves, 171
,, ,, inhibitor and augmentor fibres in, 467
,, ganglia, 176
,, abdominal nerve, 316
Spleen, the, possible formation of red corpuscles in, 38
,, its action during digestion, 436
,, blood supply to the, 737
,, lymphatic vessels of the, 738
,, hyperplastic spots of the, 738
,, Malpighian corpuscles of the, 738
,, nerves of the, 739
,, movements of the, 740
,, chemical constituents of, 742
,, uric acid in the, 780
"Spleen-curve," 740
Spleen-pulp, destruction of red corpuscles in, 36
,, ,, composition of, 737, 739
Spot, blind, of retina, iv. 110
Stagnation stage of inflammation, 338
Stapes, or stirrup bone, iv. 182
Starch, action of saliva on, 358
,, chemical composition of, 359
,, action of pancreatic juice on, 426
,, "animal," 722
,, its value in diet, 836
Starvation, its effect in checking production of glycogen, 716, 720
,, changes in body during, 789
,, fall of temperature attending, 818
Stearin, a constituent of animal fat, 772
,, its presence in blood, 50
Stereoscope, ocular movements affected by the, iv. 149
,, principle of construction, iv. 163
Stethometer of Burdon Sanderson, 540
Stimuli, defined, 56
,, various kinds of, 59
,, necessary characters of, 138
Stimulous, reflex actions varied according to nature of, 906
Stomach, structure of, 380

Stomach, cardiac glands of, 381
,, pyloric glands of, 384
,, nervous supply to, 402
,, its secretion of gastric juice 403, 404
,, movements of, 458
,, changes of food in the, 471
Stomata of cisterna in frog, 487
,, in mammals connecting serous cavities with lymphatics, 488
Storage of bile in gall-bladder, 435
,, of glycogen in the liver, 717, 725
,, of fat before hibernation, 821
Strands, spiral, of cochlea, iv. 220
Stratum granulosum of epidermis, 688
,, lucidum of epidermis, 689
Striæ acusticæ, 959
Striation, obscurity of in cardiac muscular tissue, 288
Stroma of red corpuscles, its composition, 32
,, ,, embryonic formation of from protoplasm, 36
,, of the kidney, 646
Stromuhr of Ludwig described, 220
Strychnia, reflex action as affected by, 907, 912
Submucous tissue, 378
Substance, living, compared with dead, 3, 95
,, ,, metabolic changes in, 42
,, ,, chemical composition, 43
Substances, visual, hypothetical, iv. 95
Substantia gelatinosa centralis of cord, 863
,, ,, of Rolando, 863
,, nigra in brain, 981
Succus entericus, its nature and action, 430
,, ,, how probably furnished, 449
Sugar, its presence in the blood, 51, 727
,, normally present in blood and chyle, 513
,, formed by saliva from starch, 360
,, course taken by, during digestion, 513, 524
,, in diabetic urine, 654
,, its conversion into glycogen, 722
,, a product of metabolic changes, 723
,, used up by muscle, 824
,, its value in diet, 836
Sulcus spiralis of cochlea, iv. 211
,, crucial and sigmoid, of dog's brain, 1035
Sulphur in proteids, 17, 799
,, in urine, 650
,, in keratin, 689
Suprarenal bodies, their structure, 765
,, ,, chemical composition, 766
,, ,, their functions, 766
Swallowing, mechanism of, 452
,, its action on tympanic air pressure, iv. 197

INDEX. 453

Sweat, how secreted, 695, 699
,, composition of, 696
Sweat-fibres of different animals, course of, 701
Sweat glands, their structure, 690
,, ,, action of pilocarpin on, 701
Sweat-nerves, 701
Sweating in lower animals, 700
,, nervous mechanism of, 699
,, a reflex act, 701
Sylvius, aqueduct of, 930
Sympathetic system, plain muscular fibres supplied by, 160
,, ,, its connection with spinal nerves, 171
,, ,, ganglia of the, 176
Syntonin, 97
Systole, auricular and ventricular, 228, 229, 251
,, ventricular, a simple contraction, 237
,, and diastole, comparative duration of, 249, 251
,, amount of blood driven by each, 201, 253

Tactile sensations, iv. 274
,, ,, localization of, iv. 276
Tambour, Marey's, 241
Tarsus of the eyelids, iv. 172
Taste-buds, their distribution, iv. 255
,, their nervous connections, iv. 257
Taurocholic acid, 423
Tears, secretion of, iv. 174
Tectorial membrane, iv. 221, 234
Teeth, order of their appearance, iv. 408
Tegmental system, fibres of, 991—994
Tegmentum, grey matter of, 981
,, connections of the, 992
,, nature of its functions, 1112
Temperature of living bodies, 2
,, as affecting clotting, 20
,, ,, irritability, 143, 145
,, ,, plain muscle, 161
,, ,, ciliary action, 165
,, ,, vaso-motor fibres, 315, 332
,, ,, action of gastric juice, 372
,, ,, action of rennet, 374
,, ,, point of saturation of gas, 550
,, ,, absorption of oxygen by liquids, 572
,, ,, the cutaneous vessels, 699
,, ,, perspiration, 694, 700
,, ,, storage of glycogen, 718
,, ,, sense of taste, iv. 261
,, of expired air, 550
,, regulation of, by evaporation from the skin, 699

Temperature, regulation of, by variations in loss of heat, 810, 812
,, ,, by the nervous system, 814, 815
,, of cold-blooded animals, 809
,, of warm-blooded animals, 810
,, normal, range of, 817
,, high, phenomena of death from, 818
,, low, effects of, 819
,, its relation to amount of food needed, 843
,, of body, maintenance of, 814
,, sensations of, iv. 278, 292
,, terminal organs for sensations of, iv. 291
,, sense of, in parts other than external skin, iv. 280
Tendon reflexes, "knee-jerk," 913, 926
Tenonion cavity and Tenon's capsule, iv. 167
Terminal organs, special sensations due to, iv. 288
,, ,, for sense of touch, iv. 288
,, ,, for sense of pressure, iv. 290
,, ,, for sense of heat different from those for sense of cold, iv. 291
,, ,, cutaneous, their nature, iv. 293
Testis, origin of, iv. 364
,, general structure of, iv. 365
,, lymphatics of, iv. 369
Tetanic contraction, its nature, 58
Tetanus, phenomena of, 78, 81, 139
,, carbonic acid evolved during, 103
,, exhaustion of irritability from, 149
Thalamus, optic, 971
Thermopile, various forms of, 105
Thermotaxis, centre for, 815, 816
Thirst, sensation of, iv. 885
Thoracic duct, characters of lymph from the, 497
Thorax, effect on blood-flow of pressure in the, 615, 618
Thrombi, white, their nature, 48
Thymus body, structure of the, 767
,, ,, nature and functions, 768
,, ,, its size in infancy, iv. 407
Thyroid body, 761—765
,, ,, structure of, 761
,, ,, functions of, 763
,, ,, diseases connected with, 764
,, ,, in infancy, iv. 407
Thyroid-arytenoid muscles, iv. 318, 328
Thyroid cartilage, iv. 308
Time taken by cerebral processes, 11
Tissues not indispensable for life, 3
,, classification of, 6
,, built up by the blood, 13
,, similarity of histological elements of, 41

Tissues, contractile, 54—168
,, nervous, 169—185
,, vascular, 186—352
,, digestive, 355—525
,, respiratory, 526—631
,, relative proportions of, in the body, 789
,, metabolism of, 822
,, their death gradual, iv. 418
Tone, arterial, 308
,, dependent on vaso-motor action, 319, 327
,, centre for, in medulla, 323
,, intrinsic nature of, 329
,, maintained by automatic action of cord, 922
,, of skeletal muscles, 922
,, ,, due to central nervous system, 923
Tones, musical, fundamental and partial, iv. 224
Tongue, papillæ of, iv. 255
,, localization of taste sensations in, iv. 262
Torcular Herophili, 1133
Tortoise, persistence of ventricular beat in, 285
,, heart-beat in, independent of cardiac nerves, 288
,, ,, action of atropin on, 301
Touch-cells, in epidermis and elsewhere, iv. 273
Touch-corpuscles, structure of, iv. 270
,, distribution of, iv. 271
Trabeculæ of lymphatic glands, 492
,, of spleen, 735
Trachea, structure of, 530
,, nerve supply to, 534
,, effect on respiration of its closure, 593
Tract, optic, course of, 1074
,, ascending antero-lateral, 874, 895
,, descending antero-lateral, 873
,, cerebellar, 874, 890, 950
,, ,, as to functions of, 1102
,, median posterior, 874, 891
,, ,, ,, as to functions of, 1103
,, pyramidal, crossed, 872, 888
,, ,, direct, 873, 889
,, ,, relations to volition, 1044, 1056, 1059
Tracts, afferent, in spinal cord, 1094
,, internuncial, for afferent impulses, 1104
Training, effect of in brain action, 1069
Transudation into lymph spaces, 503
,, not merely a filtration, 504
,, conditions determining, 505
,, opposite currents of, through capillary walls, 506, 520
Trapezium, 960
Traube-Hering curves, their origin, 622

Traube-Hering undulations in kidney, 660
,, variations in cerebral blood-pressure, 1136, 1137
Tricuspid valves, 230
Trypsin, a constituent of pancreatic juice, 414, 426
,, in the fœtal pancreas, iv. 389
Trypsinogen, an antecedent of trypsin, 414
Tube, Fallopian, iv. 351
Tuberculum acusticum, 959
Tubules, seminal, course of, iv. 365
,, ,, structure of, iv. 366
,, uriniferous, structure of, iv. 634, 641
,, ,, convolutions of, 636
,, ,, epithelium of the, 665
,, ,, work of the, 671
,, ,, special substances excreted by, 666
Tubuli seminiferi, *see* Tubules, seminal.
Tunicæ, intima, media and extima of arteries, 193
Tunica muscularis mucosæ, 378
,, albuginea, iv. 365
,, vaginalis, iv. 365
Tuning-fork for the measurement of velocity, 70
Tympanum of ear, iv. 179
,, conduction of sound through, iv. 181, 190
,, structure and relations, iv. 182
,, membrane of, iv. 185
,, muscles of the, iv. 194
,, its connection with sense of outwardness of sounds, iv. 242
Tyrosin, a product of pancreatic digestion, 427
,, chemical composition, 428
,, a result of proteid decomposition outside the body, 759

Umbilical cord, formation of, iv. 377
,, arteries, growth of, iv. 377
,, ,, pressure in, iv. 383
,, ,, venous blood in the, iv. 384
,, vein, pressure in, iv. 383
Undulations, respiratory, phenomena of, 614, 618
,, luminous, iv. 84
Unit, physiological, defined, 6
Urari, the nature of its action, 57, 86
,, its effect on cells of pigment epithelium of retina, iv. 70
,, diabetes in frogs produced by, 733
Urea, a constituent of the blood, 51
,, chemical relations of, 649
,, as nitrogenous waste, 102, 632, 750, 754, 759
,, absent from muscle, 102, 751
,, its relations to kreatin, 750, 752

INDEX.

Urea, its presence in the blood antecedent to kidney action, 672
,, its action on the tubules of kidney, 676
,, brought to the kidneys by the blood, 750, 760
,, its formation in the liver, 755
,, synthesis of, 756
,, its relation with cyanogen compounds, 759
,, diminished excretion of, during starvation, 790
,, excretion of, not increased by exercise, 805
,, its kinship to vegetable alkaloids, 828
,, a constituent of amniotic fluid, iv. 390
Ureter, structure of, 635, 678
,, peristaltic contractions of, 680
Uric acid, chemical composition of, 649
,, relations to urea, circumstances determining its appearance, 757
,, constant presence of in the spleen, 743
Urina hysterica, 677
Urine, composition and characters of, 648
,, normal organic constituents, 649, 653
,, inorganic salts of, 650
,, average composition of, 653
,, abnormal constituents of, 654, 734
,, secretion of, 656
,, vaso-motor mechanisms for, 657
,, its relations to the renal circulation, 664
,, albuminous, 670
,, pigments of, 673
,, discharge of, 678
,, its secretion continuous, 679
,, changes of, in the bladder, 635
,, ,, during starvation, 790
,, sugar present in diabetes, 730
,, of children, characteristics of, iv. 407
Urobilin, 652
Use, muscle substance increased by, 148
,, skilled movements facilitated by, 1069
" Uterine milk," iv. 381, 388
Uterus, the, general structure of, iv. 351
,, minute structure of, iv. 352
,, blood-vessels and lymphatics of, iv. 353
,, reception of the ovum by, iv. 360
,, changes in mucous membrane of, during menstruation, iv. 361
,, ,, after impregnation, iv. 374
,, expansion of, during pregnancy, iv. 395
,, "retraction" of, iv. 396, 398, 400

Uterus, rhythmical contractions of, during pregnancy, iv. 395
,, ,, ,, during 'labour,' iv. 396
,, nerves of, iv. 399
Utricle, primordial, 4
,, of labyrinth, iv. 199

Vagina, the, structure of, iv. 353
Vagus, see Nerve, vagus.
Valves of veins, 197, 219
,, of the heart, 197
,, ,, their action in circulation, 230
,, ,, sounds caused by their closure, 235, 236
,, ,, tricuspid, their action, 230
,, ,, semilunar, of the pulmonary artery, 230
,, ,, ,, ,, ,, their action, 231
,, ,, ,, ,, of aorta, 235, 246
,, ileo-cæcal, mechanism of, 463
,, of the thoracic duct, 482
,, of the lymphatic vessels, 483, 501
,, absence of, in pulmonary veins, 534
,, of Vicussens, 936
Valvulæ conniventes of small intestine, 442
Vapour, aqueous, in expired air, 550
Vas deferens, iv. 365
,, ,, contraction of in emission, iv. 372
Vasa vasorum of arteries, 195
,, ,, of veins, 196
,, afferentia and efferentia of kidney, 639
,, efferentia of testis, iv. 365
,, recta of testis, iv. 365
Vascular mechanism, 186—340
,, ,, structure of capillaries, 190
,, ,, ,, of minute arteries, 193
,, ,, ,, of larger arteries, 194
,, ,, ,, of veins, 196
,, ,, ,, of heart, some points in, 197
,, ,, main features of, 198
,, ,, main regulators of, 273, 306
,, walls, their action on the blood, 27
,, ,, alteration of in inflammation, 338
Vaso-motor action, 306—330
,, ,, effects of, 319
,, ,, cutaneous and splanchnic, compensatory, 350, 351
,, ,, compensatory in loss and increase of blood, 341
,, ,, in brain, 1136
,, ,, summary of, 330
,, ,, regulation of temperature by, 811
,, ,, its rhythmic tendency, 622
,, ,, centre, 1098, 322, 329
,, ,, limits of, 326

Vaso-motor centre, relations of to other centres, 327
,, fibres, constrictor, 310, 312
,, ,, ,, course of, 317, 322, 330
,, ,, ,, loss of medulla in, 317
,, ,, ,, tonic action of, 319—323
,, ,, ,, chief parts of body supplied by, 322
,, ,, dilator, 312
,, ,, ,, course of, 318
,, ,, ,, usually employed as part of reflex action, 321
,, ,, ,, retention of medulla in, 331
,, functions of the central nervous system, 321
,, nerves of veins, 333
Vegetable cell, storage of metabolic products in, 828
,, diet, results of, 839
,, ,, large amount required, 841
Veins, structure of, 196
,, minute, 197
,, valves of, 197, 219
,, their capacity as compared with arteries, 200
,, blood-pressure in, 203, 207
,, vaso-motor nerves of, 333
Velocity of nervous impulse, 75
,, of muscular contraction, 87
,, equal, of muscular current of action and nervous impulse, 124
,, comparative, of arterial, venous and capillary circulation, 209—220
,, of blood-current, 263
,, of pulse wave, 264
Venæ stellatæ of kidney, 645
Venous circulation, aids to, 219
,, pulse, 271, 613
,, sinuses of brain, 1133
Ventilation, positive and negative of lung, 594
Ventricle of heart of frog, action in heart-beat, 282
,, ,, ,, of tortoise, isolated, spontaneous beat of, 285
Ventricles of heart, synchronism of their action, 220
,, ,, ,, change of form of in cardiac cycle, 232
,, ,, ,, four stages of action of, 246
,, ,, ,, tonic contraction of, 304
,, of brain, development of, 930
Vermiform appendix, solitary follicles in the, 489
Vertigo, causes of, 1015
Vesicles, cerebral, 929
,, optic, 1140

Vesicles, otic, iv. 177
Vesiculæ seminales, iv. 370
,, ,, secretion of, iv. 370
,, ,, their action in emission, iv. 372
Vestibule of ear, iv. 176
,, ,, parts of, iv. 198
,, ,, perilymph cavity of, iv. 200
Vibrations of muscle sound, 140
,, sonorous, longitudinal and transversal, iv. 190
,, ,, of the tympanic membrane, iv. 191
,, ,, through the auditory ossicles, iv. 192
,, ,, through the bones of the skull, iv. 193
,, of sound and light compared, iv. 225
,, interference of, iv. 229
Vierordt, his hæmatachometer, 222
Vicusseus, annulus of, 209
Villi, the, of small intestine, 445
,, columnar epithelium of, 445, 516
,, goblet cells of, 446
,, structure of, 447
,, pumping action of, 519
,, of chorion, foetal, iv. 377
Vision, 1070
,, binocular, 1071, iv. 134—154
,, ,, its action in judging of distance, iv. 151
,, ,, ,, of solidity, iv. 162
,, ,, mechanism of, 1071
,, central apparatus for, 1077
,, imperfections of, 1078
,, affected by injury to cortex, 1081
,, dioptric mechanisms of, iv. 1, 6
,, astigmatism, iv. 48
,, spherical aberration, iv. 48
,, chromatic aberration, iv. 50
,, entoptic phenomena, iv. 51
,, distinct, limits of, iv. 80
,, trichromic nature of, iv. 91, 98
,, colour, Young-Helmholtz' theory of, iv. 91, 129
,, ,, Hering's theory of, iv. 93, 129
,, field of, 1071, iv. 123, 135
,, corresponding or identical points, iv. 137
,, struggle of the two fields of, iv. 155, 164
Visual areas in fovea centralis, iv. 81
,, axis, iv. 134
,, centres, lower and higher, 1084
,, impulses, development of, iv. 110
,, ,, origin of, iv. 114
,, perceptions and judgments, iv. 155—164
,, ,, psychical processes in, iv. 158
,, plane, iv. 135
,, purple, iv. 116
,, sensations, 1070—1085
,, ,, probable mode of development of, 1083
,, ,, fusion of, iv. 76, 80

INDEX. 457

Visual sensations, discrete, conditions of, iv. 79
,, ,, in relation to visual perceptions, iv. 123—133
,, ,, simultaneous, iv. 123
,, substances, hypothetical, iv.118, 119
,, units, retinal, iv. 81
Vitellin, a constituent of yolk, iv. 353
Vitreous humour, the, iv. 29, 170
Vocal cords, structure, iv. 313
,, ,, voice produced by vibration of the, iv. 306, 315
,, ,, tightening and slackening of the, iv. 322
Voice, the, iv. 306—332
,, how produced, iv. 306
,, fundamental features of the, iv. 316
,, different qualities of, iv. 326
,, chest and head voices, iv. 329
,, registers of the, iv. 331
,, breaking of the, iv. 332
Volkmann, his hæmadromometer, 220
Voluntary movements, their tetanic character, 140
,, ,, nervous mechanisms for, 914, 1034, 1056, 1066
Vomiting, mechanism of, 460
Vowel chamber, iv. 333
Vowels, how formed, iv. 334

Walking, how effected, iv. 343
Walls, vascular, their influence on transudation, 504, 506
Warmth, its effect on skin action, 675
Waste matters, their discharge from the living body, 2
,, ,, given out by amœbæ, 4
,, ,, not necessarily useless, 43
,, ,, elimination of, 632
,, nitrogenous, 102
,, ,, not increased by muscle contraction, 103, 106, 805
Water, secretion of, by the glands, 416
,, varying amount of, in living tissue, 507
,, its absorption into the portal system, 513
,, intestinal secretion of, 522
,, its discharge by the kidney, 674
,, by the skin, 699
Waves of contraction, muscular, 86
,, pulse, dicrotic, their nature, 266
,, ,, predicrotic, 270
,, ,, anacrotic, 271
,, of sound, iv. 180, 224
,, of light, iv. 84

Waves of nervous and muscle impulse,125
Wax of the ear, secretion of, 693
Web of frog, arterial changes visible in, 306
Weber's law, iv. 73, 226, 275
Weight, human, curve of, iv. 404
Wharton's jelly, iv. 377
Whispering, how effected, iv. 336, 341
White, sensation of, produced from mixing of colour sensations, iv. 90
Willis, circle of, 1131, 1134
Winking, how effected, iv. 172
,, chief use of, iv. 175
Wolffian bodies, the origin of the conducting part of the testis, iv. 364
"Word-deafness," 1089
Work, mechanical, in living body, 2
,, done by a muscle-nerve preparation, 136 et supra
,, amount done by heart, 253
,, daily, estimate of, 803
,, mechanical source of energy of, 805
,, production of heat increased by, 813

Xanthin, a constituent of urine, 650, 758
,, present in the thyroid, 762
,, ,, ,, thymus, 768
Xanthoproteic test for protein, 17, 18

Yawning, 630
Yellow elastic fibres, 190
,, spot, structure, iv. 66
,, ,, colour sensations as affected by, iv. 106
Yolk, chemical composition of, iv. 356
Young-Helmholtz, theory of primary colour sensations, iv. 91
,, ,, ,, as applied to colour blindness, iv. 101, 104
,, ,, ,, of simultaneous and succive contrasts, iv. 129

Zinn, zonule of, iv. 30
,, ,, passage of fluid by the, iv. 171
Zona pellucida or radiata, iv. 355
,, spongiosa of cord, 865
Zone, ciliary, iv. 26
,, peripheral of capillaries, 335
,, Lissauer's, 875
Zymogens, 415, 419

MACMILLAN & CO.'S PUBLICATIONS.

Just Published, with 383 Illustrations. 8vo. $5.50.

TEXT-BOOK OF COMPARATIVE ANATOMY.

By DR. ARNOLD LANG,

PROFESSOR OF ZOOLOGY IN THE UNIVERSITY OF ZURICH ; FORMERLY RITTER PROFESSOR OF PHYLOGENY IN THE UNIVERSITY OF JENA.

With Preface to the English Translation

By PROFESSOR DR. ERNST HAECKEL, F.R.S.,

DIRECTOR OF THE ZOOLOGICAL INSTITUTE IN JENA.

Translated into English by

HENRY M. BERNARD, M.A. (CANTAB.), AND MATILDA BERNARD.

Part I.

Complete, with Index and 383 illustrations. 8vo. $5.50.

This translation of the first volume of Professor Lang's *Lehrbuch der Vergleichende Anatomie* may be considered as a second edition of the original work. Professor Lang kindly placed at our disposal his notes, collected for the purposes of emendation and expansion, and they have been duly incorporated in the text. — *From the Translator's Preface.*

Professor Lang has here successfully carried out the very difficult task of selecting the most important results from the bewildering mass of new material afforded by the extensive researches of the last decades, and of combining them with great judgment. Besides this he has, more than any former writer, utilized the comparative history of development in explaining the structure of the animal body, and has endeavored always to give the phylogenetic significance of ontogenetic facts. Lastly, he has, by the clear systematic reviews of the various classes and orders which precede the anatomical account of each race, further facilitated the phylogenetic comprehension of complicated morphological problems, his wisely chosen and carefully executed illustrations assisting materially in this result. It is therefore with great pleasure that I commend this book to the English student, in the hope that the English translation will promote to as great an extent as the German original the wider study and better comprehension of animal morphology, and will attract new students to this noble science. — *From Professor Haeckel's Preface.*

MACMILLAN & CO., 112 Fourth Avenue, New York.

MACMILLAN & CO.'S PUBLICATIONS.

Just published, with 357 illustrations. 8vo. $6.00.

AN INTRODUCTION TO THE STUDY OF MAMMALS,

LIVING AND EXTINCT.

By WILLIAM HENRY FLOWER, C.B., F.R.S., D.C.L., LL.D., P.Z.S., F.L.S., F.G.S., &c.; Director of the Natural History Departments, British Museum; and RICHARD LYDEKKER, B.A., F.G.S., F.Z.S., &c.

357 Illustrations. 8vo. $6.00.

This work is based largely upon the article "Mammalia," together with forty shorter articles, written by the senior of the two authors for the ninth edition of the "Encyclopædia Britannica."

The best thanks of the authors are due to Dr. G. E. Dobson, F.R.S., Mr. Oldfield Thomas, and Dr. St. G. Mivart, F.R.S., for freely permitting the incorporation of their own work in the present volume.

Mr. Lydekker undertook the task of arranging the various articles in their proper sequence, selecting from these such portions as seemed suitable, filling up the gaps, and adding new matter where necessary, — a large amount of this new matter treating of the extinct forms, and also of the group Artiodactyla. — *From the Preface.*

CONTENTS.

CHAP.
- I. Introductory Remarks.
- II. General Anatomical Characters.
    1. Tegumentary Structures.
    2. Dental System.
    3. The Skeleton.
    4. The Digestive System.
    5. Circulatory, Absorbent, Respiratory, and Urinary Systems.
    6. Nervous System and Organs of Sense.
    7. Reproductive Organs.
- III. Origin and Classification of the Mammalia.
- IV. Geographical and Geological Distribution.
- V. The Subclass Prototheria or Ornithodelphia.
- VI. The Subclass Metatheria or Didelphia.
- VII. The Subclass Entheria and Order Edentata.
- VIII. The Orders Sirenia and Cetacea.
- IX. The Order Ungulata.
- X. The Order Rodentia.
- XI. The Order Carnivora.
- XII. The Order Insectivora.
- XIII. The Order Chiroptera.
- XIV. The Order Primates.

MACMILLAN & CO., 112 Fourth Avenue, New York.

Now ready, 8vo, with 89 illustrations, $2.25:

Lessons in Elementary Biology.

By T. JEFFREY PARKER, B.Sc., F.R.S.,

PROFESSOR OF BIOLOGY IN THE UNIVERSITY OF OTAGO DUNEDIN, NEW ZEALAND.

With 89 illustrations, a Synopsis, Index, and Glossary. 8vo. $2.25.

In his preface to the new edition of the well-known *Practical Biology*, Professor Huxley gives his reasons for beginning the study of organized nature with the higher forms of animal life, to the abandonment of his earlier method of working from the simpler to the more complex organisms. He says in effect that experience has taught him the unwisdom of taking the beginner at once into the new and strange region of microscopic life, and the advantage of making him commence his studies with a subject of which he is bound to know something, — the elementary anatomy and physiology of a vertebrate animal.

Most teachers will probably agree with the general truth of his opinion. On the other hand, the advantage to logical treatment of proceeding from the simple to the complex — of working upwards from protists to the higher plants and animals — is so immense that it is not to be abandoned without very good and sufficient reasons.

In my own experience I have found that the difficulty may be largely met by a compromise; namely, by beginning the work of the class by a comparative study of one of the higher plants (flowering plant or fern) and of one of the higher animals (rabbit, frog, or crayfish).

By the time this introductory work is over, the student of average intelligence has overcome preliminary difficulties, and is ready to profit by the second and more systematic part of the course, in which organisms are studied in the order of increasing complexity.

It is such a course of general elementary biology which I have attempted to give in the following Lessons, my aim having been to provide a book which may supply in the study the place occupied in the laboratory by "Huxley and Martin," by giving the connected narrative which would be out of place in a practical handbook. I also venture to hope that the work may be of some use to students who have studied zoology and botany as separate subjects, as well as that large class of workers whose services to English science often receive but scant recognition, — I mean amateur microscopists.

As to the general treatment of the subject I have been guided by three principles. Firstly, that the main object of teaching biology as part of a liberal education is to familiarize the student not so much with the facts as with the ideas of science. Secondly, that such ideas are best understood, at least by beginners, when studied in connection with concrete types of animals and plants. And, thirdly, that the types chosen should illustrate without unnecessary complication the particular grade of organization they are intended to typify, and that exceptional cases are out of place in an elementary course.

The types have therefore been selected with a view of illustrating all the more important modifications of structure, and the chief physiological processes in plants and animals; and by the occasional introduction of special lessons on such subjects as biogenesis, evolution, etc., the entire work is so arranged as to give a fairly connected account of the general principles of biology. It is in obedience to the last of the principles just enunciated that I have described so many of the Protozoa, omitted all but a brief reference to the development of Hydra, and to the sexual process in Penicillium, and described Nitella instead of Chara, and Polygordius instead of the earthworm. The last-named substitution is of course only made possible by the book being intended for the study and not for the laboratory, but I feel convinced that the student who masters the structure of Polygordius, even from figures and descriptions alone, will be in a far better position to profit by a practical study of one of the higher worms. — *From the Preface.*

We have nothing but praise for Mr Parker's book. In well chosen language it sets forth what is known on the subject-matter, and is in many ways an advance on any treatise we have seen. — *Anti-Jacobin, London.*

MACMILLAN & CO., 112 Fourth Avenue, New York.

MACMILLAN & CO.'S PUBLICATIONS.

THE MYOLOGY OF THE RAVEN.

(*Corvus Corax Sinuatus.*)

A Guide to the Study of the Muscular System in Birds.

By R. W. SHUFELDT.

With numerous illustrations drawn by the author, a Bibliography and Index. 8vo. $4.00.

The reviewer has gone through this valuable contribution to comparative anatomy with some care, and is glad to say that on the first of the above-mentioned points (the accuracy and exactness of the author's statements) he has no criticism to offer. Dr. Shufeldt has shown himself master of the subject, and deserves the thanks of all naturalists. Indeed, the work is one of which American scientists have reason to be proud, and it seems out of place to make any criticism. Inaccuracies of statement have been diligently sought for but with negative results.

In conclusion the reviewer wishes to reiterate the statement that the volume which he has the honor of introducing to the readers of this journal is worthy of its able, diligent author, and a credit to the scientific workers of this country.—*From the Journal of Comparative Med. and Vet. Archives.*

TEXT-BOOK OF THE PHYSIOLOGICAL CHEMISTRY OF THE BODY. Including an account of the chemical changes occurring in disease. By A. GAMGEE, M.D., F.R.S. Illustrated. 8vo. Vol. I. $4.50. Vol. II. in *Press.*

LESSONS IN ELEMENTARY ANATOMY. By ST. GEORGE MIVART, F.R.S. Illustrated. 8vo. $1.75.

LECTURES ON BACTERIA. By A. DE BARRY. Second improved edition. Authorized translation by HENRY E. F. GARNSEY, M.A. Revised by ISAAC BAYLEY BALFOUR, M.A., M.D., F.R.S. With 20 wood engravings. 12mo. $1.50.

COMPARATIVE ANATOMY OF THE VEGETATIVE ORGANS OF THE PHANEROGAMS AND FERNS. By Dr. A. DE BARRY. Translated and annotated by F. O. BOWER, M.A, F.L.S., and D. H. SCOTT, M.A., PH.D., F.L.S. With 241 Woodcuts, and an Index. 8vo. $5.50.

COMPARATIVE MORPHOLOGY AND BIOLOGY OF THE FUNGI MYCETOZA AND BACTERIA. By A. De BARRY. Translated by HENRY E. F. GARNSEY, M.A., revised by ISAAC BAYLEY BALFOUR, M.A., M.D, F.R S. With 198 Woodcuts, and Index. 8vo. $5.50.

LECTURES ON THE PHYSIOLOGY OF PLANTS. By JULIUS VON SACHS. Translated by H. MARSHALL WARD, M.A., F.L.S. With 455 Woodcuts, and an Index. 8vo. $8.00.

OUTLINES OF CLASSIFICATION AND SPECIAL MORPHOLOGY OF PLANTS. By Dr. K. GOEBEL. A new edition of Sach's Text-Book of Botany. Book III. translated by E. F. GARNSEY, M.A. Revised by ISAAC BAYLEY BALFOUR, M.A., M.D., F.R.S. With 407 Woodcuts. 8vo. $5.25.

LECTURES ON THE PHYSIOLOGY OF PLANTS. By SYDNEY HOWARD VINES, M.A., D.Sc., F.R.S. With numerous Illustrations and an Index. 8vo. $5.00.

MACMILLAN & CO., 112 Fourth Avenue, New York.

WORKS BY MICHAEL FOSTER,

M.A., M.D., LL.D., F.R.S.,

PROFESSOR OF PHYSIOLOGY IN THE UNIVERSITY OF CAMBRIDGE, AND FELLOW
OF TRINITY COLLEGE, CAMBRIDGE.

A TEXT-BOOK OF PHYSIOLOGY.

With illustrations. Fifth Edition. Largely revised.

Part I. **Comprising Book I.** Blood; The Tissues of Movement; The Vascular Mechanism. 8vo. $2.60.
Part II. **Comprising Book II.** The Tissues of Chemical Action; Nutrition. 8vo. $2.60.
Part III. **The Central Nervous System.** 8vo. $1.75.
Part IV. **The Central Nervous System (*Continued*). The Tissues and Mechanisms of Reproduction.** 8vo. $2.00.

"The present edition is more than largely revised. Much of it is re-written, and it is brought quite abreast with the latest wave of progress of physiological science. A chief merit of this work is its judicial temper, a strict sifting of fact from fiction, the discouragement of conclusions based on inadequate data, and small liking shown toward fanciful though fascinating hypotheses, and the avowal that to many questions, and some of foremost interest and moment, no satisfying answers can yet be given." — *New England Medical Journal*.

"It is in all respects an ideal text-book. It is only the physiologist, who has devoted time to the study of some branch of the great science, who can read between the lines of this wonderfully generalized account, and can see upon what an intimate and extensive knowledge these generalizations are founded. It is only the teacher who can appreciate the judicious balancing of evidence and the power of presenting the conclusions in such clear and lucid forms. But by every one the rare modesty of the author in keeping the element of self so entirely in the background must be appreciated. Reviewing this volume as a whole, we are justified in saying that it is the only thoroughly good text-book of physiology in the English language, and that it is probably the best text-book in any language." — *Edinburgh Medical Journal*.

THE ELEMENTS OF EMBRYOLOGY.

By MICHAEL FOSTER, M.A., M.D., LL.D., and the late FRANCIS M. BALFOUR, M.A., LL.D., F.R.S. Second Edition. Revised and Enlarged. Edited by ADAM SEDGWICK, M.A., and WALTER HEAPE. With illustrations. 12mo. $2.60.

"A book especially adapted to the needs of medical students, who will find in it all that is most essential for them to know in the elements of vertebrate embryology." — *Academy*.

A COURSE OF ELEMENTARY PRACTICAL PHYSIOLOGY.

By M. FOSTER, M.D., F.R.S., and J. N. LANGLEY, M.A., F.R.S. Fifth Edition. Enlarged. 12mo. $2.00.

"This work will prove of great value to the teacher of physiology, as an aid to the preparation of an eminently practical course of lectures and demonstrations of elementary experimental physiology. Its chief utility, however, will be to the intelligent student, who armed with a dissecting case, a microscope, and the book, will be enabled to pass his summer vacation in a manner at once interesting and profitable." — *Medical Record*.

MACMILLAN & CO., 112 Fourth Avenue, New York.

A TEXT-BOOK OF PHYSIOLOGY.

By JOHN GRAY MCKENDRICK, M.D., LL.D., F.R.S. Including Histology, by PHILIPP STÖHR. In two volumes.

Volume I. General Physiology. Including the Chemistry and Histology of the Tissues and the Physiology of Muscle. 8vo. $4.00.

"This volume treats of the general physiology of the tissues. . . . Taken as a whole, the first volume of Dr. McKendrick's work is a most valuable one, and we shall look for the second with great interest. If he succeeds in his treatment of special, as well as he has succeeded with general, physiology, his text-book will be entitled to a prominent place among the best text-books of physiology." — *Science*.

Volume II. Special Physiology. Including Nutrition, Innervation, and Reproduction. 8vo. $6.00.

TEXT-BOOK OF COMPARATIVE ANATOMY.

By Dr. ARNOLD LANG, Professor of Zoology in the University of Zurich; formerly Ritter Professor of Phylogeny in the University of Jena. Issued as the Ninth Edition of EDWARD OSCAR SCHMIDT'S "Handbook of Comparative Anatomy." Translated into English by HENRY H. BERNARD, M.A., Cantab., F.Z.S., and MATILDA BERNARD. With Preface by Professor ERNST HÆCKEL. 2 vols. Illustrated. Medium 8vo. Vol. I. complete, with Index, $5.50.

TEXT-BOOK OF THE DEVELOPMENTAL HISTORY OF THE INVERTEBRATES.

By Drs. KORSCHELT and HEIDER, of Berlin. Translated under the supervision of Dr. E. L. MARK, Harvard University, Mass. Fully illustrated. 8vo. *In the press.*

TEXT-BOOK OF THE DEVELOPMENTAL HISTORY OF THE VERTEBRATES.

By Dr. OXAR HERTWIG, Professor of Comparative Anatomy in the University of Berlin. Translated and Edited by Dr. E. L. MARK, Harvard University, Mass. Fully illustrated. 8vo. *In the press.*

MACMILLAN & CO., 112 Fourth Avenue, New York.

MACMILLAN & CO.'S PUBLICATIONS.

BALFOUR. A Treatise on Comparative Embryology. By F. M. BALFOUR, M.A., F.R.S., Fellow and Lecturer of Trinity College, Cambridge. With Illustrations, Second Edition, reprinted without alteration from the First Edition. In 2 vols., 8vo. Vol. I., $4.50 ; Vol. II., $5.25.

CLAUS. Elementary Text-Book of Zoology. By Dr. C. CLAUS. Translated and edited by ADAM SEDGWICK, M.A., with the assistance of F. G. HEATHCOTE, B.A. Part 1. General Part and Special Part: Protozoa to Insecta. Part 2. Special Part: Mollusca to Man. With 706 woodcuts. 2 vols., 8vo. $8.00.

ECKER. The Anatomy of the Frog. By ALEXANDER ECKER. Translated, with numerous Annotations and Additions, by G. HASLAM, M.D., and profusely illustrated with 250 wood engravings and 11 colored figures. 8vo. $5.25.

EIMER. Organic Evolution as the Result of the Inheritance of Acquired Characters according to the Laws of Organic Growth. By Dr. G. H. THEODOR EIMER, Professor of Zoology and Comparative Anatomy in Tübingen. Translated by T. J. CUNNINGHAM, M.A., F.R.S.E., late Fellow of University College, Oxford. Part I., with 6 Figures in the Text. 8vo, $3.25.

FEARNLEY. A Course of Elementary Practical Histology. By WILLIAM FEARNLEY. 12mo. $2.00.

FLOWER. Mammals, Living and Extinct. By WILLIAM HENRY FLOWER, C.B., F.R.S., D.C.L., Director of the Natural History Departments, British Museum, and RICHARD LYDECKER, B.A. 8vo, cloth, illustrated with 357 figures. $6.00.

HAMILTON (D. J.). On the Pathology of Bronchitis, Catarrhal Pneumonia, Tubercle, and Allied Lesions of the Human Lung. By D. J. HAMILTON. With Illustrations. 8vo. $2.50.

HAMILTON (D. J.) A Systematic and Practical Text-Book of Pathology. By D. J. HAMILTON. Vol. I. 8vo. $6.25.

HUXLEY. Physiography. An Introduction to the Study of Nature. By T. H. HUXLEY, F.R.S. With Illustrations and colored plates. New Edition. 12mo. $1.80.

"It is not too much to say that any one who will read through this little volume will have a clearer and more connected idea of the physical phenomena of the earth than could be obtained by the perusal of many elaborate treatises." — *Guardian.*

HUXLEY AND MARTIN. A Course of Elementary Instruction in Practical Biology. By T. H. HUXLEY, LL.D., F.R.S., assisted by H. N. MARTIN, M.A., M.D., D.Sc., F.R.S. Revised Edition, Extended and Edited by G. B. HOWES and D. H. SCOTT. With a Preface by Prof. HUXLEY. 12mo. $2.60.

MACMILLAN & CO., 112 Fourth Avenue, New York.

MACMILLAN & CO.'S PUBLICATIONS.

KLEIN. **Micro-Organisms and Disease;** an Introduction into the Study of Specific Micro-Organisms. By E. KLEIN, M.D., F.R.S. Third Edition. Revised. With 120 Illustrations. 12mo. $1.50

MIVART. **Lessons in Elementary Anatomy.** By ST. GEORGE MIVART, F.R.S., Lecturer in Comparative Anatomy at St. Mary's Hospital. With upwards of 400 Illustrations. 16mo. $1.75.

"The work is excellent, and should be in the hands of every student of human anatomy." — *Medical Times.*

"It may be questioned whether any other work on anatomy contains in like compass so proportionately great a mass of information." — *Lancet.*

PARKER. **A Course of Instruction in Zootomy** (Vertebrata). By T. JEFFERY PARKER. With seventy-four Illustrations. 12mo. $2.25.

PARKER and BETTANY. **The Morphology of the Skull.** By W. K. PARKER and G. T. BETTANY. 12mo. $2.60.

WALLACE. **Works by Alfred Russel Wallace, LL.D., F.L.S.**

Darwinism. Being a Systematic Exposition of the Theory of Natural Selection, with some of its Applications. With numerous illustrations. $1.75.

"The present work contains the conclusions upon this great subject of thirty years of thought and observation. . . . A contribution of the first importance to the literature of the subject. At the same time it would be difficult to find a book more entertaining to the general reader. He writes with the sincerity and easy mastery which comes of fulness of knowledge. There can be no more interesting guide in that great wonderland of science in which he has been so long one of the chief discoverers." — *New York Times.*

The Malay Archipelago; The Land of the Orang Utan and the Bird of Paradise. A Narrative of Travel. With Studies of Man and Nature. With Illustrations. Ninth Edition. 12mo. $1.75.

Contributions to the Theory of Natural Selection; and Tropical Nature and Other Essays. New edition. In one volume. 12mo. $1.75.

Island Life; or, The Phenomena and Causes of Insular Faunas and Floras. Including Revision and Attempted Solution of the Problem of Geological Climates. With Illustrations and Maps. New and cheaper Edition. 12mo. *In the Press.*

WIEDERSHEIM. **Elements of the Comparative Anatomy of Vertebrates.** Adapted from the German of ROBERT WIEDERSHEIM. By W. NEWTON PARKER. With Additions. Illustrated with 270 Woodcuts. 8vo. $3.00.

New and Cheaper Edition.

WEISMANN. **Essays upon Heredity and Kindred Biological Problems.** By Dr. AUGUST WEISMANN, Professor in the University of Freiburg in Breisgau. Edited by EDWARD B. POULTON, M.A., F.R.S., SELMAR SCHONLAND, and ARTHUR E. SHIPLEY. 12mo. $2.00.

MACMILLAN & CO., 112 Fourth Avenue, New York.

www.ingramcontent.com/pod-product-compliance
Lightning Source LLC
Chambersburg PA
CBHW051236300426
44114CB00011B/761